The Chemical Synthesis of Natural Products

Dedicated to the memory of
a very fine organic chemist
Dr Clive Bird
of King's College London

The Chemical Synthesis of Natural Products

Edited by

KARL J. HALE
Department of Chemistry
University College London
UK

Academic Press

CRC Press

First published 2000
Copyright © 2000 Sheffield Academic Press

Published by
Sheffield Academic Press Ltd
Mansion House, 19 Kingfield Road
Sheffield S11 9AS, England

ISBN 1-84127-039-3

Published in the U.S.A. and Canada (only) by
CRC Press LLC
2000 Corporate Blvd., N.W.
Boca Raton, FL 33431, U.S.A.
Orders from the U.S.A. and Canada (only) to CRC Press LLC

U.S.A. and Canada only:
ISBN 0-8493-9748-0

Printed on acid-free paper in Great Britain by
Bookcraft Ltd, Midsomer Norton, Bath

British Library Cataloguing-in-Publication Data:
A catalogue record for this book is available from the British Library

Library of Congress Cataloging-in-Publication Data:
The chemical synthesis of natural products / edited by Karl Hale.
    p.   cm.
  Includes bibliographical references and index.
  ISBN 0-8493-9748-0 (alk. paper)
  1. Natural products--Synthesis   I. Hale, Karl.
QD415.C46 1999
547'.2--dc21   99-40544
                                               CIP

Preface

The synthesis of complex natural products continues to occupy a central position in organic chemistry research, not only because nature keeps providing us with some of the most awe-inspiring and synthetically challenging molecules that we can ever aspire to synthesise, but also because research in this area frequently drives many important breakthroughs in methodology. Given the tremendous contribution that complex natural product synthesis makes to chemistry as a whole, it is essential that up-to-date reference volumes appear regularly on this massive subject area, to assist organic chemists engaged in this activity. Such volumes should summarise concisely the most important technological advances and research achievements that have occurred in the sub-disciplines of natural product synthesis over stated periods of time, and they should not attempt to be comprehensive.

I am happy to say that this book goes a considerable way to fulfilling this role, with all the authors having effectively highlighted most of the key advances that have occurred within their specialist sub-areas in the past decade. I believe that it will be of considerable value to all synthetic organic chemists.

Karl J. Hale
University College London

Contributors

Dr G.S. Bhatia Christopher Ingold Laboratories, Department of Chemistry, University College London, 20 Gordon Street, London WC1H 0AJ, UK

Dr C.M. Bladon Department of Chemistry, School of Physical Sciences, University of Kent, Canterbury CT2 7NH, UK

Mr J.W. Bode Laboratorium für Organische Chemie, ETH-Zentrum, Universitätstrasse 16, CH-8092 Zürich, Switzerland

Dr S. Caddick Centre for Biomolecular Design and Drug Development, The Chemistry Laboratory, University of Sussex, Falmer, Brighton BN1 9QJ, UK

Professor E.M. Carreira Laboratorium für Organische Chemie, ETH-Zentrum, Universitätstrasse 16, CH-8092 Zürich, Switzerland

Professor R.S. Coleman Department of Chemistry, Ohio State University, Columbus, OH 43210, USA

Mr M. Frigerio Christopher Ingold Laboratories, Department of Chemistry, University College London, 20 Gordon Street, London WC1H 0AJ, UK

Dr J.M. Gardiner Department of Chemistry, University of Manchester Institute of Science and Technology, PO Box 88, Manchester M60 1QD, UK

Professor K.J. Hale Christopher Ingold Laboratories, Department of Chemistry, University College London, 20 Gordon Street, London WC1H 0AJ, UK

Dr A.N. Hulme Department of Chemistry, University of Edinburgh, King's Buildings, West Mains Road, Edinburgh EH9 3JJ, UK

Dr M.L. Madaras   Department of Chemistry, Ohio State University, Columbus, OH 43210, USA

Dr C.M. Marson   Christopher Ingold Laboratories, Department of Chemistry, University College London, 20 Gordon Street, London WC1H 0AJ, UK

Dr J. Robertson   Dyson Perrins Laboratory, South Parks Road, Oxford OX1 3QY, UK

Mr S. Shanmugathasan   Centre for Biomolecular Design and Drug Development, The Chemistry Laboratory, University of Sussex, Falmer, Brighton BN1 9QJ, UK

Ms N.J. Smith   Centre for Biomolecular Design and Drug Development, The Chemistry Laboratory, University of Sussex, Falmer, Brighton BN1 9QJ, UK

Professor G.A. Sulikowski Department of Chemistry, Texas A&M University, College Station, TX 77842-3012, USA

Dr M.M. Sulikowski   Department of Chemistry, Texas A&M University, College Station, TX 77842-3012, USA

Dr A.B. Tabor   Christopher Ingold Laboratories, Department of Chemistry, University College London, 20 Gordon Street, London WC1H 0AJ, UK

Dr E. Tyrrell   Department of Chemistry, Kingston University, Penrhyn Road, Kingston-upon-Thames KT1 2EE, UK

Dr P.B. Wyatt   Department of Chemistry, Queen Mary and Westfield College, University of London, Mile End Road, London E1 4NS, UK

Contents

4 Important developments in the total synthesis of alkaloids   96
J. ROBERTSON

5 Recent developments in the chemical synthesis of naturally occurring aromatic heterocycles   128
G. A. SULIKOWSKI and M. M. SULIKOWSKI

6   Carboaromatic compounds   **144**
R. S. COLEMAN and M. L. MADARAS

7   Highlights in the chemical synthesis of terpenes and terpenoids   **180**
E. TYRRELL

8   Strategies for building and modifying naturally occurring steroids   **199**
C. M. MARSON

9 Synthesis of enediynes and dienediynes   229
S. CADDICK, S. SHANMUGATHASAN
and N. J. SMITH

10 The chemical synthesis of linear peptides and amino acids   264
C. M. BLADON and P. B. WYATT

1 Strategies for the chemical synthesis of complex carbohydrates

J.M. Gardiner

1.1 Introduction

The 1990s have seen particularly vigorous activity in the arena of oligosaccharide synthesis, and, in the broadest sense, the synthesis of oligosaccharide analogues. This chemical interest has in large part been driven by the emerging understanding of the biological roles of oligosaccharide moieties, together with the need for material for biological investigations, and the exciting opportunities for creating new therapeutic agents. There have been major advances in a variety of areas of complex carbohydrate chemistry since the late 1980s. These have encompassed new chemical glycosylation methodologies and improved strategies for using these methods in oligosaccharide assembly. Applications of enzymes in oligosaccharide synthesis have also seen major developments. The synthesis of carbohydrates on solid supports and the synthesis of saccharide libraries (solid-supported and in solution) and of multivalent saccharide assemblies have been progressing, and numerous complex oligosaccharides and related glycoconjugates have now been synthesised as a result of this array of methodological developments. And last, but not least, recent years have seen a rapidly growing interest in carbohydrate structural mimetics, where not just intersugar atoms are altered, but where structural motifs are replaced by nonsugars. In this chapter, we will briefly outline the successes of some new chemical glycosylations and also of enzymatic protocols since the late 1980s, directing readers to various more comprehensive reviews. The synthesis of carbohydrates on solid-supports, and libraries on support or in solution will be highlighted, and the growing areas of carbohydrate mimetics and multivalent carbohydrates will be reviewed.

1.2 Chemical glycosylations

The synthetic challenges presented by oligosaccharide assembly reside in the requirement for regioselective glycosylation of acceptor sugars, and in the requirement for the stereoselective introduction of one anomeric stereochemistry in the coupled product. The first issue has traditionally been dealt with by selective protection strategies. The control of anomeric stereochemistry has been tackled in a number of ways through novel chemical methodologies.

A central issue in complex oligosaccharide assembly is control of anomeric reactivity; in particular, chemoselectivity in the reactivity of different anomeric functional groups. Recent years have seen a tremendous simplification of the range of operations needed to implement each glycosidic coupling step, and an overall reduction in the number of different glycosyl donors needed to implement a given oligosaccharide synthesis. Several strategies have been devised which rely on replacement or modification of one anomeric functionality which does not serve as a donor, with another which does. (This is in addition to the phenomenon of O(2) functionality being used to modulate anomeric reactivity, by 'arming' or 'disarming' the donor).[1] These combined approaches have led to oligosaccharide syntheses being developed that have exploited a common anomeric donor group. They have also made possible the reiterative assembly of oligosaccharides, wherein identical sugar building blocks are repeatedly used as acceptor and then as donor. Good acceptor/donor intermediates for reiterative approaches include glycals and glycal epoxides (Section 1.2.4),[1] thiophenyl glycosides,[1] phenyl sulfoxides,[2] n-pentenyl glycosides and n-pentenyl dibromides,[1] p-N-acetylphenyl sulfides,[3] and O-allyl and vinyl glycosides.[4,5] In addition, Nicolaou and co-workers have employed thiophenyl and fluoro glycosides in the same way, converting thiophenyl glycosides to fluoro glycosides before using these as donors, and relying on the fact that the fluoroglycoside can couple to acceptor thioglycosides.[6] An extrapolation of this latter approach is the so-called orthogonal use of thiosulfides and fluorides, which can each act as glycosyl donors using different conditions under which the other functionality is inert. This obviates the need to modify or substitute the anomeric functionality as is required in those methods listed above. Thus, phenylthioglycosides couple to glycosylfluoride acceptors under NIS-TfOH (or AgOTf) promotion, and the product glycosyl-fluorides can then be further extended by coupling to a phenylthioglycoside acceptor using Cp_2HfCl_2-$AgClO_4$ promotion.[7] A heptasaccharide repeating β-(1 → 4) linked oligo-GlcNAc was prepared using this strategy,[7] and other systems such as an extended blood group B determinant tetrasaccharide have also been prepared.[8] The alternative of employing a spectrum of anomeric reactivities has been adapted into a one-pot sequential synthesis of a number of trisaccharides, using a combination of bromides, fluorides or trichloroacetimidates, with phenyl sulfides and methyl glycosides.[9] Additionally, phenylselenoglycosides have extended such series (section 1.2.2). For more comprehensive reviews of chemical glycosylation in general, including an explanation of the latent–active and armed–disarmed reactivity concepts, readers are directed to other sources.[1,10]

1.2.1   Glycosyl sulfoxides

Glycosyl sulfoxides, first developed by Kahne's group,[11] have proven efficient and valuable glycosyl donors. They can be selectively activated in the presence of thiophenylglycosides (from which the sulfoxides are derived), and thus can be used as donors in glycosylation reactions where the thioglycoside acts as an acceptor. Thus, all building blocks can originate from thioglycosides. The thioglycoside can then be used subsequently to extend further the saccharide structure without the need for changing anomeric functionality.

Kahne and co-workers have demonstrated the utility of this type of donor, in an attractive synthetic route to Le^a, Le^b and Le^X, all syntheses simply relying on iterative use of the same sulfoxide coupling conditions [for the synthesis of Le^X (1), see Scheme 1.1].[12] The reactivity of sulfoxides over thioglycosides has also been exploited by others. For example, Martin-Lomas's group has used this to good effect in an efficient synthesis of tetragalactoside 5 (Scheme 1.2), which retains the thioglycosidic functionality originating from starting monosaccharide 4 (the other 3 galactosides being derived from sulfoxide 3).[13]

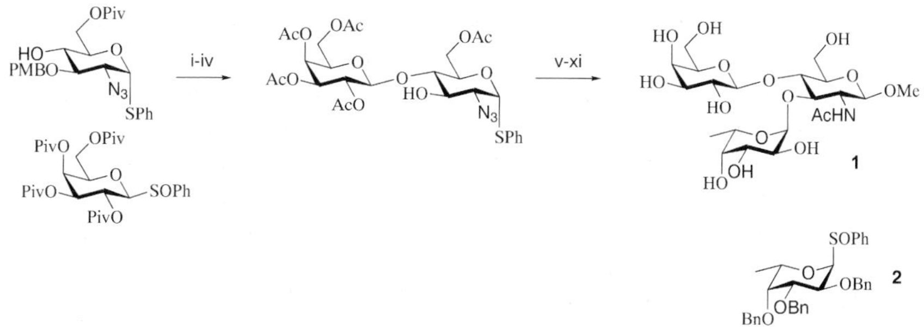

Scheme 1.1   (i) Tf$_2$O, DTBMP; (ii) LiOH, MeOH; (iii) Ac$_2$O, Pyr.; (iv) 10% TFA, CH$_2$Cl$_2$; (v) Tf$_2$O, DTBMP, **2**; (vi) HgTFA$_2$, wet CH$_2$Cl$_2$; (vii) NaH, MeI; (viii) Lindlar reduction; (ix) Pd/C. H$_2$; (x) Ac$_2$O, Pyr., DMAP; (xi) NaOMe, MeOH.

1.2.2   Selenoglycosides and telluroglycosides

Throughout the 1990s, the utility of selenoglycosides has been evidenced in a number of valuable ways. Such glycosides have conveniently extended the anomeric reactivity spectrum. They provide a valuable strategic alternative to the iterative interchange of anomeric functionality or the use of truly orthogonal reactivity.

Pinto's group were the first to demonstrate that (armed or disarmed) selenoglycosides could be activated over thioglycosides (armed or

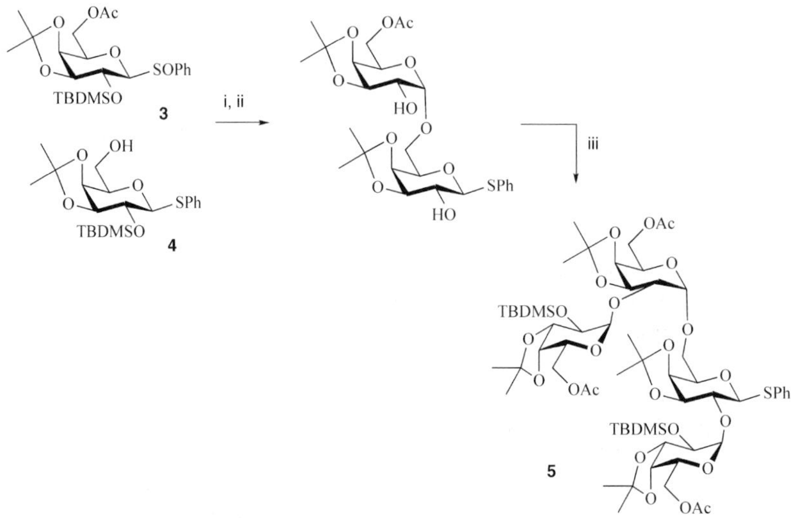

Scheme 1.2   (i) Tf$_2$O, di-t-Bu-4-Me-pyridine; (ii) TBAF; (iii) **3**, Tf$_2$O, di-t-Bu-4-Me-pyridine.

disarmed, respectively), using silver triflate as promoter.[14] Selenoglyco-sides can thus be selectively employed with thioglycoside acceptors to afford thioglycosidic disaccharide donors directly. Even *disarmed* selenoglycosides [those bearing O(2) acyl groups] such as the seleno-phenyl rhamnoside **6** can be activated over *armed* [bearing O(2) benzylic]thioglycosides such as **7** to afford specific disaccharides (e.g. **8**) (see Scheme 1.3).[15] Additionally, organic bases (collidine or 1,1,3,3-tetramethylurea) and silver triflate provide a catalyst system which allows glycosyl bromides (e.g. **9**) to be used as donors with selenoglycosidic acceptors (e.g. **10**). Furthermore, trichloroacetimidates can be activated over selenoglycosides using TMSOTf as catalyst. Similar selectivities

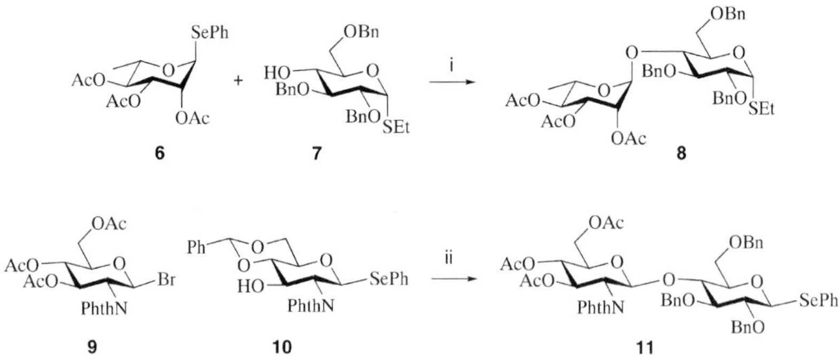

Scheme 1.3   (i) AgOTf, K$_2$CO$_3$; (ii) AgOTf, collidine or 1,1,3,3-tetramethylurea.

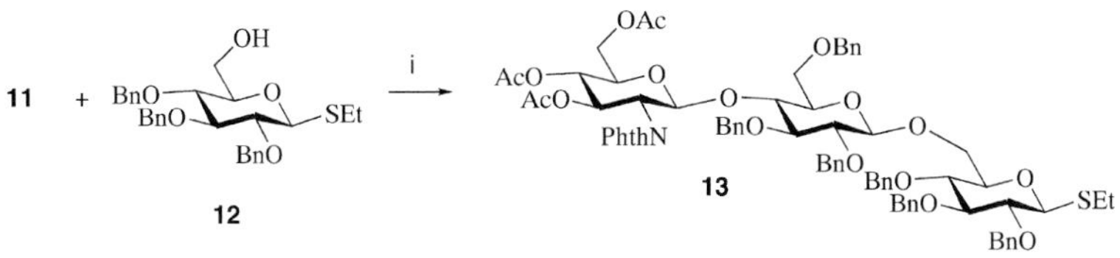

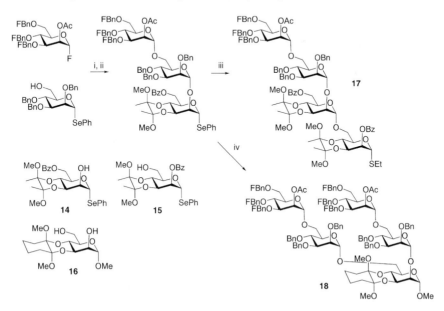

Scheme 1.3 Continued.

were observed using selenoglycosides as donors with *N*-iodosuccinimide-TfOH or iodonium di-*sym*-collidine perclorate (IDCP) as activator.[16, 17] This then completes a convenient and selective sequence of glycosylation conditions for glycosyl bromides (or trichloroacetimidates), selenides and sulfides, applied to synthesis of targets such as trisaccharide **13**. Comparable chemistry using selenoribofuranosyl donors is less efficient and controllable.[18] Iwamura and co-workers have also described activation of selenoglycosides, and glycosylation reactions using photo-induced electron transfer, and have applied this to disaccharide synthesis.[19]

Ley's group have combined the reactivity fine-tuning effects of their *trans*-1,2-cyclohexane diol acetal protection systems,[20] with the fluoro-glycoside/selenoglycoside/thioglycoside reactivity spectrum to complete a convenient, single-pot, synthesis of linear tetramannosides and branched pentamannosides (Scheme 1.4).[21] Syntheses of **17** and **18** are

Scheme 1.4 (i) 2.8 eq. AgOTf, 1.4 eq. Cp₂HfCl₂, mol. sieves; (ii) **14**, 1.05 eq. NIS/TfOH; (iii) **15**, 1.05 eq. NIS/TfOH; (iv) **16**, 1.05 eq. NIS/TfOH.

illustrative. A similar strategy provided an impressively efficient synthesis of a high mannose type nonasaccharide.[22]

Stick and co-workers have also recently reported the use of telluroglucosides (prepared from the precursor glucosyl bromides[23]) as glucosyl donors, finding that telluroglucosides are more reactive (in competition experiments) than selenoglucosides.[24] Thus, 2,3,4,6-tetrabenzoyl telluro-β-D-glucopyranoside was shown to be much more reactive than its seleno analogue in glycosylation of 1,2:3,4-di-O-isopropylidene-α-D-galactose. The extension of this reactivity study to various armed and disarmed donor–acceptor combinations is awaited, to see whether the telluro systems might add a practicable general extension to the seleno–thio–bromo series of reactivities.

1.2.3   n-*Pentenyl glycosides*

Fraser-Reid's group has pioneered the use of *n*-pentenyl glycosides for oligosaccharide synthesis.[25] These glycosides are effective glycosyl donors when activated by NIS-TESOTf. These workers have also introduced pentenyl dibromides as 'latent' derivatives of 'active' pentenyl glycosides. The utility of *n*-pentenyl glycosides and their derived dibromides is illustrated in a recent synthesis of the proteoglycan linkage region precursor **19** (Scheme 1.5).[26] An *n*-pentenyl glycosyl disaccharide donor reacts with an *n*-pentenyl dibromide glycosyl acceptor to construct the tetrasaccharide, and release of the *n*-pentenyl glycoside from the latent dibromide then provides for glycosylation of the linker serine. Earlier examples of syntheses of important oligosaccharide structures include an approach to the glycophosphatidylinositol membrane anchor core hexasaccharide-inositide.[27]

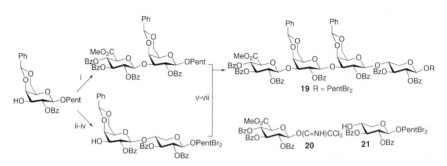

Scheme 1.5   (i) **20**, TESOTf; (ii) ClAcCl, Pyr; (iii) NIS, TESOTf, **21**, (iv) CS$_2$·H$_4$N$_2$, *i*-Pr$_2$EtN; (v) NIS, TESOTf; (vi) NaI, MEK; (vii) HOCH$_2$CH(CO$_2$PTP)NHFmoc, NIS, TESOTf.

1.2.4 Glycals and glycal epoxides

Danishefsky's group have extensively used glycals and their derived epoxides (1,2-anhydrosugars) and epimines for the synthesis of many complex carbohydrates. Central to many of their syntheses was the efficient epoxidation of glycals by dimethyldioxirane (**24** to **25**),[28] and the efficient iodosulfonamidation of glycals followed by conversion to N-sulfonyl glycal azidines (**22**). Fortuitously, the DMDO epoxidation of glucal and galactal derivatives is routinely highly α-stereoselective. The iodosulfona-midation process also generally generates the α-N-glycosylsulfonamide.

The primary value of glycal epoxides lies in their ability to function as efficient glycal donors under Lewis acid (typically zinc chloride) catalysis. The free 2-OH thereby generated in **26** can then be used for further glycosidation, or be specifically deoxygenated, or modified in some other way. Partly-protected glycals can themselves act as acceptors in glyco-sylations with glycal epoxides, thus opening the way for reiterative appli-cations (see section 1.4). It is the N-sulfonyl glycal azidines that function as the active glycosyl donors in the iodosulfonamidation protocol, this method affording 2-N-sulfonyl β-glycosides. This is the stereochem-istry in many biologically important 2-amino sugar derivatives. All these applications are summarised in Scheme 1.6.

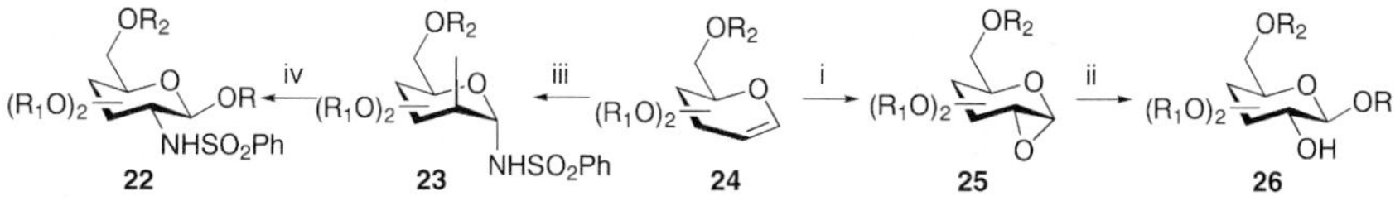

Scheme 1.6 (i) DMDO; (ii) ROH, Lewis acid; (iii) I(sym-coll)$_2$Cl)$_4$, PhSO$_2$NH$_2$; (iv) base then ROH.

Combination of these general methods with other glycosylation protocols has allowed Danishefsky's group to report the synthesis of numerous oligosaccharide targets.[29] Amongst the most significant have been the total syntheses of the GM$_3$ and GM$_4$ gangliosides[30, 31] the A and B blood group antigens, the Lewis x,[32] Lewis y[33, 34] and Lewis b[35] determinants and also of the Globo H antigen.[36] (The Lewis X unit had also been sialylated to afford SLe^X, described in a previous report, and indeed extended to a hexasaccharide derivative.)

The synthesis[37] of MBr1 nicely exemplifies many of the aforementioned applications of glycals in oligosaccharide synthesis (Scheme 1.7). Glyco-syl fluoride **27** was obtained from a precursor D-galactal. It coupled with **28**, itself a product of a galactal epoxide–glucal coupling, to afford, after removal of the PMB group, the trisaccharide glycal **29**. The sec-ond trisaccharide unit, **33**, was assembled from the D-galactal epoxide

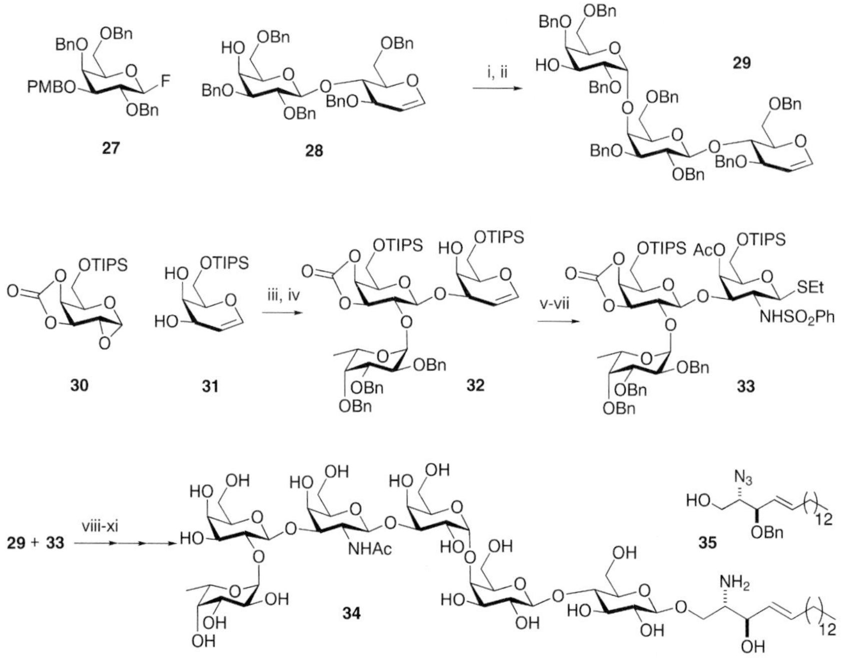

Scheme 1.7 (i) SnCl$_2$, AgClO$_4$, DTBP; (ii) deprotect PMB; (iii) ZnCl$_2$; (iv) SnCl$_2$, AgClO$_4$, fucosyl fluoride; (v) I(sym-coll)$_2$ClO$_4$, PhSO$_2$NH$_2$; (vi) Ac$_2$O, base; (vii) EtSH, LiHMDS; (viii) MeOTf, mol. sieves; (ix) DMDO; (x) ZnCl$_2$, **35**.

30 and the D-galactal **31**, followed by regioselective fucosylation using the appropriate fluoride. Iodosulfamidation, and reaction with lithium ethanethiolate as nucleophile, provided the 2-sulfamido thioglycoside **33**. Coupling of acceptor **29** with donor **33**, followed by elaboration of the glucal terminus via the 1,2-anhydrosugar, and subsequent standard manipulations, afforded the target **34**.

All four unusual monosaccharides (**A–D**) present in the calicheamicin saccharide unit **36** were also derived from various glycals (**37–40**). Glycal coupling methods were again used to couple these together, with the trichloroacetimidate of **36** (R = C(=NH)CCl$_3$) ultimately being used to attach the eneyne core to complete the natural product synthesis.[38]

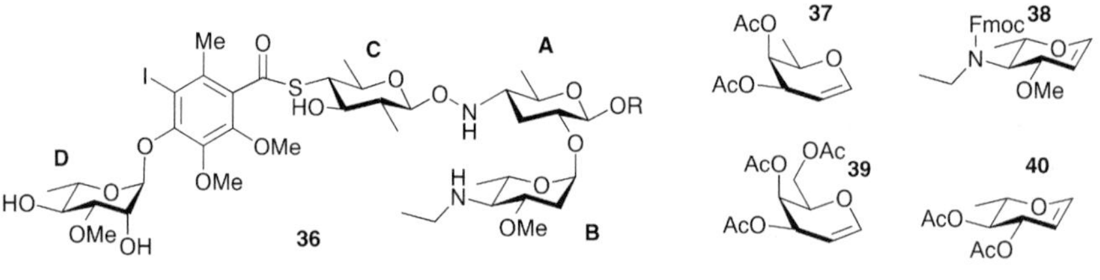

1.2.5 Glycosylidene carbenes and P-containing glycosides

Vasella's group has published a series of papers thoroughly exploring applications of glycosylidene carbenes as glycosyl donors, generated from glycosylidene diazirines.[39] These glycosyl donors have been used for glycosylation of a wide range of alcohol targets, including inositols and ginkgolides as well as disaccharides.[40–43]

Phosphorus-containing anomeric functionalities are also emerging as potentially valuable for the provision of glycosyl donor reactivity.[44–46]

1.3 Enzymes in oligosaccharide synthesis[47–51]

Harnessing the regioselectivity (with respect to glycosyl acceptor) and stereoselectivity (with respect to glycosyl donor anomeric stereochemistry) of enzymes involved in the *in vivo* assembly (or degradation) of oligosaccharides offers substantial benefits. The aforementioned native selectivities allow for oligosaccharide synthesis with unprotected sugars, and side step the anomeric stereochemical control issues which often plague most chemical methods available for glycosidic bond formation. Glycosyltransferases and glycosidases are both now widely employed in oligosaccharide synthesis. In addition to the regio- and stereochemical advantages inherent in employing enzymatic methods, the potential of sequential and/or multienzyme processes for assembling more complex oligosaccharides in a single process is particularly exciting. The utility of some enzymatic processes for solid-supported synthesis is also of significance, as this may well play a role in the development of a viable automated solid-phase synthesis of oligosaccharides.

1.3.1 Glycosyltransferases

Leloir-type glycosyltransferases have been of most interest and utility because of their high level of anomeric stereocontrol and high regioselectivity with respect to glycosyl acceptors. These types of enzymes are responsible for the majority of glycoconjugate oligosaccharide syntheses in mammals, and although there are estimated to be several hundred glycosyltransferases,[52] until recently, only a few were generally available. The enzymes use sugar nucleotide diphosphates (or in the case of neuraminic acid, its cytosine monophosphate) as activated glycosyl donors. Earlier problems with enzyme availability are now being addressed, with a number now being commercially available and a significant number becoming accessible through genetic engineering methods.[53, 54]

The basic process of a transferase-catalysed glycoside bond forming reaction is given in Scheme 1.8 (for the nucleotide diphosphate-dependent cases).

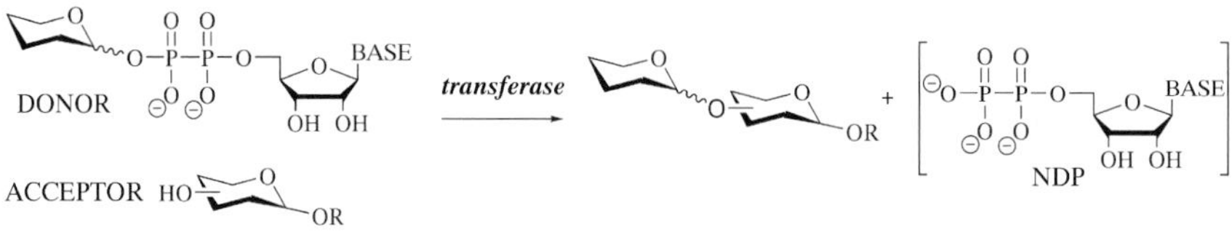

Scheme 1.8.

An attractive feature of transferases is that, although there are numerous transferases, each catalysing the formation of a different glycosidic linkage, they all use a specific monosaccharide derivative as donor. The common monosaccharide glycosyl donors employed by transferases are α-UDP-Glc, α-UDP-Gal, α-GDP-Man, α-UDP-GlcNAc, α-GDP-GalNAc, β-GDP-Fuc, α-TDP-Rha and α-CMP-NeuAc. Thus, for example, all galactosyltransferases use α-UDP-Gal as substrate, different enzymes combining this with different acceptors.

1.3.1.1 Regeneration methods

The high cost of sugar nucleotides means that stoichiometric use is undesirable for scale synthesis. In addition, product inhibition from the released nucleotide phosphates can prove problematic. *In situ* generation methods have thus been developed for the use of α-UDP-Gal, α-CMP-NeuAc,[55] α-UDP-GlcNAc,[56] α-GDP-Man and β-GDP-Fuc,[57] with those reported in the past decade or so referenced.

These methods are based on coupling the nucleotide sugar phosphate to nucleotide phosphate turnover (i.e. glycosidation side-products), with enzymes allowing the synthesis of more nucleotide sugar phosphate from inexpensive sugar sources. Typically, the sugar or sugar 1-phosphate is employed along with the use of PEP-pyruvate interconversion by pyruvate kinase (as a phosphate source, sometimes repeatedly). Examples are illustrated by the regeneration of α-UDP-Gal (Scheme 1.9) and of α-CMP-NeuAc (Scheme 1.10).

The former relies on a pyruvate-kinase-driven UDP to UTP reaction and the latter on phosphorylation of Glc-α-1-P, followed by enzymatic C(4) epimerisation. The latter uses pyruvate kinase twice, first to drive ADP to ATP synthesis providing the ATP needed to take CMP to CDP, followed by a second pyruvate kinase coupled process taking the CDP to CTP, which is then employed for monophosphorylation of NeuAc.

The glycosyltransferases which have seen most applications are β-(1,4)-GalT, α-(2,3)- and α-(2,6)-SialylT and α-(1,3)-FucT, with applications of

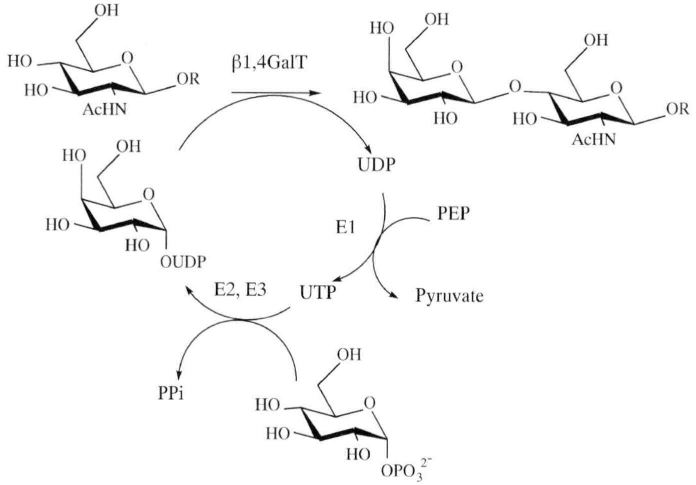

Scheme 1.9   E1: pyruvate kinase, E2: UDP-Glc pyrophosphorylase, E3: UDP-Glc epimerase.

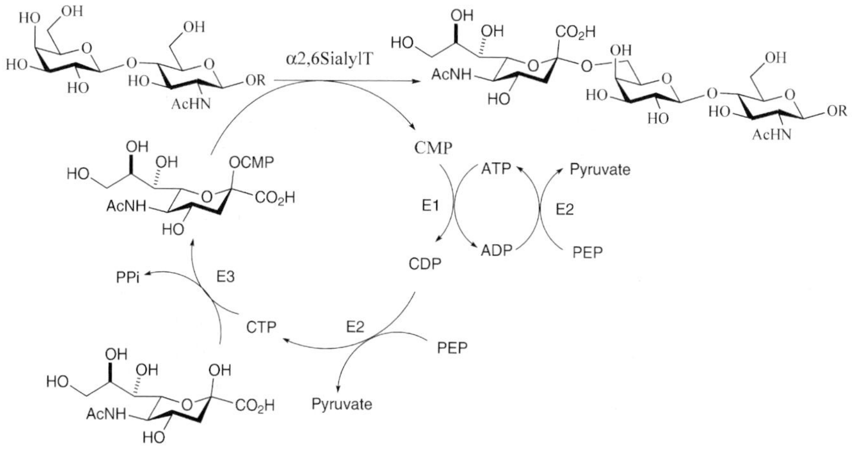

Scheme 1.10   E1: nucleoside monophosphate kinase or adenylate kinase, E2: pyruvate kinase, E3: CMP-NeuAc synthase.

N-acetylglucosaminyltransferase and α-(1,2)-mannosyltransferase also being notable. The past decade has seen demonstration that some of these enzyme systems are able to accept a range of nonnatural substrates, and to transfer non-native donors. The β-(1,4)-GalT system has been particularly well explored, and it has proven useful for galactosylation of a range of systems, including thio and amino sugars, and glycals. It has even been used to galactosylate the anomeric hydroxyl of

3-acetamido-3-deoxyglucose to obtain a 1,1-linked disaccharide (although rates may be as much as two or three orders of magnitude slower).

The past decade has witnessed particularly notable applications of combinations of multienzyme glycosyltransferase reactions, where each transferase is coupled to a suitable enzymatic sugar nucleotide regeneration system *in situ*. Significant examples include the synthesis of Sialyl Lewis X using the UDP-Gal, GDP-Fuc and CMP-NeuAc recycling systems, which employs three transferases [β-(1,4)-GalT, α-(2,3)-SialylT and α-(1,3)-FucT] for sequential assembly of the tetrasaccharide.[57] The efficiency resulting from such a multiprocess approach, without any requirements for protection or deprotection, enables this to be employed for kilogram-scale SLe^X synthesis.

Transferases have also facilitated the synthesis of a useful ^{13}C-labelled analogue of SLe^X, **41** (Scheme 1.11),[57] and also of bivalent SLe^X analogues **42** relying on simultaneous enzymatic assembly of two SLe^X residues on a bis-GlcNAc bearing template. The structure shown in Scheme 1.11 has five-fold higher affinity for E-selectin than SLe^X itself (suggesting possible multivalency; see section 1.5).[58, 59] Sequential use of transferases has been taken a step further and applied to reiterative synthesis of specific polysaccharides, illustrated by Wong and co-workers' synthesis of hyaluronic acid.[60]

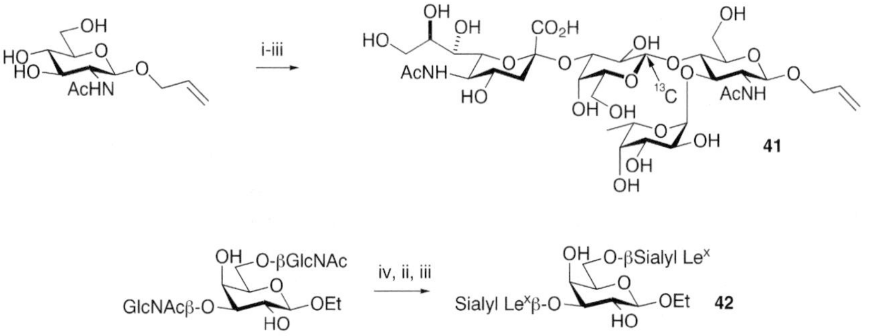

Scheme 1.11   (i) UDP-[1-^{13}C]-Gal, β1,4GalT; (ii) CMP, NeuAc, α2,3-sialylT; (iii) GDP-Fuc, α1,3FucT; (iv) UDP-Gal, β1,4GalT.

1.3.1.2   *Engineered transferases*

The group of Flitsch has described the use of a recombinant and modified (genetically engineered) mannosyltransferase to synthesise the trisaccharide **43** (Scheme 1.12) which is the common core unit of all *N*-linked oligosaccharides.[61] This core trisaccharide contains a β-1,4-linked mannose residue, with the reducing terminal unit attached to Asn. The Flitsch group employed the precursor pyrophosphate—an analogue

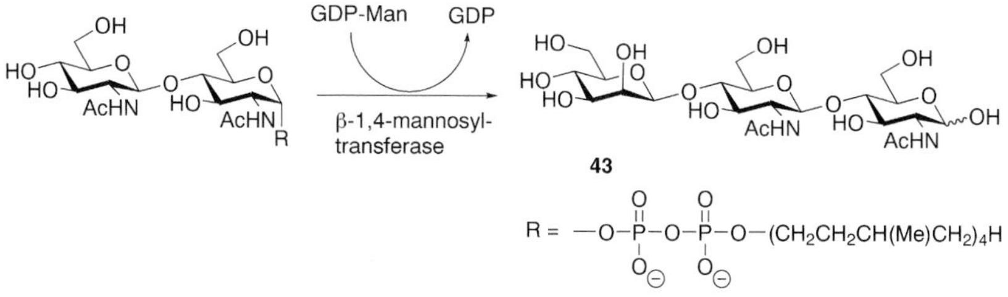

Scheme 1.12.

of the natural dolichyl pyrophosphate used for Asn attachment—as acceptor and GDP-Man as standard donor. The expressed enzyme was then immobilised on an affinity column charged with zinc(II). The trisaccharide was obtained in high yield.

1.3.2   Glycosidases

There are a wide array of enzymes which have evolved to catalyse the hydrolysis of glycosidic bonds. As with proteases, there is the potential to drive these enzymatic reactions in the direction of synthesis, by suitable engineering of conditions and/or substrate structures. In the case of glycosidases, ready availability and low cost make this attractive. In particular, the availability of glycosidases able to assist construction of disaccharide links for which a suitable transferase is not available is of importance. This approach is typified by the use of a glycoside donor converted by the enzyme into a reactive intermediate which normally proceeds to react with water but which is instead trapped out by a glycoside acceptor. In other words, kinetically-controlled conditions are used. Disaccharides or higher oligosaccharides and reactive glycosides (aryl glycosides, glycosyl fluorides) have generally been employed. Competition from hydrolysis is always a limiting consideration. Linking a glycosidase-catalysed glycoside forming reaction to a subsequent transferase-catalysed process can be advantageous, since such a strategy uses the irreversible transferase process to pull the glycosidase equilibrium through to a final product which is no longer a glycosidase substrate. An example of this is the synthesis of trisaccharide **44** via a β-galactosidase-catalysed disaccharide synthesis and a 2,6-sialyltransferase-catalysed completion of the trisaccharide assembly process (Scheme 1.13).[62]

The donor may also be a monosaccharide with a reactive anomeric group, such as nitrophenyl. The synthesis of Lewis a in Scheme 1.14 illustrates this, and, in addition, involves chemical elaboration of a

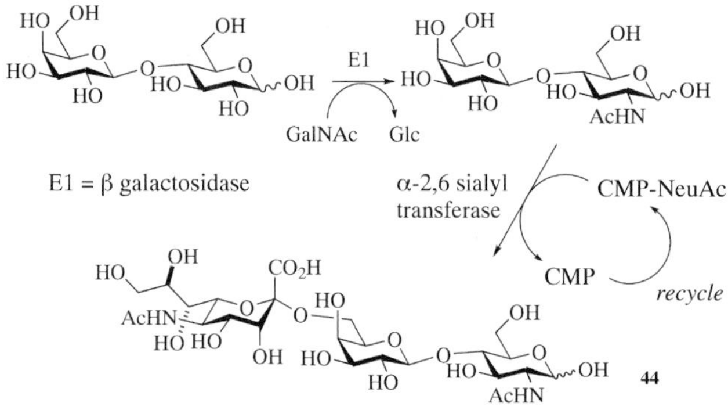

Scheme 1.13.

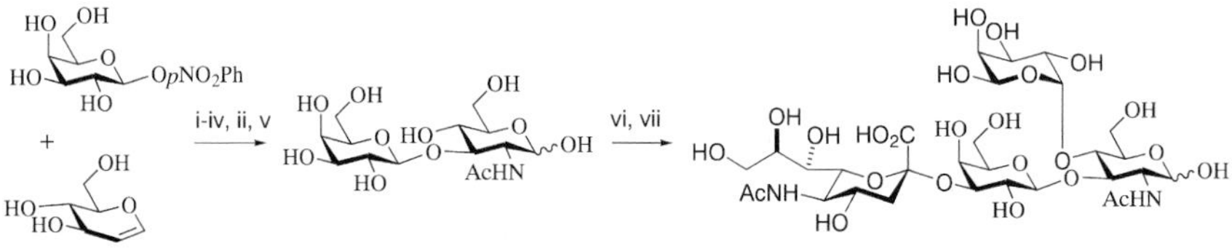

Scheme 1.14 (i) β-galactosidase; (ii) Ac_2O; (iii) NaN_3, CAN; (iv) [H]; (v) saponify; (vi) CMP-NeuAc, SialylT; (vii) GDP-Fuc, FucT.

glycal intermediate to a 2-NAc sugar followed by transferase-catalysed elaboration.

Crout and co-workers have described the use of α-galactosidases to prepare several thioglycoside trisaccharides, using *p*-nitrophenylglycosides as substrate donors. Substrate **45** and acceptor **46** provided trisaccharide **47** when exposed to *A. niger* α-galactosidase, and a mixture of **47** and **48** when *A. oryzae* α-galactosidase I was used. The slightly differing yields depended on the nature of the thio aglycon (Scheme 1.15).[63]

The same group has used sequential glycosidase-catalysed reactions for the synthesis of trisaccharides. Thus, the core trisaccharide of the

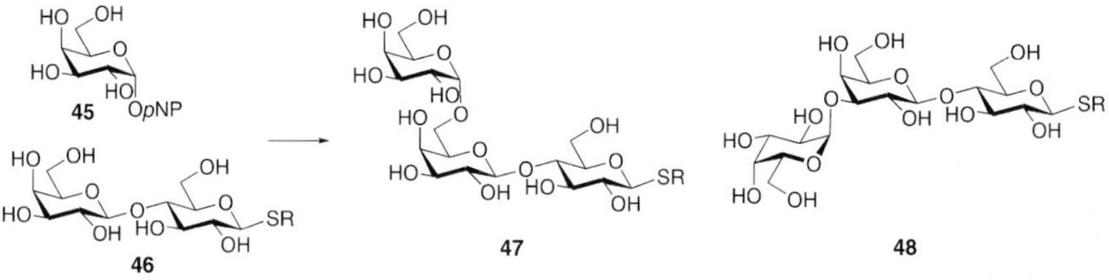

Scheme 1.15.

N-glycans (synthesised using a modified mannosyltransferase by Flitsch *et al.*; see section 1.3.1) was prepared from constituent monosaccharides using two different glycosidase-catalysed reactions (Scheme 1.16).[64]

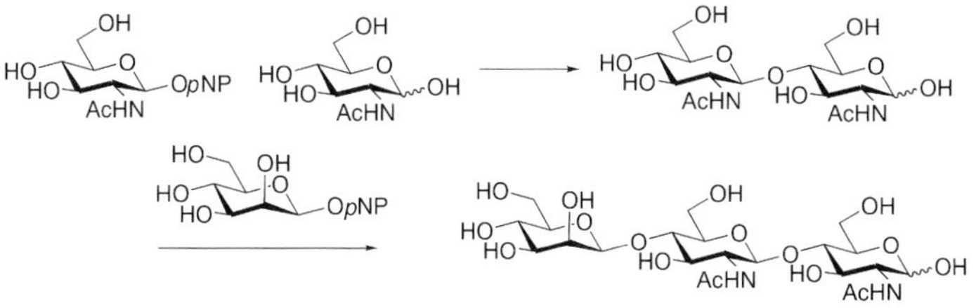

Scheme 1.16.

1.3.2.1   Turning glycosidases into transferases

The group of Withers at the University of British Columbia (UBC) has recently developed glycosidase enzymes with engineered active sites. These effectively function as glycosyltransferases, and are termed 'glycosynthases'.[65] Such enzymes allow glycosyl fluorides (glucosyl or galactosyl) to be reacted with acceptor mono- or disaccharides and can be used for gram scale synthesis.

It is likely that the next few years will begin to see the further application of sulfotransferases[66] as a contributor to the synthesis of bioactive regiospecifically sulfated oligosaccharides. To date, sulfotransferases have seen some valuable uses,[67, 68] such as being critical to the synthesis of SLe^X 6-sulfate.

1.4   Solid-phase synthesis of oligosaccharides and synthesis of oligosaccharide libraries

Both DNA and peptide sequences have succumbed to automated synthesis in part because of their 'linear' structures, and because a single bond-forming step is usually required to establish each inter-residue linkage. In addition, stereochemical problems are often not encountered during sequence assembly. The much more diverse requirements for linking carbohydrates together, in terms of reactivity differences, stereocontrol, and the occurrence of branched structures, makes the development of solid-phase synthesis (and ultimately automation) far more challenging but nonetheless a very formidable potential technology. A number of groups have focused efforts on developing solid-supported oligosaccharide synthesis. There are two directions of interest: first, the development of methods for sequential solid-phase synthesis of target saccharides, and second, the use of solid-support immobilisation as a means of facilitating the generation of carbohydrate libraries.

Danishefsky's group has extended its glycal/glycal epoxide/glycal aziridine methodology (section 1.2.4) to solid-supported synthesis and has prepared a series of saccharide targets, both linear and branched.[29, 69] Synthesis of the pentasaccharide **49** (Scheme 1.17) is illustrative of the iterative protocol developed by Danishefsky and co-workers for glycals.[70]

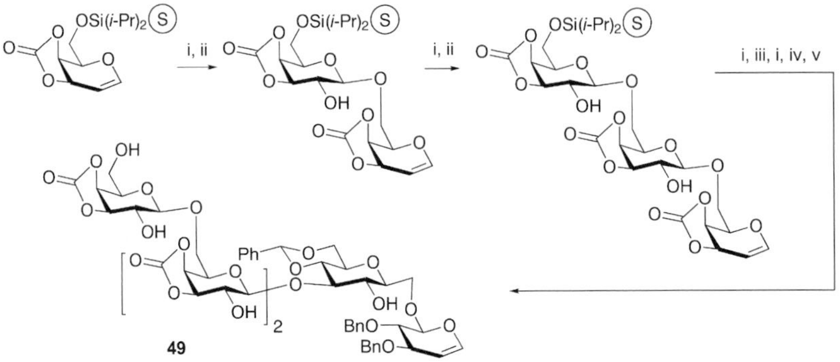

Scheme 1.17   (i) DMDO; (ii) ZnCl$_2$, 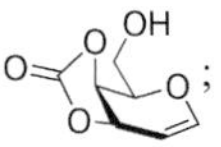 ; (iii) ZnCl$_2$, 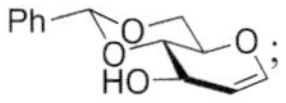 ; (iv) ZnCl$_2$, 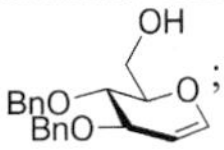 ; (v) TBAF, AcOH.

Other notable examples include the synthesis of a Le^b glucal precursor (the synthesis from tetrasaccharide to hexasaccharide was completed post-cleavage). The previously discussed iodosulfamidation reaction of glycals was also adapted for use on solid-supports (Scheme 1.18), with a view to facilitating the solid-supported synthesis of N-linked glycopeptides.[71] Additionally, these workers have recently shown that polymer-bound glycals can be directly converted to supported thioethylglycosides and thioethyl-2-amidoglycosides, which can each then function as supported glycosyl donors.

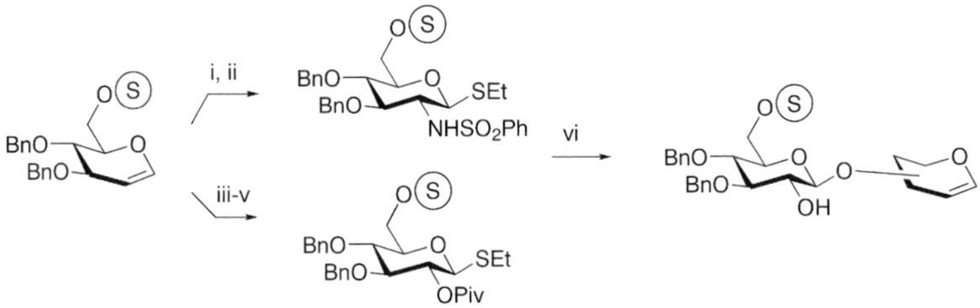

Scheme 1.18   (i) I(sym-coll)$_2$ClO$_4$, PhSO$_2$NH$_2$; (ii) LiHMDS·EtSiH; (iii) DMDO; (iv) EtSH, Tf$_2$O; (v) PivCl, DMAP; (vi) 3-, 4- or 6-OH glycal, ZnCl$_2$.

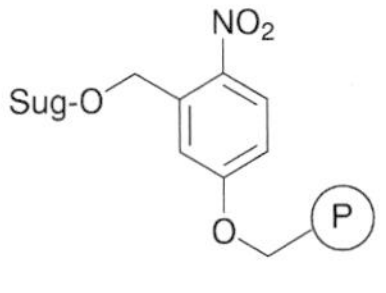

Significantly, Nicolaou's group has described a solid-supported approach to oligosaccharide assembly, which is potentially applicable to the generation of libraries.[72] It exploits thiophenyl glucosides as donors and DMTST (dimethylthiomethylsulfonium triflate) as a promoter. Very high β-anomeric selectivity is observed, and combinations of TBDPS protection at O(6), and Fmoc protection at O(3), facilitate branching. The growing carbohydrate is immobilised by using a nitroaromatic linker to polystyrene (**50**), affording a simple, mild, photolytic cleavage of the final oligosaccharide. The approach was demonstrated by the synthesis of the heptasaccharide phytoalexin elicitor (HPE) **51**.

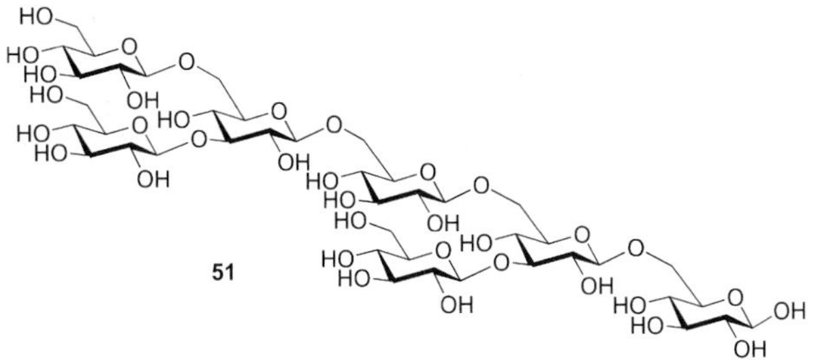

Ito and Ogawa have reported an interesting use of a solid-supported glycosyl donor to prepare β-mannoside saccharides (Scheme 1.19).[73] A thiomannoside was attached to a solid-support via a linker at O(2) (**52**).

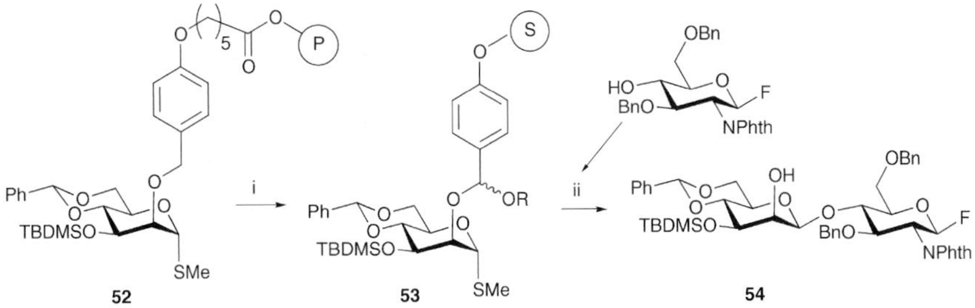

Scheme 1.19   (i) ROH, DDQ, mol. sieves; (ii) MeOTf, DBMP, mol. sieves.

An oxidative synthesis of a mixed acetal **53** using another sugar alcohol allowed for a subsequent intramolecular β-selective glycosylation and concomitant release from the support. Orthogonality of thio- and fluoroglycoside functionality allowed the second acceptor alcohol to be a fluoroglycoside providing, for example, disaccharide **54**. The synthesis of β-mannosides is always problematic and readers are directed to a review of other tethered strategies to provide β-mannosides.[74]

Schmidt's group has successfully used controlled-pore glass (CPG) as a support for attachment of a 3,4,6-tribenzyl-2-phenoxyacetyl (R=PhOCH$_2$CO) thiomannoside (iteratively employing the trichloroacetimidate **55**) via a thiopropylsilyl linker connected to the CPG surface (**56**). Iterative use of the trichloroacetimidate donor and O(2) deprotection of the immobilised terminal mannose residue allowed construction of a protected α-1,2-linked trimannoside **57** (Scheme 1.20).[75] Previously, this group had reported the solid-supported synthesis of a (1 → 6)-linked pentaglucoside using an analogous thioglycoside attachment strategy, in that case to a modified polystyrene support.[76]

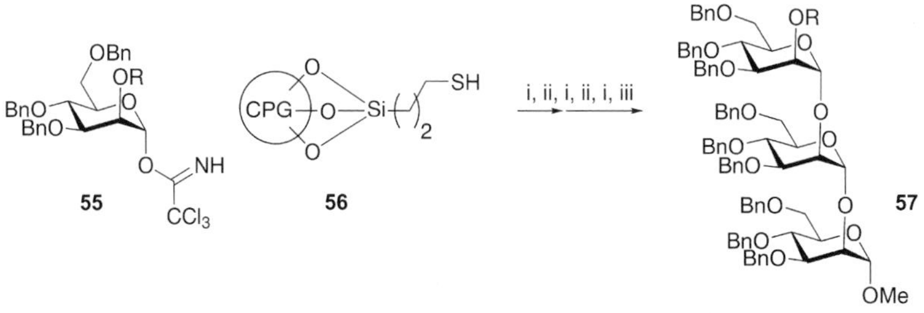

Scheme 1.20   (i) TMSOTf, −40°C; (ii) guanidine, DMF; (iii) NBS, MeOH.

De Napoli and co-workers[77] have also used trichloroacetimidates to extend polymer-supported oligosaccharides, in this case immobilising the first residue through a 3-OH linked succinimidyl linkage to the amino group of a Tentagel resin.

Roush's group has constructed trisaccharide **60** using trichloroacetimidate (**59**) and glycosyl acetate **61** as donors. Immobilisation of the glycal residue was achieved via the sulfonate group of a modified polystyrene resin (from Merrifield resin), giving supported glycal **58** (Scheme 1.21).[78] By inclusion of 6-bromo and 2-iodo functions in the building blocks, and by substitution of the resin-cleaved leaving group,

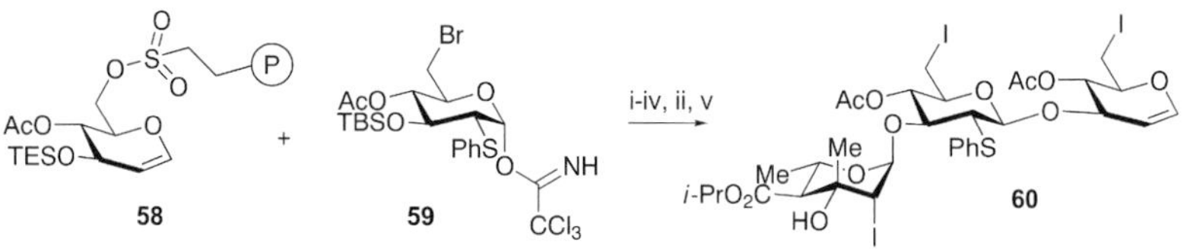

Scheme 1.21   (i) TMSOTf, −78°C, mol. sieves; (ii) Ac$_2$O, Pyr; (iii) Et$_3$N · (HF)$_3$; (iv) TMSOTf, −78°C, *i*-PrO$_2$C... (**61**); v. NaI, 2-butanone.

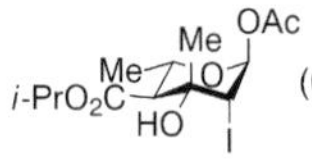

each residue ultimately contained an iodo group. These iodo functions were then removed reductively to afford **60**. The method should be capable of yielding many other deoxy oligosaccharide analogues.

The high reactivity and good stereoselectivity of glycosyl sulfoxide donors (section 1.2.1) make them ideally suited for use in solid-supported oligosaccharide synthesis. Iterative use of such donors provided a $(1 \rightarrow 6)$-linked trigalactoside and other disaccharides relevant to Lewis antigens.[79] Kahne's group has also reported the synthesis (and screening against a target lectin) of solid-supported carbohydrate libraries.[80] A library of approximately 1300 disaccharides and trisaccharides was prepared on Tentagel resin, with the first sugar immobilised through a thioaryl glycosidic link. Six first-unit sugars were employed, and 12 sulfoxide donors (10 monosaccharides and 2 disaccharides). A further combinatorial step involved N-acylation with 17 different acylating agents. These workers were also able to extend the solid-supported work to probe the role of multivalency (see section 1.5) on polymer supported arrays of carbohydrate ligands.[81]

Fraser-Reid and co-workers have recently extended their n-pentenyl glycoside methodology to the solid-supported synthesis of trisaccharides and have applied the technique for the synthesis of libraries.[82] The synthesis of triglucoside **62** (Scheme 1.22) is illustrative. Gal-$\beta(1 \rightarrow 2)$Man-$\alpha(1 \rightarrow 6)$glucosamine was also prepared on support, using NPhth, benzoyl and acetyl as intermediate protecting groups in an analogous synthesis.

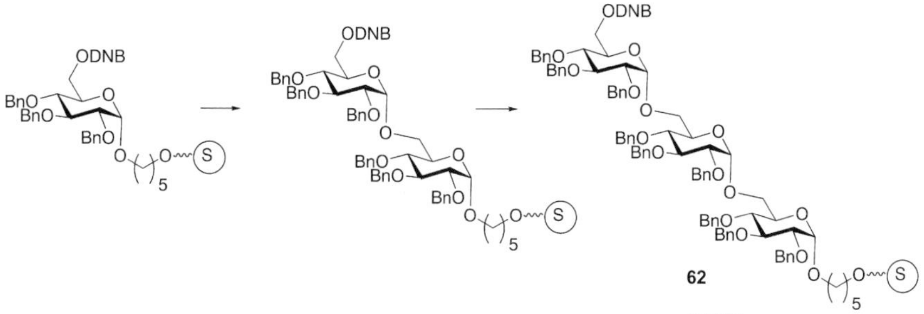

Scheme 1.22   (i) NaOMe, MeOH, THF; (ii) NIS, TESOTf,

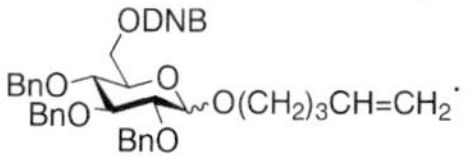

The group of Boons has recently described a 'two-directional' strategy for trisaccharide library generation, based on use of a Tentagel-linked (fully protected) thioglucoside using a split-and-mix type protocol [a succinic linker from the sugar O(6) is employed] (Scheme 1.23).[83] This is able to act as a donor with various methyl glycoside acceptors (**63–65**)

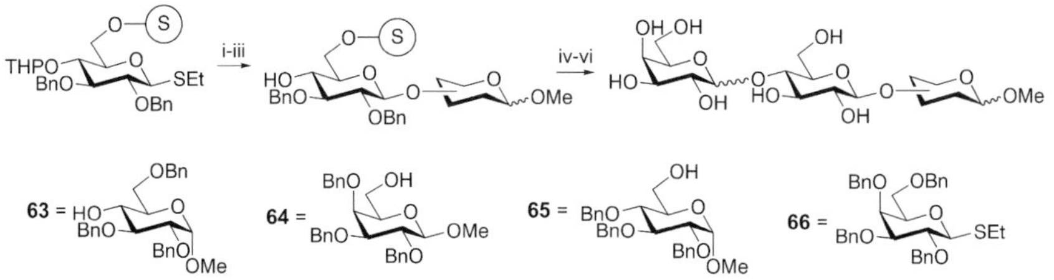

Scheme 1.23   (i) split, react with **63, 64** or **65**, NIS, TMSOTf; (ii) mix; (iii) HOAc, THF/H$_2$O; (iv) **66**, NIS, TMSOTf; (v) MeONa, MeOH then Dowex 50; (vi) Pd(OAc)$_2$, H$_2$, EtOH.

(after splitting batches of the solid-supported thioglycoside) which each have a single free hydroxyl. The resulting disaccharides are mixed, and in the subsequent steps, release of one hydroxyl [by removal of a uniquely located THP protecting group on O(4)] from the original solid-linked glucoside now allows this to act as donor when a new thioglycoside (**66**) is introduced in solution.

Other progress has been made by Krepinsky's group,[84, 85] and others.[86] The former has introduced synthesis on a 'soluble solid-support' consisting of a DOX (α,α'-dioxyxylyl diether) linker attached to an MPEG [poly(ethylene glycol) ω-monomethyl ether] support. Trichloroacetimidate donors are employed in this method, activated by TES triflate[85] or dibutyl boron triflate.[84] The soluble supported growing saccharide is precipitated after each glycosylation round, before being redissolved for subsequent glycosylation. This technique offers the advantages of solution-phase glycosylation and the purification advantages of being able to wash excess reagents and by-products off the precipitated solid-support. The synthesis of penta-(α1 → 2) mannose, as well as of mixed di- and trisaccharides, was demonstrated.

The use of allylglycosides, and vinyl glycosides derived from these, has been advocated by Boons's group for the *unsupported* solution-phase synthesis of trisaccharide libraries.[87] This approach utilises the group's previously reported Wilkinson's catalyst isomerisation technology[4, 5] and the knowledge that the vinyl glycosides can act as good glycosyl donors, and allyl glycosides as acceptors, under TMS-triflate catalysis. A basis set of glucosides, galactosides and fucosides was used which rely on having one specific hydroxyl acetylated (and thus selectively deprotectable) and the others benzyl protected. These components can then be provided as either or both acceptor (**c**) and donor (**b**) using the isomerisation methodology. As an illustration, this was applied to the synthesis of a 20-compound trisaccharide library **72** (Scheme 1.24).

Compounds **67–70** are converted to corresponding donors and acceptors and all combinations of disaccharide independently prepared.

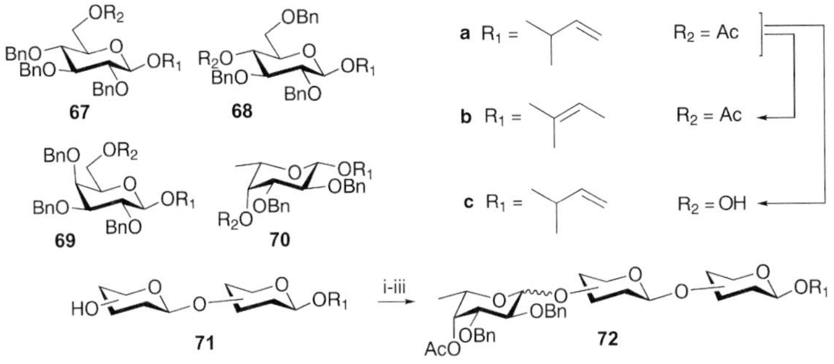

Scheme 1.24 (i) TMSOTf, MeCN, fucosyl donor; (ii) NaOMe; (iii) Pd/C, H$_2$.

Deacylation provides a library of acceptors, which can then be reacted with a donor, for example **70b**, to provide, after reduction, a library (**72**) of trisaccharides. Boons's group has also used a thioglycoside as a donor then as an acceptor (with a glycosyl fluoride donor) in a two-directional synthesis of individual trisaccharides.[88] Hindsgaul's group has also generated trisaccharide libraries in solution, essentially employing random fucosylation of Gal-β(1 → 3)GlcNAc. All 18 possible trisaccharides were generated.[89]

1.5 Multivalent carbohydrates

Protein–carbohydrate binding is generally weak, and simultaneous binding of multiple sugar residues, multivalent binding, is implicated in a number of biologically important recognition events. Multivalent recognition may use multiple recognition sites on one protein, on several receptor proteins or may involve clustering of several separate receptor proteins.[90, 91] There is growing interest in the development of multivalent carbohydrate ligand analogues as potential therapeutics.[92] The cell-dependent nature of oligosaccharide binding proteins also means synthetic multivalent carbohydrate conjugates offer potential for cell-directed delivery of therapeutic agents, a concept which has already seen applications to non-viral gene transfection. The past few years have seen a number of examples of synthetic multivalent carbohydrates reported employing a variety of synthetic linear or branched/dendritic synthetic backbones or scaffolds on which carbohydrate molecules are attached to provide a three-dimensional carbohydrate array. A number of these have shown significantly increased binding affinities and promising preclinical studies.

A variety of polymeric backbones or dendritic-type structures have been used as scaffolds. Peptidic scaffolds (generally based on lysine) have

been employed for di- and trivalent galactosides,[93–96] and more complex oligolysine structures have provided 8-mer (**73**) and 16-mer mannosides,[97] di-, tetra-, octavalent,[98] and nonavalent[99] N-acetyllactosamine-bearing dendrimers and divalent to 16-mer multivalent sialosides (**74** being the dodecavalent system).[100, 101] A polyglycine scaffold was used to prepare multiple Lewis^X structures.[102] Roy's group has reported libraries of N-linked lactosyl oligoglycines of various valencies,[103] and poly(Ala2Thr)-bearing Gal-β(1 → 4)-GalNAc-β attached to the threonine side-chain has been reported.[104] A series of oligoasparagine-based multivalent SLe^X analogues have been prepared from GlcNAc bearing a side-chain N-linked Asn. The SLe^X units were assembled on the mutlivalent system using enzymatic methods.[105] A related polyaspartimide bearing SLe^X units was described by Thoma's group using a different protocol employing

SLe^X bearing a propylamino glycoside function.[106] A Tyr-Glu-Glu 'side-chain-linked' tripeptide was the scaffold for a trigalactoside reported by Findeis, with galactosides attached by hexamine spacers to form amide links to the carboxylates of Thr, the central Glu and the side-chain acid of the terminal Glu.[107]

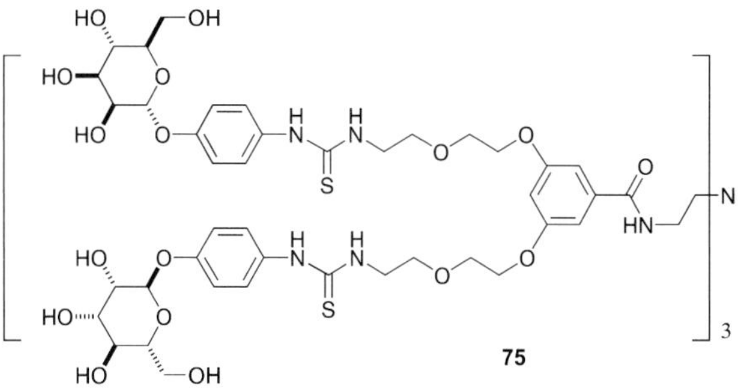

75

A number of nonpeptidic dendritic systems have been described, including the hexavalent mannoside **75**,[108] the PAMAM-dendrimer-based system **76** bearing 24 glucose or galactose residues,[109] other PAMAM-based dendritic mannosides of valency 4 through to 32,[110] and trivalent sialosides.[111] Stoddart's group has also reported a variety of dendritic carbohydrates, based on a poly(proylene imine) unit (including *S*-linked systems), and a convergent approach using amino [tris(hydroxymethyl)]-methane as a building block.[112–116] Tetra-, hexa- and octavalent thiourea linked *N*-glycosidic dendrimers,[117] and related trivalent mannosides [using tris-(2-aminoethyl)amine, tris-(3-hydroxypropyl)nitromethane or tris-(2-carboxyethyl)nitromethane as scaffolds][118] have been prepared (from glycosyl isothiocyanates).

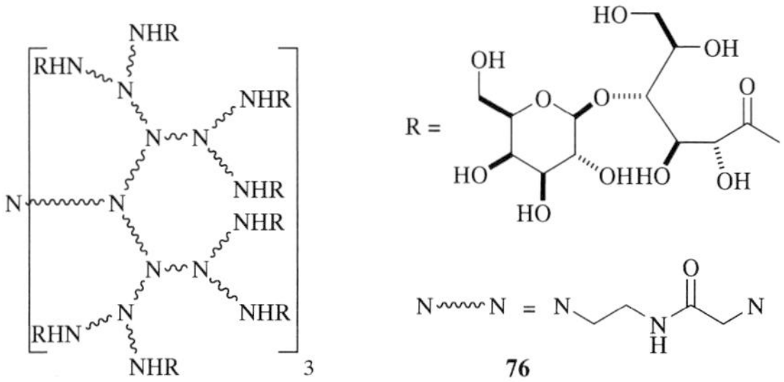

76

Another series of polymeric supported systems include polyacrylamides bearing *C*-linked sialosides (**77**).[119] These were inhibitors of virally-mediated (influenza A) hemagluttination, and their inhibitory effect was two-fold to twenty-fold greater in the presence of a strong monomeric

C-sialoside linker

R_1 = H, or C-sialidase linker

77

inhibitor of influenza neuraminidase (NA). A polyacrylamide backbone has also been used to attach the GM_3 trisaccharide (as a thioglycoside [Neu5Acα(2 → 3)-Gal-β(1 → 4)-Glcβ(1 → SR)]).[120] A range of polyamide-supported *N*-linked glycosides were prepared using chemoenzymatic methods.[121] Ring-opening metathesis polymerisation has proven valuable for the synthesis of a range of *O*- and *C*-glycosidic glucose or mannose-bearing polymers (**78**)[122, 123] and sulfated galactosides (**79**).[124]

A calixarene-based tetravalent system bearing four spacer-linked α-thiasialosides has been described.[125] A review of multivalent glyco-conjugates has recently appeared elsewhere,[126] as has a shorter review specifically of polymer-supported polyvalent glycosides.[127] There has also been a review of synthetic carbohydrate vaccine research, which covers in part examples of multivalent agents.[128]

1.6   Synthesis of oligosaccharide analogues and glycomimetics

The synthesis of complex carbohydrates whose oligomerisation is based not on *O*-glycosidic linkages, but on other bond types between the constituent monosaccharides, and the synthesis of more structurally modified glycomimetics has been the focus of a wide range of efforts. Such analogues offer the possibility of mimicking the binding potential of native oligosaccharides, whilst increasing stability to (chemical or enzymatic) glycosidic bond cleavage. This is thus a potentially vital area for the development of bioactive carbohydrate-based therapeutics.

1.6.1   C-Glycosides and biologically active conformational studies

Replacing the glycosidic oxygen with carbon could provide conformationally similar but glycosidase-stable analogues. There are now a considerable array of strategies for the synthesis of C-glycosides, and an exhaustive review is beyond the scope of this chapter. The area is the subject of a recent book.[129] We will therefore confine ourselves here to discussing some recent examples where the introduction of the C-glycosidic link has been used to probe whether the O- to C-substitution is valid with regard to bioactive conformations and thereby to carbohydrate mimetic drug design. Some comment will be added on other less common oligo-C-saccharide designs.

Kishi and co-workers were the first to undertake conformational analysis of C-linked saccharides.[130] Their results indicated a close conformational analogy with the native saccharides, and this greatly enhanced the case in favour of targeting C-glycosidic analogues.[131] Kishi's group has reported the synthesis and conformational assessment of C-trisaccharides (**80**) related to blood-group antigens. On the basis of nuclear magnetic resonance (NMR) coupling constants and nuclear Overhauser effect (NOE) experiments, conformations were proposed that were similar to the O-glycosidic systems. A series of deoxy analogues were prepared, and significant conformational changes predicted. Changes in binding affinities for the O-glycoside analogues and the corresponding C-glycosides were found to mirror each other closely, and this was taken as an indication that the binding conformations, and changes imparted by deoxygenation, were analogous.[132] This has generally led to the supposition that this could be globally extrapolated to other C-glycosides.

However, NMR and modelling studies of C-lactose (**81b**) indicate that the C-glycoside significantly populates an energy minimum which is not significantly populated by the O-glycoside.[133, 134] This important result means that it is not always possible to extrapolate free O-glycoside conformations even to free C-glycosidic conformations. The results also indicated that the C-glycoside was significantly more flexible than the O-lactose (which has likely consequences in terms of entropy loss on binding).

This analysis of free C-disaccharide conformations was extended to a comparison of protein binding by C- and O-saccharides, and provided an unexpected outcome, which has major significance for the design, and understanding, of *C*-glycosidic saccharide analogues as potential carbohydrate-based therapeutics. NMR studies show that *C*-lactose and *O*-lactose clearly adopt entirely different conformations on binding to the galactose-binding protein, ricin B. The *O*-glycoside is bound in the *syn* conformation (consistent with the crystal structure) but the *C*-glycoside is bound in the *anti* conformation.[135, 136] This is of additional note as there had until recently been relatively little previous evidence for the occurrence of the *anti* conformer in *O*-glycosides, and it has never been observed in hard-sphere, exo-anomeric (HSEA) and related calculations. However, Dabrowski's group has also recently reported that the *anti* conformer of the Galβ(1 → 3)-Glcβ1-OMe disaccharide does exist in equilibrium with the predominant *syn* form.[137] That this is indeed an energetically available conformation raises the possibility that there may be protein recognitions that select natural carbohydrate *anti* conformers. Evidence for the existence of the *anti* conformer in other carbohydrate analogues has recently been reported in the case of aminothio acetal analogues.[138] These studies demonstrate not only that *C*-glycosides may in fact adopt different unbound conformations, but that protein binding can actually select out (*C*-glycosidic) conformations which represent minimal contributions to analogous *O*-glycoside conformations and which are not bound by the same protein.

There have also been examples of *C*-glycosides based on azasugars. An approach to aza-*C*-disaccharides, such as D-azaMan-β-(1 → 5)-D-Man, was reported using Pd-catalysed coupling followed by azasugar ring elaboration.[139] This general method was extended to D-azaMan-β-(1 → 5)-D-Gal, D-azaMan-β-(1 → 6)-D-Glc, D-azaMan-β-(1 → 4)-D-Talo and D-azaMan-β-(1 → 1)-D-Glc. Johnson's group has previously reported the first example of this type of system.[140] Other groups have reported other examples of aza-*C*-disaccharides.[141–144]

1.6.2   Acetylenes

The Vasella group has published a series of papers over the past few years describing the synthesis of acetylenic *C*-glycoside building blocks for constructing acetylene-linked oligosaccharide analogues. Pd-catalysed coupling of acetylenic bromides and silanes (or germanes) is the key synthetic method employed. Their general methodology allows for the synthesis of a variety of acyclic as well as cyclic oligosaccharides of this type, illustrated by **82**[145] and **83**.[146] Cellobiose with 1-β and 4' acetylenes have also been prepared as components suitable for alternating *O*-glycosidic and acetylenic *C*-glycosidic structures.[147]

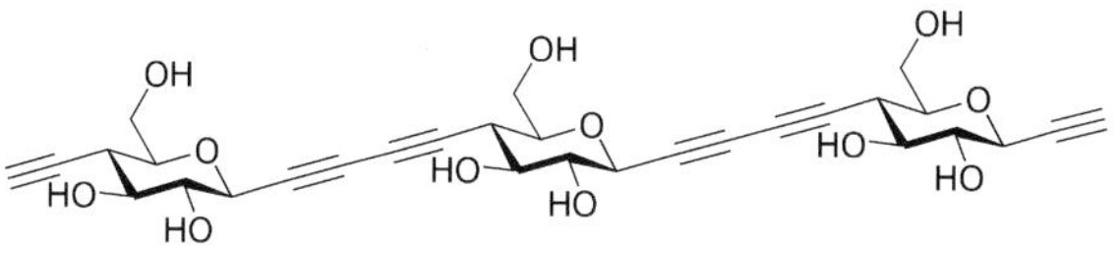

82

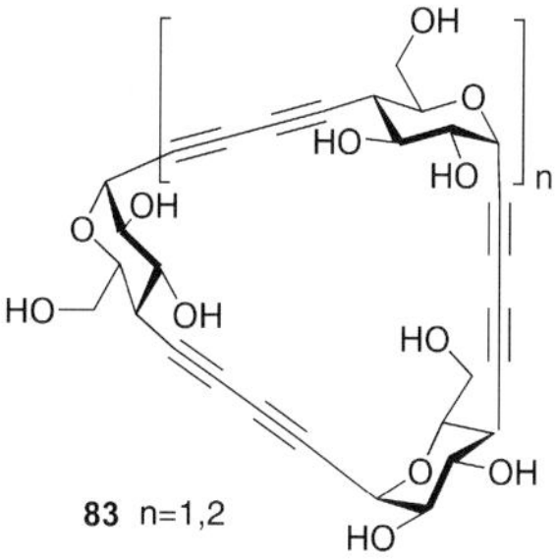

83 n=1,2

1.6.3   Thio-linked analogues

Despite the heritage of thioglycosides as players in chemical glycosylation reactions, there are fewer examples of using this anomeric functionality as a retained alternative to the glycosidic oxygen. The thio-linked disaccharide **84** (an analogue to the Fuc-GlcNac component of SLe^x) was prepared by ring opening of a cyclic sulfamidate with a free thiosugar.[148] Schmidt's group reported the synthesis of **85**, a fully thio-linked analogue using appropriate sugar thiols (prepared via regiospecific tosylate reaction with KSAc) as glycosyl acceptors.[149] Earlier examples of sulfur-linked saccharides have been reviewed.[150]

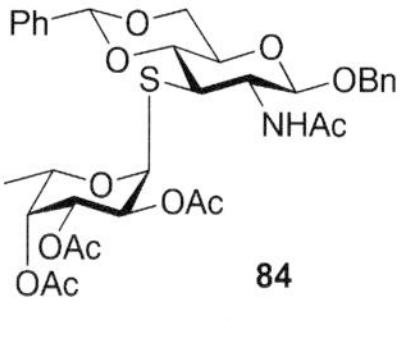

84

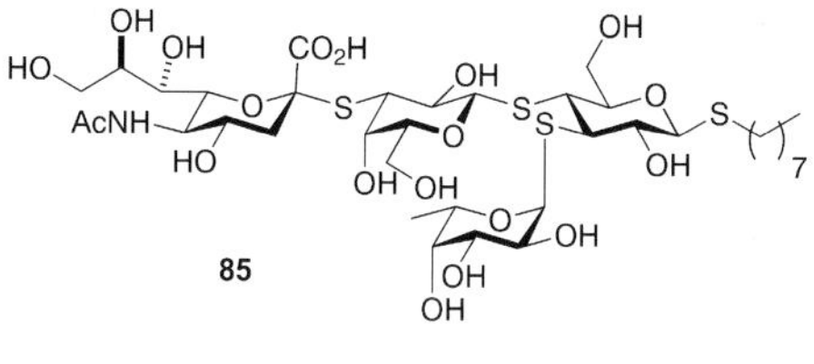

85

1.6.4   Amide- and phosphate-linked oligosaccharides

Syntheses of oligosaccharides linked by amide bonds or phosphate diesters have been described. The attraction of amide linkages lies in the facility of their construction from amine-carboxylate precursors and the stability of the resulting amide bond. Wessel and co-workers have reported 2-amino sugars linked through amide bonds to O(3) of the adjacent monomer (e.g. **86**),[151] and also direct amide links to C(6) (from 6-carboxyl sugars) giving structures such as **87**.[152] In the latter case, the acid function was also used as the handle for attachment to a solid support (via a benzhydrylamine polystyrene resin) providing thereby the means to prepare tetrasaccharide analogues on solid phase. This methodology is clearly well set up for application to the synthesis of libraries, as it employs well-established amide bond forming reactions for linking the sugar units, and there are no concerns about anomeric stereocontrol. A third variation of the use of amide links is the use of an anomeric carboxylate to link 2-amino sugars providing β1,2-linked 'carbopeptoids' of type **88**.[153] The synthesis of **89** has also been reported, which is a 1,6-linked, and sulfated, analogue of **88**; compound **89** was shown to be a potent inhibitor of human immunodeficiency virus (HIV) replication at ca. 1 μM.[154]

There have also been described several examples of amide-linked sialic acid oligomer mimetics, mostly notably those from Gervay's group. This group reported, first, the (solution-phase) preparation of the amide-linked dimer **90**,[155] and subsequently, the synthesis of dimeric (e.g. **91**) through to octameric amide-linked sialooligomers, using solid-supported peptide synthesis chemistry on Rink amide resin.[156] The latter study used circular dichroism to demonstrate ordered secondary structures in the higher oligomers. An amide-linked sialic-acid–galactose dimer, **92**, has also been synthesised.[157] Some other galactose-, glucose- and fucose-containing amide-linked disaccharides have likewise been prepared.

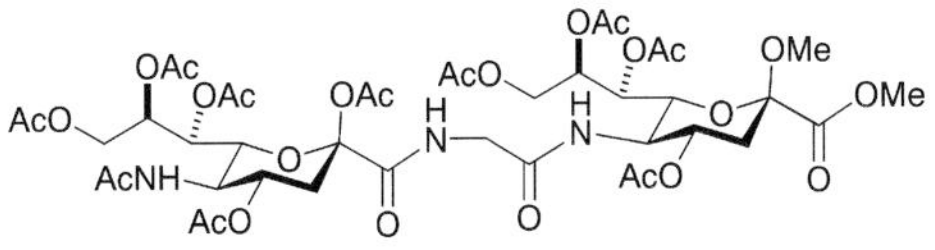

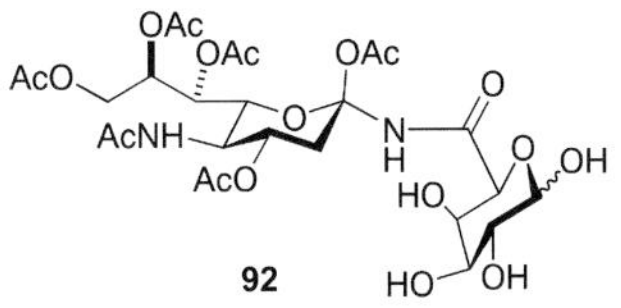

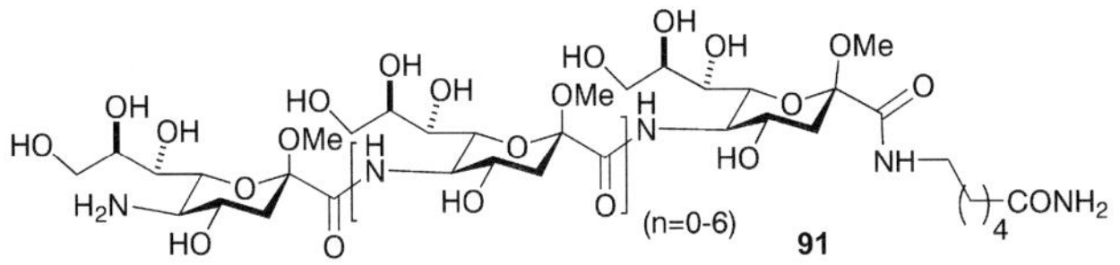

Nicolaou's group has reported the synthesis of phosphate diester oligosaccharide analogues, nicknamed 'carbonucleotoids' (Scheme 1.25). These were prepared via standard nucleic acid type iterative coupling chemistry. All monomers were derived from **94**, prepared from the known *C*-glucoside **93**.[158,159]

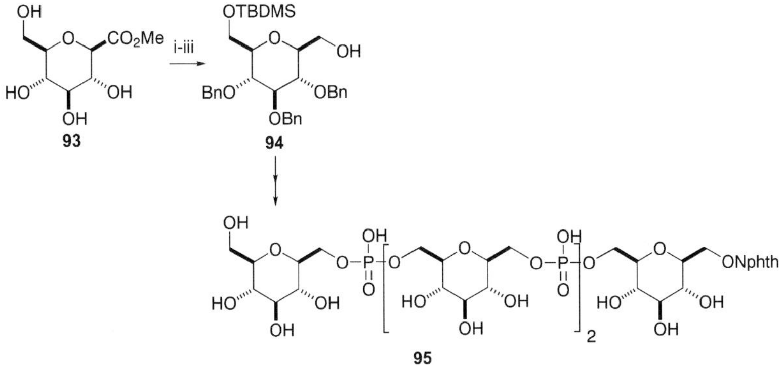

Scheme 1.25   (i) TBDMSCl, Imidazole, DMF; (ii) BnBr (9 equiv.), Ag₂O (6 equiv.), DMF; (iii) DIBAL, THF.

1.6.5   Structural mimetics

We will now discuss examples where the analogy extends beyond merely replacing the glycosidic oxygen with another functionality to substitution of whole sugar units by nonsugar units, providing true glycomimetics.

The selectins have been targets for the synthesis of various mimetics and have provided a testing ground for the development of some interesting structural replacements of sugar units. A SLe^X mimetic, in which the GalNAc linking the Fuc and the Gal residues was replaced with a *trans*-1,2-cyclohexane diol, **96**, showed about four times stronger E- and P-selectin affinity than SLe^X itself.[160] Further simplified analogues, **97**, with the galactose replaced by simple alkyl or substituted alkyl chains, also showed mM IC_{50}s against E- and/or P-selectin. A further related structural mimetic of SLe^X also replaced the sialic acid with a carboxylate and a methylene cyclohexane to generate **98**.[161] An NMR conformational comparison of **98** with SLe^X was suggestive of the mimetic adopting a solution conformation very similar to that of the native ligand.[162] In other work the GlcNAc has also been replaced.[163]

96

97

a R = R_1 = H, n =1
b R = R_1 = CH_2OH, n =1
c R = OH, R_1 = H, n =1
d R = R_1 = H, n = 2
e R = R_1 = H, n = 3

Hanessian and co-workers have described the preparation of two series of SLe^X mimetics. In the first the sialic acid residue was replaced with a methylene carboxylate, and the galactose and/or *N*-Ac glucosamine units by a (−)-quinic-acid-derived carbocycle and/or a conformationally biased ethylene dioxy tether, as in compounds **99**–**102**.[164] Hanessian's group has also described the synthesis of a series of mimetics (**103**–**106**) which utilise mono- and bicyclic lactams as β-turn scaffold-type mimetics replacing the GlcNAc unit.[165]

98

99

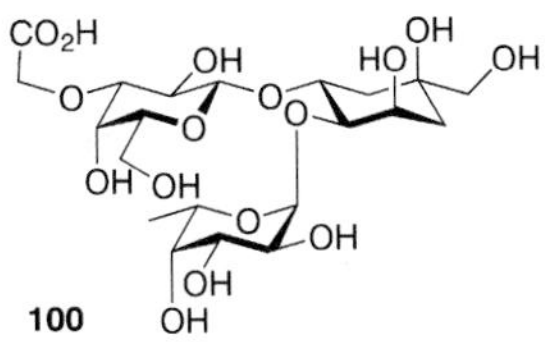

100 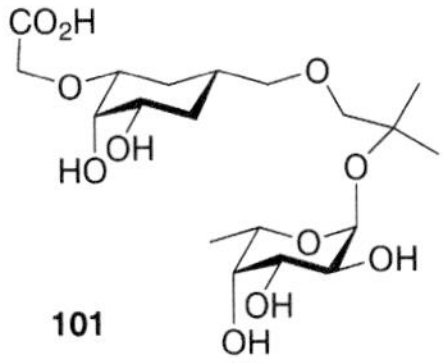   101

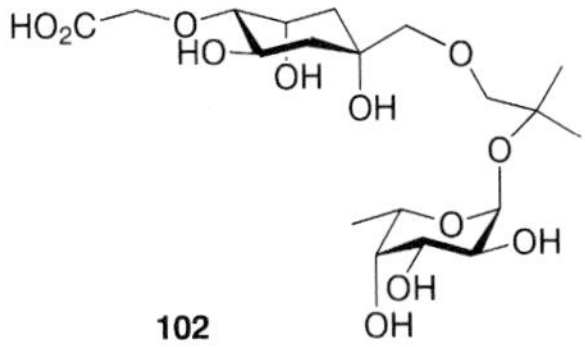

102 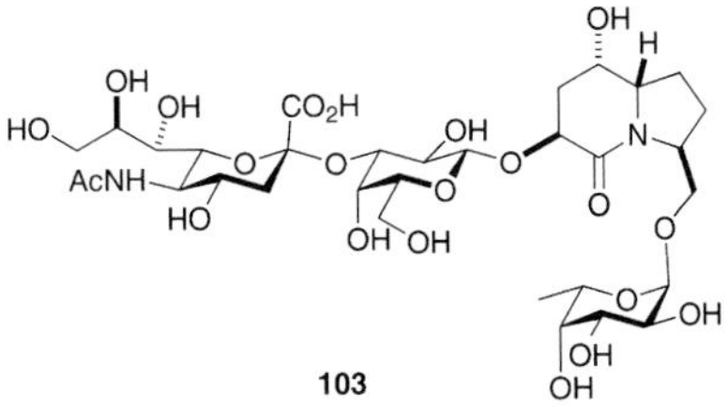   103

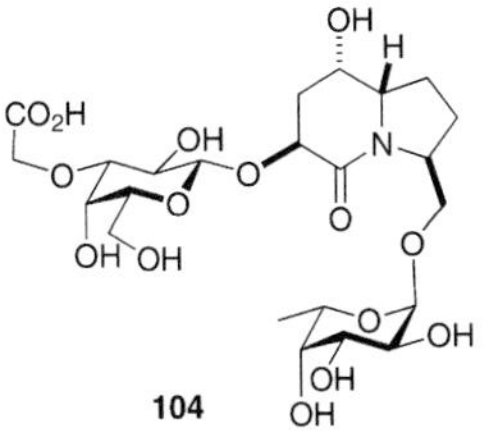

104 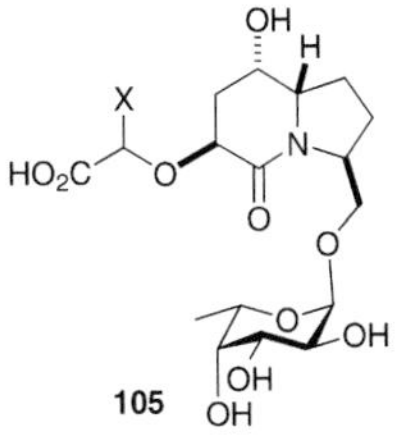   105 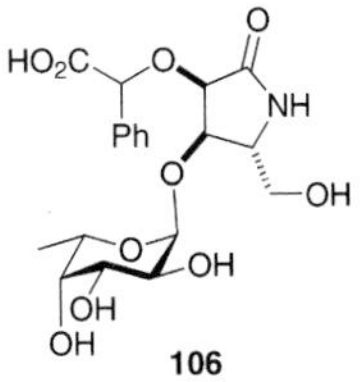   106

Wong's group have reported a considerable number of SLe^X mimetics, several with good selectin affinity. A range of *O*- and *C*-fucosidic analogues **107–111** showed IC$_{50}$s of 1.3–20 mM.[166] Lipid attachment evidently leads to some increases in activity.[167] Further analogues based on **111** but using an aminocyclohexanol (e.g. **112** and **113**) or a threonine (**114–118**) as the linking core section were subsequently described. Only those with IC$_{50}$s < 5 mM are shown here.[168, 169] The diol side-chain of the peptidic component is proposed to mimic the essential 4-OHs and 6-OHs of the galactose present in sialyl Lewis X.

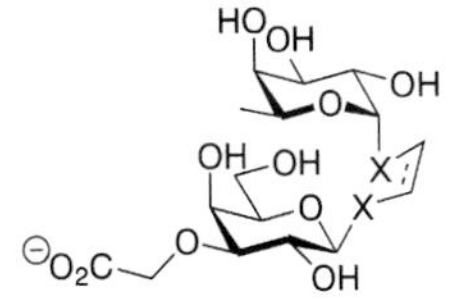

107 X = X = O
108 X = X = CH$_2$
109 X = X = CH$_2$, alkene

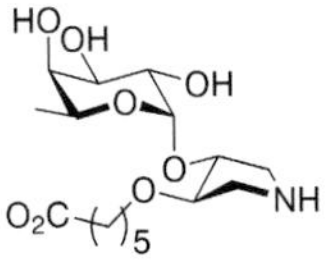

110

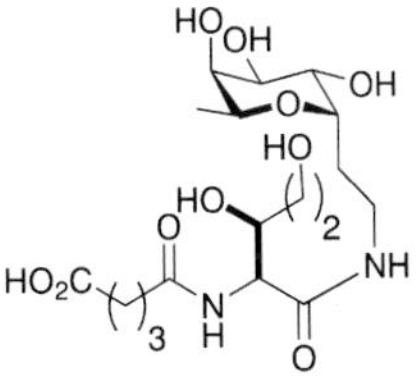

111

The same group has also prepared a series of mannose-based analogues along similar lines.[170] In addition, various fucose-based and mannose-based analogues have been described containing a range of different non-carbohydrate components designed to attach functionality suitable to the binding site of E-, P- or L-selectin (based on models of the binding site).

Wong's group has also announced the synthesis of the 1,1-linked disaccharide (3-*O*-carboxymethyl)-β-D-galactopyranosyl α-D-mannopyranoside **119**, from a precursor silyl galactoside and a mannosyl fluoride. The disaccharide has five times the E-selectin binding affinity of SLe^X.[171] Examples of other SLe^X mimetics binding to E-, P- or L-selectins (and in many cases all three) with affinties of 0.3 μM to 3 mM are **120**–**123**.[172–179]

Another interesting series of conformational mimetics have been reported by Richard Taylor's group at York, which utilise functionalised tetralin or naphthyl ring systems as scaffolds, simultaneously replacing all three sugars apart from the fucose (**124**–**126**).[180] A series of biaryl mannosides (**127**–**131**) rationally designed based on a model of SLe^X–E-selectin binding and prior structure–activity relationship (SAR) information, have been described. Amongst these were several mimetics with better affinity than SLe^X itself (inhibition of SLe^X-expressing HL60 cells binding to E-selectin).[181] The IC_{50}s for **127** and **128** were 2 mM and 2.6 mM, respectively, that of **129** was 4.9 mM, and **130** and **131** 4.6 mM and 3.8 mM, respectively (SLe^X, 3 mM). The biaryl systems were constructed by Pd(0)-catalysed boronic acid-halide coupling and attached to the sugar by BF_3-catalysed glycosylation of the mannose peracetate.

References

1. G.-J. Boons, 1996, *Tetrahedron*, 52, 1095-1121; K.J. Hale and A.C. Richardson, 1993, in *The Chemistry of Natural Products*, 2nd edn, ed. R.H. Thomson, Blackie, London, 1.
2. L.A.J.M. Sliedrecht, G.A. van der Marel, and J.H. van Boom, 1994, *Tetrahedron Lett.*, 35, 4015-4018.
3. R. Roy, F.O. Andersson, and M. Letellier, 1992, *Tetrahedron Lett.*, 33, 6053-6056.
4. G.-J. Boons, A. Burton, and S. Isles, 1996, *J. Chem. Soc., Chem. Commun.*, 141-142.
5. G.-J. Boons and S. Isles, 1994, *Tetrahedron Lett.*, 35, 3593-3596.
6. K.C. Nicolaou, N.J. Bockovich, and D.R. Carcanague, 1993, *J. Am. Chem. Soc.*, 115, 8843-8844, and references therein.
7. O. Kanie, Y. Ito, and T. Ogawa, 1994, *J. Am. Chem. Soc.*, 116, 12073-12074.
8. O. Kanie, Y. Ito, and T. Ogawa, 1996, *Tetrahedron Lett.*, 37, 4551-4553.
9. H. Yamada, T. Haada, H. Miyazaki, and T. Takahashi, 1994, *Tetrahedron Lett.*, 35, 3979-3982.
10. S.H. Khan and R.A. O'Neill, 1996, *Modern Methods in Carbohydrate Synthesis*, Harwood, Amsterdam.
11. D. Kahne, S. Walker, Y. Cheng, and D. Van Engen, 1989, *J. Am. Chem. Soc.*, 111, 6881-6882.
12. L. Yan and D. Kahne, 1996, *J. Am. Chem. Soc.*, 118, 9239-9248.
13. N. Khiar and M. Martin-Lomas, 1995, *J. Org. Chem.*, 60, 7017-7021.
14. S. Mehta and B.M. Pinto, 1991, *Tetrahedron Lett.*, 32, 4435-4438.
15. S. Mehta and B.M. Pinto, 1993, *J. Org. Chem.*, 58, 3269-3276.
16. H.M. Zuurmond, P.A.M. van der Klein, P.H. van der Meer, G.A. van der Marel, and J.H. van Boom, 1992, *Rec. Trav. Chim. Pays-Bas*, 111, 365-366.
17. H.M. Zuurmond, P.A.M. van der Klein, P.H. van der Meer, G.A. van der Marel, and J.H. van Boom, 1993, *J. Carbohydr. Chem.*, 12, 1091-1103.
18. L.A.J.M. Sliedregt, H.J.G. Broxterman, G.A. van der Marel, and J.H. van Boom, 1994, *Carbohydr. Lett.*, 1, 61-68.
19. T. Furuta, K. Takeuchi, and M. Iwamura, 1996, *J. Chem. Soc., Chem. Commun.*, 157-158.
20. S.V. Ley, R. Downham, P.J. Edwards, J.E. Innes, and M. Woods, 1995, *Contemp. Org. Synth.*, 365.
21. M.-K. Cheung, N.L. Douglas, B. Hinzen, S.V. Ley, and X. Pannecoucke, 1997, *Synlett*, 257-260.
22. P. Grice, S.V. Ley, J. Pietruszka, H.M.I. Osborn, H.W.M. Priepke, and S.L. Warriner, 1997, *Chem. Eur. J.*, 3, 431-440.
23. R.V. Stick, D.M.G. Tilbrook, and S.J. Williams, 1997, *Aust. J. Chem.*, 50, 233-235.
24. R.V. Stick, D.M.G. Tilbrook, and S.J. Williams, 1997, *Aust. J. Chem.*, 50, 237-240.
25. B. Fraser-Reid, J.R. Merritt, A.L. Handlon, and W.C. Andrews, 1993, *Pure Appl. Chem.*, 65, 779-786.
26. J.G. Allen and B. Fraser-Reid, 1999, *J. Am. Chem. Soc.*, 121, 468-469.
27. R. Madsen, U.E. Udodong, C. Robeerts, D.R. Mootoo, P. Konradsson, and B. Fraser-Reid, 1995, *J. Am. Chem. Soc.*, 117, 1554-1565, and preceding papers in series.
28. S.J. Danishefsky and R.L. Halcomb, 1989, *J. Am. Chem. Soc.*, 111, 6661-6666.
29. S.J. Danishefsky and M.T. Bilodeau, 1996, *Angew. Chem., Int. Ed. Engl.*, 35, 1380-1419.
30. K.K.-C. Liu and S.J. Danishefsky, 1993, *J. Am. Chem. Soc.*, 115, 4933-4934.
31. J. Gervay, J.M. Peterson, T. Oriyama, and S.J. Danishefsky, 1993, *J. Org. Chem.*, 58, 5465-5468.
32. S.J. Danishefsky, J. Gervay, J.M. Peterson, F.E. McDonald, K. Koseki, D.A. Griffith, T. Oriyama, and S.P. Marsden, 1995, *J. Am. Chem. Soc.*, 117, 1940-1953.
33. V. Behar and S.J. Danishefsky, 1994, *Angew. Chem., Int. Ed. Engl.*, 33, 1468-1470.

34. S.J. Danishefsky, V. Behar, J.T. Randolph, and K.O. Lloyd, 1995, *J. Am. Chem. Soc.*, 117, 5701-5711.

35. J.T. Randolph and S.J. Danishefsky, 1994, *Angew. Chem., Int. Ed. Engl.*, 33, 1470-1473.

36. G. Ragupathi, T.K. Park, S. Zhang, I.J. Kim, L. Graber, S. Adluri, K.O. Lloyd, S.J. Danishefsky, and P.O. Livingston, 1997, *Angew. Chem., Int. Ed. Engl.*, 36, 125-128.

37. M.T. Bilodeau, T.K. Park, S. Hu, J.T. Randolph, S.J. Danishefsky, P.O. Livingston, and S. Zhang, 1995, *J. Am. Chem. Soc.*, 117, 7840-7841.

38. J. Aiyar, S.A. Hitchock, D. Denhart, K.K.-C. Liu, S.J. Danishefsky, and D.M. Crithers, 1994, *Angew. Chem., Int. Ed. Engl.*, 33, 855-858, and references therein.

39. A. Vasella, 1993, *Pure Appl. Chem.*, 65, 731-752.

40. K. Briner, B. Bernet, J.-L. Maloisel, and A. Vasella, 1994, *Helv. Chim. Acta*, 77, 1969-1984.

41. E. Bozo and A. Vasella, 1994, *Helv. Chim. Acta*, 77, 745-753.

42. P.R. Muddasani, E. Bozo, B. Bernet, and A. Vasella, 1994, *Helv. Chim. Acta.*, 77, 257-290.

43. P.R. Muddasani, B. Bernet, and A. Vasella, 1994, *Helv. Chim. Acta.*, 77, 334-350.

44. S.-I. Hashimoto, H. Sakamoto, T. Honda, and S. Ikegami, 1997, *Tetrahedron Lett.*, 38, 5181-5184.

45. H. Kondo, S. Aoki, Y. Ichikawa, R.L. Halcomb, H. Ritzen, and C.-H. Wong, 1994, *J. Org. Chem.*, 59, 864-877.

46. H. Schene and H. Waldmann, 1998, *Eur. J. Org. Chem.*, 1227-1230.

47. S.C. Crawley and M.M. Palcic, 1996, in *Modern Methods in Carbohydrate Synthesis* (S.H. Khan and R.A. O'Neill, eds), Harwood, Amsterdam, pp. 492-517.

48. H.J.M. Gijsen, L. Qiao, W. Fitz, and C.-H. Wong, 1996, *Chem. Rev.*, 96, 443-473.

49. G.J. McGarvey and C.-H. Wong, 1997, *Liebigs Ann. Recueil*, 1059-1074.

50. C.-H. Wong, 1996, in *Modern Methods in Carbohydrate Synthesis* (S.H. Khan and R.A. O'Neill, eds), Harwood, Amsterdam, pp. 467-491.

51. C.-H. Wong and G.M. Whitesides, 1994, in *Enzymes in Synthetic Organic Chemistry*, Pergamon, Oxford, 252-297.

52. K. Drickamer and M.E. Taylor, 1998, *Trends in Biochem. Sci.*, 321-324.

53. B.W. Murray, S. Takayama, J. Schultz, and C.-H. Wong, 1996, *Biochemistry*, 35, 11183-11195.

54. H. Schacter, 1994, in *Molecular Glycobiology* (O. Hindsgaul, ed), IRL Press, New York, pp. 88-162.

55. Y. Ichikawa, J.J.-C. Liu, G.-J. Shen, and C.-H. Wong, 1991, *J. Am. Chem. Soc.*, 113, 6300-6302.

56. G.C. Look, Y. Ichikawa, G.-J. Shen, G.-J. Cheng, and C.-H. Wong, 1993, *J. Org. Chem.*, 58, 4326-4330.

57. Y. Ichikawa, Y.-C. Lin, D.P. Dumas, G.-J. Shen, E. Garcia-Junceda, M.A. Williams, R. Baker, C. Ketcham, L.E. Walker, J.-C. Paulson, and C.-H. Wong, 1992, *J. Am. Chem. Soc.*, 114, 9283-9289.

58. S.A. DeFrees, W. Kosch, W. Way, J.C. Paulson, S. Sabesan, R.L. Halcomb, D.-H. Huang, Y. Ichikawa, and C.-H. Wong, 1995, *J. Am. Chem. Soc.*, 117, 66-79.

59. S.A. DeFrees, F.C.A. Gaeta, Y.C. Lin, Y. Ichikawa, and C.-H. Wong, 1993, *J. Am. Chem. Soc.*, 115, 7549-7550.

60. C. Deluca, M. Lansing, I. Martin, F. Crescenzi, G.-J. Shen, M. O'Regan, and C.-H. Wong, 1995, *J. Am. Chem. Soc.*, 117, 5869-5870.

61. G.M. Watt, L. Revers, M.C. Webberley, I.B.H. Wilson, and S.L. Flitsch, 1997, *Angew. Chem., Int. Ed. Engl.*, 36, 2354-2355.

62. G.F. Herrmann, Y. Ichikawa, C. Wandrey, F.C.A. Gaeta, J.C. Paulson, and C.-H. Wong, 1993, *Tetrahedron Lett.*, 34, 3091-3094.

63. G. Vic, M. Scigelova, J.J. Hastings, O.W. Howarth, and D.H.G. Crout, 1996, *J. Chem. Soc., Chem. Commun.*, 1473-1474.

64. S. Singh, M. Scigelova, and D.H.G. Crout, 1996, *J. Chem. Soc., Chem. Commun.*, 993-994.
65. L.F. Mackenzie, Q.P. Wang, R.A.J. Warren, and S.G. Withers, 1998, *J. Am. Chem. Soc.*, 120, 5583-5584.
66. K.G. Bowman and C.R. Bertozzi, 1999, *Chem. & Biol.*, 6, R9-R22.
67. P.R. Scudder, K. Shailubhai, K.L. Duffin, P.R. Streeter, and G.S. Jacob, 1994, *Glycobiology*, 4, 929-933.
68. E.V. Chandrasekaran, R.K. Jain, R.D. Larsen, K. Wlaichuk, and K.L. Matta, 1995, *Biochemistry*, 34, 2925-2936.
69. P.H. Seeberger and S.J. Danishefsky, 1998, *Acc. Chem. Res.*, 31, 685-695.
70. K.F. McClure, J.T. Randolph, R. Ruggeri, and S.J. Danishefsky, 1993, *Science*, 260, 1307-1309.
71. J.Y. Roberge, X. Beebe, and S.J. Danishefsky, 1995, *Science*, 269, 202-204.
72. K.C. Nicolaou, N. Winssinger, J. Pastor, and F. DeRoose, 1997, *J. Am. Chem. Soc.*, 119, 449-450.
73. Y. Ito and T. Ogawa, 1997, *J. Am. Chem. Soc.*, 119, 5562-5566.
74. F. Barressi and O. Hindsgaul, 1996, in *Modern Methods in Carbohydrate Synthesis*, (S.H. Khan and R.A. O'Neill, eds) Harwood, Amsterdam, pp. 251-276.
75. A. Heckel, E. Mross, K.-H. Jung, J. Rademann, and R.R. Schmidt, 1998, *Synlett.*, 171-173.
76. J. Rademann and R.R. Schmidt, 1996, *Tetrahedron Lett.*, 37, 3989-3990.
77. M. Adinolfi, G. Barone, L. De Napoli, A. Iadonisi, and G. Piccialli, 1996, *Tetrahedron Lett.*, 37, 5007-5010.
78. J.A. Hunt and W.R. Roush, 1996, *J. Am. Chem. Soc.*, 118, 9998-9999.
79. L. Yan, C.M. Taylor, R. Goodnow, Jr, and D. Kahne, 1994, *J. Am. Chem. Soc.*, 116, 6953-6954.
80. R. Liang, L. Yan, J. Loebach, M. Ge, Y. Uozumi, K. Sekanina, N. Horan, J. Gildersleeve, C. Thompson, A. Smith, K. Biswas, W.C. Still, and D. Kahne, 1996, *Science*, 274, 1520-1522.
81. R. Liang, J. Loebach, N. Horan, M. Ge, C. Thompson, L. Yan, and D. Kahne, 1997, *Proc. Natl. Acad. Sci.*, 94, 10554-10559.
82. R. Rodebaugh, S. Joshi, B. Fraser-Reid, and H.M. Geysen, 1997, *J. Org. Chem.*, 62, 5660-5661.
83. T. Zhu and G.-J. Boons, 1990, *Angew. Chem., Int. Ed. Engl.*, 37, 1898-1900.
84. Z.-G. Wang, S.P. Douglas, and J.J. Krepinsky, 1996, *Tetrahedron Lett.*, 37, 6985-6988.
85. S.P. Douglas, D.M. Whitfield, and J.J. Krepinsky, 1995, *J. Am. Chem. Soc.*, 117, 2116-2117.
86. L. Jiang, R.C. Hartley, and T.-K. Chan, 1996, *J. Chem. Soc., Chem. Commun.*, 2193-2194.
87. G.-J. Boons, B. Heskamp, and F. Hout, 1996, *Angew. Chem., Int. Ed. Engl.*, 35, 2845-2847.
88. T. Zhu and G.-J. Boons, 1998, *Tetrahedron Lett.*, 39, 2187-2190.
89. O. Kanie, F. Barresi, Y. Ding, J. Labbe, A. Otter, L.S. Forsberg, B. Ernst, and O. Hindsgaul, 1995, *Angew. Chem., Int. Ed. Engl.*, 34, 2720-2722.
90. L.L. Kiessling and N.A. Pohl, 1996, *Chem. & Biol.*, 3, 71-77.
91. K. Drickamer, 1995, *Nature Struct. Biol.*, 2, 437-439.
92. J.M. Gardiner, 1998, *Expert Opinion in Investigational Drugs*, 7, 405-411.
93. J. Haensler and F.C. Szoka, Jr, 1993, *Bioconjugate Chem.*, 4, 85-93.
94. E.A.L. Biessen, H. Broxterman, J.H. van Boom, and T.J.C. van Berkel, 1995, *J. Med. Chem.*, 38, 1846-1852.
95. R.T. Lee and Y.C. Lee, 1997, *Bioconjugate Chem.*, 8, 762-765.
96. A.R.P.M. Valentijn, G.A. van der Marel, L.A.J.M. Sliedregt, T.J.C. van Berkel, E.A.L. Biessen, and J.H. van Boom, 1997, *Tetrahedron*, 53, 759-770.
97. D. Pagé, D. Zanini, and R. Roy, 1996, *Bioorg. Med. Chem.*, 4, 1949-1961.
98. D. Zanini, W.K.C. Park, and R. Roy, 1995, *Tetrahedron Lett.*, 36, 7383-7386.

99. R. Roy, W.K.C. Park, Q. Wu, and S.-N. Wang, 1995, *Tetrahedron Lett.*, 36, 4377-4380.
100. D. Zanini and R. Roy, 1997, *J. Am. Chem. Soc.*, 119, 2088-2095.
101. R. Roy, 1996, *Current Opin. Struct. Biol.*, 6, 692-702.
102. K. von dem Bruch and H. Kunz, 1994, *Angew. Chem., Int. Ed. Engl.*, 33, 101-103.
103. R. Roy and U.K. Saha, 1996, *J. Chem. Soc., Chem. Commun.*, 201-202.
104. T. Tsuda and S.-I. Nishimura, 1996, *J. Chem. Soc., Chem. Commun.*, 2779-2780.
105. G. Baisch and R. Ohrlein, 1996, *Angew. Chem., Int. Ed. Engl.*, 35, 1812-1815.
106. G. Thoma, B. Ernst, F. Schwarzenbach, and R.O. Duthaler, 1997, *Bioorg. Med. Chem. Lett.*, 7, 1705-1708.
107. M. Findeis, 1994, *Int. J. Pep. Prot. Res.*, 48, 477-485.
108. D. Pagé, S. Aravind, and R. Roy, 1996, *J. Chem. Soc., Chem. Commun.*, 1913-1914.
109. K. Aoi, K. Itoh, and M. Okada, 1995, *Macromolecules*, 28, 5391-5393.
110. D. Pagé and R. Roy, 1997, *Bioconjugate Chem.*, 8, 714-723.
111. G. Kretzschmar, U. Sprengard, H. Kunz, E. Bartnik, W. Schmidt, A. Toepfer, B. Horsch, M. Krause, and D. Seiffge, 1995, *Tetrahedron*, 51, 13015-13030.
112. P.R. Ashton, S.E. Boyd, C.L. Brown, S.A. Nepogodiev, E.W. Meijer, H.W.I. Peerlings, and J.F. Stoddart, 1997, *Chem. Eur. J.*, 3, 974-984.
113. P.R. Ashton, S.E. Boyd, C.L. Brown, N. Jayaraman, S.A. Nepogodiev, and J.F. Stoddart, 1996, *Chem. Eur. J.*, 2, 1115-1128.
114. P.R. Ashton, S.E. Boyd, C.L. Brown, N. Jayaraman, and J.F. Stoddart, 1997, *Angew. Chem., Int. Ed. Engl.*, 36, 732-735.
115. N. Jayaraman and J.F. Stoddart, 1997, *Tetrahedron Lett.*, 38, 6767-6770.
116. H.W.I. Peerlings, S.A. Nepogdiev, J.F. Stoddart, and E.W. Meijer, 1998, *Eur. J. Org. Chem.*, 1879-1886.
117. T.K. Lindhorst and C. Kieburg, 1996, *Angew. Chem., Int. Ed. Engl.*, 35, 1953-1956.
118. S. Kotter, U. Krallmann-Wenzel, S. Ehlers, and T.K. Lindhorst, 1998, *J. Chem. Soc. Perkin Trans 1*, 2193-2200.
119. S.-K. Choi, M. Mammen, and G.M. Whitesides, 1996, *Chem. & Biol.*, 3, 97-104.
120. S. Cao and R. Roy, 1996, *Tetrahedron Lett.*, 37, 3421-3424.
121. C. Unverzagt, S. Kelm, and J.C. Paulson, 1994, *Carbohydr. Res.*, 251, 285-301.
122. M. Kanai, K.H. Mortell, and L.L. Kiessling, 1997, *J. Am. Chem. Soc.*, 119, 9931-9932.
123. K.H. Mortell, R.V. Weatherman, and L.L. Kiessling, 1996, *J. Am. Chem. Soc.*, 118, 2297-2298.
124. D.D. Manning, L.E. Strong, X. Hu, P.J. Beck, and L.L. Kiessling, 1997, *Tetrahedron*, 53, 11937-11952.
125. S.J. Meunier and R. Roy, 1996, *Tetrahedron Lett.*, 37, 5469-5472.
126. R. Roy, 1997, *Topics in Current Chemistry*, 187, 241-274.
127. R. Roy, 1996, *Trends in Glycosci, Glycotech.*, 8, 79-99.
128. T. Toyokuni and A.K. Singhal, 1995, *Chem. Soc. Rev.*, 95, 231-242.
129. D.E. Levy and C. Tang, 1995, in *The chemistry of C-glycosides* (J.E. Baldwin and P.D. Magnus, eds), Pergamon, Oxford.
130. Y. Kishi, 1993, *Pure Appl. Chem.*, 65, 771-778.
131. A. Wei, A. Haudrechy, C. Audin, H.-S. Jun, N. Haudrechy-Bretel, and Y. Kishi, 1995, *J. Org. Chem.*, 60, 2160-2169.
132. A. Wei, K.M. Boy, and Y. Kishi, 1995, *J. Am. Chem. Soc.*, 117, 9432-9436.
133. J.-F. Espinosa, M. Martín-Pastor, J.L. Asensio, H. Dietrich, M. Martín-Lomas, R.R. Schmidt, and J. Jiménez-Barbero, 1995, *Tetrahedron Lett.*, 36, 6329-6332.
134. J.-F. Espinosa, H. Dietrich, M. Martín-Lomas, R.R. Schmidt, and J. Jiménez-Barbero, 1996, *Tetrahedron Lett.*, 37, 1467-1470.
135. J.-F. Espinosa, F.J. Cañada, J.L. Asensio, M. Martín-Pastor, H. Dietrich, M. Martín-Lomas, R.R. Schmidt, and J. Jiménez-Barbero, 1996, *J. Am. Chem. Soc.*, 118, 10862-10871.

136. J.-F. Espinosa, F.J. Cañada, J.L. Asensio, H. Dietrich, M. Martín-Lomas, R.R. Schmidt, and J. Jiménez-Barbero, 1996, *Angew. Chem., Int. Ed. Engl.*, 35, 303-306.
137. J. Dabrowski, T. Kozar, H. Grosskurth, and N.E. Nifant'ev, 1995, *J. Am. Chem. Soc.*, 117, 5534-5539.
138. K. Bock, J.Ø. Duus, and S. Refu, 1994, *Carbohydr. Res.*, 253, 51-67.
139. B.A. Johns, Y.T. Pan, A.D. Elbein, and C.R. Johnson, 1997, *J. Am. Chem. Soc.*, 119, 4856-4865.
140. C.R. Johnson, M.W. Miller, A. Golebiowski, H. Sundram, and M.B. Ksebati, 1994, *Tetrahedron Lett.*, 35, 8991-8994.
141. O.M. Saavedra and O.R. Martin, 1996, *J. Org. Chem.*, 61, 6987-6993.
142. O.R. Martin, L. Liu, and F. Yang, 1996, *Tetrahedron Lett.*, 37, 1991-1994.
143. E. Frérot, C. Marquis, and P. Vogel, 1996, *Tetrahedron Lett.*, 37, 2023-2026.
144. A. Baudat and P. Vogel, 1996, *Tetrahedron Lett.*, 37, 483-484.
145. C.Z. Cai and A. Vasella, 1996, *Helv. Chim. Acta.*, 79, 255-268.
146. R. Burli and A. Vasella, 1997, *Helv. Chim. Acta.*, 80, 1027-1052; ibid. 2215-2237.
147. A. Ernst and A. Vasella, 1996, *Helv. Chim. Acta.*, 79, 1279-1294.
148. B. Aguilera and A. Fernandez-Mayoralas, 1996, *J. Chem. Soc., Chem. Commun.*, 127-128.
149. T. Eisele, A. Toepfer, G. Kretzxhmar, and R.R. Schmidt, 1996, *Tetrahedron Lett.*, 37, 1389-1392.
150. J. Defaye and J. Gelas, 1991, in *Studies in Natural Product Chemistry*, ed. A. Rahman, Elsevier, Amsterdam, 315-357.
151. H.P. Wessel, C.M. Mitchell, C.M. Labato, and G. Schmid, 1994, *Angew. Chem., Int. Ed. Engl.*, 34, 2712-2713.
152. C. Muller, E. Nitas, and P. Wessel, 1995, *J. Chem. Soc., Chem. Commun.*, 2425-2426.
153. Y. Suhara, J.E.K. Hildreth, and Y. Ichikawa, 1996, *Tetrahedron Lett.*, 37, 1575-1578.
154. Y. Suhara, M. Ichikawa, J.E.K. Hildreth, and Y. Ichikawa, 1996, *Tetrahedron Lett.*, 37, 2549-2552.
155. P.S. Ramamorthy and J. Gervay, 1997, *J. Org. Chem.*, 62, 7801-7805.
156. L. Szabo, B.L. Smith, K.D. McReynolds, A.L. Parrill, E.R. Morris, and J. Gervay, 1998, *J. Org. Chem.*, 63, 1074-1078.
157. S. Sabesan, 1997, *Tetrahedron Lett.*, 38, 3127-3130.
158. E.-F. Fuchs and J. Lehman, 1975, *Chem. Ber.*, 108, 2254.
159. K.C. Nicolaou, H. Florker, M.G. Egan, T. Barth, and V.A. Estevez, 1995, *Tetrahedron Lett.*, 36, 1775-1778.
160. A. Toepfer, G. Kretschmar, and E. Bartnik, 1995, *Tetrahedron Lett.*, 36, 9161-9164.
161. H.C. Kolb and B. Ernst, 1997, *Chem. Eur. J.*, 3, 1571-1578.
162. W. Jahnke, H.C. Kolb, M.J.J. Blommers, J.L. Magnani, and B. Ernst, 1997, *Angew. Chem., Int. Ed. Engl.*, 36, 2603-2607.
163. J.C. Prodger, M.J. Bamford, M.I. Bird, P.M. Gore, D.S. Holmes, R. Priest, and V. Saez, 1996, *Bioorg. Med. Chem. Lett.*, 6, 793-801.
164. S. Hanessian, G.V. Reddy, H.K. Huynh, J. Pan, and S. Pedatella, 1997, *Bioorg. Med. Chem. Lett.*, 7, 2792-2734.
165. S. Hanessian, H.K. Huynh, G.V. Reddy, G. McNaughton-Smith, B. Ernst, H.C. Kolb, J. Magnani, and C. Sweeley, 1998, *Bioorg. Med. Chem. Lett.*, 8, 2803-2808.
166. T. Uchiyama, V.P. Vassilev, T. Kajimoto, W. Wong, H. Huang, C.-C. Lin, and C.-H. Wong, 1995, *J. Am. Chem. Soc.*, 117, 5395-5396.
167. T.J. Woltering, G. Weitz-Schmidt, and C.-H. Wong, 1996, *Tetrahedron Lett.*, 37, 9033-9036.
168. R. Wang and C.-H. Wong, 1996, *Tetrahedron Lett.*, 37, 5427-5430.
169. C.-C. Lin, M. Shimazaki, M.P. Heck, S. Aoki, R. Wang, T. Kimura, H. Ritzen, S. Takayama, S.-H. Wu, G. Weitz-Schmidt, and C.-H. Wong, 1996, *J. Am. Chem. Soc.*, 118, 6826-6840.

170. T.G. Marron, T.J. Woltering, G. Weitz-Schmidt, and C.-H. Wong, 1996, *Tetrahedron Lett.*, 37, 9037-9040.
171. K. Hiruma, T. Kajimoto, G. Weitz-Schmidt, I. Ollmann, and C.-H. Wong, 1996, *J. Am. Chem. Soc.*, 118, 9265-9270.
172. S.A. DeFrees, L. Philips, L. Guo, and S. Zalipsky, 1996, *J. Am. Chem. Soc.*, 118, 6101-6104.
173. D. Dupre, H. Bui, I.L. Scott, R.V. Market, K.M. Keller, P.J. Beck, and T.P. Kogan, 1996, *Bioorg. Med. Chem. Lett.*, 6, 569-572.
174. M.J. Bamford, M. Bird, P.M. Gore, D.S. Holmes, R. Priest, J.C. Prodger, and V. Saez, 1996, *Bioorg. Med. Chem. Lett.*, 6, 239-244.
175. A. Liu, K. Dillon, R.M. Campbell, D.C. Cox, and D.M. Huryn, 1996, *Tetrahedron Lett.*, 37, 3785-3788.
176. W. Spevak, C. Foxall, D.H. Charych, F. Dasgupta, and J.O. Nagy, 1996, *J. Med. Chem.*, 39, 1018-1020.
177. B.N.N. Rao, M.B. Anderson, J.H. Musser, J.H. Gilbert, M.E. Schaefer, C. Foxall, and B.K. Brandley, 1994, *J. Biol. Chem.*, 269, 19663-19666.
178. C.-H. Wong, F. Morris-Varras, C.-C. Lin, and G. Weitz-Schmidt, 1997, *J. Am. Chem. Soc.*, 119, 8152-8158.
179. S.-H. Wu, M. Shimazaki, C.-C. Lin, L. Qiao, W.J. Moree, and C.-H. Wong, 1996, *Angew. Chem., Int. Ed. Engl.*, 35, 88-90.
180. P.V. Murphy, R.E. Hubbard, D.T. Mallanack, R.E. Wills, J.G. Montana, and R.J.K. Taylor, 1998, *Bioorg. Med. Chem.*, 6, 2421-2439.
181. T.P. Kogan, B. Dupre, K.M. Keller, I.L. Scott, H. Bui, R.V. Market, P.J. Beck, J.A. Vuytus, B.M. Revelle, and D. Scott, 1995, *J. Med. Chem.*, 38, 4976-4984.

2 Total synthesis of macrolides

J.W. Bode and E.M. Carreira

2.1 Introduction

The structural and stereochemical complexity of macrolide antibiotic natural products has piqued the imagination of organic chemists and fuelled the discovery of innovative reaction methodology. As a consequence, the phenomenal advances in the development of synthetic methods for the stereoselective synthesis of this class of natural products have resulted in a progressive evolution in the efficiency with which these structures are assembled. The breathtaking pace of progress in this area makes it difficult to present comprehensively in a single collection the many superb syntheses that mirror the state-of-the-art. Bound by this limitation, we nonetheless attempt to present in this chapter a wide perspective of the field by discussing four macrolides—oleandolide, Sch 38516 (fluviricin B_1), macrolactin A and lankacidin C—whose syntheses exemplify the remarkable advances in the area. Although all are grouped under the rubric of macrolide antibiotics, the diversity of strategies crafted to assemble the macrocycle are noteworthy. Additionally, these targets incorporate the various synthetic challenges representative to this general class of natural products, including densely functionalised structures wherein the synthetic challenge arises from the multitudinous, continuous stereocentres as well as sparsely functionalised structures in which the remote relationship between few stereocentres defines the inherent difficulty in asymmetric synthesis.

2.2 Oleandolide

The erythromycins (**1**) and related macrolides, unlike few other natural products over the past three decades, have provided a testing ground for the design of numerous asymmetric reaction processes: aldol addition, aldehyde allylation, ketone reduction, macrolactonisation, alkene hydroboration, and reduction. The various synthetic tools available to the organic chemist are well documented in a number of excellent reviews.[1] A close structural and pharmacological relative of the erythromycins, oleandomycin **2**, differs only in the extent of hydroxylation at C(8) and C(12).[2] Since 1990 the three syntheses reported independently by Tatsuda,[3] Paterson[4] and Evans[5] provide an excellent

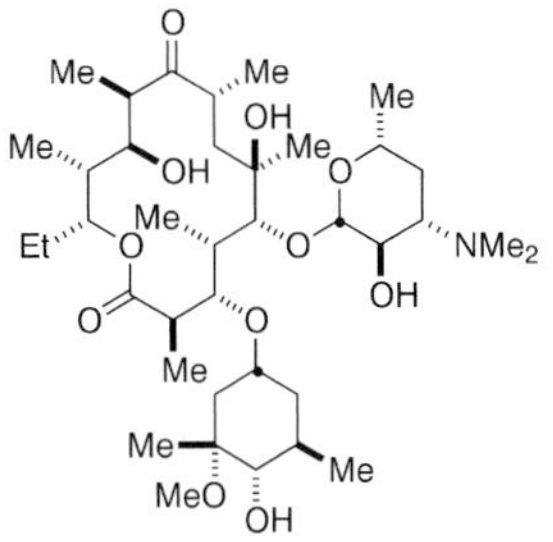

Erythromycin B
1

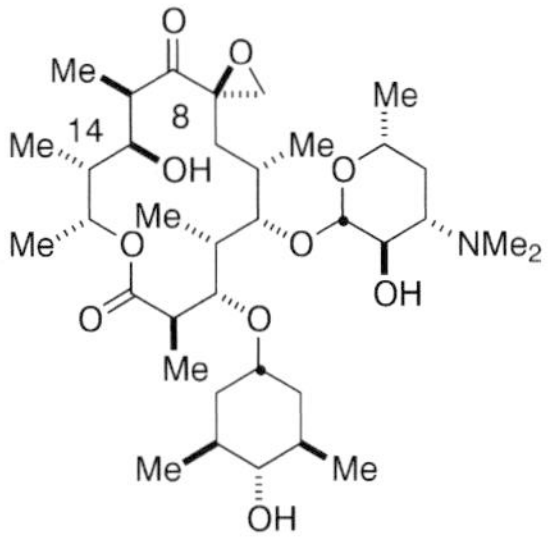

Oleandomycin
2

compilation of modern stereoselective methods and strategies for the asymmetric synthesis of polyketide-derived macrolides.

The first total synthesis of oleandomycin was reported by Tatsuda, commencing with methyl α-L- and D-rhamnosides. These sugars were converted to idopyranoside derivatives **3** and **4** through a sequence of synthetic manipulations developed previously by these investigators (Scheme 2.1).[6] The methyl pyranosides were efficiently employed as

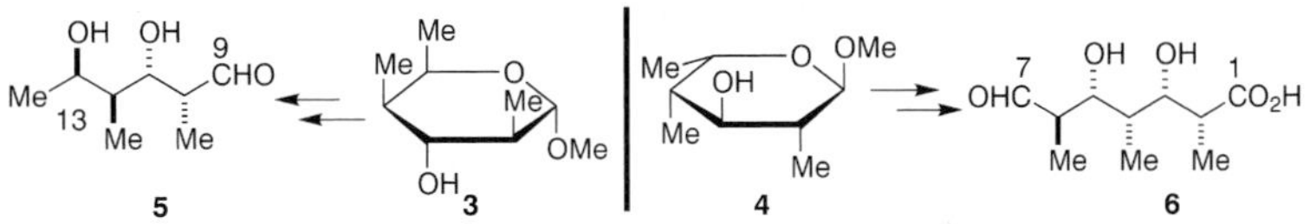

Scheme 2.1.

building blocks for the C(1)–C(7) and C(9)–C(13) segments, **5** and **6**, respectively. Three of the stereogenic centres in polyketide **6** were derived directly from **4**, while the C(5) and C(6) stereocentres were installed through reagent-controlled aldehyde addition reaction (Scheme 2.2). Treatment of aldehyde **7** with the (S,S)-tartrate-derived (E)-crotylboronate **8** afforded threo adduct **9** as the only isolated product in 90% yield.[7]

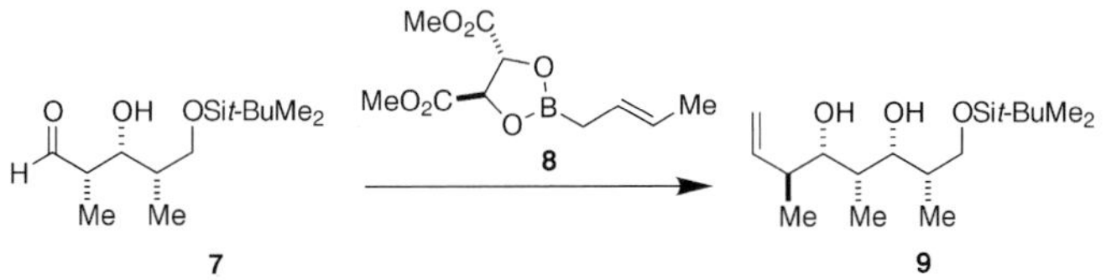

Scheme 2.2.

Previous synthetic investigations of this class of macrolide antibiotics have underscored the capricious nature of cyclisations that convert the seco acids to the requisite macrolactones. Most notably, Woodward's total synthesis of erythronolide highlighted the critical role of protecting groups in facilitating macrocyclisation.[8] In the Tatsuda oleandolide synthesis, cyclisation to the 14-membered macrocycle was conducted via an intramolecular Horner–Wadsworth–Emmons condensation reaction 10 → 11 (Scheme 2.3). In this strategy installation of the C(8) epoxide was

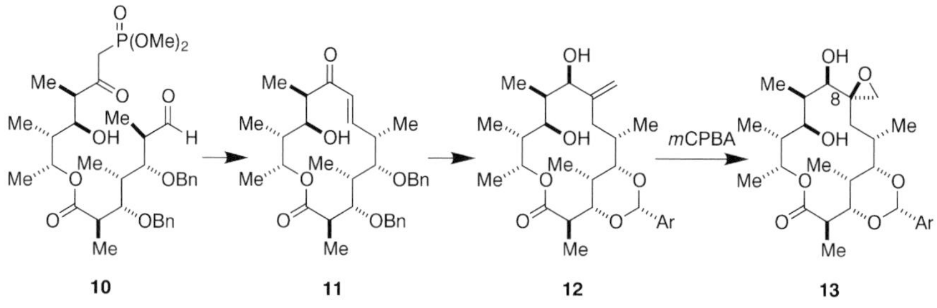

Scheme 2.3.

relegated until late in the synthetic route. In this regard, treatment of allylic alcohol 12 with m-CPBA gave epoxide 13 as the only isolable product. The synthesis of oleandolide was subsequently completed following chemoselective oxidation of the C(9) alcohol and hydrolysis of the benzylidene acetal.

The second synthesis of oleandolide was reported in 1994 by Paterson and co-workers. With a longest linear sequence of 20 steps, oleandolide was prepared in a stereocontrolled fashion in an impressive overall yield of 8%.[9] The stereoselective synthesis of the polyketide fragments exploited state-of-the-art aldol additions[10] involving both reagent- and substrate-controlled methods developed by Paterson (Scheme 2.4). The versatility

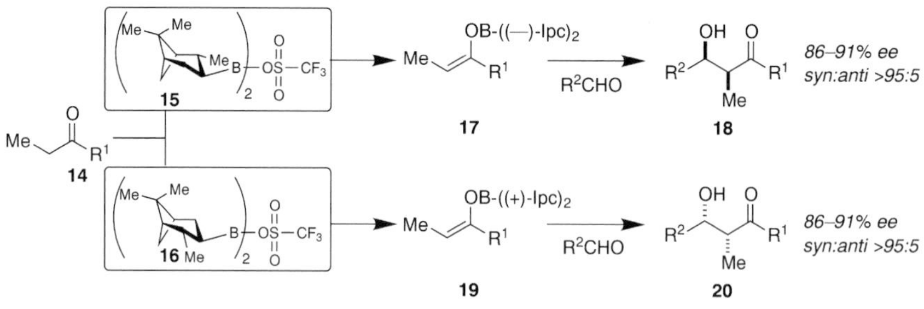

Scheme 2.4.

of such methodology has been powerfully demonstrated in a series of elegant syntheses.[11] The synthetic chemistry has considerable generality;

thus, enolisation of ethyl ketones **14** with (−)- or (+)-diisopinylcam-pheylboron triflate and i-Pr$_2$NEt affords (Z)-enol borinates **17** and **19** which participate in stereoselective aldol addition reactions with a range of aldehyde substrates. Ready access to both stereochemical families of *syn* adducts **18** and **20** is provided with each isolated in greater than 95:5 diastereoselectivity (*syn:anti*) and up to 91% enantiomeric excess (ee). Following a plan relying on reagent control, Paterson assembled the constituent fragments that led to **21**, an advanced intermediate in

21 X = O
22 X = CH$_2$

the initial approach to oleandolide. A meticulous examination of the structural and reaction chemistry of this macrocycle allowed Paterson to identify critical issues relevant to the installation of the requisite epoxide at C(8).

The results of initial investigations revealed that olefination of ketone **21** to give **22** could only be effected in low yields despite the fact that a wide range of reagents were examined, such as Zn/CH$_2$I$_2$/TiCl$_4$, Cp$_2$TiCH$_2$ClAlMe$_2$, Me$_3$SiCH$_2$Li/CeCl$_3$, Me$_2$S=CH$_2$, Me$_2$S(O)=CH$_2$, CH$_2$N$_2$ and Ph$_3$P=CH$_2$. A computational study of this lactone (**21**) using the MM2 force field in MacroModel revealed that the diastereotopic *re*- and *si*- faces of the C(8) ketone were shielded by the macrocycle and the 9-(S)-OSiMe$_2t$-Bu moieties, respectively. The investigators integrated these observations along with the results from modelling to fashion a synthetic route leading to oleandolide. Implementation of this strategy necessitated the preparation of a synthetic target possessing the diastereomeric 9(R) alcohol along with the C(7) and C(9) diols protected as the corresponding cyclic acetal (see **23**). The revised route required a different approach to the synthesis of the fragments; importantly, the

23 X = O
24 X = CH$_2$

improved synthesis that ensued is an elegant illustration of substrate-controlled stereoselective C—C bond formation.[12]

Treatment of **25** with dicyclohexylboron chloride (**26**) and Hünig's base afforded the *E*-enolate **27** that subsequently provided *anti* adduct **28** (Scheme 2.5). In a complementary fashion, enolisation of the same ketone

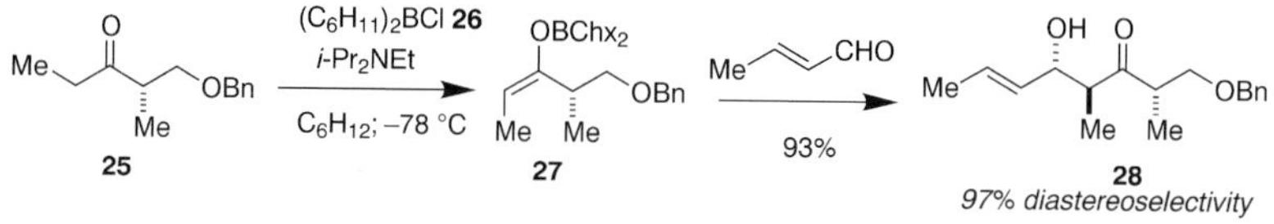

Scheme 2.5.

with stannous triflate and triethyl amine yielded the *Z*-enolate **30** which participated in a substrate-controlled aldol addition reaction with methacrolein to afford the *syn* product **32** in 93% diastereoselectivity (Scheme 2.6).

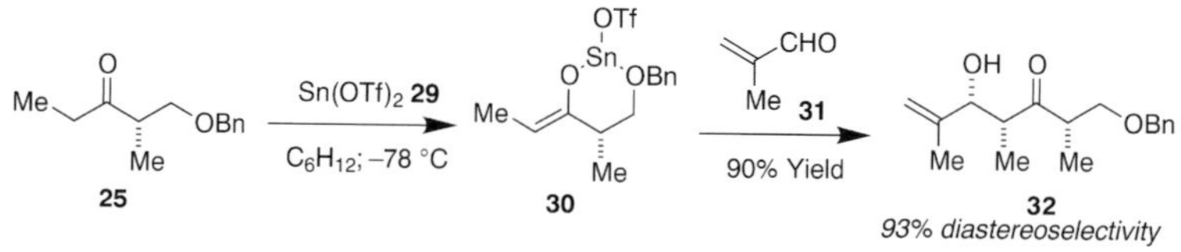

Scheme 2.6.

One of the remarkable transformations carried out by Paterson in the synthesis of oleandolide was the conversion of olefinic hydroxyketone **32** to triol **36** (Scheme 2.7). In a single step, stereoselective reduction of the hydroxyketone and hydroboration of the alkene were effected using excess Ipc$_2$BH. Chelate **33** is ideally poised for subsequent intermolecular

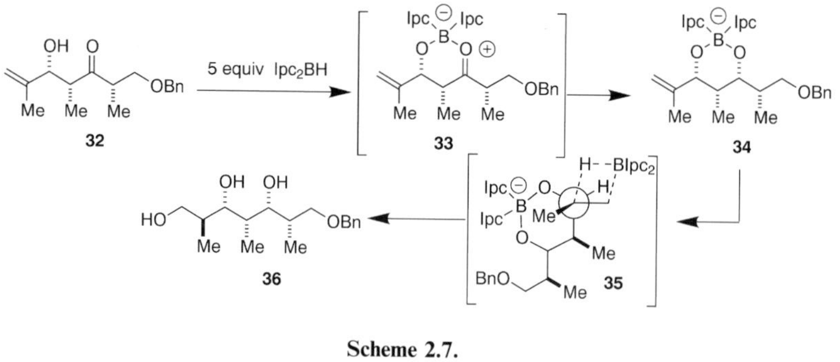

Scheme 2.7.

syn-selective reduction by a second equivalent of borane reagent affording **34**. Under the same reaction conditions, the disubstituted terminal alkene succumbed to regio- and stereoseletive hydroboration which after oxidative work-up provided **36**. The stereochemical outcome of the hydroboration reaction is consistent with a transition-state model **35** in which unfavourable steric effects, such as 1,3-allylic and reagent–substrate interactions, are minimised.

An efficient, convergent, series of transformations allowed for the preparation of ketone lactone **23**, setting the stage for the installation of the C(8) epoxide. Treatment of **23** with trimethylsulfonium methylide **37** afforded epoxide **38** as the only observed product (Scheme 2.8). Subsequent completion of this efficient synthesis was effected following a short sequence of steps to provide synthetic oleandolide.

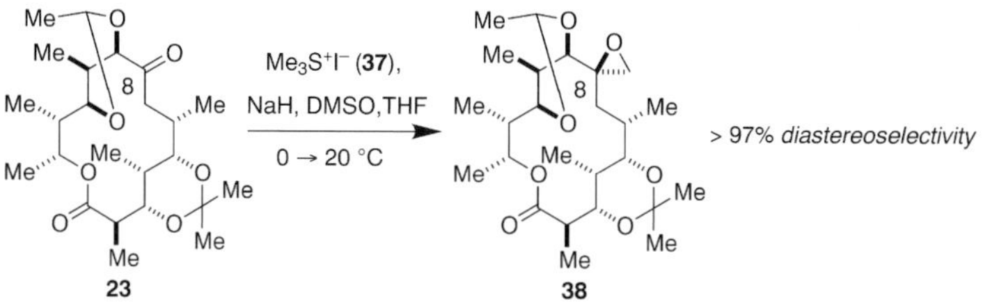

Scheme 2.8.

In the most recent synthesis of oleandolide Evans followed a strategy differing in several aspects. First, chiral auxiliary based ketoimide aldol addition reactions were utilised to assemble the desired constituents rapidly; and, second, installation of the C(8) epoxide was conducted on an acyclic fragment with the stereochemistry established on the basis of the principles of acyclic allylic stereocontrol.

Evans has pioneered the use of carboximide auxiliaries for the asymmetric synthesis of polypropionate-derived adducts.[13] A remarkably efficient and elegant illustration of the construction of dipropionate fragments is the aldol addition reaction of a β-keto imide such as **39** with aldehydes (Scheme 2.9).[14] In this regard, the synthesis of the C(1)–C(8) sector of oleandolide commenced by treatment of **39** with Ti(*i*-PrO)Cl$_3$

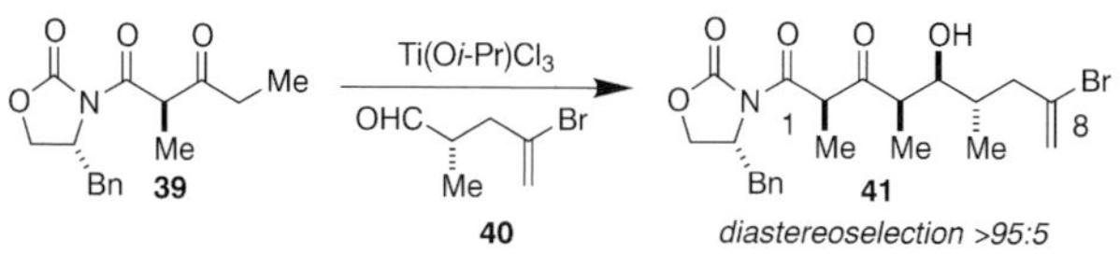

Scheme 2.9.

and triethylamine, resulting in chemoselective enolisation of the ketone at C(4). This (Z)-enolate participated in a selective aldol addition reaction with **40** to give polyketide **41**. For the preparation of the C(9)–C(14) sub-unit, the complementary stereochemical outcome was accessed through the use of stannous triflate in the enolisation protocol (Scheme 2.10).

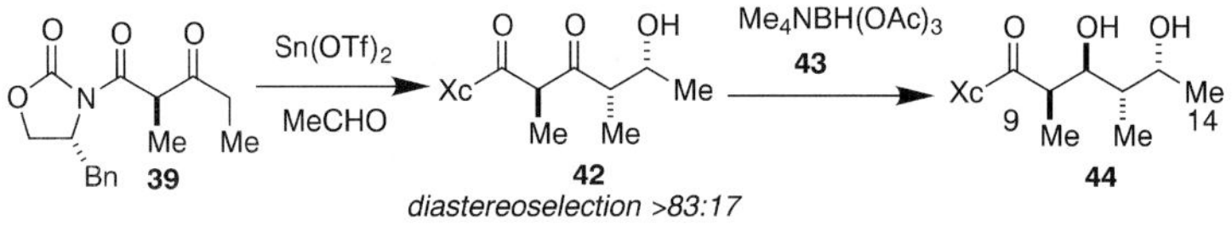

Scheme 2.10.

Enolisation of **39** followed by addition to acetaldehyde afforded **42** as an 83:17 mixture of diastereomers. In contrast to the reaction of Ti(IV) enolates the adduct was isolated possessing the *anti* stereochemical relationship relative to the preexisting stereocentre. The structural param-eters that led to this noteworthy process have been discussed in a number of landmark publications by Evans.[15] Directed reduction of the ensuing hydroxyketone with Me$_4$NBH(OAc)$_3$ **43** produced the 1,3-*anti* diol **44**. This reagent was developed by Evans for the diastereoselective, directed reduction of β-hydroxy ketones to 1,3-*anti* diols.[16]

A strategically efficient step in the Evans synthesis is the convergent coupling of two highly functionalised fragments, **45** and **46**, which was carried out in an innovative manner employing Pd(0)-catalysis (Scheme 2.11). Treatment of vinylstannane **45** with acid chloride **46** [Pd$_2$(dba)$_3$,

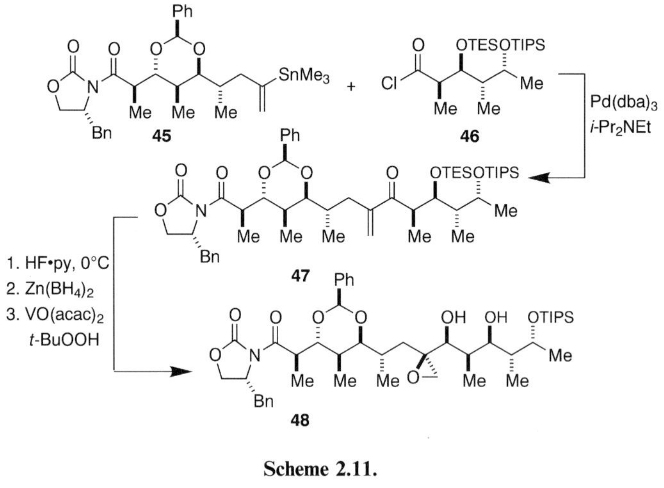

Scheme 2.11.

i-Pr$_2$NEt, C$_6$H$_6$, 80°C] gave enone **47** in 85% yield. These conditions were identified after screening procedures that minimised unwanted reactions such as protodestannylation, dimerisation and alkene isomerisation. A remarkable aspect of this C—C bond-forming reaction is that the coupling was conducted without epimerisation in the acid chloride **46**. With the C(1)–C(14) fragment in hand, the investigators chose the 9-(*S*) hydroxyl to control stereochemically the installation of the C(8) epoxide via a hydroxyl-directed Sharpless epoxidation reaction of the corresponding acyclic precursor. In this manner, the complete acyclic fragment **48**, possessing all of the requisite stereocentres, was assembled in 11 steps commencing from the propionyl-derived oxazolidinone auxiliary.

Prominent amongst the important advances that have resulted from the wealth of synthetic studies on these and related macrolides are the reagents and reaction conditions that have been developed for macrolactonisation. The Paterson and Evans syntheses both made use of one of the more successful methods that was developed by Yamaguchi to give the macrolide in excellent yield.[17] In the Evans route, epoxy seco acid **49** underwent macrocyclisation to give **50** (Scheme 2.12) which, following deprotonation and oxidation, provided synthetic oleandolide.

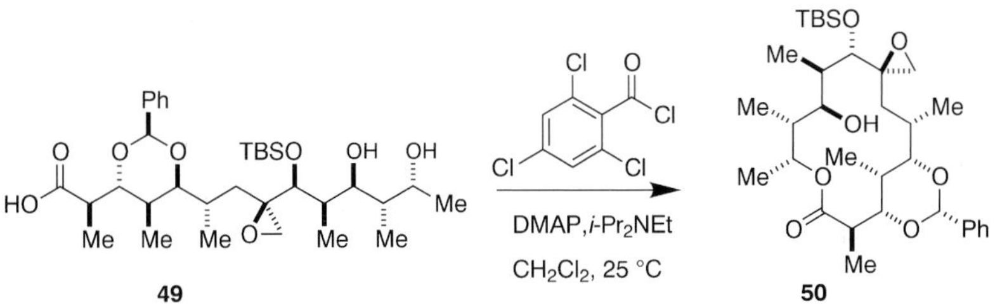

Scheme 2.12.

The three syntheses of oleandolide by Tatsuda, Paterson and Evans corroborate the phenomenal advances that have been born over the past two decades of synthesis in the polyketide area. Thus, the effectiveness with which continuous stereocentres can be assembled through allylation, aldol addition, reduction and epoxidation methods permits the rapid construction of stereochemically complex subunits that lead to highly convergent syntheses.

2.3   Fluvirucin B$_1$ (Sch 38516)

In contrast to the stereochemically congested propionate-derived macrolides, the synthetic challenges posed by Sch 38516, or fluviricin B$_1$ **51**, stem from the paucity and remoteness of its stereogenic centres.[18] These effectively preclude substrate-controlled methodologies, which renders the construction of the molecule a challenge (Scheme 2.13). Moreover, the deceptive simplicity of this macrolide leads to potential difficulties

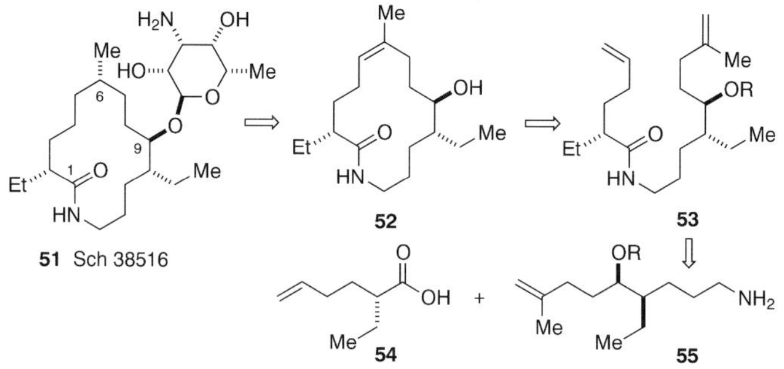

Scheme 2.13.

in macrocycle formation, as the highly flexible ring encompasses few conformational constraints that would facilitate cyclisation. Two research groups have relied on modern transition-metal catalysis for the construction of the remote chiral centres and have significantly extended the utility of methods employed for the formation of the 14-membered ring.

The general approach used by Hoveyda is outlined in the retrosynthetic analysis shown in Scheme 2.13.[19] The stereoselective preparation of **55** commenced with chiral allylic alcohol **56**, prepared by Ti(IV)-catalysed Sharpless kinetic resolution of the racemate (Scheme 2.14).

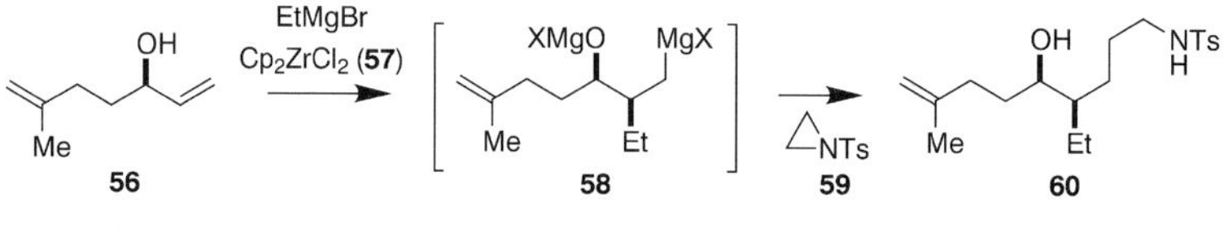

Scheme 2.14.

Diastereoselective ethylmagnesiation of **56** catalysed by Cp_2ZrCl_2 **57** (5 mol%), as developed by Hoveyda, provided **58** which was then reacted *in situ* with excess *N*-tosyl aziridine **59** to give the fully functionalised C(6)–C(13) subunit **60** in 42% yield and 97:3 diastereoselectivity.[20]

The synthesis of carboxylic acid **54** also made use of an Hoveyda catalytic, enantioselective ethylmagnesiation reaction, this time on 3,5-dihydrofuran.[21] Thus, treatment of ethyl magnesium bromide with excess (1.3 equiv.) of **61** and catalytic (*R,R*)-ethylene-1,2-bis(η^5-4,5,6,7-tetra-hydro-1-indenyl)titanium (*R*)-1,1'-binaphth-2,2'-diolate **62** gave **63** in

greater than 99% ee. Hydromagnesiation of this olefinic alcohol with
n-PrMgBr and 3 mol% Cp$_2$TiCl$_2$ afforded dianion **64** which was then
subjected to *in situ* coupling with vinyl bromide [3 mol% (Ph$_3$P)$_2$NiCl$_2$] to
provide **65** in 72% overall yield (Scheme 2.15). An interesting modification of the Ley oxidation permitted direct conversion of the primary
alcohol **65** to the corresponding acid **54**.[22]

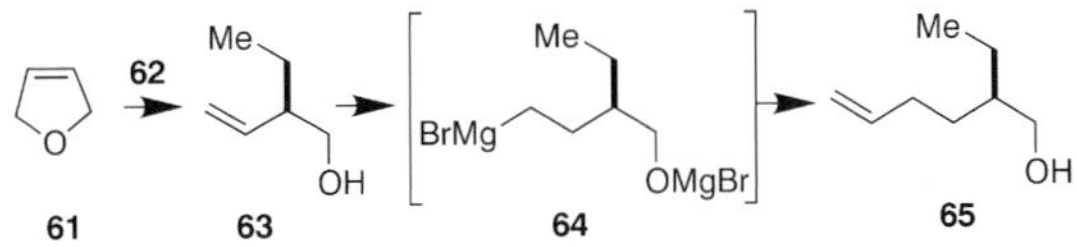

Scheme 2.15 **62** = Catalytic R,R-ethylene-1,2-bis(η^5-4,5,6,7-tetrahydro-1-indenyl)titanium (R)-
1,1'-binapth-2,2'-diolate.

Amide **66** constituted the key intermediate for the successful
construction of the macrocycle. In the presence of 20 mol% of the
Schrock catalyst Mo(CHCMe$_2$Ph)[N(2,6-(i-Pr)$_2$C$_6$H$_3$)][OCMe(CF$_3$)$_2$]$_2$ **67**
at 22°C, **66** smoothly participated in ring-closing metathesis to provide **68**
in 90% yield as a single olefin stereoisomer (Scheme 2.16).[23] Investigations into the origin of the efficiency of this reaction revealed the
importance of the remote stereochemical stereocentres for successful ring
closure; in their absence, the metathesis reaction preferentially afforded
dimeric adducts.

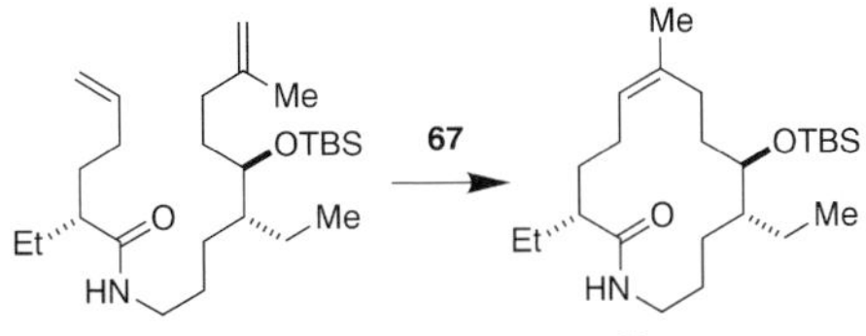

Scheme 2.16 **67** = Schrock catalyst, Mo(CHCMe$_2$Ph)[N(2,6-(i-Pr)$_2$C$_6$H$_3$)][OCMe(CF$_3$)$_2$]$_2$.

The use of olefin metathesis reactions has become increasingly common
for the construction of both medium and large ring compounds in natural
and nonnatural products.[24,25] Of considerable interest are the recent
findings that successful macrocyclisation by ring-closing metathesis may
depend not so much on conformational preorganisation of the substrate
but rather on the proximity between the olefins and polar functionality.
In demonstrating that ring-closing metathesis was viable in conformationally unbiased systems, Fürstner elegantly extended this methodology to the preparation of several macrocyclic natural products
including (−)-gloeosporone, (+)-ricinelaidic acid and various odiferous

compounds.[26] These and other studies led to the successful application of this reaction to the total synthesis of the epothilone macrolide natural products by several groups.[27]

Although direct glycosylation of aglycone **52** did not give the desired product, Hoveyda found that the saccharide could be appended prior to ring closure. Treatment of **53** with fluoroglycoside **69** in the presence of SnCl$_2$ (**70**) gave a 92% yield of **71** as a single anomer.[28, 29] Subsequent ring-closing metathesis cleanly furnished the macrocycle **72** in 92% yield as a single olefin isomer. The C(6) stereocentre in Sch 38516 was established by macrocyclic stereocontrol, by catalytic hydrogenation of **72** in 84% yield and greater than 98% diastereoselectivity (Scheme 2.17). Following deprotection of the sugar the natural product **51** was isolated.

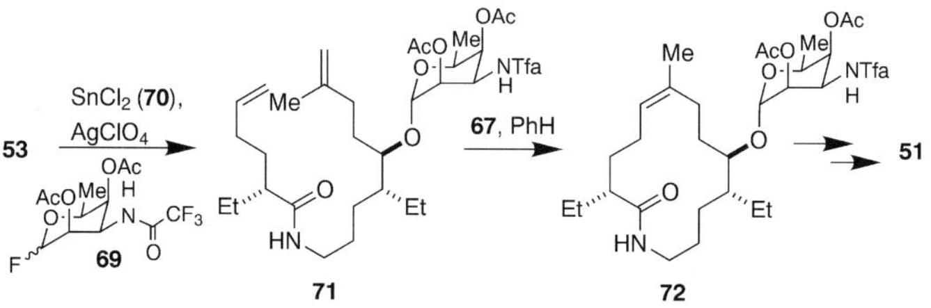

Scheme 2.17.

The Trost strategy for constructing the remote chiral centres in Sch 38516 highlights the use of modern Pd(0) catalysis (Scheme 2.18).[30]

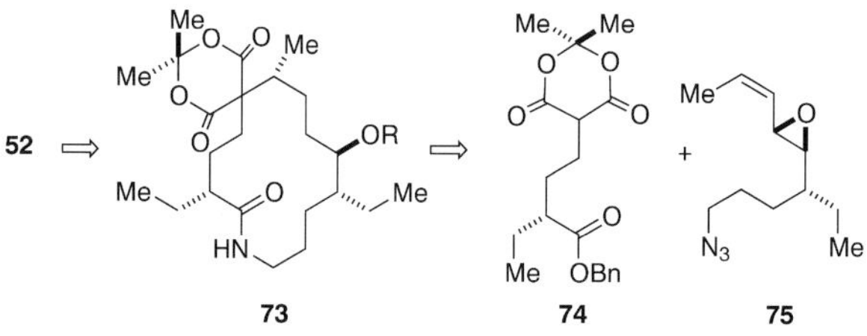

Scheme 2.18.

In the key coupling reaction, the addition of vinyl epoxide **75** to substituted Meldrum's acid derivative **74** was catalysed by 1.5 mol% [(dba)$_3$Pd·CHCl$_3$] and 20 mol% of phosphine ligand (Scheme 2.19). This remarkably convergent and efficient transformation documented the first use of a substituted epoxy olefin and nucleophile in an *intermolecular* coupling reaction, giving **77** as a single diastereomer in 75% yield. The stereochemical outcome of this reaction is noteworthy. In this regard,

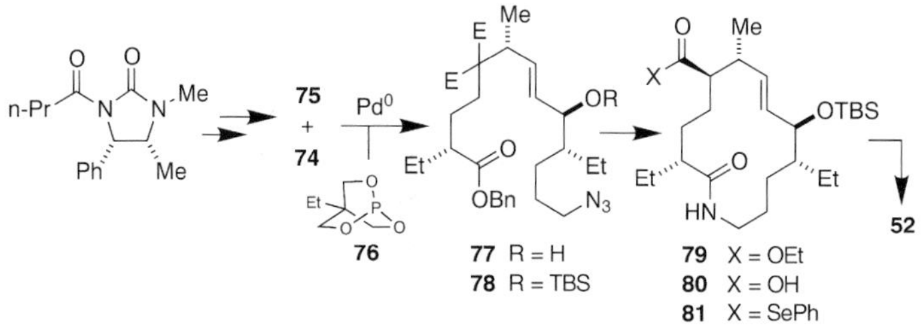

Scheme 2.19.

Trost has proposed a transient π-allylpalladium intermediate **83** as responsible for the transfer of stereochemical information, allowing for

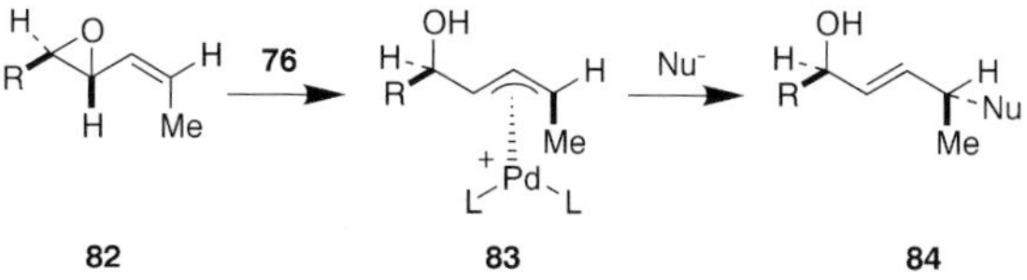

Scheme 2.20.

effective relay of asymmetry by the remote stereocentres (Scheme 2.20).

Hydrogenolysis of the benzyl ester and reduction of the azide in the presence of Pd/C, followed by macrolactamisation, provided the macrocyclic lactam. It is worth noting that attempts to form the macrolactam by more traditional methods, including those utilising Yamaguchi's reagent, did not provide the desired product. Trost effected macrolactam formation by treatment of the amino acid intermediate generated during hydrogenolysis with bromotripyrrolidinophosphonium hexafluorophosphate, a protocol that may find general utility for the synthesis of other macrocyclic natural products.[31] Successful completion of the synthesis required subsequent hydrogenation of the internal olefin, removal of the dicarboxylates, followed by a final deprotection step.

The two expedient syntheses of Sch 38516 by Hoveyda and Trost attest to the importance of transition-metal catalysis in addressing synthetic challenges in modern organic chemistry. Additionally, the target structure serves to illustrate the problems posed in macrolide construction by acyclic precursors that lack entropic biases facilitating cyclisation. In both cases, the total synthesis has demanded that established methodologies, catalytic ring-closing metathesis and catalytic nucleophilic addition to alkenyl epoxides be carefully explored and pushed to their limits.

2.4 Macrolactin A

A study by Fenical of a taxanomically unclassified deep-sea bacterium culminated in the isolation and characterisation of macrolactins A–F as a novel class of polyene macrolide antibiotics.[32] In preliminary studies these were shown to have a prophylactic effect on T-lymphoblast cells against human immunodeficiency virus (HIV) replication. The fact that these novel macrolides are not readily available from natural sources necessitates that biological research of this medicinally interesting class of structures rely on a *de novo* laboratory synthesis. This feature coupled with the fact that these structures represent an unusual synthetic challenge have led a number of groups to study efficient synthetic approaches to a representative member of this family of macrolides. Three different groups have independently reported syntheses of the macrolactin A core with two groups documenting the total synthesis of the natural product.[33–35]

An early report by Pattenden outlines the general strategy that was independently pursued by numerous investigators in this area. The study delineated an approach based on Pd(0)-catalysed transformations such as Stille and Suzuki cross-coupling reactions to assemble convergently three subunits of roughly equal stereochemical complexity (Scheme 2.21).

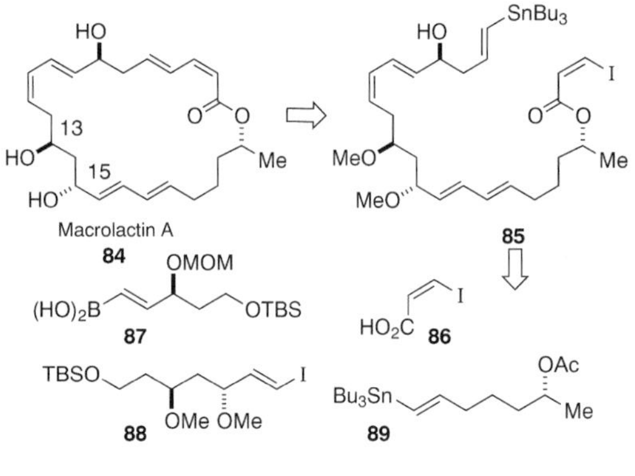

Scheme 2.21.

Moreover, this general strategy relied on Pd(0)-catalysis to effect macrocyclisation. Thus, Pattenden reported that **85** underwent macrocyclisation (58% unoptimised yield) upon subjection to modified Stille conditions. Although the route proved successful for the preparation of the macrolactin core, subsequent deprotection of the bismethylethers in the corresponding macrocycle proved prohibitive to the preparation of the natural product.

The first total synthesis of the natural product was reported by Smith. The investigators carried out a meticulous study of the macrocyclisation reaction chemistry by examining construction of the C(9)—C(10) bond through both possible permutations involving vinyl stannane and iodide reaction partners in Pd(0)-catalysed cross-coupling reactions. The syntheses of the constituent fragments incorporating the C(7), C(13) and C(15) stereogenic centres commenced with enantioselective allylation of propynal with both (R)- and (S)-diisopinylcampheyl allyl boranes to afford alkynes **90** and **96** in 90% ee. Following protection of the secondary alcohol and ozonolytic cleavage of the terminal alkene in **90**, homologation of aldehyde **91** (CrCl$_2$, LiI, Bu$_3$SnCHBr$_2$)[36, 37] furnished **92**. This vinylstannane served as the starting point for a series of chemoselective Pd(0)-mediated transformations that led to triene **95** (Scheme 2.22), namely, cross-coupling with (Z)-3-iodopropenoic acid **93**

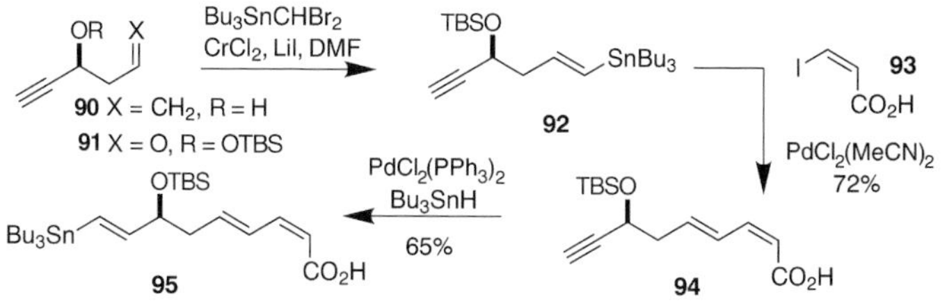

Scheme 2.22.

and alkyne hydrostannylation. This set of reactions nicely attests to the power of state-of-the-art Pd(0)-catalysed C(sp^2)–C(sp^2) bond-forming processes which proceeded with unmasked polar, protic functionality, such as alcohols and carboxylic acids.

The enantiomeric propargylic alcohol **96** served as the starting material for the preparation of the second fragment (Scheme 2.23). Alkyne **97** was

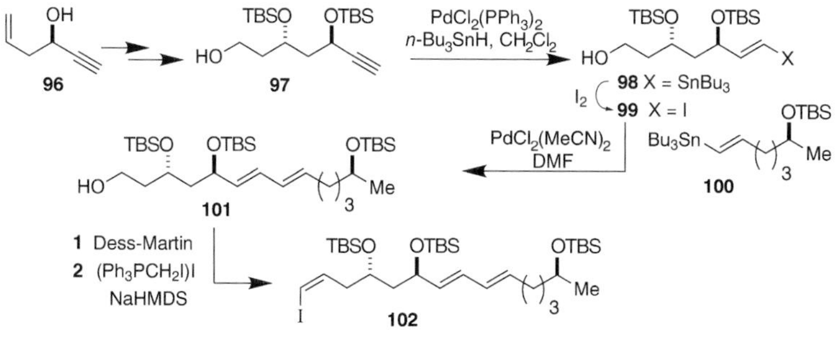

Scheme 2.23.

stereoselectively hydrostannylated to give vinylstannane **98**, and this subjected to tin–halogen exchange to obtain **99**. Coupling of this vinylic iodide with vinyl stannane **100** furnished the C(9)–C(24) subunit **102**. The aldehyde isolated from oxidation of **101** was subjected to homologation by utilising the Stork reagent[38] to furnish vinyl iodide **102**, an intermediate which maps onto the C(10)–C(24) region of macrolactin A.

The final cyclisation reaction in the Smith synthesis was effected by closure along the C(9)—C(10) bond, employing both possible Stille precursors: the C(9) stannane/C(10) iodide **103** and the C(9) iodide/C(10) stannane **104** (Scheme 2.24). Ring closure involving the latter proved to be optimal (R = Et), affording protected macrolactin A **105** in 56% yield which upon desilylation then afforded synthetic macrolactin A.

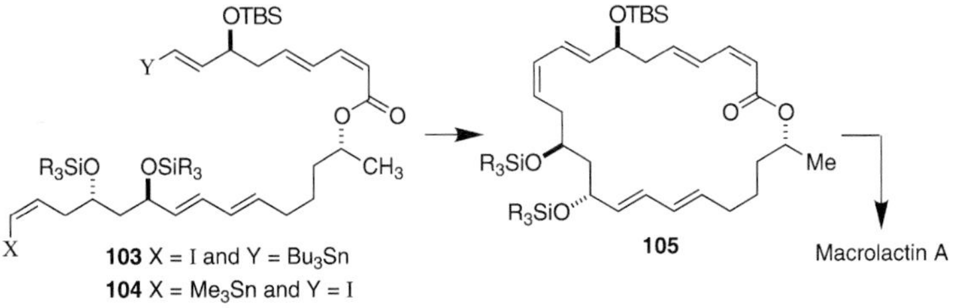

Scheme 2.24.

The Carreira route to macrolactin A also utilised a series of Stille cross-coupling reactions to assemble convergently the constituent fragments as well as to effect macrocycle formation. For the synthesis of subunits incorporating stereogenic skipped polyol stereocentres, the approach utilised the recently developed catalytic, enantioselective aldol addition reaction of silylketene acetal **107**.[39, 40] The route benefits from

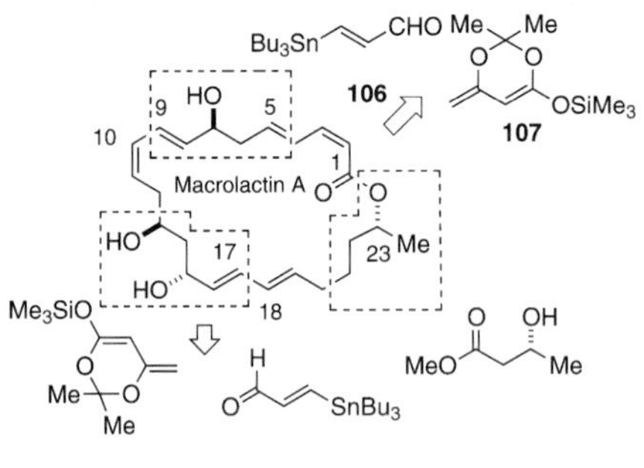

the ability to access highly functionalised diketide fragments through the aldol addition reaction to β-tributylstannyl propenal **106**. These adducts are ideally functionalised with electrophilic and nucleophilic reaction centres and can be directly employed as substrates in subsequent $C(sp^2)$–$C(sp^2)$ cross-coupling reactions. Moreover, the synthetic strategy underscores the application of modern catalytic methods for the expeditious synthesis of macrolides through C—C bond-forming reactions: aldol addition and Stille cross-coupling reactions.

Construction of the key chiral diketide subunits commenced with aldehyde **106**, masked dienolate **107**, and the enantiomeric Ti(IV) catalysts (S)- and (R)-**108/109**, respectively (Scheme 2.25).[41] The starting

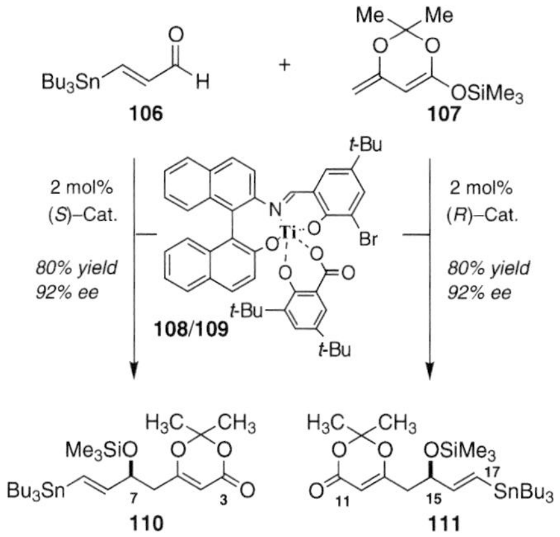

aldehyde was readily prepared by the method of Quintard, wherein propynal diethyl acetal was treated with $Bu_3SnH/BuLi/CuCN$, and the acetal cleaved by acidic work-up.[42] In separate experiments, with 2 mol% of (S)- and (R)-**108/109**, protected acetoacetate aldol adducts **110** and **111** were each isolated in 80% yield and 92% ee.

The Pd(0)-catalysed coupling that joined stannane **112** with vinyl iodide **113** was effected with $Pd_2(dba)_3$, i-Pr_2NEt, and $CdCl_2$ and provided diene **114** in 69% yield (Scheme 2.26). The use of catalytic quantities of

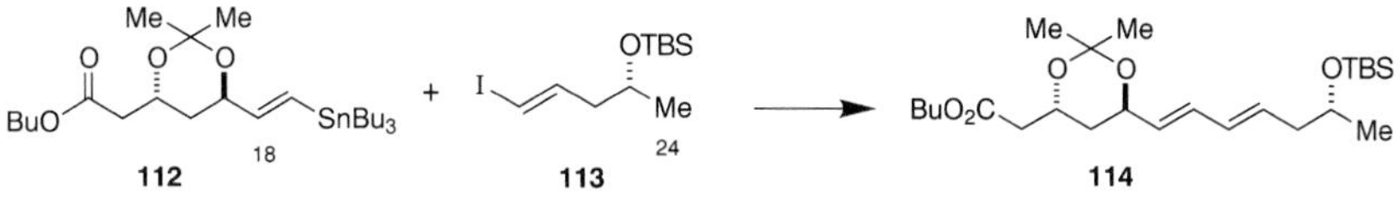

Scheme 2.26.

$CdCl_2$ was necessary in order to minimise the reaction of **112** with itself to give the undesired diene.[43] The C(1)–C(10) and C(11)–C(24) portions of the macrolide were joined via a Still-modified Horner–Wadsworth–Emmons condensation reaction to afford cis-enoate in good yield, with the adduct isolated as a 13:1 mixture of geometrical isomers

(Scheme 2.27).[44] Subsequent intramolecular Stille cross-coupling of **117** afforded the requisite macrocyclic lactone which, following deprotection, yielded macrolactin A.

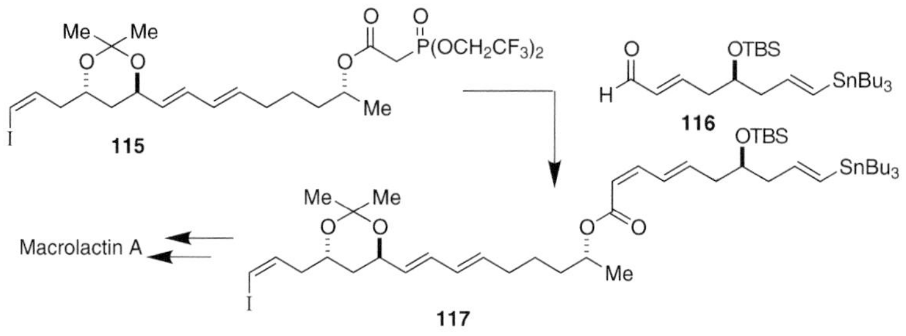

Scheme 2.27.

2.5   Lankacidin

To conclude this survey of modern syntheses of macrolide antibiotics, Kende's total synthesis of Lankacidin C **(118)** is discussed.[45] This

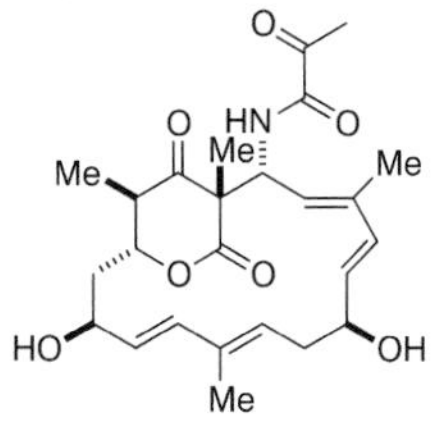

Lankacidin C (**118**)

molecule presents all the problems posed by the natural products detailed above and posseses several unique and highly sensitive structural features that mandated careful synthetic planning and experimentation.[46] The synthesis illustrates current methods for stereoselective substituted diene construction as well as the elegant application of catalytic, asymmetric reduction with reagent control to introduce a remote stereocentre on the macrocyclic framework. Taken together, these and other features of the Kende lankacidin C synthesis make it a noteworthy addition to modern macrolide assembly.[47]

Fragments **120** and **121** were prepared from readily available dithioacetal **119** and a known chiral β-lactam, respectively (Scheme 2.28).[48, 49] In initial model studies, the coupling reaction of a β-lactam

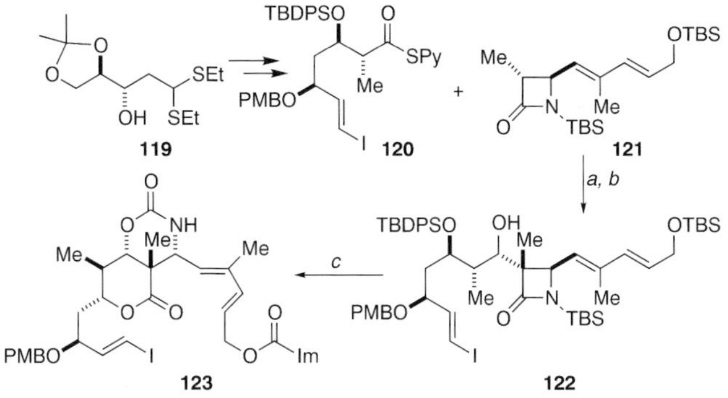

Scheme 2.28.

enolate with an aldehyde afforded products in a nonselective manner. This led to the implementation of an acylation–reduction sequence which provided the product with the required stereochemistry. The enolate generated from deprotonation of **121** gave a single Claisen adduct when reacted with 2-thiopyridyl ester **120**. Reduction of the resulting ketone by KEt₃BH at −78°C provided the desired product **122** stereoselectively. This stereochemical outcome appears to be a result of the presence of the proximal quaternary methyl group in the β-lactam ring. The enhanced facial selectivity of the acylation step in comparison with the analogous aldol reaction is unclear but may be related to greater steric demands posed by the acylation chemistry.

Having served admirably in the stereoselective coupling reaction, the β-lactam was rearranged to provide in one pot the fully functionalised δ-lactone. Treatment of **122** with Bu₄NF, followed by catalytic MsOH, and Et₃N/1,1-carbonyldiimidazole delivered **123** in 85% overall yield. This impressive step was carefully worked out in model studies prior to its application to the synthesis of lankacidin, and demonstrates a novel N/O intramolecular acyl transfer approach to stereochemically complex δ-lactones. It is worth noting that Thomas has utilised a similar, independently-conceived approach towards this region of the lankacidins.[50]

Selective hydrolysis and oxidation of **123** afforded **124** which participated in a Stille cross-coupling reaction with vinyltin **125** to furnish the corresponding allylic alcohol (Scheme 2.29). Although initial experiments to effect ring closure through a lithiated sulfone were successful in providing the macrocyclic product, subsequent attempts at desulfurisation resulted only in decomposition. Thus, these researchers turned to the Stork–Takahashi cyanohydrin chemistry to effect macro-cyclic ring-closure. Transformation of Stille product **126** to cyanohydrin

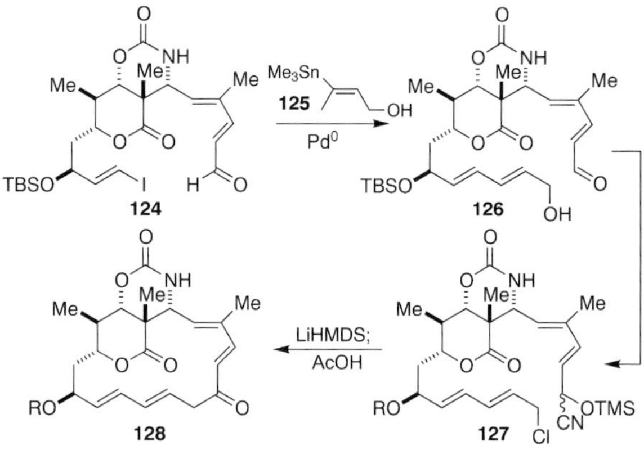

Scheme 2.29.

127 and treatment with LiHMDS, followed by hydrolysis with acetic acid, gave macrocyclic ketone 128 in 61% overall yield.

Faced with the problem of chemoselectively reducing the unsaturated ketone on the macrocycle, Kende found that the traditional method of NaBH$_4$ and CeCl$_3$ gave a complex mixture of products. L-selectride provided few advantages, giving only a 1:1 ratio of products, the conjugate reduction product and an epimeric mixture of the secondary alcohol, with the undesired isomer prevailing. Ultimately, reduction of 128 mediated by 10 mol% of Corey (R)-oxazaborolidine 129 gave the desired product 130 in 89% yield and 10:1 diastereoselectivity (Scheme 2.30).[51]

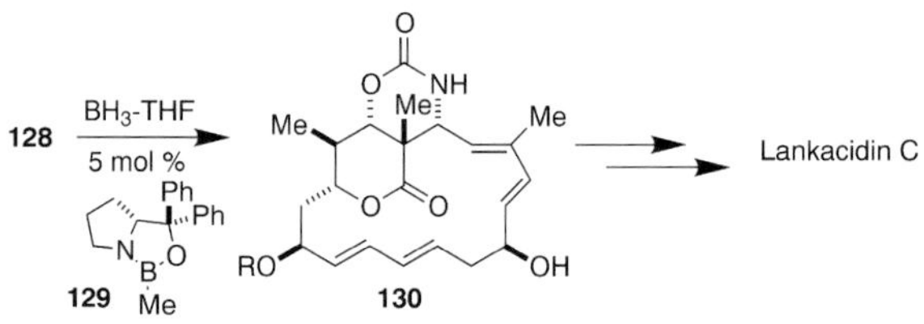

Scheme 2.30.

The success of this reaction illustrates the recent advances in the use of reagent-controlled stereochemical induction, in which a chiral catalyst can overcome the inherent stereochemical biases of the macrocyclic system. Indeed, when the corresponding (S)-CBS catalyst was employed, only the undesired isomer was obtained. Silylation of 130 gave a compound previously prepared by degradation studies on the natural product. This material was used to work out a careful sequence of steps compatible with the highly sensitive structure that subsequently permitted the total synthesis of lankacidin C.

2.6 Conclusions

The syntheses of oleandolide, fluviricin, macrolactin and lankacidin illustrate the application of contemporary synthetic methodology for the construction of stereochemically and structurally complex structures. The critical role that C—C bond formation plays in the development of efficient and elegant synthetic strategies is noteworthy. In this regard, Pd(0)-mediated couplings, olefin metathesis and alkene carbometallation have expanded the strategic options available in the design and execution of multistep syntheses. Additionally, contemporary advances in catalytic, enantioselective and diastereoselective aldehyde addition reactions enables highly convergent transformations wherein C—C bond formation is coupled with concomitant incorporation of a stereogenic centre.

The challenges posed by target-orientated syntheses are far from monolithic; rather, the synthetic inquiry (or problem) is a continuously evolving one. Indeed, analyses of even the most efficient, elegant and artful of syntheses always demonstrates gaps in the science of synthesis. Continued discovery and innovation in synthetic methods stemming from research at several interfaces (inorganic and organometallic chemistry, as well as biocatalysis) will lead to an expanding landscape of synthetic reactions leading to progressive advances in the synthesis of molecules.

References

1. (a) S. Masamune and P.A. McCarthy, 1984, in *Macrolide Antibiotics, Chemistry, Biology and Practice*, ed. S. Omura, Academic Press, Orlando, FL; (b) I. Paterson and M.M. Mansuri, 1985, *Tetrahedron Lett.*, 41, 3569; (c) I. Paterson and M.M. Mansuri, 1985, *Tetrahedron*, 45, 3569; (d) M. Bartra, F. Urpi, and J. Vilarassa, 1993, in *Recent Progress in the Chemical Synthesis of Antibiotics and Related Microbial Products*, ed. G. Kukacs, Springer-Verlag, Berlin, vol. 2.
2. (a) N.K. Kochetkov, A.F. Sviridov, and M.S. Ermolenko, 1981, *Tetrahedron Lett.*, 22, 4315; (b) ibid, 1981, 22, 4319; (c) S.S. Costa, A. Olesker, T.T. Thang, and G. Lukacs, 1984, *J. Org. Chem.*, 49, 2338; (d) Y. Kobayashi, H. Uchiyama, H. Kanbara, and F. Sato, 1985, *J. Amer. Chem. Soc.*, 107, 5541; (e) I. Paterson and P. Arya, 1988, *Tetrahedron*, 44, 253; (f) N.K. Kochetkov, D.V. Yashunsky, A.F. Sviridov, and M.S. Ermolenko, 1990, *Carbohydr. Res.*, 200, 209.
3. K. Tatsuda, T. Ishiyama, S. Takima, Y. Koguchi, and H. Gunji, 1990, *Tetrahedron Lett.*, 31, 709.
4. I. Paterson, R.D. Norcross, R.A. Ward, P. Romea, and M.A. Lister, 1994, *J. Amer. Chem. Soc.*, 116, 11287.

5. D.A. Evans, A.S. Kim, R. Metternich, and V.J. Novack, 1998, *J. Amer. Chem. Soc.*, 120, 5921.

6. K. Tatsuda, Y. Koguchi, and M. Kase, 1988, *Bull. Chem. Soc. Jpn.*, 61, 2525.

7. W.R. Roush and R.L. Halterman, 1986, *J. Am. Chem. Soc.*, 108, 294.

8. R.B. Woodward, E. Loguschp, K.P. Nambiar, K. Sakan, D.E. Ward, B.-W. Au-Yueng, P. Balaram, L.J. Browne, P.J. Card, C.H. Chen, R.B. Chenevert, A. Fliri, K. Frobel, H.-J. Gais, D.G. Garratt, K. Hayakawa, W. Heggie, D.P. Hesson, D. Hoppe, I. Hoppe, J.A. Hyatt, D. Ikeda, P.A. Jacobi, K.S. Kim, Y. Kobuke, K. Kojima, K. Krowicki, V.J. Lee, T. Leutert, S. Malchenko, J. Martens, R.S. Matthews, B.S. Ong, J.B. Press, T.V. Rajan Babu, G. Rousseau, H.M. Sauter, M. Suzuki, K. Tatsuda, L.M. Tolbert, E.A. Truesdale, I. Uchida, Y. Ueda, T. Uyehara, A.T. Vasella, W.C. Vladuchick, P.A. Wade, R.M. Williams, and H.N.C Wong, 1981, *J. Amer. Chem. Soc.*, 103, 3213.

9. For preliminary studies, see: (a) I. Paterson, M.A. Lister, and R.D. Norcross, 1992, *Tetrahedron Lett.*, 33, 1767; (b) I. Paterson, R.A. Ward, P. Romea, and R.D. Norcross, 1994, *J. Amer. Chem. Soc.*, 116, 3623; (c) I. Paterson and C.K. McClure, 1987, *Tetrahedron Lett.*, 28, 1229.

10. K. Ganesan and H.C. Brown, 1994, *J. Org. Chem.*, 59, 2336.

11. (a) I. Paterson, M.A. Lister, and C.K. McClure, 1986, *Tetrahedron Lett.*, 27, 4787; (b) I. Paterson, J.M. Goodman, and M. Isaka, 1989, *Tetrahedron Lett.*, 30, 7121; (c) I. Paterson, J.M. Goodman, M.A. Lister, R.S. Schumann, C.K. McClure, and R.D. Norcross, 1990, *Tetrahedron*, 46, 4663; (d) I. Paterson, 1992, *Pure Appl. Chem.*, 64, 1821; (e) I. Paterson and J.A. Channon, 1992, *Tetrahedron Lett.*, 33, 797; (f) I. Paterson and R.D. Tillyer, 1992, *Tetrahedron Lett.*, 33, 4233; (g) I. Paterson and J.A. Channon, 1991, *Tetrahedron Lett.*, 32, 797.

12. (a) H.C. Brown, R.K. Dhar, R.K. Bakshi, P.K. Pandiarajan, and B. Singaram, 1989, *J. Amer. Chem. Soc.*, 111, 3441; (b) H.C. Brown, R.K. Dhar, K. Ganesan, and B. Singaram, 1992, *J. Org. Chem.*, 57, 499; (c) H.C. Brown, R.K. Dhar, K. Ganesan, and B. Singaram, 1992, *J. Org. Chem.*, 57, 2716; (d) H.C. Brown, K. Ganesan, and R.K. Dhar, 1992, *J. Org. Chem.*, 57, 3767; (e) H.C. Brown, K. Ganesan, and R.K. Dhar, 1993, *J. Org. Chem.*, 58, 147; (f) K. Ganesan and H.C. Brown, 1993, *J. Org. Chem.*, 58, 7162.

13. (a) D.A. Evans, J.V. Neslon, E. Vogel, and T.R. Taber, 1981, *J. Amer. Chem. Soc.*, 103, 3099; (b) D.A. Evans, 1982, *Aldrichim. Acta*, 15, 23.

14. (a) D.A. Evans, J.S. Clark, R. Metternich, V.J. Novack, and G.S. Sheppard, 1990, *J. Amer. Chem. Soc.*, 112, 866; (b) D.A. Evans, H.P. Ng, J.S. Clark, and D.L. Rieger, 1992, *Tetrahedron*, 2127.

15. Lonomycin: D.A. Evans, A.M. Ratz, B.E. Huff, and G.S. Sheppard, 1995, *J. Amer. Chem. Soc.*, 117, 3448; (b) Rutamycin: D.A. Evans, H.P. Ng, and D.L. Rieger, 1993, *J. Amer. Chem. Soc.*, 115, 11446; (c) Calyculin: D.A. Evans, J.R. Gage, and J.L. Leighton, 1992, *J. Amer. Chem. Soc.*, 114, 9434.

16. D.A. Evans, K.T. Chapman, and E.M. Carreira, 1988, *J. Amer. Chem. Soc.*, 110, 3560.

17. J. Inanaga, K. Hirata, H. Saeki, T. Katsuki, and M. Yamaguchi, 1979, *Bull. Chem. Soc., Jpn.*, 52, 1989.

18. V.R. Hegde, M.G. Patel, V.P. Gullo, A.K. Ganguly, O. Sarre, M.S. Puar, and A.T. McPhail, 1990, *J. Amer. Chem. Soc.*, 112, 6403.

19. (a) A.F. Houri, Z. Xu, D.A. Cogan, and A.H. Hoveyda, 1995, *J. Amer. Chem. Soc.*, 117, 2943; (b) Z. Xu, C.W. Johannes, S.S. Salman, and A.H. Hoveyda, 1996, *J. Amer. Chem. Soc.*, 118, 10926; (c) Z. Xu, Z.W. Johannes, A.F. Houri, D.S. La, D.A. Cogan, G.E. Hofilena, and A.H. Hoveyda, 1997, *J. Amer. Chem. Soc.*, 119, 1032.

20. (a) A.H. Hoveyda and Z. Xu, 1991, *J. Amer. Chem. Soc.*, 113, 5079; (b) A.H. Hoveyda, Z. Xu, J.P. Morken, and A.F. Houri, 1991, *J. Amer. Chem. Soc.*, 113, 8950; (c) A.F. Houri, M.T. Didiuk, Z. Xu, N.R. Horan, and A.H. Hoveyda, 1993, *J. Amer. Chem. Soc.*, 115, 6614; (d) J.P. Morken and A.H. Hoveyda, 1993, *J. Org. Chem.*, 4237.

21. (a) J.P. Morken, M.T. Didiuk, and A.H. Hoveyda, 1993, *J. Amer. Chem. Soc.*, 115, 6697; (b) A.H. Hoveyda and J.P. Morken, 1996, *Angew. Chem., Int. Ed. Engl.*, 35, 1262.

22. S.V. Ley, J. Norman, W.P. Griffith, and S.P. Marsden, 1994, *Synthesis*, 639.

23. (a) R.R. Schrock, J.S. Murdzek, G.C. Bazan, J. Robbins, M. DiMare, and M. O'Regan, *J. Amer. Chem. Soc.*, 1990, 112, 3875; (b) G.C. Bazan, R.R. Schrock, H.-N. Cho, and V.C. Gibson, 1991, *Macromolecules*, 24, 4495.

24. R.H. Grubbs, S.J. Miller, and G.C. Fu, 1995, *Acc. Chem. Res.*, 28, 446.

25. For early examples of macrocyclic ring closure by olefin methasis, see: (a) B.C. Borer, S. Deerenberg, H. Bieräugel, and U.K. Pandit, 1994, *Tetrahedron Lett.*, 3191; (b) S.F. Martin, Y. Liao, H.J. Chen, M. Pätzel, and M.N. Ramser, 1994, *Tetrahedron Lett.*, 6005.

26. (a) A. Fürstner and K. Langemann, 1996, *J. Org. Chem.*, 61, 3942; (b) A. Fürstner and K. Langemann, 1997, *J. Amer. Chem. Soc.*, 119, 9130; (c) A. Fürstner and K. Langemann, 1997, *Synthesis*, 792.

27. For a recent review, see K.C. Nicolaou, F. Roschangar, and D. Vouroumis, 1998, *Angew. Chem., Int. Ed. Eng.*, 37, 2015.

28. (a) T. Mukaiyama, Y. Murai, and S. Shoda, 1981, *Chem. Lett.*, 431; (b) T. Mukaiyama, Y. Hashimoto, and S. Shoda, 1983, *Chem. Lett.*, 935.

29. (a) K.C. Nicolaou, R.E. Dolle, D.P. Papahatjis, and J.L. Randall, 1984, *J. Amer. Chem. Soc.*, 106, 4189; (b) K.C. Nicolaou, J.L. Randall, and G.T. Furst, 1985, *J. Amer. Chem. Soc.*, 107, 5556.

30. B.M. Trost, M.A. Ceschi, and B. König, 1997, *Angew. Chem., Int. Ed. Engl.*, 36, 1486.

31. J. Coste, E. Frérot, and P. Jouin, 1994, *J. Org. Chem.*, 59, 2437.

32. K. Guftafson, M. Roman, and W. Fenical, 1989, *J. Amer. Chem. Soc.*, 111, 7519.

33. G. Pattenden and R.J. Boyce, 1996, *Tetrahedron Lett.*, 37, 3501.

34. Y. Kim, R.A. Singer, and E.M. Carreira, 1998, *Angew. Chem., Int. Ed. Engl.*, 37, 1261.

35. A.B. Smith, III and G.R. Ott, 1998, *J. Amer. Chem. Soc.*, 120, 3935.

36. D.M. Hodgson, 1992, *Tetrahedron Lett.*, 33, 5603.

37. K. Takai, K. Nitta, and K. Utimoto, 1986, *J. Amer. Chem. Soc.*, 108, 7408.

38. G. Stork and K. Zhao, 1986, *Tetrahedron Lett.*, 25, 508.

39. (a) R.A. Singer and E.M. Carreira, 1995, *J. Amer. Chem. Soc.*, 117, 12360; (b) E.M. Carreira and R.A. Singer, 1996, *Drug Discovery Today*, 1, 145.

40. J. Krüger and E.M. Carreira, 1998, *J. Amer. Chem. Soc.*, 120, 837.

41. E.M. Carreira, R.A. Singer, and W. Lee, 1994, *J. Amer. Chem. Soc.*, 116, 8837.

42. (a) B.H. Lipshutz, E.L. Ellsworth, S.H. Dimock, and D.C. Reuter, 1989, *Tetrahedron Lett.*, 30, 2065; (b) R. Ostwald, P.-Y. Chavant, H. Stadtmüller, and P. Knochel, 1994, *J. Org. Chem.*, 59, 4143.

43. (a) D.A. Evans and W.C. Black, 1993, *J. Amer. Chem. Soc.*, 111, 4497; (b) E. Negishi, N. Okukado, A.O. King, D.E. Van Horn, and B.I. Spiegel, 1978, *J. Amer. Chem. Soc.*, 100, 2254; (c) J.K. Stille and B.L. Groh, 1987, *J. Amer. Chem. Soc.*, 109, 813.

44. W.C. Still and C. Gennari, 1983, *Tetrahedron Lett.*, 24, 4405.

45. (a) E. Gaumann, R. Huter, W. Keller-Schierlein, L. Neipp, V. Prelog, and H. Zahner, 1964, *Helv. Chim. Acta*, 47, 78; (b) J.M.J. Sakamoto, S. Kondo, S. Yumoto, and M. Arishima, 1962, *J. Antibiot., Ser. A*, 15, 98; (c) S. Harada, E. Higashide, T. Fugono, and T. Kishi, 1969, *Tetrahedron Lett.*, 27, 2239; (d) E. Higashide, T. Hugono, K. Hatano, and M. Shibata, 1971, *J. Antibiot.*, 24, 1; (e) S. Harada, T. Kishi, and K. Mizuno, 1971, *J. Antibiot.*, 24, 13; (f) T. Fugono, S. Harada, E. Higashide, and T. Kishi, 1971, *J. Antibiot.*, 24, 23.

46. (a) S. Harda and T. Kishi, 1974, *Chem. Pharm. Bull.*, 22, 99; (b) S. Harada, 1975, *Chem. Pharm. Bull*, 23, 2201; (c) For an illustration of this instability, see: J.W. McFarland, D.K. Pirie, J.A. Retsema, and A.R. English, 1984, *Antimicrob. Agents Chemother.*, 25, 226.

47. (a) A.S. Kende, K. Koch, G. Dorey, I. Kaldor, and K. Liu, 1993, *J. Amer. Chem. Soc.*, 115, 9842; (b) A.S. Kende, K. Liu, I. Kaldor, G. Dorey, and K. Koch, 1995, *J. Amer. Chem. Soc.*, 117, 8258.

48. (a) M.Y.H. Wong and R.G. Gray, 1978, *J. Amer. Chem. Soc.*, 100, 3548; (b) H. Maehr, A. Perrotta, and J. Smallheer, 1988, *J. Org. Chem.*, 53, 832.
49. (a) R. Labia and C. Morin, 1984, *Chem. Lett.*, 1007; (b) T.H. Salzmann, R.W. Ratcliffe, B.G. Christensen, and F.A. Bouffard, 1980, *J. Amer. Chem. Soc.*, 102, 6161.
50. (b) E.J. Thomas and A.C. Williams, 1987, *J. Chem. Soc., Chem. Commun.*, 992; (e) J.M. Roe and E.J. Thomas, 1990, *Synlett*, 727.
51. (a) E.J. Corey, R.K. Bakshi, S. Shibata, C.-P. Chen, and V. Singh, 1987, *J. Amer. Chem. Soc.*, 109, 7925.

3 Polyether total synthesis

A.N. Hulme

3.1 Introduction

Interest in the synthesis of polyethers[1] was sparked by the discovery that many such compounds are able to function as ionophores, transporting cations (Na^+, K^+, Ca^{2+}, Mg^{2+}) through lipophilic membranes. This ionophore activity causes an imbalance in the intracellular ion concentration and hence disrupts the normal growth patterns of bacteria. Throughout the 1970s and 1980s many members of this class were isolated, including the antibiotics monensins A and B (**1** and **2**),[2] etheromycin (**3**)[3] and salinomycin (**4**).[4] Several polyether antibiotics are used commercially for the treatment of coccidial infections in poultry, and as growth promoters in ruminant animals.[1]

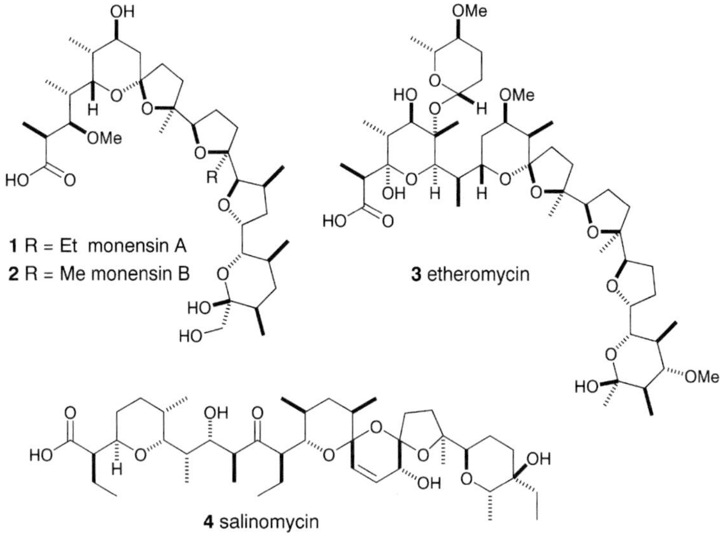

1 R = Et monensin A
2 R = Me monensin B

3 etheromycin

4 salinomycin

Whilst crown ethers generally rely upon a simple repeating structure to give the molecule a well-defined ligation pattern, polyether ionophores have a three-dimensional structure that is defined by the minimisation of steric interactions along a complex carbon backbone as well as hydrogen bonding interactions. Natural polyether antibiotics typically possess a large number of five- or six-membered cyclic ether rings, (e.g. monensin A

(1), the first of this class to be structurally characterised).[2] A further feature of polyether antibiotics is the frequent appearance of a rigid spiroketal unit and a terminal carboxyl group. For the monensins, this spiroketal unit works alongside several intramolecular hydrogen bonds to impart a cyclic structure to both the free acid and the metal-ion complexed states (Scheme 3.1).[5] Hence, the exterior surface of the

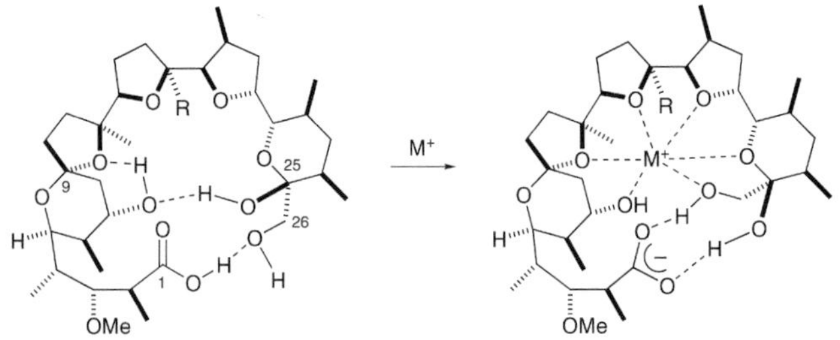

Scheme 3.1   Metal-ion complexation by the monensins.

monensins is almost entirely lipophilic, whilst the interior is highly polar in nature and thus ideally suited for the complexation of metal cations, particularly Na^+. It is this bifunctional character that enables the monensins and many other polyether antibiotics to transport cations across biological membranes.

Ever since the discovery of the first polyether antibiotics, a large number of other polyether-containing natural products have been isolated and characterised. Moreover, many have been shown to exhibit not just antibiotic activity but also antitumour,[6] antifungal[7] and neurotoxic properties.[8] This structurally complex, and synthetically challenging, class of natural products has therefore been the subject of much interest over the past decade, and many novel synthetic strategies have been developed to control both the absolute and relative stereochemistry observed in the highly functionalised polyether rings.

Traditionally, the strategies employed in the synthesis of the cyclic ether rings themselves have fallen into five broad categories:

- epoxide and polyepoxide cyclisation reactions;
- stepwise ether formation through intermediate metal-oxo species;
- acid-catalysed acetal formation;
- cyclisation through Williamson ether formation reactions;
- cyclic ether formation through C—C bond coupling of acyclic ether precursors.

Many of these approaches are said to be 'biomimetic' inasmuch as the synthetic chemist has attempted to mimic in the laboratory the route by which the natural product is formed in the living organism. This can sometimes allow the proposed biosynthetic pathway to be tested and can also lead to the development of new synthetic methodology. With the isolation of ever more complex polyether natural products, new strategies for ether ring formation will continue to be required. Some of the new methodologies that have been developed in recent years will be touched upon in this chapter.

The landmark total syntheses of monensin A (1) by Kishi[9] and Still,[10] amongst others, have been the subject of several reviews[11, 12] and so will not feature in this chapter. Newer members of the polyether class contain larger ether ring sizes (e.g. seven-, eight- and nine-membered rings) than the simple five- and six-membered rings first encountered, and these medium rings have provided significant synthetic challenges in themselves. Several new members of this class are also macrolides, for example spongistatin 1 (5).[13] Hence, synthetic strategies towards these polyethers must also tackle issues of macrolactonisation. Other polyether natural products have been identified with multiple contiguous tetrahydrofuran rings. Structures of this sort are exemplified by parviflorin (6).[14] Further

5 spongistatin 1

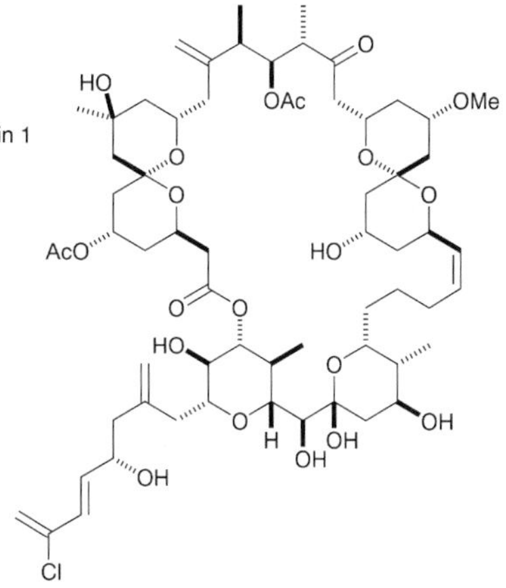

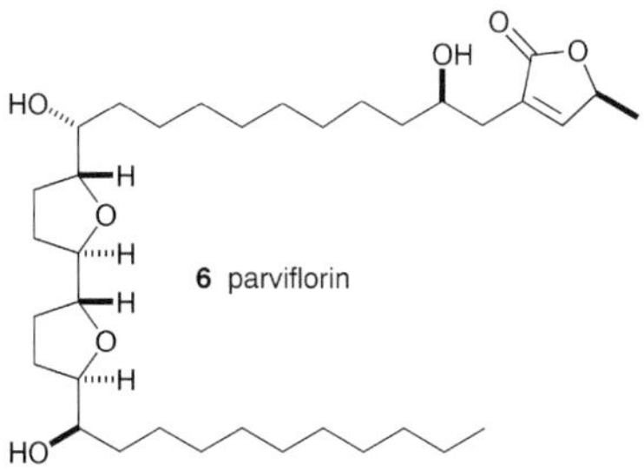

6 parviflorin

hurdles are raised by the *trans*-fused cyclic ether systems found in brevetoxin A (7).[15] In this chapter, recent approaches to several polyether

7 brevetoxin A

natural products will be discussed. The molecules we will address will include monensin A (1), etheromycin (3), salinomycin (4), spongistatin 1 (5), parviflorin (6) and brevetoxin A (7).

3.2 Biomimetic approaches to the synthesis of polyketide polyether antibiotics

Through elegant labelling studies, Cane has shown that monensin A (1) is a polyketide-derived natural product composed of acetate, propionate and butyrate units. These studies have also confirmed that the oxygen atoms of the C, D and E rings of the monensins arise from atmospheric oxygen and not from the polyketide carbonyl groups.[16] This finding prompted Cane, Celmer and Westley to propose that the biosynthesis of many polyether natural products takes place in two distinct stages.[17] For monensin, the first event is an enzymatic epoxidation of an acyclic polyene precursor 8. The second is a cascade of intramolecular epoxide ring-openings on 9 to form the polyether framework (Scheme 3.2).

Based upon this hypothesis, biomimetic approaches to the synthesis of monensin B (2) have been devised by both Still[18] and Schreiber.[19] These have focused on the introduction of the epoxide units in a single step by the epoxidation of a macrolide diene or triene, followed by lactone hydrolysis and cyclisation to polyether fragments. Whilst these approaches have demonstrated the efficiency of the cascade cyclisation process, the required relative stereochemistry for only two out of three epoxides was established. More recent work by Paterson[20] has led to a successful polyepoxide cascade cyclisation for the synthesis of the ABCD-ring subunit of etheromycin (3), which will be discussed shortly.

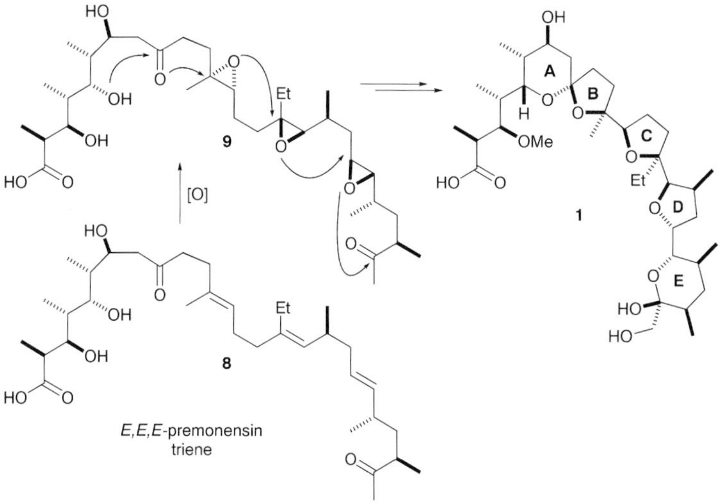

Scheme 3.2 Cane–Celmer–Westley model for the biosynthesis of monensin (**1**).

The first step in the Cane–Celmer–Westley model for polyether biosynthesis remains the focus of some controversy.[21] In the only reported feeding studies with *Streptomyces cinnamonensis*, no conversion of radiolabelled *E, E, E*-premonensin triene (**8**) to monensin A (**1**) was detected.[22] McDonald[23] and others[24] have therefore proposed that an alternative sequential *syn* oxidative polycyclisation of a *Z, Z, Z*-premonensin triene precursor **10** might be operating (Scheme 3.3). Such a biosynthetic hypothesis would also allow for the late-stage introduction of the oxygen atoms in the C, D and E rings.

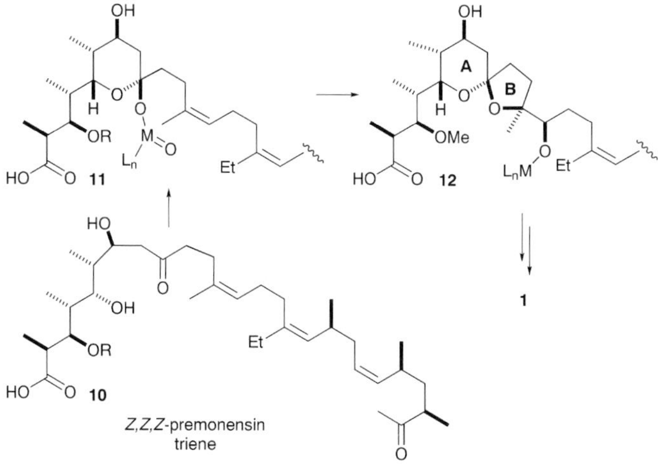

Scheme 3.3 *Syn* oxidative polycyclisation hypothesis.

A key tenet of the McDonald cyclisation theory is the formation of an alkoxy-bound metal-oxo intermediate such as **11**. Intramolecular [2+2] cycloaddition, to form a metallaoxetane, and subsequent reductive elimination of the metal then generates compound **12**. Oxidation of the metal would activate the system for a further cycle of the *syn* oxidative cyclisation process. Since the oxygen atoms are added to the double bonds in a *syn* fashion by [2+2] cycloaddition, a Z, Z, Z-polyene precursor **10** would be required to construct the correct stereochemistry for the polyether framework of monensin. However, this type of reaction is not currently known in nature, and the hypothesis remains untested by feeding studies.

The polyether antibiotic etheromycin (**3**) is characterised by a complex array of six ether rings (A–F) and a multitude of associated stereocentres. The Cane–Celmer–Westley hypothesis suggests that it might be derived biosynthetically from the cyclisation of polyepoxide **13**. This would require the C(7) hydroxyl of **13** to form the A-ring hemiacetal, and the C(9) hydroxyl to trigger the epoxide ring-opening cascade [shown in Scheme 3.4 as a solid curved arrow from C(9)] to give **14**.[25] It is worth noting that for this to be synthetically viable, the alternative cyclisation mode for polyepoxide **13**, where the C(7) and C(9) hydroxyls are involved in bicyclic acetal formation (dotted curved arrow), must be less favourable.

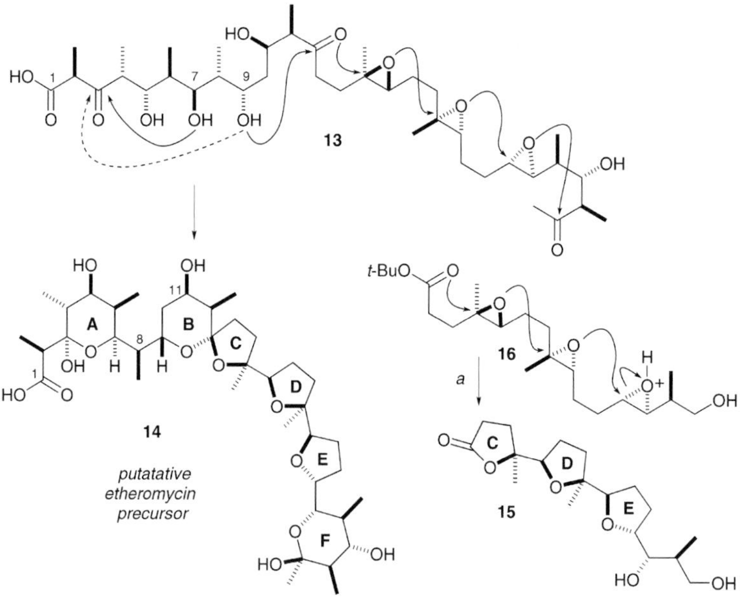

Scheme 3.4 Putative cyclisation mode for etheromycin and CDE ring synthesis. (*a*) CSA, CH₂Cl₂, 0°C, 18 h.

Through a series of studies Paterson has shown that the aforementioned polyepoxide ring-opening strategy is feasible for the construction of the CDE fragment **15** (Scheme 3.4)[26] and similar BCD fragments.[27] In contrast to the approaches used by Still[18] and Schreiber[19] for monensin B, stepwise use of the Sharpless asymmetric epoxidation reaction upon suitably functionalised acyclic allylic alcohol precursors was found to be successful for the assembly of the required tri-epoxide **16**. The biomimetic acid-catalysed cyclisation reaction of **16** was conducted in the presence of camphorsulfonic acid (CSA) to give the CDE ring fragment **15** (Scheme 3.4).

In the most complex studies to date, more heavily functionalised precursors have been synthesised. This has allowed the construction of an ABCD ring model **17**.[20] Control of the stereochemistry in the A-ring fragment **18**, was achieved through a series of aldol reactions (Scheme 3.5). The first of these was an Evans propionate *syn*-aldol process which

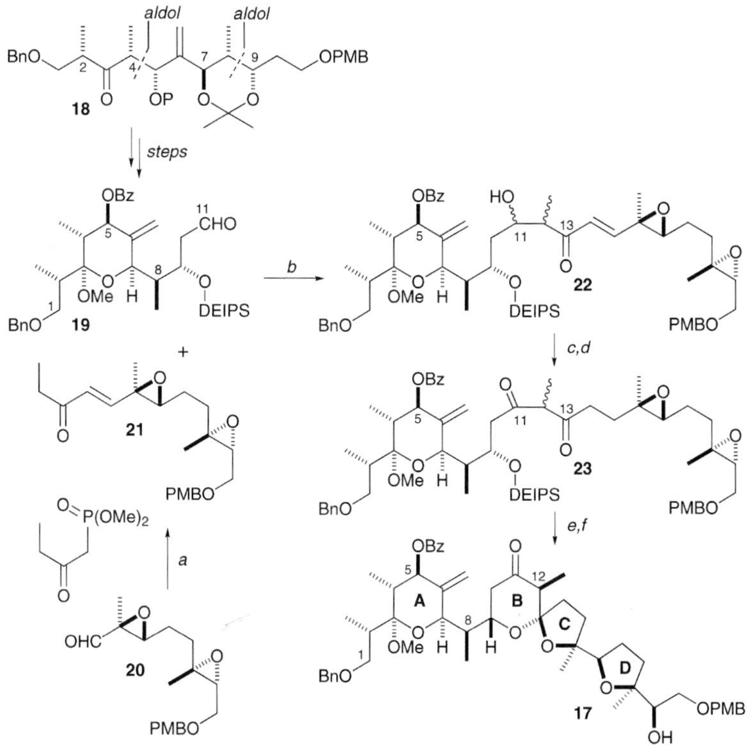

Scheme 3.5 Construction of an ABCD ring model of etheromycin. (*a*) Ba(OH)$_2$•8H$_2$O, THF (aq.); (*b*) LDA, THF, $-78°$C; ZnCl$_2$; **19**; (*c*) H$_2$, Rh/Al$_2$O$_3$, THF; (*d*) Dess–Martin periodinane, CH$_2$Cl$_2$, $-78°$C; (*e*) 0.5 M HCl, THF; (*f*) PPTS, (MeO)$_3$CH, MeOH.

set the C(8)–C(9) stereochemistry with high stereoselectivity. In the second key aldol reaction, the *syn–syn* relationship between the C(2), C(4) and C(5) stereocentres was controlled using a tin(II) enolate derived from a C(1)–C(4) dipropionate fragment. Previous studies[25] had shown that by positioning a benzoate protecting group on the C(5) hydroxyl, the A-ring tetrahydropyran could be formed efficiently as its methyl acetal **19** [leaving the C(9) hydroxyl free to form the etheromycin B-ring].

Once again, Sharpless asymmetric epoxidation gave excellent control over the required diepoxide stereochemistry in **20** (Scheme 3.5). The latter was then converted into enone **21** by using a mild, Ba(OH)$_2$-mediated Horner–Wadsworth–Emmons reaction. The zinc enolate derived from enone **21** coupled to aldehyde **19** to give an 84% yield of the enone adducts **22**. Reduction of the enone double bond and oxidation of the free hydroxyl gave the β-diketone cyclisation precursor **23**. This was treated directly with 0.5 M HCl in THF to remove the DEIPS group from the C(9) hydroxyl. The latter event allowed the BCD-ring poly-epoxide cyclisation cascade to occur. A 70:30 mixture of diastereomers was obtained at C(12). Concomitant hydrolysis of the A-ring acetal was observed under these reaction conditions. Isolation of the fully cyclised ABCD-ring fragment **17** was achieved after the major C(12) diastereo-mer was subjected to acetal formation to reform the A-ring [PPTS, (MeO)$_3$CH]. Paterson's work underscores the utility of polyepoxide cascade cyclisations for the synthesis of polyether antibiotics of this sort.

Support for an alternative *syn* oxidative cyclisation mechanism in the biosynthesis of the polyether antibiotics has been provided by McDonald.[23, 28, 29] By using triene **24** (Scheme 3.6) as a model for the putative premonensin *Z, Z, Z*-triene **10**, McDonald investigated the viability of this cyclisation mode for the synthesis of the CD ring system of monensin.

Triene **24** underwent Sharpless asymmetric dihydroxylation (AD-mix β), with unexpectedly high selectivity for a terminal trisubstituted double bond to give diol **25** in moderate yield and 89% enantiomeric excess (ee). PCC-induced *syn* oxidative cyclisation of **25** led to the *cis*-tetrahydro-furan diol **26** as a single stereoisomer.[28] However, this reaction proceeded in poor yield (24%). Hence, an alternative *syn* cyclisation was investigated with Collins reagent [CrO$_3$(py)$_2$]. This provided ketone **27** in moderate yield. Acetate protection of the C(12) hydroxyl and reduction of the C(17) ketone next furnished the secondary alcohol **28** (consistent with Felkin–Anh stereocontrol) with high efficiency. A *syn* oxidative cyclisa-tion was conducted on hydroxyalkene **28** using dichloroacetylperrhenate to establish the second tetrahydrofuranyl ring in the conformationally preferred *trans* arrangement. Hence, a synthesis of the *cis,trans*-bis-tetrahydrofuranyl alcohol **29** was achieved, which represents a model

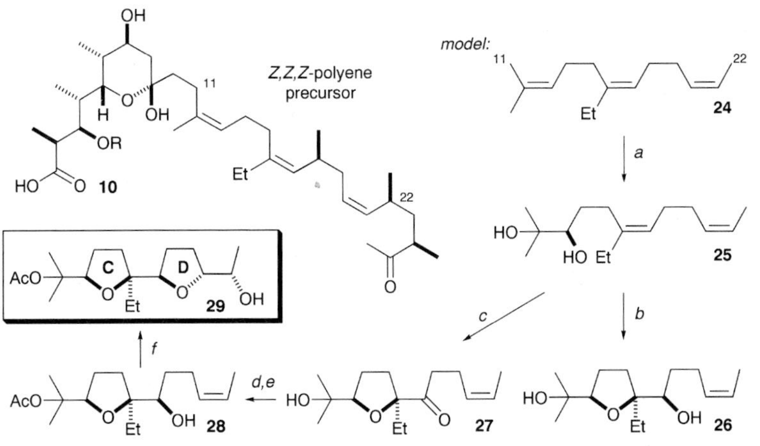

Scheme 3.6 Sequential *syn* oxidative cyclisation of a CD ring model of monensin. (*a*) AD-mix β, CH$_3$SO$_2$NH$_2$, *t*-BuOH/H$_2$O (1:1), 0°C; (*b*) PCC, CH$_2$Cl$_2$; (*c*) CrO$_3$(py)$_2$, CH$_2$Cl$_2$; (*d*) Ac$_2$O, DMAP (cat.), Et$_3$N; (*e*) NaBH$_4$, CeCl$_3$, EtOH, −78°C; (*f*) (Cl$_2$CHCO$_2$)ReO$_3$, (ClCHCO)$_2$O, CH$_2$Cl$_2$, 0°C.

system having the correct stereochemical oxygenation pattern for the CD-rings (C(12)–C(21)) of monensin A (**1**).

Although McDonald's *syn* dihydroxylation, *syn* oxidative cyclisation sequence requires the use of three metal-oxo reagents (Os, Cr, Re), it has been speculated that a single biosynthetic enzyme containing a metal-oxo site might catalyse all three transformations if coupled to an external oxidising agent.[29] Hence, these synthetic studies have provided evidence for a viable alternative to the Cane–Celmer–Westley model for polyether antibiotic biosynthesis.

3.3 Strategies for the synthesis of the polyether antibiotic salinomycin

The polyether ionophore antibiotic salinomycin (**4**) is produced by a strain of *Streptomyces albus*. It has been found to exhibit marked anti-microbial activity against Gram-positive bacteria, mycobacteria and fungi.[4] Salinomycin is used commercially as a growth promoter for cattle and for the treatment of coccidial infections in poultry.[1] The structural determination of salinomycin by Kinashi in 1973[30] showed that it contains a tricyclic dispiroketal ring system built around a central unsaturated six-membered ring. Similar ring systems have since been encountered in the polyether antibiotics noboritomycin,[31] 20-deoxysalinomycin,[32] and X-14766A,[33] amongst others.

In degradation studies by Kishi, a retroaldol reaction to the fragments **30a** and **31** was observed when C(20) *O*-acetate protected salinomycin methyl ester **32** was heated at 200°C *in vacuo* (Scheme 3.7).[34] Subsequently, Yonemitsu noted that when ethyl ketone **31** was treated with

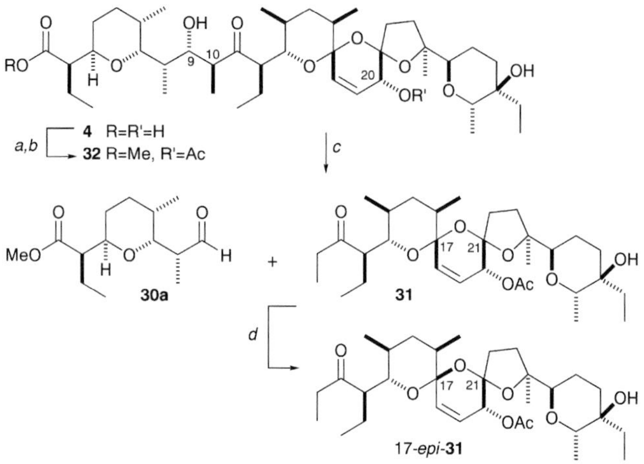

Scheme 3.7 Salinomycin degradation studies. (*a*) CH$_2$N$_2$; (*b*) Ac$_2$O; (*c*) 200°C, 0.5 mmHg; (*d*) CSA, CH$_2$Cl$_2$.

camphorsulfonic acid (CSA) it underwent spiroketal rearrangement to become epimeric at C(17).[35] In the synthetic approaches to salinomycin that are reviewed in this section this retroaldol reaction was conceptually exploited in the retrosynthetic strategy, with the first disconnection being made at the C(9)—C(10) bond.

A number of different approaches have been recorded to the synthesis of the tricyclic core. The first total synthesis of salinomycin, by Kishi, in 1982,[34] capitalised on an acid-catalysed intramolecular ketalisation, an approach that was mirrored by Yonemitsu (section 3.3.1). More recent approaches by Kocienski (section 3.3.2) and Brimble[36] have made use of the oxidative rearrangement of a 2-acyl furan and the oxidative cyclisation of a hydroxyspiroacetal, respectively.

3.3.1 *The Yonemitsu total synthesis of salinomycin*

Yonemitsu relied upon the cheap 'chiral pool' starting materials, D-glucose, D-mannitol and (*S*)-lactic acid for installing much of the chirality present in the carbon backbone of the target.[37] As in Kishi's synthesis of salinomycin, a late-stage aldol coupling between the C(1)–C(9) aldehyde **30b** and a protected C(10)–C(30) fragment **33** was exploited (Scheme 3.8).

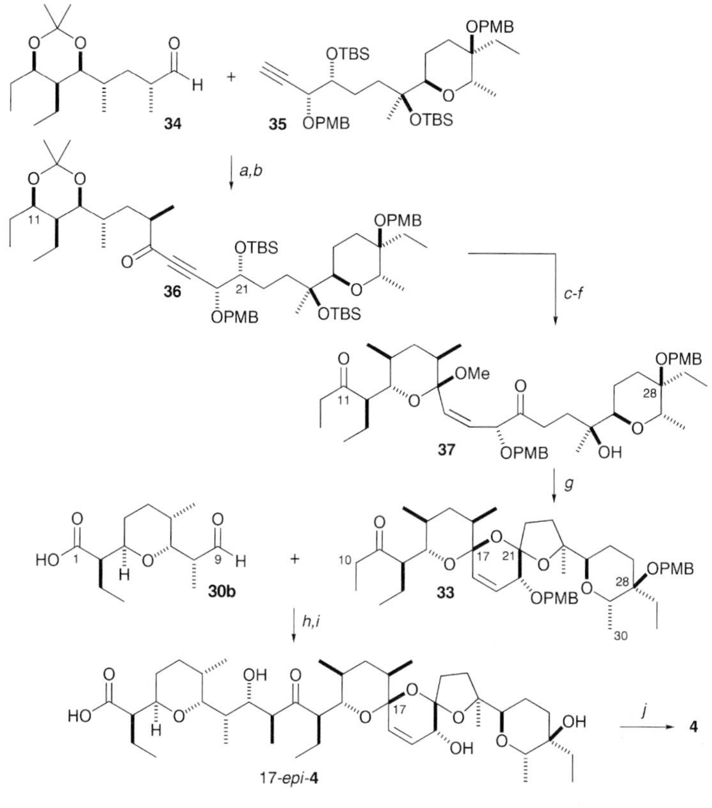

Scheme 3.8 Yonemitsu's approach to the synthesis of salinomycin. (*a*) **35**, *n*-BuLi, THF, −78°C; **34**; (*b*) [O]; (*c*) CSA, MeOH; (*d*) TBAF, dioxane–THF, 65°C; (*e*) H₂, Lindlar cat., MeOH; (*f*) DMSO, (COCl)₂, CH₂Cl₂; Et₃N; (*g*) CSA, CH₂Cl₂; (*h*) (ᶜHex)₂NMgBr, THF, −55°C; (*i*) DDQ, CH₂Cl₂-buffer; (*j*) TFA, CH₂Cl₂, molecular sieves.

Coupling of aldehyde **34**[38] with the alkynyllithium derived from acetylene **35**[37] allowed assembly of the C(10)–C(30) carbon backbone (Scheme 3.8). The resulting alcohol was readily oxidised to ynone **36**. Treatment with camphorsulfonic acid in methanol enabled intramolecular acetalisation to occur, to produce the β-methyl glycoside. Desilylation was achieved with TBAF. Partial reduction of the triple bond to the *cis* alkene and Swern oxidation of the secondary alcohols at C(11) and C(21) finally afforded the dispiroacetal cyclisation precursor **37**. Acid-catalysed intramolecular ketalisation generated a 1:2 mixture of two diastereomers which were readily separable. Unfortunately, the desired C(21) stereochemistry was present only in the minor product **33**, both diastereomers being epimeric with salinomycin at C(17). (The same cyclisation conducted with a benzyl protecting group at C(28) led to a more favourable 1:1 mixture of diastereomers, but the benzyl protecting group later proved too difficult to remove in the final stages of the synthesis.)

Coupling of the C(1)–C(9) and C(10)–C(30) fragments was achieved in moderate yield (23%) by an aldol reaction which utilised the magnesium enolate derived from ethyl ketone **33** at −55°C (Scheme 3.8). Rapid and efficient deprotection of both *p*-methoxybenzyl groups was then effected with DDQ. Upon treatment with TFA the 17-*epi*-salinomycin produced by this route was found to epimerise at the C(17) ketal position, to generate salinomycin (**4**).

That the correct acetal stereochemistry at C(17) should now be favoured under acid-catalysed equilibration conditions should not be surprising since the vast majority of natural spiroacetal systems are generated under thermodynamic control with the relative configuration determined by a combination of anomeric and steric effects.[39] However, we will return to the stereochemical disparity displayed by the spiro-acetals in **33** and **4** in section 3.3.2 below.

3.3.2   Kocienski's route to salinomycin

Kocienski's synthesis of salinomycin[40] exploited the aldol strategy shown in Scheme 3.9 to assemble the natural product, and closely followed the precedent set by both Kishi[34] and Yonemitsu.[37] Kocienski's synthesis of the C(1)–C(9) aldehyde **30a** made use of the stereoselective addition of a pyranosyl cuprate to an η^3-allyl cationic complex. Specifically, Faller's η^3-allyl cationic molybdenum complex **38** (Scheme 3.10)[41] was selected as the electrophile partner because of its ease of preparation in enantio-merically enriched form by Sharpless asymmetric epoxidation. The addition of cuprate **39** (derived from **40**) to the η^3-allyl cationic com-plex **38** proceeded in moderate yield (44%) but with high regio- and diastereofacial selectivity to give **41**. Oxidative destruction of the η^2-molybdenum intermediate **41**, conversion of the primary alcohol to the corresponding methyl ester and ozonolysis of the double bond revealed aldehyde **30a**, the desired C(1)–C(9) fragment.

For the synthesis of the C(11)–C(30) fragment, Kocienski made extensive use of furan chemistry. First, he selectively lithiated furan (**42**) and added the resulting furanyllithium to lactone (**43**) to obtain **44**, after selective desilylation (Scheme 3.11). Furan (**44**) was then treated with

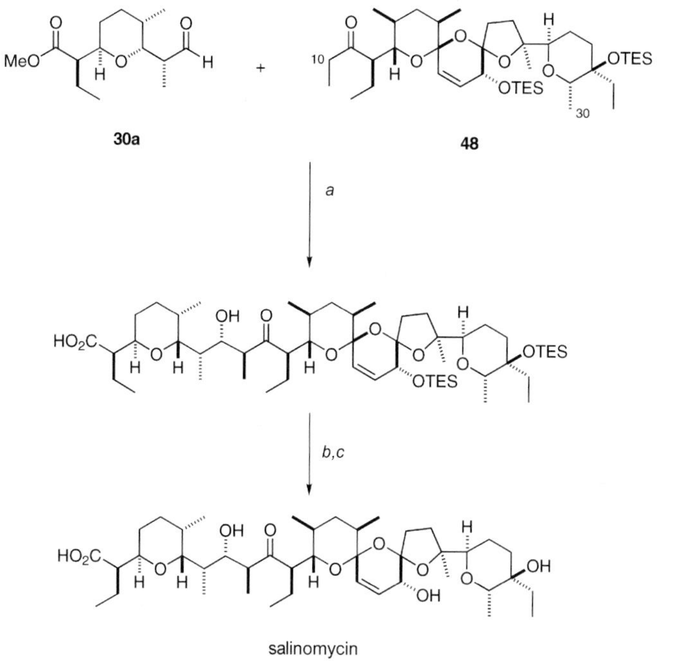

Scheme 3.9 (*a*) (^cHex)₂NMgBr (0.95 M in THF), **48**, THF, −65°C to −50°C; **30a**, −65°C to −50°C; (*b*) n-Bu₄NF, THF; (*c*) CF₃CO₂H, CH₂Cl₂.

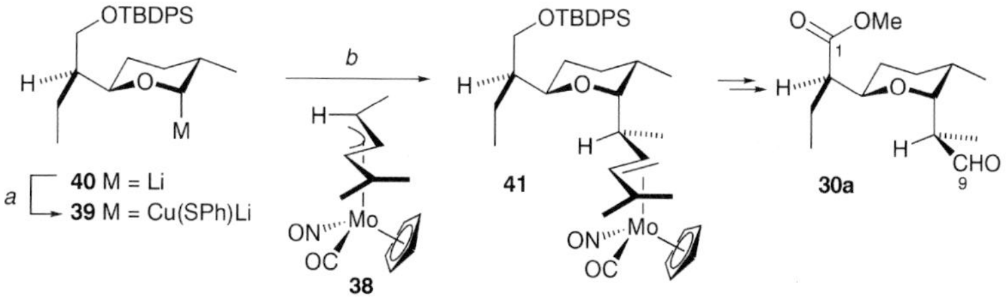

Scheme 3.10 Synthesis of the C(1)–C(9) aldehyde fragment of salinomycin. (*a*) CuBr•SMe₂, THF, −80°C; (*b*) **38**, THF, −80°C.

N-bromosuccinimide (NBS) to release the latent enedione functionality. Aqueous HF promoted a rearrangement of spiroacetal **45** to **46** and **47**. The latter were isolated in 44% and 15% yield, respectively. Kocienski has suggested that the reaction sequence shown in Scheme 3.11 operates during the formation of intermediate **45**. The acid-catalysed rearrangement of **45** to the tricyclic acetals **46** and **47** is thought to be kinetically controlled.

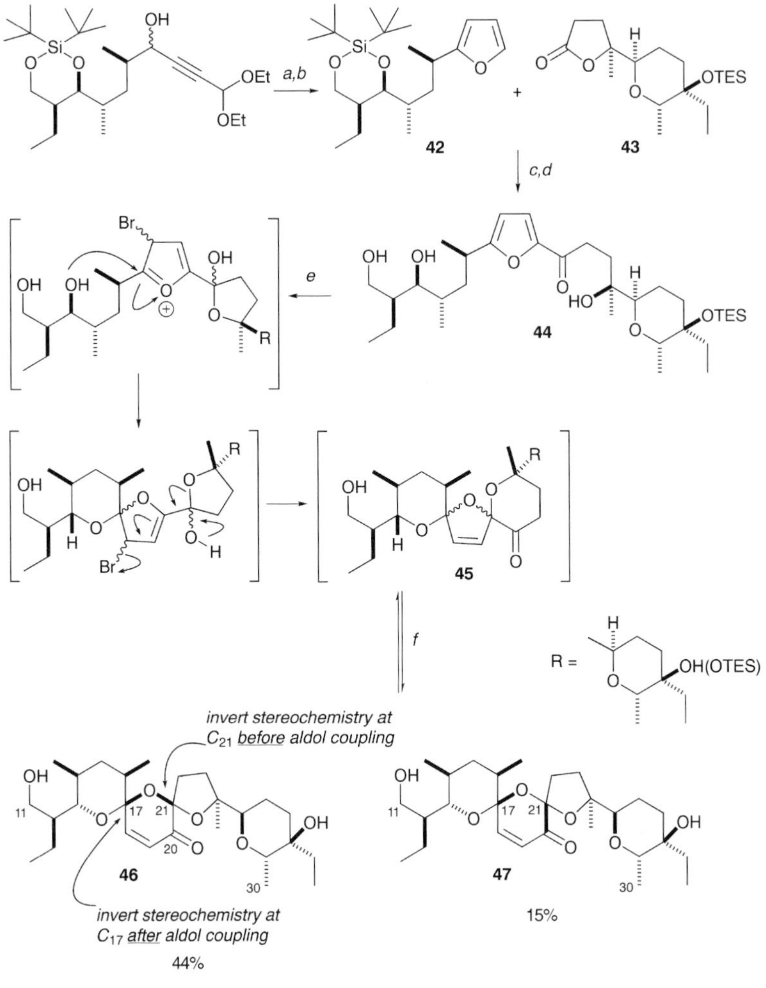

Scheme 3.11 Construction of the dispirocyclic core of salinomycin. (*a*) H$_2$, Lindlar cat., MeOH; (*b*) PPTS, CH$_2$Cl$_2$; (*c*) **42**, ^{n}BuLi, THF, $-60°$C; **43**, $-65°$C; (*d*) HF•pyridine, THF–pyridine; (*e*) NBS, THF/H$_2$O, $-10°$C; (*f*) HF, MeCN/H$_2$O.

Unfortunately, the major tricyclic acetal **46** had incorrect stereochemistry at both C(17) and C(21). After silylation with TESOTf, reduction of the enone carbonyl at C(20) with NaBH$_4$ in the presence of CeCl$_3$•7H$_2$O gave a 7:1 mixture of diastereomers in favour of the incorrect stereochemistry at C(20). Fortunately, a Mitsunobu inversion could be used to correct the stereochemistry at C(20). However, rectification of the stereochemistry at C(17) and C(21) was more demanding. An acid-catalysed rearrangement of the C(20)-inverted tricyclic core (CSA) eventually produced the desired C(21) stereochemistry, it giving the C(10)–C(30) segment **48** shown in Scheme 3.9, after protecting group adjustment. An *anti* aldol reaction was then used to join the C(1)–C(9) and C(10)–C(30) fragments together (Scheme 3.9). Inversion of the C(17) stereochemistry was finally accomplished by acid-catalysed rearrangement with TFA. This delivered salinomycin.

These results lend weight to the argument of Kishi,[34] that the repulsive dipolar interactions between two axially disposed oxygen atoms in the salinomycin spirocyclic core are compensated for by a hydrogen bond between the C(9) and C(20) hydroxy groups. Systems in which a remote hydrogen bond is precluded (e.g. 20-deoxysalinomycin) have been shown to adopt the 17-*epi* configuration at equilibrium. Further support for the hydrogen bonding argument has been provided by X-ray data for salinomycin derivatives, which reveal a folded conformation in which the C(9) and C(20) hydroxyl groups are in proximity.[30, 42] Hence the synthesis of salinomycin (**4**) has served to illustrate the delicate balance of steric and electrostatic interactions which determine the stereochemistry of spiroacetals in the polyether antibiotics.

3.4 The spongistatins

In 1993 three groups independently reported the isolation and structure elucidation of a new class of antimitotic macrolide polyether derived from marine sponges of the genus *Spongia*[13, 43] and *Spirastrella* (spongistatins 1–9),[44] *Cinachyra* (cinachyrolide A)[45] and *Hyrtios* (altohyrtins A–C).[46] These compounds were all highly cytotoxic towards a wide variety of cancer cell lines. All the members of this class were found to contain six pyran rings and a 42-membered lactone ring. Structurally, altohyrtin A appeared to be identical to spongistatin 1, and cinachyrolide A to spongistatin 4. The relative and absolute configurations of these molecules have been the subject of debate. However, detailed nuclear magnetic resonance (NMR) studies[47] and recent total syntheses have established the structures for spongistatin 1 (**5**)[48] and spongistatin 2 (**49**)[49] to be as shown in Scheme 3.12. Spongistatin 1 (**5**) is the most cytotoxic member of this family. It possesses a most unusual chlorodiene functionality in its side-chain.

Clearly, the synthesis of any molecules belonging to this class, with their vast array of stereocentres, their complex functionalities and their large macrocyclic rings, is a serious undertaking. The first groups to attain success in this area were those of Evans (section 3.4.1) and Kishi (section 3.4.2). Numerous other approaches to their synthesis have also been reported.[50]

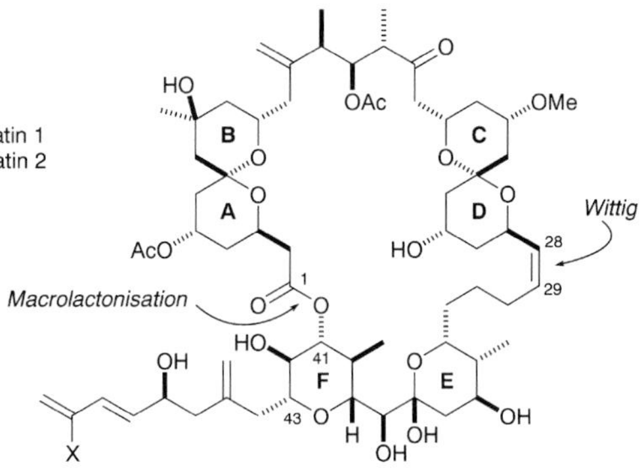

Scheme 3.12   Retrosynthetic analysis of spongistatins 1 and 2.

In constructing the polyether macrolactone ring, Kishi and Evans both chose to utilise the powerful disconnections of a Wittig reaction to form the C(28)–C(29) double bond, and a macrolactonisation to form the 42-membered ring (Scheme 3.12). In both studies the Wittig reaction was used to couple two major fragments effectively and to set the C(28)–C(29) *cis* alkene geometry with greater than 95:5 selectivity. In Kishi's study the side-chain at C(43) was introduced *prior* to the Wittig coupling [i.e. a fully developed C(29)–C(51) fragment was coupled to the C(1)–C(28) fragment], whereas in Evans's study the side-chain was introduced in the dihydropyran F-ring precursor after Wittig coupling. Both groups found that subjecting a cyclisation precursor with unprotected hydroxyl groups at both C(41) and C(42) to Yamaguchi's macrolactonisation conditions (2,4,6-trichlorobenzoyl chloride, R$_2$NEt, PhH *or* PhMe; DMAP, reflux) gave the desired macrolactonisation product with a high degree of regioselectivity.

The use of many aspects of acyclic stereochemical control, such as the regioselective opening of chiral epoxides to control the absolute stereochemistry of hydroxyl substituents along the carbon backbone, was a common feature of both syntheses. However, the strategies used by Evans and Kishi to control the array of stereochemistry along the spongistatin backbone did differ significantly in their construction of the C(1)–C(28) sector and it is this divergence that will be the focus of sections 3.4.1 and 3.4.2.

3.4.1   The Evans total synthesis of spongistatin 2

In his total synthesis of spongistatin 2 (**49**), Evans chose to construct the C(1)–C(28) segment via an *anti* aldol coupling between ketone **50** and aldehyde **51** (Scheme 3.13).[49] Aldol coupling of a C(8)–C(15) methyl ketone and a C(1)–C(7) aldehyde gave β-hydroxy ketones **52** [diastereomeric ratio (dr) 50:50]. Oxidation of the diastereomeric mixture and

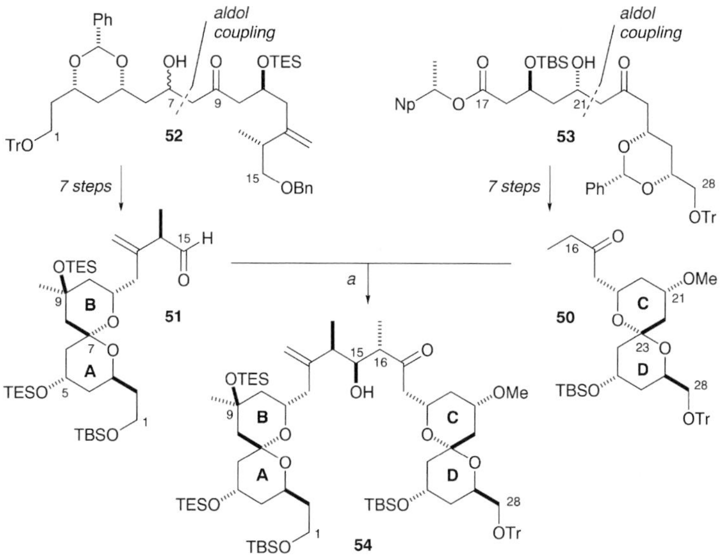

Scheme 3.13 Evans's approach to the C(1)–C(28) subunit of spongistatin 2. (*a*) **50**, (*c*Hex)₂BCl, Et₃N, pentane, 0°C; −78°C, **51**.

deprotection with concomitant spirocyclisation in the presence of acid [HF (aq.), MeCN] generated a single C(7) spiroketal diastereomer (the thermodynamically favoured product) in high yield. This was elaborated to the desired aldehyde **51** by conventional manipulations.

Construction of the carbon backbone of ethyl ketone **50** followed a similar pattern, but, in this case, the methyl ketone/aldehyde aldol coupling was highly diastereoselective (dr 96:4), favouring the desired C(21) stereochemistry (**53**; Scheme 3.13). Acid-catalysed (CSA) spirocyclisation produced an unfavourable mixture of epimers at C(23) (1:6), but the composition of this mixture could be pushed towards the desired stereoisomer (2.2:1 selectivity) by equilibration with magnesium trifluoroacetate. The hydroxyl groups at C(25) and C(28) were protected, and ethyl ketone **50** was formed by converting the C(17) ester to the Weinreb amide and treating the latter with ethyl magnesium bromide.

There is considerable precedent for an ethylketone derived *E*-boron enolate giving rise to an *anti* selective aldol reaction, through a Zimmerman–Traxler six-membered ring transition state.[51] However, the aldol union between C(15) and C(16) involves the reaction of two chiral partners **50** and **51**, and hence the stereochemical influence of each reacting substrate must be considered seriously. An investigation into the intrinsic diastereofacial bias of the *E*-boron enolate derived from **50** revealed that the remote chirality present in this spiroketal fragment would not significantly influence the stereochemical outcome of the aldol coupling (dr 67:33 with isobutyraldehyde). However, the reaction of a model α-methyl aldehyde (mimicking aldehyde fragment **51**) with the *E*-boron enolate derived from 3-pentanone indicated that the aldehyde component would exert a strong influence over the outcome of the reaction (dr 97:3). The preference for a Felkin transition state **TS-I**, as opposed to an *anti*-Felkin transition state **TS-II**, as frequently observed in the reaction of *E*-enolates with α-chiral aldehydes, has been attributed to destabilisation of the *anti*-Felkin transition state by a developing *gauche*-pentane interaction.[52]

When the C(15)–C(16) aldol coupling was performed using fragments **50** and **51** (Scheme 3.13), a mixture of diastereomers was isolated in good

TS-I   **TS-II**

Felkin (TS-I) and *anti*-Felkin (TS-II) transition states for the reaction of an *E*-boron enolate.

yield with a strong bias for the desired Felkin adduct **54** with C(14)–C(15) *syn*, C(15)–(16) *anti* stereochemistry (dr 90:10). This was converted to the required C(28)-aldehyde through a series of standard manipulations. Wittig coupling with a functionalised EF-ring C(29)-phosphonium salt, introduction of the side-chain at C(43) and macrolactonisation allowed the successful completion of the synthesis of spongistatin 2 (**49**).[49]

3.4.2   Kishi's total synthesis of spongistatin 1

In their total synthesis of spongistatin 1 (**5**), Kishi and co-workers elected to construct the C(1)–C(28) fragment via a Ni(II)/Cr(II)-mediated coupling between aldehyde **55** and vinyl iodide **56** (Scheme 3.14).[48] Compound **55** was obtained from dithiane **59**, which in turn was assembled from the opening of epoxide **57** with the HO cuprate derived from vinyl iodide **58** (Scheme 3.14). Removal of the silyl protecting groups and cleavage of the dithiane moiety generated a suitable precursor for acid-catalysed (PPTS) spiroketal formation at C(7). Standard manipulation of the protecting groups and oxidation at C(17) completed the synthesis of aldehyde **55**.

Kishi's route to the C(17)–C(28) *trans* vinyl iodide **56** made use of a series of regioselective epoxide ring-opening reactions to introduce much of the required stereochemistry. The vinyl iodide functionality was installed by a stereo- and regioselective alkyne hydrostannation reaction late in the synthesis followed by treatment with NIS.

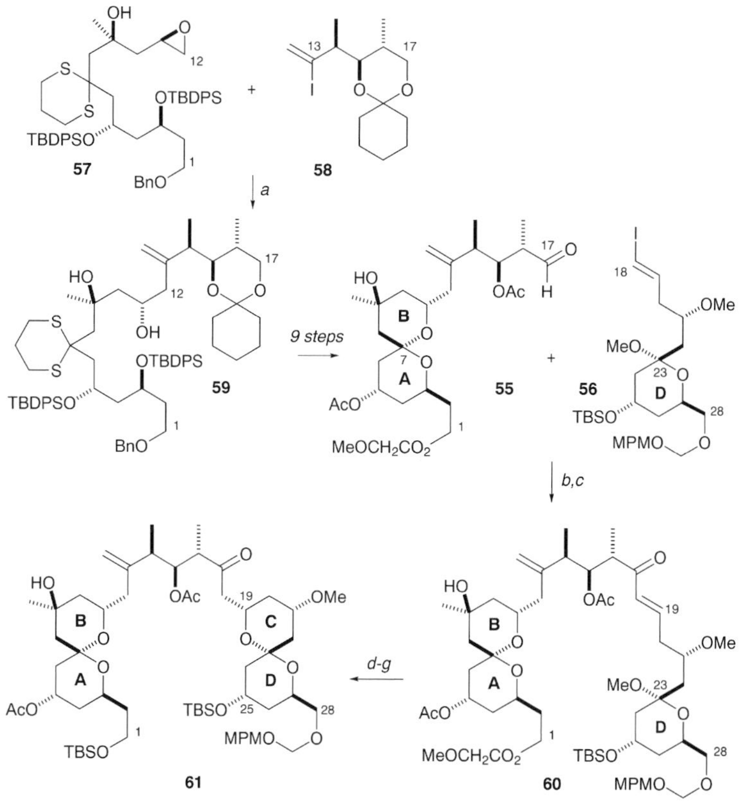

Scheme 3.14 Kishi's approach to the C(1)–C(28) subunit of spongistatin 1. (*a*) **58**, *t*-BuLi, THF, −78°C; 2-thienyl-CuCNLi; **57**, −78°C → 0°C; (*b*) NiCl$_2$/CrCl$_2$, bipy, THF; (*c*) Dess–Martin periodinane, py, CH$_2$Cl$_2$; (*d*) PPTS, acetone/H$_2$O; (*e*) Triton-B, MeOH/MeOAc, 0°C; (*f*) HF•py, CH$_3$CN; (*g*) 2,6-lutidine, TBSOTf, −78°C.

The Ni(II)/Cr(II)-mediated coupling of **55** with **56** proceeded smoothly to give a mixture of epimeric allylic alcohols. Subsequent oxidation with Dess–Martin periodinane gave α, β-unsaturated ketone **60**. Hydrolysis of the methyl glycoside with aqueous PPTS next set the stage for intramolecular Michael addition (Triton-B) to set up the CD-ring spiroketal, which was isolated as a single diastereomer with the correct stereochemistry at C(19), but incorrect stereochemistry at the C(23) spirocentre. The PPTS hydrolysis conditions also effected deprotection of the C(1) methoxyacetate. Because the TBS protecting group at O(25) conferred good stability upon the C(23)-spirocentre, Kishi decided to remove this group with aqueous HF in acetonitrile. The mildly acidic conditions nicely corrected the stereochemistry at C(23). Reprotection with TBSOTf finally provided **61**. This was then taken forward towards the natural product.

3.5   Other polyether natural products

It is not possible within the scope of this chapter to review all the polyether macrolide natural product syntheses that have been reported. Notable recent targets that have succumbed to synthesis are the bryo-statins **62**,[53] and halichondrin B **63**[54] and much of the early work in this area has already been covered in an excellent review.[55]

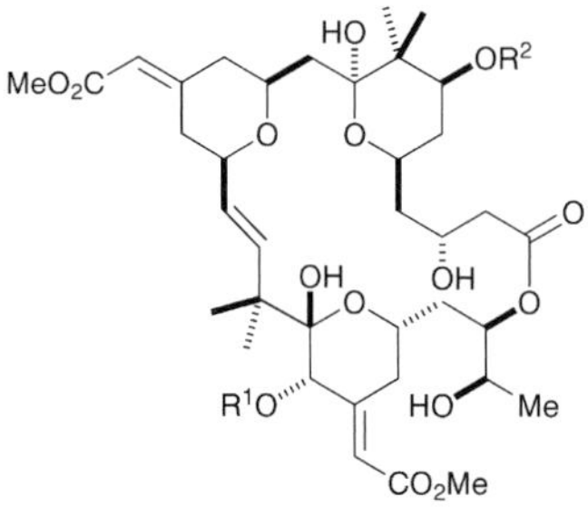

62 bryostatins

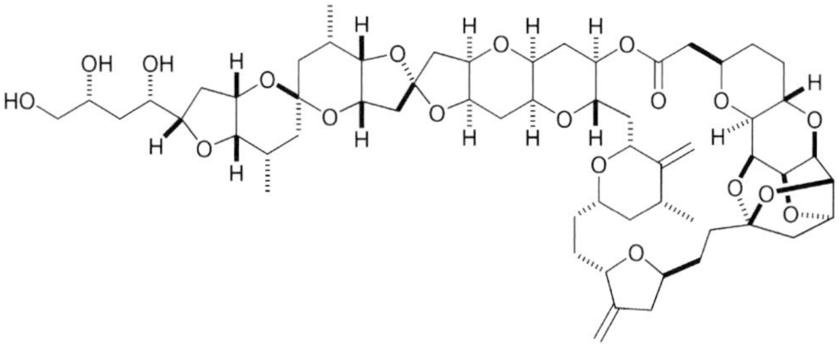

63 halichondrin B

3.5.1   Annonaceous acetogenins

The annonaceous acetogenins are a new class of polyethers isolated from the plant family Annonaceae. Uvaricin **64** (Scheme 3.15) was the first

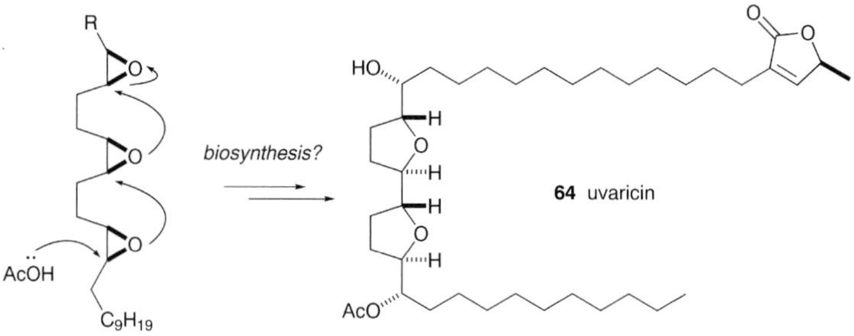

Scheme 3.15   Uvaricin.

family member to be isolated, in 1982.[56] Since that time over 220 related molecules have been encountered.[57] The annonaceous acetogenins have wide-ranging biological effects, which include antifungal, antiprotozoal, anthelminthic and anticancer activity.[57]

The annonaceous acetogenins are derived from C(32) or C(34) long-chain fatty acids and exhibit as a key structural motif a series of 2,5-linked oligo-tetrahydrofurans. Representative examples of this class include the tris-tetrahydrofuran goniocin (65)[58], the bis-tetrahydrofurans parviflorin (6)[14] and uvaricin (64),[56] as well as the mono-tetrahydrofuran longifolicin (66).[59] A polyepoxide cascade cyclisation process has been proposed for the biosynthesis of these natural products.[57] Thus it has been suggested that uvaricin (64) might be derived from the reaction of a tri-epoxide precursor with acetic acid, in the manner shown in Scheme 3.15.[56] The butenolide head-group of the molecule is likely to be introduced via an aldol-type reaction with a three-carbon unit. The presence of remote hydroxyl groups along the carbon backbone of certain annonaceous acetogenins (e.g. 65) has meant that Mosher ester methodology has been used extensively for the determination of the absolute chirality of these molecules.[60] The discovery of the acetogenins has thus led to considerable refinement of this technique.

65 goniocin

66 longifolicin

The stereoselective synthesis of the oligo-tetrahydrofurans present in the annonaceous acetogenins has been the subject of several reviews.[61] Hence, this section will concentrate solely on the more recent strategies employed by two of the key players in this area, Hoye and Marshall.

3.5.2 An asymmetric dihydroxylation/epoxidation based approach to the total synthesis of parviflorin

If one examines the central core of the annonaceous acetogenin parviflorin 6 (Scheme 3.16) one quickly recognises that it possesses C_2 symmetry. In a recent approach towards the synthesis of this fatty-acid

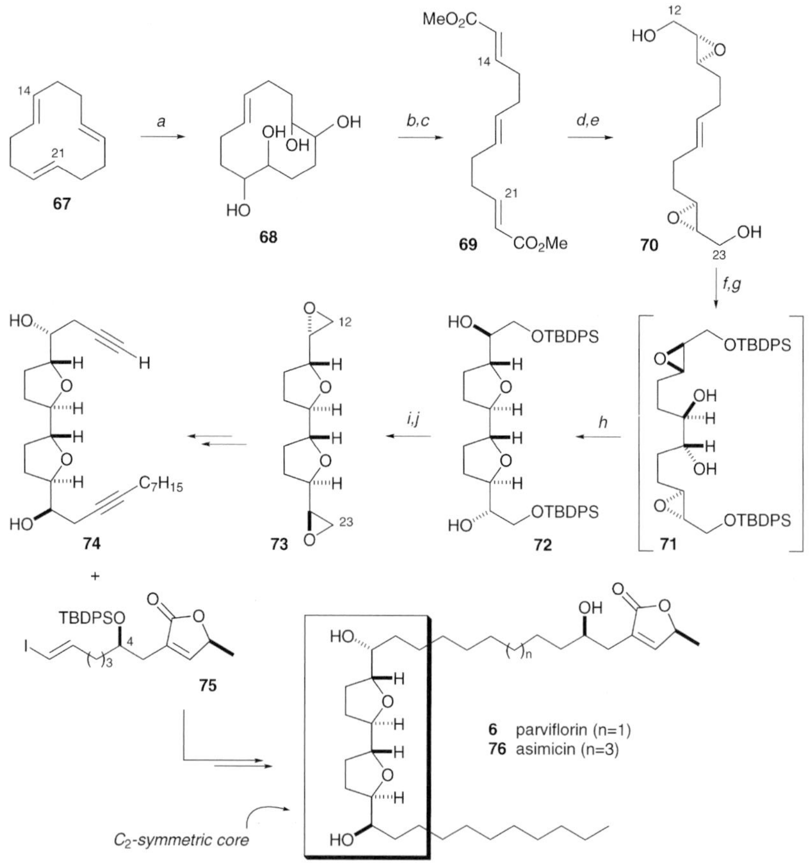

Scheme 3.16 Hoye's two-directional synthesis of the parviflorin core. (*a*) OsO$_4$ (cat.), NMO, acetone/H$_2$O; (*b*) NaIO$_4$, Et$_2$O/H$_2$O; (*c*) Ph$_3$P=CHCO$_2$Et, CH$_2$Cl$_2$; (*d*) DIBAL-H, CH$_2$Cl$_2$, −78°C; (*e*) Ti(*i*-PrO)$_4$, L-(+)-DET, *t*-BuOOH, CH$_2$Cl$_2$, −15°C; (*f*) TBDPSCl, Et$_3$N, DMAP, CH$_2$Cl$_2$; (*g*) AD-mix-β, *t*-BuOH/H$_2$O, MeSO$_2$NH$_2$, 0°C; (*h*) TFA, CH$_2$Cl$_2$; (*i*) TsCl, Et$_3$N, DMAP, CH$_2$Cl$_2$; (*j*) TBAF, THF.

derived natural product,[62] Hoye has exploited this feature to develop a highly efficient two-directional strategy.[63] In his synthesis, the C(14)– C(21) carbons of the central core were derived from cyclododecatriene (**67**) via selective dihydroxylation of two of the double bonds [NMO, OsO$_4$ (cat.)] (Scheme 3.16). Oxidative cleavage of the resultant tetrol (**68**), oxidation to the bis-aldehyde and chain extension using a Wittig reaction provided the bis-enoate **69**, which was used to create the C(12)–C(23) backbone of parviflorin. Reduction to the bis-allylic alcohol (DIBAL-H) and double Sharpless asymmetric epoxidation gave **70** in greater than

97% ee. Protection of the terminal alcohols and dihydroxylation of the remaining double bond gave an intermediate diol **71** which underwent epoxide cyclisation upon treatment with acid (TFA) to provide the C_2 symmetric bis-tetrahydrofuran core **72**.

The correct stereochemistry for the parviflorin core was set up by inverting the stereocentres at C(13) and C(22) via formation of the C_2 symmetric bis-epoxide **73**. Sequential opening of the terminal epoxides, first with 1-lithio-2-trimethylsilylethyne and subsequently with 1-lithio-1-nonyne, led to desymmetrisation of the core (Scheme 3.16). Palladium(0)-catalysed coupling of the deprotected terminal alkyne in **74** with a fully functionalised C(1)–C(9) vinyl iodide **75** gave the full parviflorin backbone. The synthesis was completed by hydrogenation of the eneyne and alkyne bonds in the presence of Wilkinson's catalyst. These were reduction conditions which left the butenolide unit intact and which allowed the C(4) hydroxyl to be unmasked to give parviflorin (**6**).

3.5.3 A nonracemic α-alkoxy stannane based approach to the total synthesis of asimicin

Marshall's approach (Scheme 3.17) to asimicin (**76**) (Scheme 3.16), the 34-carbon fatty-acid homologue of parviflorin (**6**), exploited the reaction of a tartrate-derived bis-aldehyde **77** and a chiral nonracemic α-alkoxy allylic stannane **78** to build up the C(2)-symmetric bis-tetrahydrofuran central core.[64] The preparation of **77** involved a two-directional extension of diisopropyl L-tartrate using Horner–Wadsworth–Emmons chemistry. The absolute stereochemistry of the tartrate unit was retained throughout this protocol.[65] In the presence of indium trichloride α-alkoxy allylic stannanes such as **78** undergo 1,3-transmetallation to give dichloroin-dium–allyl intermediates such as **79**.[66] Subsequent S_E2' addition to an aldehyde results in the formation of a monoprotected *anti* 1,2-diol. Hence, reaction of the α-alkoxy allylic stannane **78** with bis-aldehyde **77** under these conditions resulted in a two-directional extension of the carbon chain to give diol **80** with the absolute stereochemical control as shown.[64] Double Williamson ether ring closure resulted in the formation of the C_2-symmetric core **81** of asimicin.

Desymmetrisation of diol **81** was achieved by monotosylation (BuLi; TsCl) and subsequent reduction with superhydride (Scheme 3.17). Hydrogenation of the resultant dienol over Rh/Al_2O_3 gave alcohol **82**. After the conversion of the alcohol to the corresponding iodide, two-carbon homologation using a vinyl cuprate and asymmetric dihydroxyla-tion (AD-mix-β) allowed the introduction of the hydroxyl at C(4) with the correct stereochemistry. Straightforward chain extension, butenolide

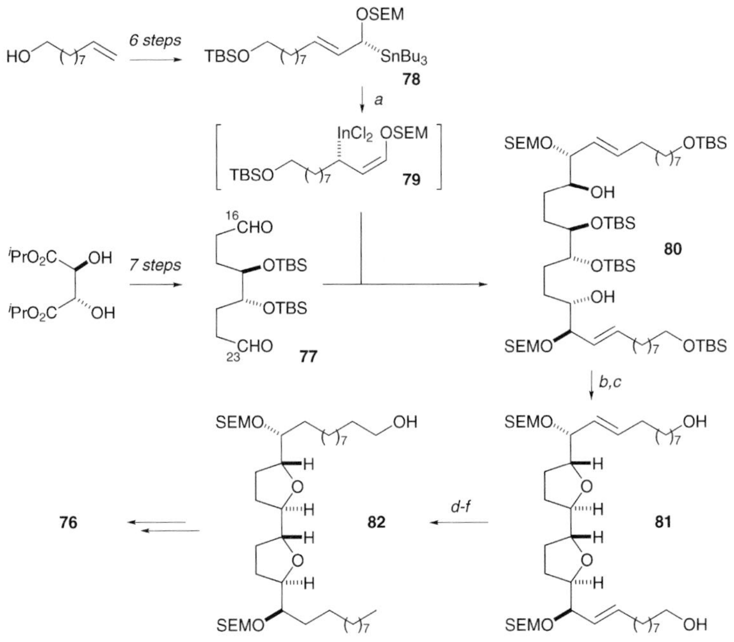

Scheme 3.17 Marshall's two-directional synthesis of the asimicin core. (*a*) InCl₃, EtOAc, $-78°C \rightarrow 23°C$; (*b*) TsCl, pyridine; (*c*) TBAF, THF, 40°C; (*d*) BuLi, THF/DMSO, 40°C; TsCl, 23°C; (*e*) LiBEt₃H, THF; (*f*) H₂, Rh-Al₂O₃ (cat.), EtOAc.

formation and SEM deprotection (PPTS, EtOH) allowed the successful completion of the synthesis of asimicin **76**.

3.5.4 Structure and biological activity of the trans fused polyether marine neurotoxins

Brevetoxins A (**7**)[15] and B (**83**)[67] are complex *trans* fused polyether marine neurotoxins associated with the deadly red tide dinoflagellate blooms that periodically occur around the world. These blooms kill fish

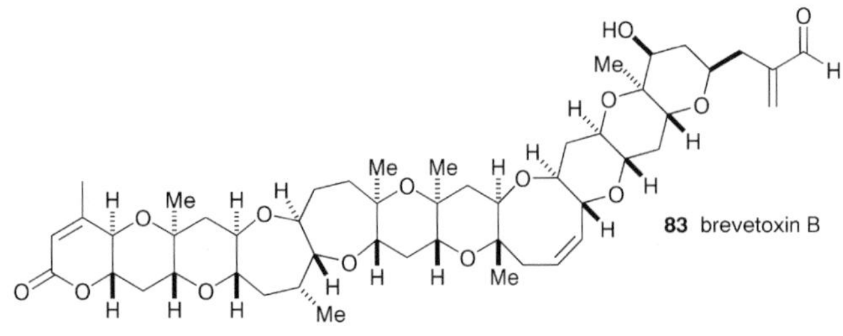

83 brevetoxin B

and marine life and have also resulted in human poisoning. The structurally related polyethers, ciguatoxin **84**[68] and maitotoxin,[69] have

84 ciguatoxin

likewise been shown to be responsible for food poisoning after the ingestion of coral reef fish. These neurotoxins are thought to be produced by an epiphytic dinoflagellate, with the neurotoxins being transferred through the food chain and accumulated at dangerous levels in carnivorous fish.

Brevetoxin B (**83**) was the first member of this family of neurotoxins to be discovered. Its structure was elucidated in 1981 through a combination of spectroscopic and X-ray crystallographic methods.[67a] The structure of ciguatoxin (**84**) was determined in 1990.[68a] However, confirmation of the C(2)-hydroxyl stereochemistry and absolute configuration was not achieved until 1997, when chemical synthesis of appropriate derivatives allowed CD studies to provide conclusive proof of the (2S,5R) stereochemistry.[68b]

As with other polyether natural products, various pathways involving polyepoxide cyclisation cascades have been proposed for the biosynthesis of these *trans* fused polyether ladders. These proposals rely upon successive *trans* epoxide openings in an *endo* fashion (contrary to Baldwin's rules).[70]

The *trans* fused polyether marine neurotoxins are lipid-soluble and have been shown to be selective activators of voltage channels in nerves, heart and muscle.[71] Toxicological studies indicate that the potency of these molecules is associated with their molecular size and their degree of conformational flexibility.[72] Although brevetoxin B contains two contiguous seven-membered rings which confer some flexibility on the structure, the molecule overall appears quite rigid. It is less potent than the less conformationally restricted polyether brevetoxin A (**7**), or the larger neurotoxins, ciguatoxin (**84**) and maitotoxin.

The most striking structural feature of these marine neurotoxins is the ladder-like arrangement of the *trans* fused polyether rings, which provides a particularly exciting challenge for total synthesis. As a consequence, a wide range of strategies have been instituted for the synthesis of the required seven-, eight- and nine-membered ether rings, and these have been reviewed.[73] The classic synthesis of brevetoxin B (**83**) by Nicolaou in 1995 has been the subject of several accounts.[74] Hence, this section will focus solely on some of the more recent strategies for the synthesis of the *trans* fused polyether marine neurotoxins.

3.5.5   Nicolaou's total synthesis of brevetoxin A

The recent synthesis of brevetoxin A (**7**) by Nicolaou represents a landmark achievement in the field of polyether synthesis.[75] The retrosynthetic strategy adopted by Nicolaou for brevetoxin A (**7**) is outlined in Scheme 3.18. This strategy relies upon a Horner–Wittig coupling to set-up the Z-alkene found in the eight-membered F-ring. This is also used simultaneously to link two complex fragments, the BCDE-ring system **85** and GHIJ-ring system **86**. (Fragments **85** and **86** themselves were ultimately derived from the sugars D-glucose and D-mannose.) Synthesis of the polyether rings in brevetoxin A (**7**) was achieved by a

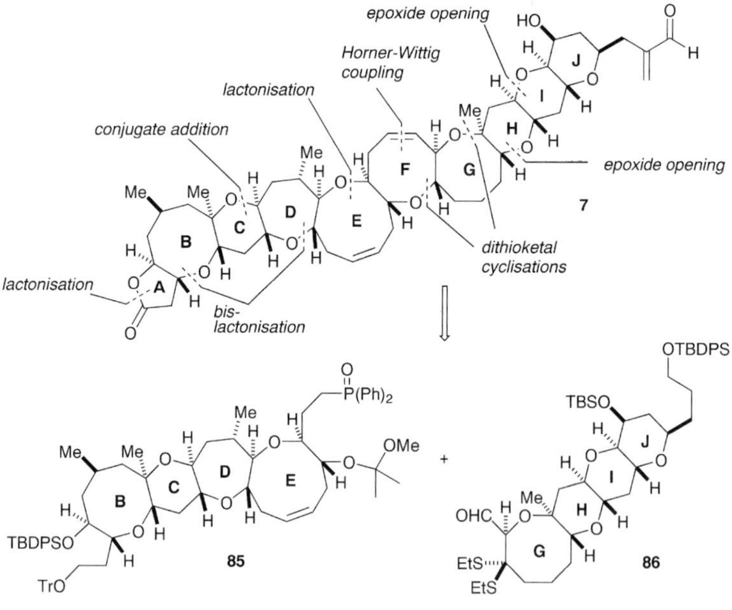

Scheme 3.18   Nicolaou's retrosynthetic analysis of brevetoxin A.

variety of different strategies (epoxide opening, conjugate addition, lactonisation with concomitant reduction and dithioketal cyclisation), as outlined in Scheme 3.18.

In the synthesis of the BCDE-ring component **85** we again see use of a two-directional strategy[63] to ensure the rapid development of an advanced precursor (see sections 3.5.2 and 3.5.3).[75] Thus, synthesis of the BCD-ring fragment **87** was achieved from a dihydroxy acid precursor **88** by a Yamaguchi bis-lactonisation to form the seven- and eight-membered rings (Scheme 3.19). Deprotection of the hydroxyl in **89**, de-

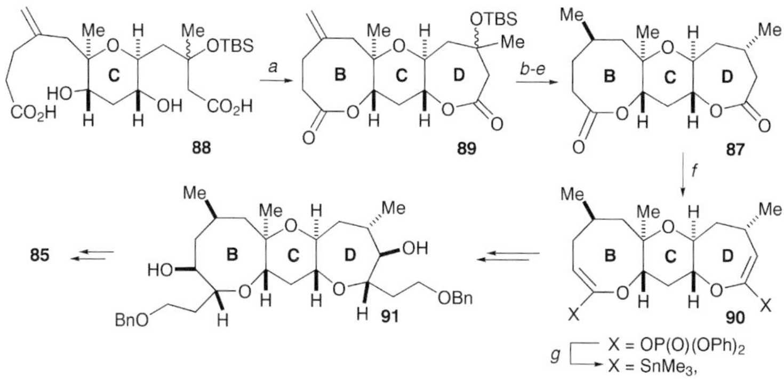

Scheme 3.19 Nicolaou's lactonisation-based ether cyclisation strategy: synthesis of the BCDE-ring fragment of brevetoxin A. (*a*) 2,4,6-trichlorobenzoyl chloride, Et$_3$N, THF, 0°C; DMAP, PhH, 75°C; (*b*) HF • py, CH$_2$Cl$_2$, 0°C; (*c*) [PhC(CF$_3$)$_2$O]$_2$SPh, CH$_2$Cl$_2$, 0°C; (*d*) H$_2$, (Ph$_3$P)$_3$RhCl (cat.), PhH; (*e*) H$_2$, Pd/C (cat.), EtOAc; (*f*) KHMDS, (PhO)$_2$P(O)Cl, HMPA, THF −78°C; (*g*) Me$_3$SnSnMe$_3$, LiCl, Pd(PPh$_3$)$_4$ (cat.), THF, 70°C.

hydration to the enone and sequential hydrogenation of the two double bonds (Wilkinson's catalyst, H$_2$; Pd/C, H$_2$) gave the bis-lactone **87** with a reasonable degree of control over the stereochemistry at the two new methyl stereocentres (dr 75:25; desired diastereomer:other diastereomers) in excellent yield. Two-directional functionalisation of **87** relied upon conversion of the lactones to their cyclic ketene acetal phosphates **90** and palladium-catalysed coupling with hexamethyl distannane.[76] Trans-metallation (Sn → Li → Cu) and alkylation resulted in the formation of a bis-enol ether which was subjected to hydroboration to give the correct diol **91**. End-differentiation was achieved by selective silyl protection of the B-ring hydroxyl. Although this process was not particularly selective, recycling the bis-silylated material allowed the isolation of a 90% yield of the monosilylated material overall. Introduction of the final ether ring in the BCDE-ring fragment **85** made use of a further lactonisation/ketene acetal phosphate/palladium-catalysed coupling, ether formation sequence.

The construction of the GHIJ-ring component **86** illustrates Nicolaou's use of a dithioketal cyclisation. The HIJ-ring precursor **92** was constructed using traditional methodology, utilising two epoxide opening reactions to create the required stereochemistry at the HI- and GH-ring fusions.[77] This was coupled to the remaining carbon backbone of the G-ring (fragment **93**) using Wittig chemistry to create the Z-alkene cyclisation precursor **94** (Scheme 3.20). Hydroxydithioketal cyclisation of

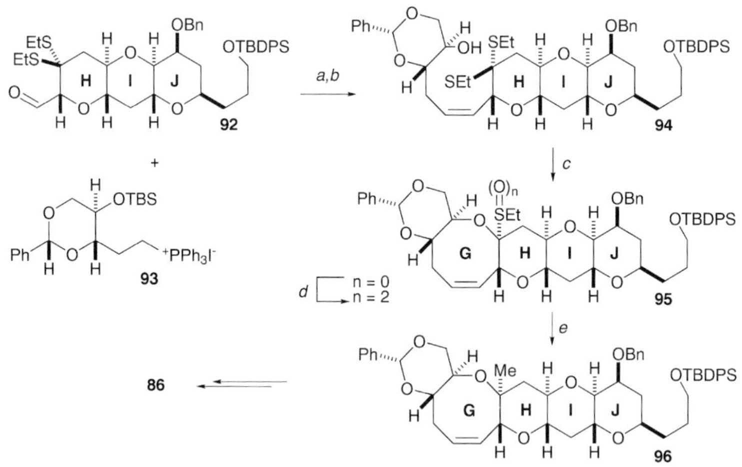

Scheme 3.20 Nicolaou's dithioketal-based ether cyclisation strategy: synthesis of the GHIJ-ring fragment of brevetoxin A. (*a*) **93**, *n*-BuLi, HMPA, THF; **92**, THF, $-78°C$; (*b*) TBAF, THF; (*c*) AgClO$_4$, NaHCO$_3$, SiO$_2$, 4 Å MS, MeNO$_2$; (*d*) *m*CPBA, CH$_2$Cl$_2$, $0°C$; (*e*) AlMe$_3$, CH$_2$Cl$_2$, $-78°C$.

94, facilitated by AgClO$_4$, resulted in the formation of the mixed eight-membered ring thioketal **95**. The introduction of the α-methyl group at the G/H ring junction was accomplished by oxidation of **95** to the corresponding sulfone and reaction with AlMe$_3$ to give **96** in 94% yield. Nicolaou has proposed that the well-defined conformation of the intermediate oxonium species is responsible for the observed stereocontrol. Standard manipulation of the protecting groups in **96** and hydrogenation to remove the double bond in the G-ring allowed the conversion of **96** to the desired GHIJ-ring fragment **86** in a further nine steps.

Generation of the anion of phosphine oxide **85** (Scheme 3.18) and Horner–Wittig coupling with aldehyde **86** allowed the selective formation of the F-ring Z-alkene. The eight-membered F-ring was formed using a dithioketal cyclisation, in a similar fashion to that used in the development of the G-ring (**94 → 95**; Scheme 3.20). Straightforward manipulation of the functionality at either terminus of the polyether backbone allowed the successful completion of the synthesis of brevetoxin A (**7**). The development of a number of strategies for the synthesis of *trans* fused medium ether rings by Nicolaou in the pursuit of this impressive total synthesis will undoubtedly assist in the future design and synthesis of analogues of this potent neurotoxin for biological testing.

3.5.6 Other approaches to the marine neurotoxins

It is perhaps not surprising that these challenging polyether marine neurotoxins have forced the introduction of a plethora of new methods for the synthesis of *trans* fused polyethers of varying sizes.[73, 78] The use of organometallic reagents to effect C—C bond-forming reactions on acyclic ether precursors is an approach that is gaining ascendancy in this area.

The viability of a metathesis-based approach towards the synthesis of the polyether marine neurotoxins was first demonstrated by Nicolaou and co-workers.[79] Their strategy utilised the reaction of an ester precursor **97** in the presence of excess Tebbe reagent to generate the enol ether **98** *in situ* (Scheme 3.21). Ring-closing metathesis led to the formation of cyclic enol ether **99**, which was used to construct the OPQ-ring system of maitotoxin via hydroboration of the enol ether double bond.

Clark's approach to the construction of the *trans* fused polyether ladders of brevetoxin A (**7**) and B (**83**) made use of the ring-closing metathesis reactions of preformed enol ethers[80] and allyl ethers.[81] Thus enol ether precursors **100** were treated with the Schrock molybdenum catalyst **101** to give the cyclised products **102** in high yield (Scheme 3.21). Direct conversion of the metathesis product to alcohols **103** suitable for further development was achieved with moderate to excellent selectivity (**103:104**) by hydroboration of the requisite di- or trisubstituted glycals **102**.[80] This methodology has been shown to be effective for the synthesis of *trans* fused six- and seven-membered rings. However, when the methodology was applied to the synthesis of larger eight- or nine-membered rings it failed to produce the required glycals cleanly, because of competitive intermolecular metathesis reactions. Hence, an alternative strategy was pursued by Clark for the synthesis of these rings, where ring-closing metathesis was achieved by using the allyl ether rather than the enol ether.[81] Ring-closing metathesis of allyl ethers **105** in the presence of catalyst **101** was found to give high yields of the eight- and nine-membered rings **106** under moderately high dilution conditions

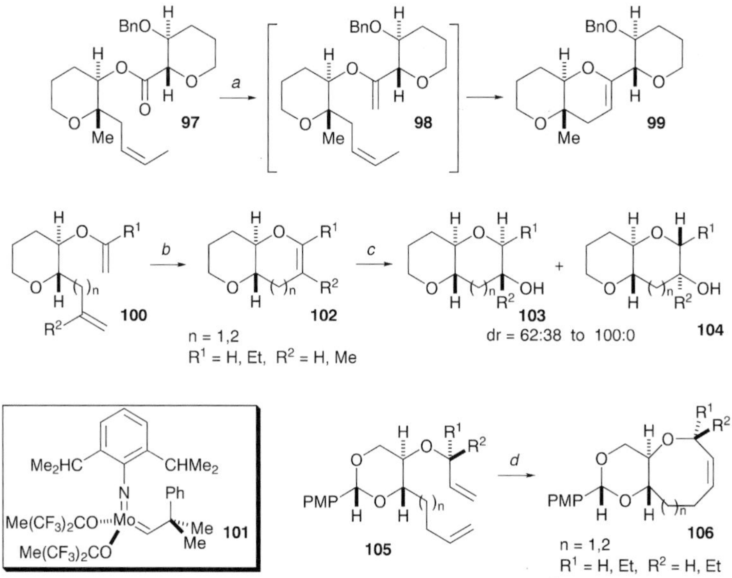

Scheme 3.21 A metathesis-based approach to the synthesis of the marine neurotoxins. (a) Tebbe reagent, THF, 25°C → reflux; (b) **101** (13 mol%), C_5H_{12}; (c) thexylBH$_2$, THF, −20°C; NaOH, H$_2$O$_2$; (d) **101** (25 mol%), PhH, 60°C.

(Scheme 3.21). Other investigators in this area have also made use of sugar-derived metathesis precursors to generate products which are highly oxygenated in the left-hand ring.[82] Since this chemistry is in the early stages of development, the application of this methodology to the total synthesis of polyether neurotoxins, such as brevetoxin A (**7**), remains to be investigated. However, the future of such strategies for the rapid and efficient construction of polyether natural product targets would appear to be very bright.

References

1. J.W. Westley (ed.), 1983, *Polyether Antibiotics: Naturally Occurring Ionophores*, vols 1–2, Marcel Dekker, New York.
2. (a) A. Agtarap, J.W. Chamberlin, M. Pinkerton, and L. Steinrauf, 1967, *J. Am. Chem. Soc.*, 89, 5737; (b) M. Gorman, J.W. Chamberlin, and R.L. Hamill, 1968, *Antimicrob. Agents Chemother.*, 363.
3. W.D. Celmer, P.W. Cullen, M.T. Jefferson, C.E. Moppett, J.B. Routien, and F.C. Sciavolino, 1979, *Chem. Abs.*, 90, 150295x.
4. Y. Miyazaki, M. Shibuya, H. Sugawara, O. Kawaguchi, C. Hirose, and J. Nagatsu, 1974, *J. Antibiot.*, 27, 814.

5. W.L. Duax, G.D. Smith, and P.D. Strong, 1980, *J. Am. Chem. Soc.*, 102, 6725.

6. G.R. Pettit, 1994, *Pure Appl. Chem.*, 66, 2271.

7. R.K. Pettit, S.C. McAllister, G.R. Pettit, C.L. Herald, J.M. Johnson, and Z.A. Cichacz, 1998, *Int. J. Antimicrobial Agents*, 9, 147.

8. T. Yasumoto and M. Murata, 1993, *Chem. Rev.*, 93, 1897.

9. (a) G. Schmid, T. Fukuyama, K. Akasaka, and Y. Kishi, 1979, *J. Am. Chem. Soc.*, 101, 259; (b) T. Fukuyama, C.-L.J. Wang, and Y. Kishi, 1979, ibid., 101, 260; (c) T. Fukuyama, K. Akasaka, D.S. Karanewsky, C.-L.J. Wang, G. Schmid, and Y. Kishi, 1979, ibid., 101, 262.

10. (a) D.B. Collum, J.H. McDonald III, and W.C. Still, 1980, *J. Am. Chem. Soc.*, 102, 2117; (b) D.B. Collum, J.H. McDonald III, and W.C. Still, 1980, ibid., 102, 2118.

11. T.L.B. Boivin, 1987, *Tetrahedron*, 43, 3309.

12. K.C. Nicolaou and E.J. Sorensen, 1996, *Classics in Total Synthesis*, chapters 12 and 15, VCH, Weinheim, New York, Basel, Cambridge, and Tokyo.

13. G.R. Pettit, Z.A. Cichacz, F. Gao, C.L. Herald, M.R. Boyd, J.M. Schmidt, and J.N.A. Hooper, 1993, *J. Org. Chem.*, 58, 1302.

14. S. Ratnayake, Z. Gu, L.R. Miesbauer, D.L. Smith, K.V. Wood, D.R. Evert, and J.L. McLaughlin, 1994, *Can. J. Chem.*, 72, 287.

15. (a) Y. Shimizu, H.-N. Chou, H. Bando, G. Van Duyne, and J.C. Clardy, 1986, *J. Am. Chem. Soc.*, 108, 514; (b) Y. Shimuzu, H. Bando, H.-N. Chou, G. Van Duyne, and J.C. Clardy, 1986, *J. Chem. Soc., Chem. Commun.*, 1656.

16. D.E. Cane, T.-C. Liang, and H. Hasler, 1982, *J. Am. Chem. Soc.*, 104, 7274.

17. D.E. Cane, W.D. Celmer, and J.W. Westley, 1983, *J. Am. Chem. Soc.*, 105, 3594.

18. W.C. Still and A.G. Romero, 1986, *J. Am. Chem. Soc.*, 108, 2105.

19. S.L. Schrieber, T. Sammakia, B. Hulin, and G. Schulte, 1986, *J. Am. Chem. Soc.*, 108, 2106.

20. I. Paterson, R.D. Tillyer, and J.B. Smaill, 1993, *Tetrahedron Lett.*, 34, 7137.

21. U. Koert, 1995, *Angew. Chem., Int. Ed. Engl.*, 34, 298.

22. D.S. Holmes, J.A. Sherrington, U.C. Dyer, S.T. Russell, and J.A. Robinson, 1990, *Helv. Chim. Acta*, 73, 239.

23. F.E. McDonald and T.B. Towne, 1994, *J. Am. Chem. Soc.*, 116, 7921.

24. C.A. Townsend and A. Basak, 1991, *Tetrahedron*, 47, 2591.

25. I. Paterson, R.D. Tillyer, and G.R. Ryan, 1993, *Tetrahedron Lett.*, 34, 4389.

26. I. Paterson, I. Boddy, and I. Mason, 1987, *Tetrahedron Lett.*, 28, 5205.

27. I. Paterson and P.A. Craw, 1989, *Tetrahedron Lett.*, 30, 5799.

28. F.E. McDonald, T.B. Towne, and C.C. Schultz, 1998, *Pure Appl. Chem.*, 70, 355.

29. F.E. McDonald and C.C. Schultz, 1997, *Tetrahedron*, 53, 16435.

30. H. Kinashi, N. Otake, H. Yonehara, S. Sato, and Y. Saito, 1973, *Tetrahedron Lett.*, 4955.

31. C. Keller-Julsen, H.D. King, M. Kuhn, H.R. Loosli, and A. Von Wartburg, 1978, *J. Antibiot.*, 31, 820.

32. J.W. Westley, J.F. Blount, R.H. Evans, and C.-M. Liu, 1977, *J. Antibiot.*, 30, 610.

33. J.W. Westley, L.H. Sello, N. Troupe, C. Liu, J.F. Blount, R.G. Pitcher, T.H. Williams, and P.A. Miller, 1981, *J. Antibiot.*, 34, 139.

34. Y. Kishi, S. Hatakeyama, and M.D. Lewis, 1982, in *Frontiers of Chemistry*, ed. K.J. Laidler, Pergamon Press, Oxford, p. 287.

35. K. Horita, S. Nagato, Y. Oikawa, and O. Yonemitsu, 1989, *Chem. Pharm. Bull.*, 37, 1726.

36. (a) M.A. Brimble and H. Prabaharan, 1998, *Tetrahedron*, 54, 2113; (b) P.R. Allen, M.A. Brimble, and F.A. Fares, 1998, *J. Chem. Soc., Perkin Trans. I*, 2403.

37. K. Horita, Y. Oikawa, S. Nagato, and O. Yonemitsu, 1988, *Tetrahedron Lett.*, 29, 5143.

38. K. Horita, S. Nagato, Y. Oikawa, and O. Yonemitsu, 1987, *Tetrahedron Lett.*, 28, 3253.

39. F. Perron and K.F. Albizati, 1989, *Chem. Rev.*, 89, 1617.

40. P.J. Kocienski, R.C.D. Brown, A. Pommier, M. Procter, and B. Schmidt, 1998, *J. Chem. Soc., Perkin Trans. I*, 9.

41. J.W. Faller and D. Linebarrier, 1988, *Organometallics*, 7, 1670.
42. J.W. Westley, R.H. Evans, L.H. Sello, N. Troupe, C.-M. Liu, J.F. Blount, R.G. Pitcher, T.H. Williams, and P.A. Miller, 1984, *J. Antibiot.*, 34, 139.
43. G.R. Pettit, Z.A. Cichacz, F. Gao, C.L. Herald, and M.R. Boyd, 1993, *J. Chem. Soc., Chem. Commun.*, 1166.
44. (a) G.R. Pettit, C.L. Herald, Z.A. Cichacz, F. Gao, J.M. Schmidt, M.R. Boyd, N.D. Christie, and F.E. Boettner, 1993, *J. Chem. Soc., Chem. Commun.*, 1805; (b) G.R. Pettit, C.L. Herald, Z.A. Cichacz, F. Gao, M.R. Boyd, N.D. Christie, and J.M. Schmidt, 1993, *Nat. Prod. Lett.*, 3, 239; (c) G.R. Pettit, Z.A. Cichacz, C.L. Herald, F. Gao, M.R. Boyd, J.M. Schmidt, E. Hamel, and R. Bai, 1994, *J. Chem. Soc., Chem. Commun.*, 1605.
45. N. Fusetani, K. Shinoda, and S. Matsunaga, 1993, *J. Am. Chem. Soc.*, 115, 3977.
46. (a) M. Kobayashi, S. Aoki, H. Sakai, K. Kawazoe, N. Kihara, T. Sasaki, and I. Kitagawa, 1993, *Tetrahedron Lett.*, 34, 2795; (b) M. Kobayashi, S. Aoki, H. Sakai, N. Kihara, T. Sasaki, and I. Kitagawa, 1993, *Chem. Pharm. Bull.*, 41, 989; (c) M. Kobayashi, S. Aoki, and I. Kitagawa, 1994, *Tetrahedron Lett.*, 35, 1243.
47. M. Kobayashi, S. Aoki, K. Gato, and I. Kitagawa 1996, *Chem. Pharm. Bull.*, 44, 2142.
48. (a) J. Guo, K.J. Duffy, K.L. Stevens, P.I. Dalko, R.M. Roth, M.M. Hayward, and Y. Kishi, 1998, *Angew. Chem., Int. Ed. Engl.*, 37, 187; (b) M.M. Hayward, R.M. Roth, K.J. Duffy, P.I. Dalko, K.L. Stevens, J. Guo, and Y. Kishi, 1998, ibid., 37, 192.
49. (a) D.A. Evans, P.J. Coleman, and L.C. Dias, 1997, *Angew. Chem., Int. Ed. Engl.*, 36, 2738; (b) D.A. Evans, B.W. Trotter, B. Côté, and P.J. Coleman, 1997, ibid., 36, 2741; (c) D.A. Evans, B.W. Trotter, B. Côté, P.J. Coleman, L.C. Dias, and A.N. Tyler, 1997, ibid., 36, 2744.
50. (a) A.B. Smith III, Q. Lin, K. Nakayama, A.M. Boldi, C.S. Brook, M.D. McBriar, W.H. Moser, M. Sobukawa, and L. Zhuang, 1997, *Tetrahedron Lett.*, 38, 8675; (b) I. Paterson, D.J. Wallace, and R.M. Oballa, 1998, ibid., 39, 8545; (c) E. Fernandez-Megia, N. Gourlaouen, S.V. Ley, and G.J. Rowlands, 1998, *Synlett*, 991; (d) M.T. Crimmins and D.G. Washburn, 1998, *Tetrahedron Lett.*, 39, 7487 (e) S.A. Hermitage, S.M. Roberts, and D.J. Watson, 1998, ibid., 39, 3567; (f) S. Lemaire-Audoire and P. Vogel, 1998, ibid., 39, 1345.
51. D.A. Evans, E. Vogel, and J.V. Nelson, 1979, *J. Am. Chem. Soc.*, 101, 6120.
52. (a) D.A. Evans, J.V. Nelson, and T.R. Taber, 1982, *Topics in Stereochem.*, 13, 1; (b) W.R. Roush, 1991, *J. Org. Chem.*, 56, 4151.
53. (a) D.A. Evans, P.H. Carter, E.M. Carreira, J.A. Prunet, A.B. Charette, and M. Lautens, 1998, *Angew. Chem., Int. Ed. Engl.*, 37, 2354; (b) K.J. Hale, J.A. Lennon, S. Manaviazar, M.H. Javaid, and C.J. Hobbs, 1995, *Tetrahedron Lett.*, 36, 1359; (c) M. Kalesse and M. Eh, 1996, *Tetrahedron Lett.*, 37, 1767; (d) J. De Brabander, B.A. Kulkarni, R. Garcia-Lopez, and M. Vandewalle, 1997, *Tetrahedron Asymm.*, 8, 1721; (e) S. Kiyooka and H. Maeda, 1997, ibid., 8, 3371; (f) J.M. Weiss, H.M.R. Hoffmann, 1997, ibid., 8, 3913; (g) T. Obitsu, K. Ohmori, Y. Ogawa, H. Hosomi, S. Ohba, S. Nishiyama, and S. Yamamura, 1998, *Tetrahedron Lett.*, 39, 7349.
54. (a) S.D. Burke, K.J. Quinn, and V.J. Chen, 1998, *J. Org. Chem.*, 63, 8626; (b) K. Horita, M. Nagasawa, Y. Sakurai, and O. Yonemitsu, 1998, *Chem. Pharm. Bull.*, 46, 1199; (c) D.P. Stamos and Y. Kishi, 1996, *Tetrahedron Lett.*, 37, 8643; (d) K. Horita, S. Hachiya, K. Ogihara, Y. Yoshida, M. Nagasawa, and O. Yonemitsu, 1996, *Heterocycles*, 42, 99.
55. R.D. Norcross and I. Paterson, 1995, *Chem. Rev.*, 95, 2041.
56. S.D. Jolad, J.J. Hoffmann, K.H. Schram, J.R. Cole, M.S. Tempesta, G.R. Kriek, and R.B. Bates, 1982, *J. Org. Chem.*, 47, 3151.
57. L. Zheng, Q. Ye, N.H. Oberlies, G. Shi, Z.-M. Gu, K. He, and J.L. McLaughlin, 1996, *Nat. Prod. Rep.*, 275.
58. Z.-M. Gu, X.-P. Fang, L. Zeng, and J.L. McLaughlin, 1994, *Tetrahedron Lett.*, 35, 5367.
59. Q. Ye, D. Alfonso, D. Evert, and J.L. McLaughlin, 1996, *Bioorg. Med. Chem.*, 4, 537.

60. (a) M.J. Rieser, Y. Hui, J.K. Rupprecht, J.F. Kozlowski, K.V. Wood, J.L. McLaughlin, P.R. Hanson, Z. Zhuang, and T.R. Hoye, 1992, *J. Am. Chem. Soc.*, 114, 10203; (b) Z.-M. Gu, L. Zeng, X.-P. Fang, T. Colman-Saizarbitoria, M. Huo, and J.L. McLaughlin, 1994, *J. Org. Chem.*, 59, 5162.

61. (a) U. Koert, 1995, *Synthesis*, 115; (b) R. Hoppe and H.-D. Scharf, 1995, ibid., 1447.

62. T.R. Hoye and Z. Ye, 1996, *J. Am. Chem. Soc.*, 118, 1801.

63. C.S. Poss and S.L. Schreiber, 1994, *Acc. Chem. Res.*, 27, 9.

64. J.A. Marshall and K.W. Hinkle, 1997, *J. Org. Chem.*, 62, 5989.

65. J.A. Marshall and K.W. Hinkle, 1996, *J. Org. Chem.*, 61, 4247.

66. J.A. Marshall and K.W. Hinkle, 1995, *J. Org. Chem.*, 60, 1920.

67. (a) Y.-Y. Lin, M. Risk, S.M. Ray, D. Van Engen, J. Clardy, J. Golik, J.C. James, and K. Nakanishi, 1981, *J. Am. Chem. Soc.*, 103, 6773; (b) M. S. Lee, D.J. Repeta, K. Nakanishi, and M.G. Zagorski, 1986, ibid., 108, 7855.

68. (a) M. Murata, A.M. Legrand, Y. Ishibashi, M. Fukui, and T. Yasumoto, 1990, *J. Am. Chem. Soc.*, 112, 4380; (b) M. Satake, A. Morohashi, H. Oguri, T. Oishi, M. Hirama, N. Harada, and T. Yasumoto, 1997, *J. Am. Chem. Soc.*, 119, 11325.

69. (a) M. Murata, H. Naoki, S. Matsunaga, M. Satake, and T. Yasumoto, 1994, *J. Am. Chem. Soc.*, 116, 7098; (b) M. Sasaki, N. Matsumori, T. Maruyama, T. Nonomura, M. Murata, K. Tachibana, and T. Yasumoto, 1996, *Angew. Chem., Int. Ed. Engl.*, 35, 1672.

70. K. Fujiwara, N. Hayashi, T. Tokiwano, and A. Murai, 1999, *Heterocycles*, 50, 561.

71. K.S. Rein, D.G. Baden, and R.E. Gawley, 1994, *J. Org. Chem.*, 59, 2101.

72. R.E. Gawley, K.S. Rein, G. Jeglitsch, D.J. Adams, E.A. Theodorakis, J. Tiebes, K.C. Nicolaou, and D.G. Baden, 1995, *Chemistry & Biology*, 2, 533.

73. E. Alvarez, M.-L. Candenas, R. Perez, J.L. Ravelo, and J.D. Martín, 1995, *Chem. Rev.*, 95, 1953.

74. (a) K.C. Nicolaou and E.J. Sorenson, 1996, in *Classics in Total Synthesis*, chapter 37, VCH, Weinheim, New York, Basel, Cambridge and Tokyo; (b) K.C. Nicolaou, 1996, *Angew. Chem., Int. Ed. Engl.*, 35, 589; (c) K.C. Nicolaou, E.J. Sorenson, and N. Wissinger, 1998, *J. Chem. Ed.*, 75, 1225.

75. K.C. Nicolaou, Z. Yang, G.-Q. Shi, J.L. Gunzner, K.A. Agrios, and P. Gärtner, 1998, *Nature*, 392, 264.

76. K.C. Nicolaou, G.-Q. Shi, J.L. Gunzner, P. Gärtner, and Z. Yang, 1997, *J. Am. Chem. Soc.*, 119, 5467.

77. K.C. Nicolaou, C.A. Veale, C.-K. Hwang, J. Hutchinson, C.V.C. Prasad, and W.W. Ogilvie, 1991, *Angew. Chem., Int. Ed. Engl.*, 30, 299.

78. K.J. Ivin, 1998, *J. Mol. Cat. A*, 133, 1.

79. (a) K.C. Nicolaou, M.H.D. Postema, and C.F. Claiborne, 1996, *J. Am. Chem. Soc.* 118, 1565; (b) K.C. Nicolaou, M.H.D. Postema, E.W. Yue, and A. Nadin, 1996, *J. Am. Chem. Soc.* 118, 10335.

80. J.S. Clark and J.G. Kettle, 1997, *Tetrahedron Lett.*, 38, 123.

81. J.S. Clark and J.G. Kettle, 1997, *Tetrahedron Lett.*, 38, 127.

82. (a) M.A. Leeuwenburgh, H.S. Overkleeft, G.A. van der Marel, and J.H. van Boom, 1997, *Synlett*, 1263; (b) J.D. Rainer and S.P. Allwein, 1998, *J. Org. Chem.*, 63, 5310.

4 Important developments in the total synthesis of alkaloids

J. Robertson

4.1 Introduction

The literature describing the chemical synthesis of alkaloids is truly vast. There can be no question of providing comprehensive coverage, and this review is of necessity subjective. The discussion is restricted to 'complex' alkaloids, that is, to those molecules in which the nitrogen is present not merely as a minor appendage, nor as an amino acid function, nor as a simple amide. Additionally, alkaloids in which the nitrogen atom is present in a monocyclic ring system or as part of a wholly aromatic ring have been largely omitted. This still leaves a wide variety of fascinating molecules, particularly so in a decade in which old 'solved' problems have been tackled anew in the light of recent synthetic developments and where many long-standing 'resistant' molecules have finally succumbed to total synthesis. Because of the wide variation in biogenetic origin, natural source and chemical structure meaningful groupings are limited. Therefore the synthetic coverage has been given in broadly chronological order to try to convey a sense of the trends that have developed. Although this may have resulted in a somewhat disjointed presentation it is hoped that an appreciation of the combination of ingenuity and insight that underpins most of these syntheses will be conveyed.

4.2 Reserpine

Although four total syntheses of reserpine had been described prior to the review period, none had been successful in controlling the C(3) stereochemistry in the C-Ring. In 1989, however, Stork adumbrated his group's solution to this problem.[1] His stereospecific synthesis (Scheme 4.1) was based on the assembly of a fully functionalised ring-E synthon, with control of relative stereochemistry via a tandem Michael–Michael reaction wherein a furyldimethylsilyl group was used as a hydroxyl surrogate. The conditions chosen for the Tamao–Kumada oxidative cleavage of the silyl unit also effected a Baeyer–Villiger rearrangement to allow unmasking of the E-ring precursor with DIBAL. The key to the successful stereocontrol in forming ring-C was rationalised by Stork in terms of ion pairing. Thus, under conditions where the axial cyanide group remained tightly bound to the iminium ion (heating in acetonitrile), approach of the indole nucleus occurred from the more accessible β face to give the isoreserpine stereochemistry. However, under conditions in which the cyanide–iminium ion pair suffered disruption (either by adding AgBF₄ or by effecting the reaction in a THF/ hydrochloric acid mixture) a favourable stereoelectronically controlled axial attack enabled indole alkylation to proceed on the α face of the DE-substructure to provide reserpine after esterification (Scheme 4.1; no yields were reported).

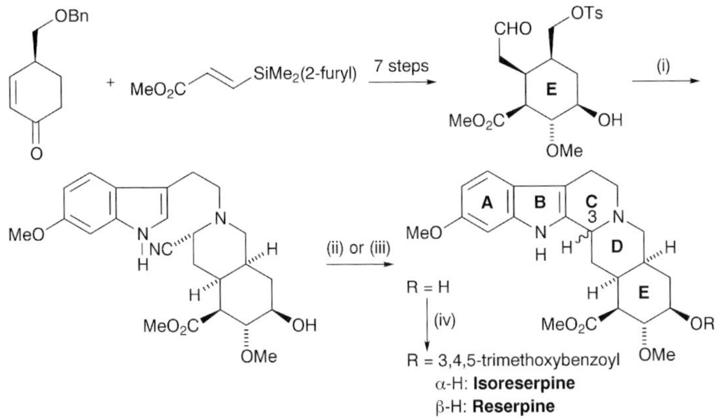

Scheme 4.1   (i) 6-methoxytryptamine, H⁺, CN⁻; (ii) CH₃CN, heat (→ α-H); (iii) e.g., aq. HCl, THF (→ β-H); (iv) 3,4,5-trimethoxybenzoyl chloride.

Aside from the Stork synthesis there have been formal total syntheses by Fraser-Reid[2] and by Chu.[3] More recently Hanessian has achieved a stereospecific synthesis in 20 steps and 2.6% overall yield from (−)-quinic acid, the paper including a good overview of the problems associated with controlling the C(3) stereochemistry of this target.[4]

4.3   Histrionicotoxin

Histrionicotoxin, isolated from the Colombian frog *Dendrobates histrionicus*, is of biological interest mainly for its ability to inhibit the function of certain membrane ion channels. From a chemical standpoint however, it is the unusual combination of a spiropiperidine ring system and bis-enyne functionality that has stimulated most endeavours towards its synthesis. To date only two total syntheses of the natural product have appeared, with only one of these (which is the sole enantioselective synthesis)[5] falling within the period under review. Stork's route

(Scheme 4.2) is noteworthy for its elegant application of his allylic epoxide cyclisation methodology[6] to build the functionalised cyclohexane ring found in the target, in a single step. Stork's route also introduced a new method for Z-iodoalkene synthesis[7] which was developed largely with this target in mind.

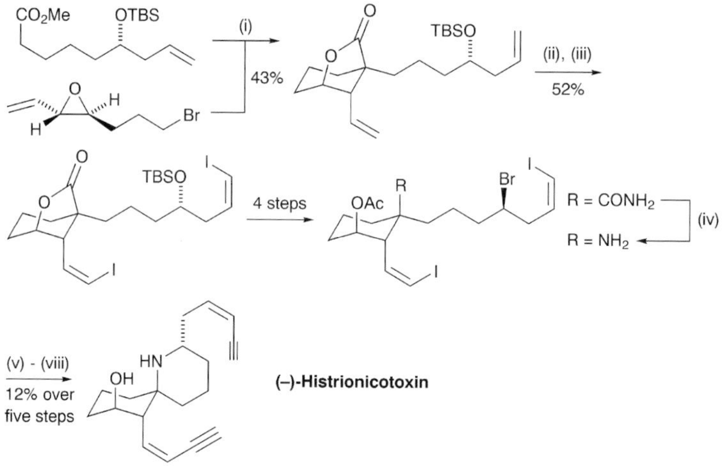

Scheme 4.2 (i) LDA; (ii) O_3; PPh_3; (iii) $(Ph_3P^+CH_2I)I^-$, NaHMDS; (iv) $(CF_3CO_2)_2IPh$, H_2O; (v) Et_3N; (vi) $Pd(PPh_3)_4$, CuI, TMSC≡CH; (vii) TBAF; (viii) aq. K_2CO_3, MeOH.

Most of the remaining synthetic output relating to this class has been directed towards the synthesis of the simpler analogue, perhydrohistrionicotoxin. Tanner's route to the latter[8] provides a good illustration of the tremendous increase in complexity one can obtain after an intramolecular Staudinger reaction and an imino ene-type cyclisation have been carried out on an acyclic precursor (Scheme 4.3).

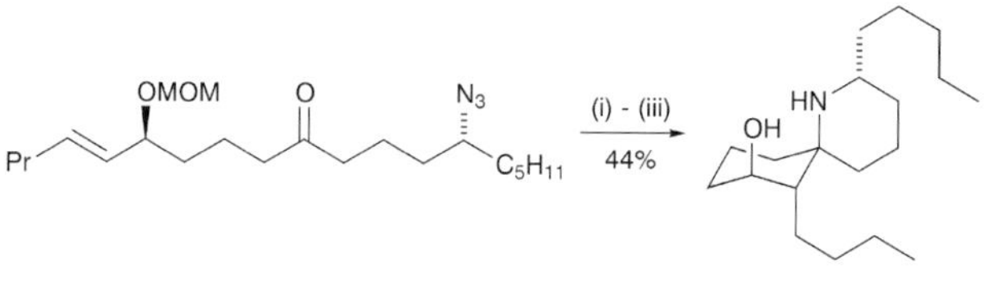

Perhydrohistrionicotoxin

Scheme 4.3 (i) PPh_3; (ii) $TiCl_2(Oi-Pr)_2$; (iii) H_2, PtO_2.

4.4   Atisine

The first formal synthesis of the natural enantiomer of atisine, the major diterpene alkaloid in *Aconitum heterophyllum*, was recorded by Fukumoto's group[9]. It built on their earlier synthesis of the racemate[10] in which key features were a double Mannich approach to assemble the AE-ring system and an intramolecular Michael–Michael reaction for constructing the bicyclo[2.2.2]octane system. The latter methodology gave access to a late-stage pentacyclic intermediate that was taken on to a compound that Pelletier had previously converted to atisine itself (Scheme 4.4).

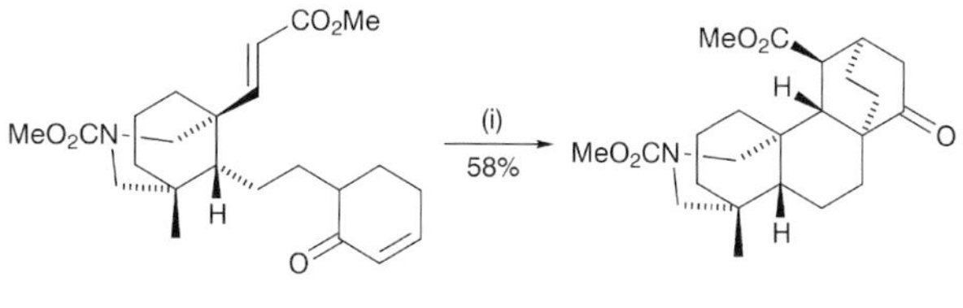

Scheme 4.4   (i) LHMDS.

4.5   *Daphniphyllum* alkaloids

Heathcock's series of reports on the synthesis of the *Daphniphyllum* alkaloids has possibly had the greatest impact of any of the chemistry discussed in this chapter. The successful realisation of a spectacular biomimetic conversion, that resulted in the formation of a pentacyclic ring system from an acyclic precursor, has superbly exemplified what can be achieved when an excellent proposal is aided and abetted by a little luck (a cylinder of 'ammonia' in this instance). Heathcock has chronicled[11] all his achievements in this area, starting from his first attempts at classical synthesis, with the bulk of the chemical details being contained in a series of full papers, *vide infra*.

Our starting point is his formation of the secodaphnane core structure, 'protodaphniphyllene', the proposed first pentacyclic intermediate—from which all members of the family derive—from the squalene dialdehyde derivative shown (Scheme 4.5).[12] The use of methylamine rather than ammonia proved to be critical to the success of this process since the first enamine cyclisation in the methylamine case was much more efficient, and the N-methyl substituent provided an effective hydride source to quench the late-stage isopropyl cation.

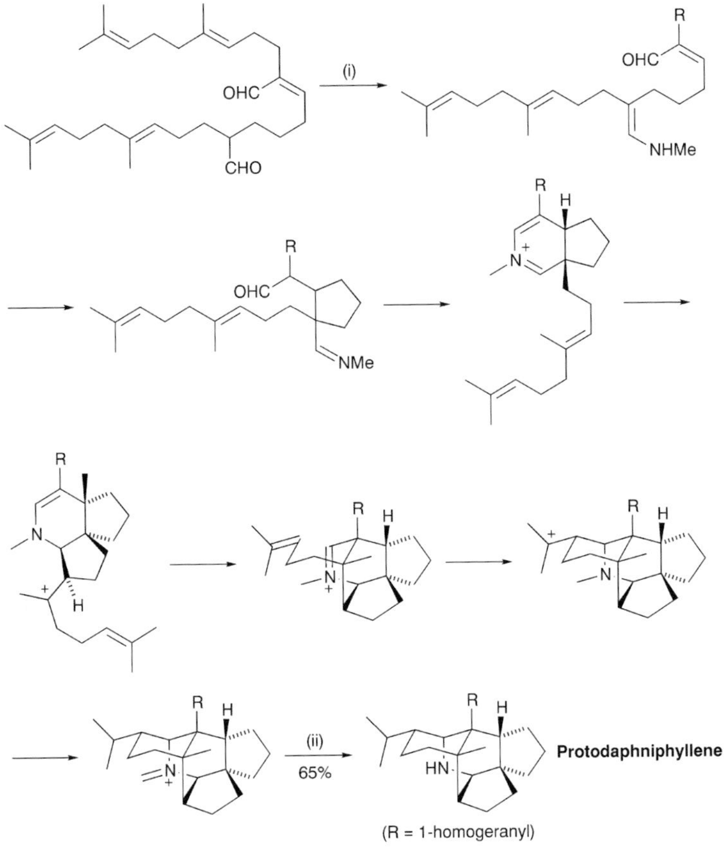

Scheme 4.5   (i) MeNH$_2$, HOAc; (ii) H$_2$O.

In further probes of the biosynthetic interrelationship of these alkaloids, the conversion of the secodaphnane skeleton into that of daphnilactone A was achieved[13] by a fragmentation–oxidation–ring-closure sequence (Scheme 4.6), and in later work[14] the daphnane core structure of (+)-codaphniphylline was accessed by direct pyrrolidine ring closure (without incorporation of a methylene group) in a closely related sequence (Scheme 4.7).

The major part of the related heptacyclic structure of (±)-bukittinggine was assembled by similar procedures and the pyrrolidine ring introduced by a novel N-cyclisation onto a π-allyl palladium intermediate.[15] Five further steps were required to effect stereoselective reduction of the exocyclic alkene and formation of the lactone ring (Scheme 4.8).

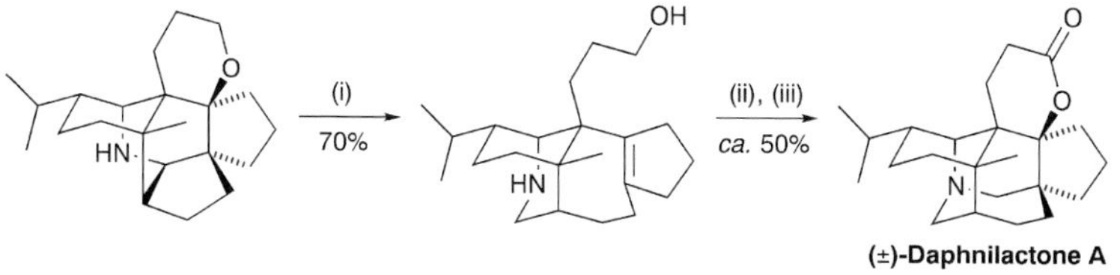

Scheme 4.6   (i) DIBAL; (ii) CrO$_3$, H$_2$SO$_4$; (iii) HCHO, pH 7.

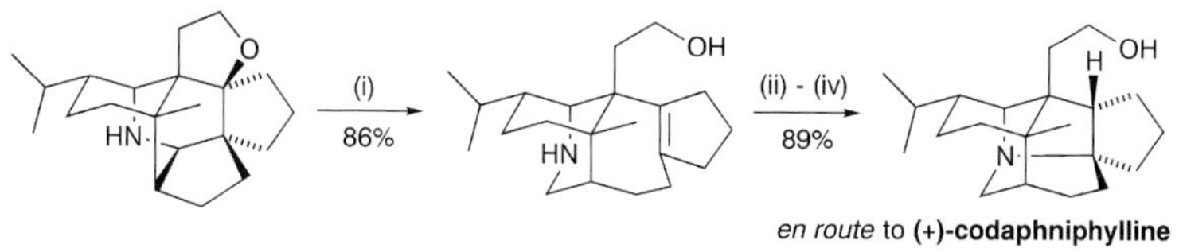

en route to **(+)-codaphniphylline**

Scheme 4.7   (i) DIBAL; (ii) PhNCO; (iii) HCO$_2$H; (iv) KOH, MeOH.

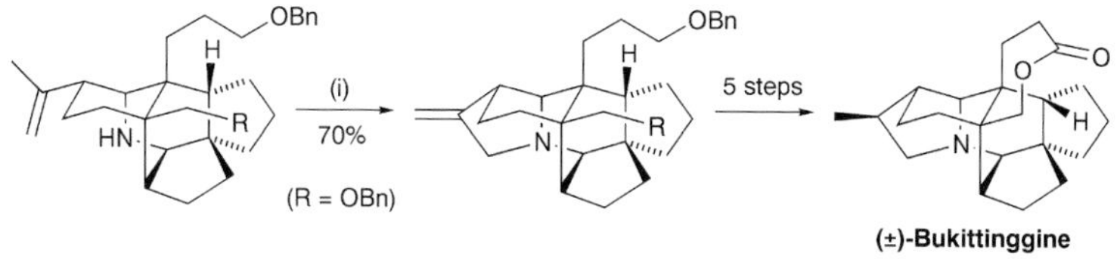

(±)-Bukittinggine

Scheme 4.8   (i) Pd(O$_2$CCF$_3$)$_2$, Ph$_3$P, benzoquinone.

4.6   Dendrobine

Dendrobine, obtained from a Japanese orchid and a key constituent of a Chinese folk medicine, has been a popular target of synthesis throughout the decade, and some excellent chemistry has appeared as a result. Sha's stereospecific synthesis effected skeletal assembly by an α-carbonyl radical cyclisation and was otherwise noteworthy for the excellent stereo-chemical control it recorded during a key hydroboration reaction en route to the (+)-enantiomer.[16] However, Livinghouse's synthesis of the racemate[17] was perhaps the most instructive, as it provided an excellent example of an approach to heterocycle synthesis that was based on the trapping of nitrilium ions generated by C-acylation of alkyl isocyanides. In this example application of such a sequence resulted in a pyrroline intermediate that was selectively reduced from the less hindered *exo* face after N-methylation. Closure of the third ring proceeded through a ketyl radical-anion conjugate cyclisation, with only four straightforward steps being needed to complete the synthesis (Scheme 4.9).

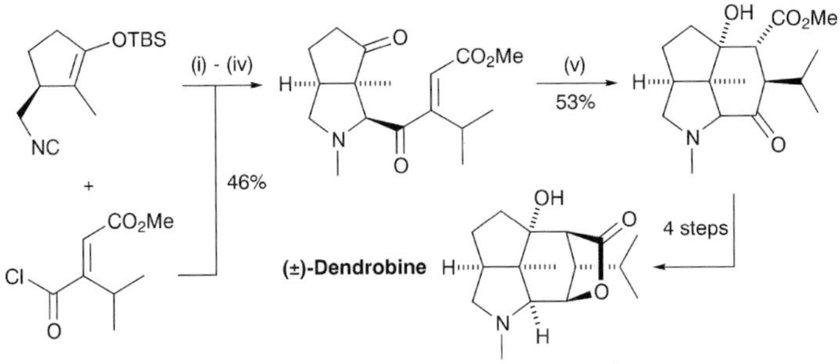

Scheme 4.9   (i) MS4Å; (ii) AgBF$_4$; (iii) MeOTf; (iv) KHB(Ot-Bu)$_3$; (v) SmI$_2$.

Formal syntheses of (−)-dendrobine and the racemate were reported by Mori[18] and Martin,[19] respectively, the latter relying on Oppolzer's methodology for N-acyldienamine synthesis.

4.7   Quinocarcin

The absolute stereochemistry of (−)-quinocarcin was confirmed by the first enantioselective synthesis of Garner.[20] A synthesis of the racemate had been reported some years earlier by Fukuyama.[21] Terashima's group have also been active in this area and have published routes to both enantiomers.[22] Key features of Garner's synthesis include an asymmetric azomethine ylid cycloaddition—with Oppolzer's acryloyl camphor sultam—to generate the diazobicyclo[3.2.1]ring system and an unprecedented intramolecular Wittig olefination on an imide carbonyl to generate the dihydropyridine ring. Reduction of the amide and conversion to the α-cyanoamine enabled aminal formation to complete the synthesis (Scheme 4.10).

4.8   Magellanine

The first total synthesis of the *Lycopodium* alkaloid (−)-magellanine and the first of an alkaloid having the magellanane skeleton were reported by Overman,[23] this report being rapidly followed by details of Paquette's total synthesis which proceeded by threefold annulation of the central cyclopentane ring.[24] The Overman synthesis highlights the Prins-terminated pinacol rearrangement which had been developed for the synthesis of angularly-fused polycyclic compounds. Thus a vinylcyclopentanol derivative was assembled from (1R, 5S)-bicyclo[3.2.0]heptenone

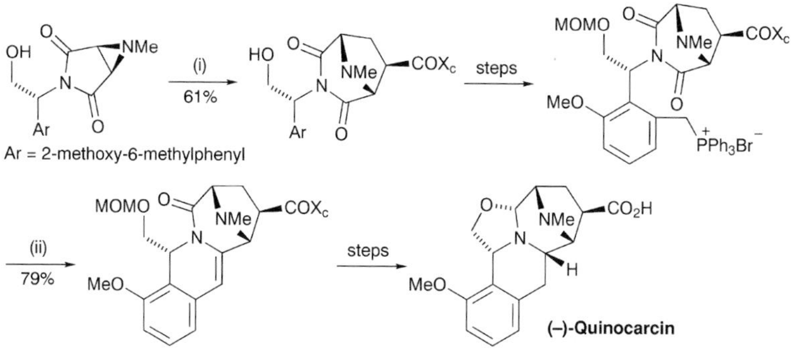

Scheme 4.10   (i) CH$_2$=CHCOX$_c$ (see text), $h\nu$; (ii) KOt-Bu, DMF, heat.

and oxonium ion formation and Prins-pinacol rearrangement were initiated with tin(IV) chloride. The tetracyclic product, containing all of the ring carbons with the natural product stereochemistry, was taken on to (−)-magellanine via alkene cleavage and reductive amination followed by elaboration of the cyclohexanone ring (Scheme 4.11).

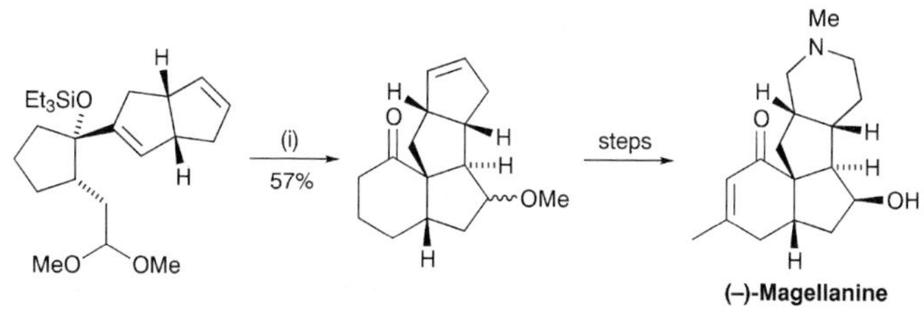

Scheme 4.11   (i) SnCl$_4$.

4.9   Ervitsine

A powerful biomimetic approach to ervitsine, a minor indole alkaloid isostructural with the biogenetically related ervatamine group, has been described by Bosch, who established the tetracyclic ring system in a single step from a simple 2-acyl indole.[25] Deprotonation and enolate addition to the 4-position of the pyridinium component were followed by activation of the resultant dihydropyridine with Eschenmoser's salt and cyclisation to give a low yield of product which merely required elimination and reduction to generate the racemic alkaloid (Scheme 4.12).

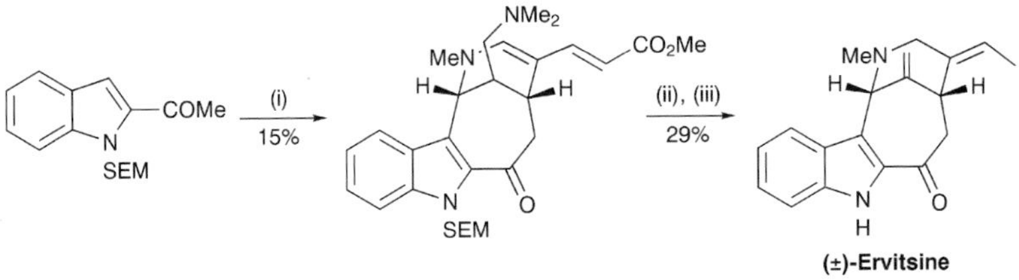

Scheme 4.12 (i) LICA then 3-(2-methoxycarbonyl)vinyl-1-methylpyridinium iodide then $CH_2=N^+Me_2I^-$; (ii) MCPBA, heat; (iii) HCl, ethanol then $NaBH_4$.

4.10 Gelsemine

Ever since the structure of gelsemine was first reported in 1959 numerous research groups have grappled with the dual problems of assembling the tetracyclic core and introducing the spiro-fused oxindole functionality. Finally, in 1992, the first of four racemic total syntheses was recorded[26] in which an acyliminium ion, within a heavily functionalised azadecalin ring system, was trapped by a suitably disposed silyl enol ether (Scheme 4.13).

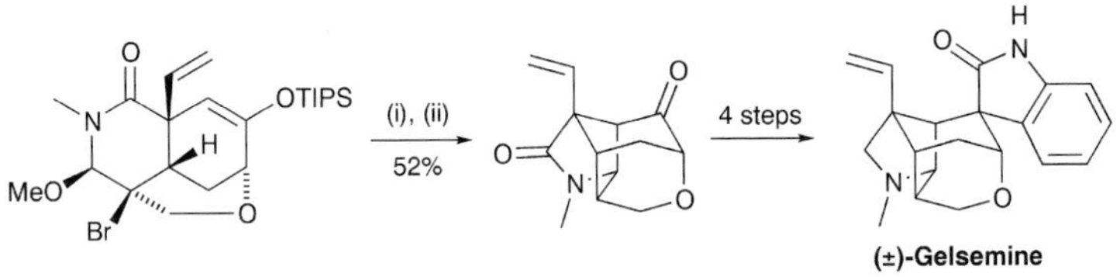

Scheme 4.13   (i) TFA; (ii) Bu₃SnH.

Johnson's synthesis[27] required 27 steps in all, proceeding in 0.26% overall yield. Of the subsequent syntheses, that of Speckamp[28] was the most efficient (1.1% yield over a mere 19 steps). The key chemistry here employed an alternative silyl enol ether capture of an acyliminium ion, the stereochemical features in the precursor having been established by Diels–Alder chemistry (as in the early stages of Hart's synthesis[29]). From that point the synthesis progressed rapidly—via allylic oxidation, and carbonylative coupling of a derived enol triflate—with the oxindole system being accessed through a Heck reaction; five more steps resulted in (±)-gelsemine (Scheme 4.14).

Fukuyama's very different approach[30] enabled control of the stereochemistry at the spiro-ring junction by introducing the oxindole at an early stage and effecting carbon–carbon bond formation to this centre in an unusual divinylcyclopropane–cycloheptadiene rearrangement.

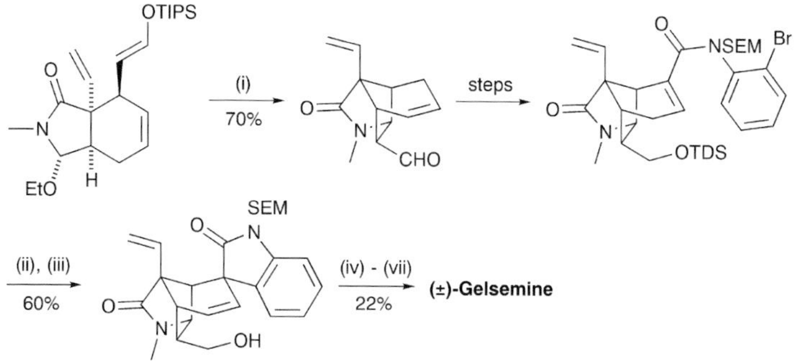

Scheme 4.14   (i) $BF_3 \cdot OEt_2$; (ii) $Pd_2(dba)_3$, Et_3N; (iii) TBAF; (iv) HgO, Tf_2O, $PhNMe_2$; (v) $NaBH_4$, NaOH; (vi) TBAF; (vii) AlH_3.

4.11   Papuamine

Just within the period under review an antifungal metabolite, papuamine, composing 1.3% of the dry weight of a Papua New Guinean sponge, was isolated and shown to have an unusual C_2-symmetric pentacyclic structure centred on a 13-membered diamine ring.[31] Three total syntheses of both the natural (−)-enantiomer and the unnatural enantiomer have appeared since that time,[32] all three sharing certain features in common, particularly in the use of palladium alkenyl-coupling chemistry to effect macrocyclisation at the central bond of the conjugated diene. The Weinreb synthesis[32c, d] deserves particular note as it provides an excellent example of the power of his recently developed intramolecular allenyl-imine ene cyclisation. In this example, one diastereomer of the key silyl-allene (absolute stereochemistry arising from a lipase-mediated partial hydrolysis of dimethyl 4-cyclohexene-1,2-dicarboxylate) was found to undergo ene cyclisation, in the presence of tin(IV) chloride, to give the desired *trans,cis,cis* relative stereochemistry. On this basis the same aldehyde was treated with 0.5 equivalents of 1,3-diaminopropane where-upon imine formation and double ene cyclisation ensued to afford the stereochemically correct macrocyclisation precursor in good yield. De-silylation, free-radical stannylation and a comparatively efficient coupling gave the natural product, which also confirmed the absolute configuration as that drawn (Scheme 4.15).

4.12   Ecteinascidin 743

Interest in the complex marine alkaloid ecteinascidin 743 centres on its potential as a clinically useful antitumour agent which, for the moment,

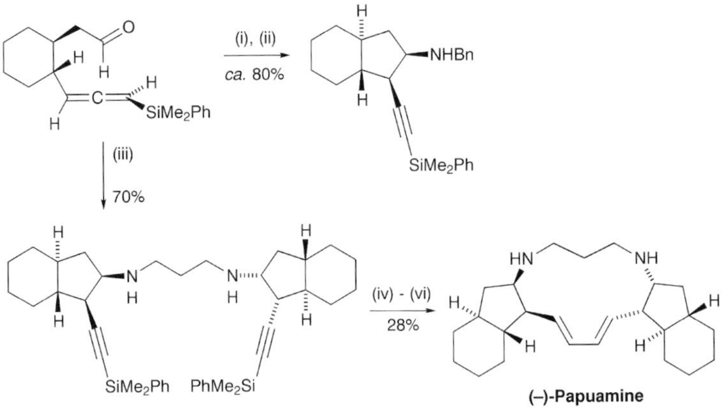

Scheme 4.15 (i) BnNH$_2$; (ii) SnCl$_4$; (iii) 0.5 equiv. 1,3-diaminopropane, heat; (iv) TBAF; (v) Bu$_3$SnH, AIBN; (vi) PdCl$_2$(PPh$_3$)$_2$, O$_2$.

remains underexplored because of its low availability from the natural source. A single total synthesis has been reported, by Corey's group,[33] in a route involving the application of three Pictet–Spengler reactions to establish the isoquinoline rings. The route is complex and notable, particularly for its careful orchestration of protecting groups and for its bold endgame strategy. One of two oxygenated phenylalanine sub-structures, prepared by asymmetric hydrogenation of parent enamide precursors, was taken through the first Pictet–Spengler reaction after unmasking and condensation of the α-acyloxyaldehyde and amino groups to provide a key bridged lactone intermediate. This was coupled with a second phenylalanine derivative and another Pictet–Spengler reaction used to form the second isoquinoline unit to give the major features of the target molecule. A sequence of functional group interconversions followed (including selective palladium-catalysed methylation of the tri-flate derived from the less-hindered phenolic oxygen). The 10-membered lactone ring was introduced from an α-hydroxycyclohexadienone pre-cursor in an unusual one-pot sequence in which the liberated thiolate nucleophile was trapped by conjugate addition to an intermediate quinone methide (Scheme 4.16, step iii). The final Pictet–Spengler reaction was achieved, after oxidative deamination, to provide the natural product once two deprotection steps had been performed.

4.13   Croomine

The structurally interesting homoindolizidine alkaloid croomine has stimulated two total syntheses. The first racemic synthesis of Williams[34]

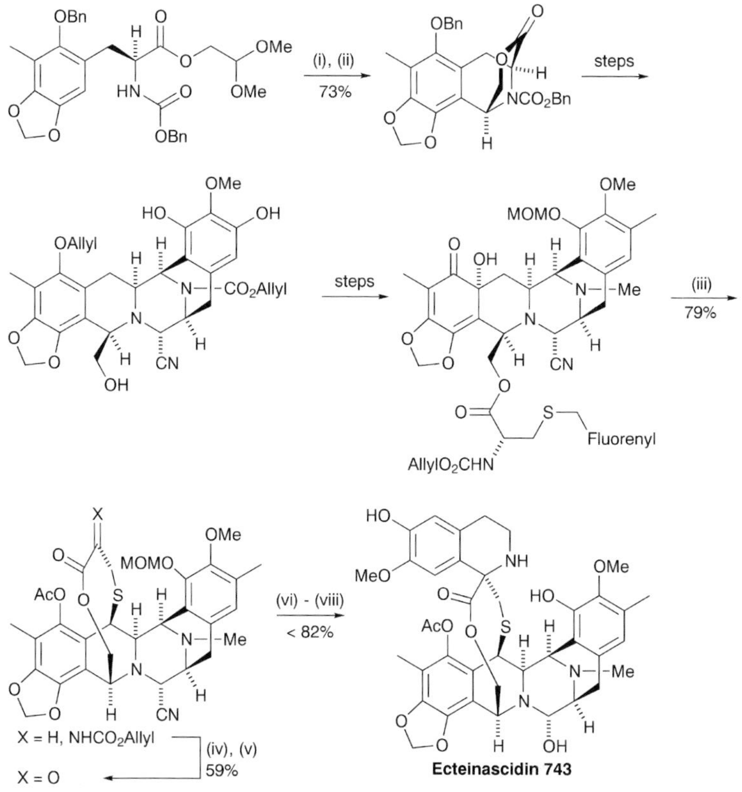

Scheme 4.16   (i) BF$_3$·OEt$_2$, H$_2$O; (ii) BF$_3$·OEt$_2$, MS 4Å; (iii) DMSO, Tf$_2$O; *i*-Pr$_2$NEt; *t*-BuOH; (Me$_2$N)$_2$C=N*t*-Bu; Ac$_2$O; (iv) PdCl$_2$(PPh$_3$)$_2$, Bu$_3$SnH, HOAc; (v) *N*-methyl-pyridinium-4-carboxaldehyde iodide, DBU, (CO$_2$H)$_2$; (vi) (3-hydroxy-4-methoxyphenyl)ethyl-amine, SiO$_2$; (vii) aq. TFA; (viii) aq. AgNO$_3$.

relied upon a highly stereocontrolled iodoaminocyclisation, which proceeded in only 25% yield but with 50%–60% recovery of starting material. The stereochemical outcome of this key reaction was ascribed to the intermediacy of a spiroaziridinium ion that was intercepted with inversion by the ester carbonyl group (Scheme 4.17). Martin's concise and generally efficient route to the natural (+)-enantiomer made use of his vinylogous Mannich methodology, in this case two furan moieties taking the part of a dienol.[35] Key features include the high stereocontrol in both the vinylogous Mannich reactions and the novel application of the Rapoport procedure for iminium ion formation (for another application of this procedure, see Oppolzer's synthesis of (+)-3-isorauniticine[36]).

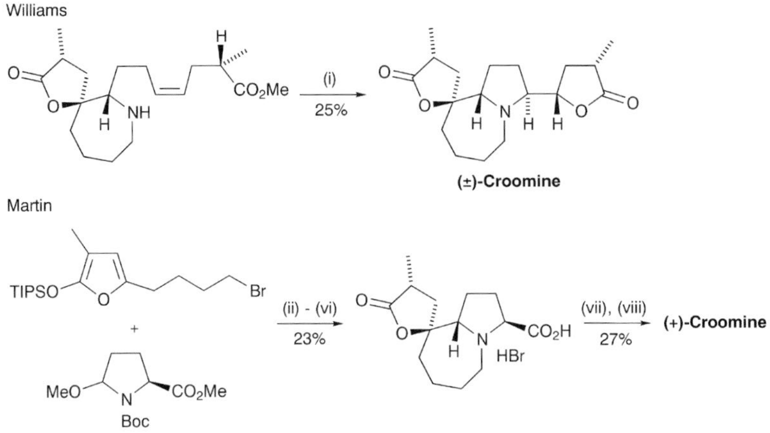

Williams

(±)-Croomine

Martin

+

(+)-Croomine

Scheme 4.17   (i) I_2; (ii) TIPSOTf; (iii) TFA; (iv) H_2, Rh-C; (v) NMM, DMF; (vi) aq. HBr; (vii) POCl$_3$, DMF then 3-methyl-2-triisopropylsilyloxyfuran; (viii) H_2, Pd-C, aq. HCl.

4.14   Strychnine

The 1990s have seen a resurgence of interest in the synthesis of strychnine after a 40-year hiatus since the classic first synthesis of Woodward.[37] Six total syntheses have been recorded.[38] Of these, Rawal's concise racemic synthesis[38g] and Overman's enantioselective route to the (−)-enantio-mer[38e] provide excellent illustrations. In the Rawal route (Scheme 4.18), *exo*-selective intramolecular Diels–Alder cycloaddition of a precursor prepared from *o*-nitrophenylacetonitrile constructed the ABCE-ring system in excellent yield. Lactamisation and Heck cyclisation gave (±)-isostrychnine, which Prelog had shown could be isomerised to strychnine under basic conditions.

The Overman route proceeded via the Wieland–Gumlich aldehyde in an excellent demonstration of the power of the aza–Cope–Mannich sequence wherein 3-acyl pyrrolidines are generated in a single step. From the so-formed A-CDE-ring system, five efficient steps led to the Wieland–Gumlich aldehyde and thence (−)-strychnine. The whole route produced the heptacyclic alkaloid in only 20 steps and in a respectable 3% overall yield (Scheme 4.19).

Martin's recent (formal) synthesis[38h] built on his earlier vinylogous Mannich–Diels–Alder methodology and previous biosynthetic propo-sals. The key conversion to *O*-benzyl akuammicine was proposed to proceed through electrophilic chlorination, iminium ion capture and [1,2]-alkyl shift, although the postulated chlorinated intermediate was not fully characterised (Scheme 4.20).

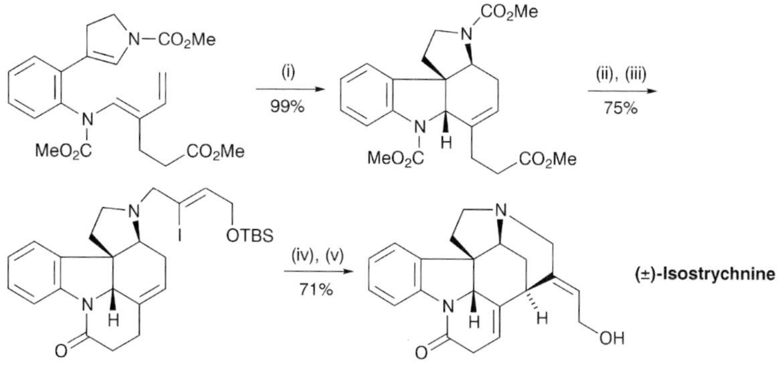

Scheme 4.18 (i) 185°C; (ii) TMSI then MeOH, heat; (iii) 1-bromo-2-iodo-4-(*tert*-butyldimethylsilyl)oxybut-2-ene, K_2CO_3; (iv) Pd(OAc)$_2$, Bu$_4$NCl, K_2CO_3; (v) 2N HCl.

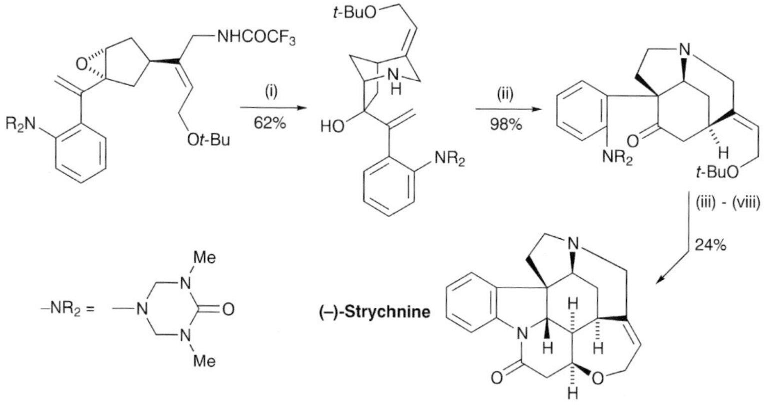

Scheme 4.19 (i) NaH, heat; KOH, aq. EtOH; (ii) (HCHO)$_n$, Na$_2$SO$_4$; (iii) LDA, MeO$_2$CCN; (iv) methanolic HCl; (v) Zn, H$_2$SO$_4$; (vi) NaOMe, MeOH; (vii) DIBAL; (viii) CH$_2$(CO$_2$H)$_2$, Ac$_2$O, NaOAc, HOAc.

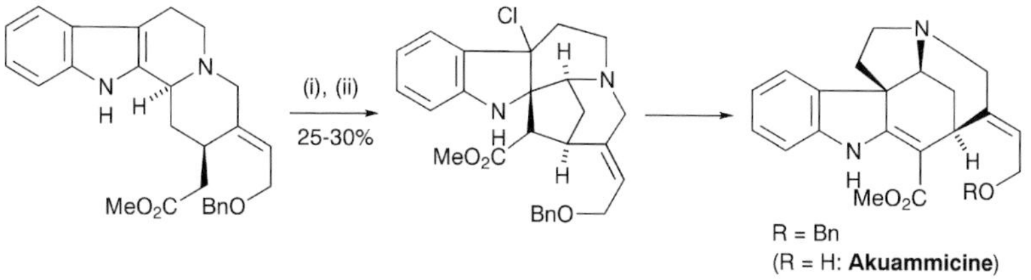

Scheme 4.20 (i) SnCl$_4$, *t*-BuOCl; (ii) LHMDS.

4.15   Huperzine B

The first total synthesis of the acetylcholinesterase-inhibiting *Lycopodium* alkaloid (±)-huperzine B, by Bai,[39] was based on the Schumann synthesis of obscurine.[40] It proceeded through a series of consecutive 1,4- and 1,2- enamine additions to an α,β-unsaturated bicyclic imine that was, in turn, generated *in situ* by acetal–imine interchange. This reaction led to the complete huperzine ring system which merely required oxidation, elimination and alkene isomerisation to deliver the target (Scheme 4.21).

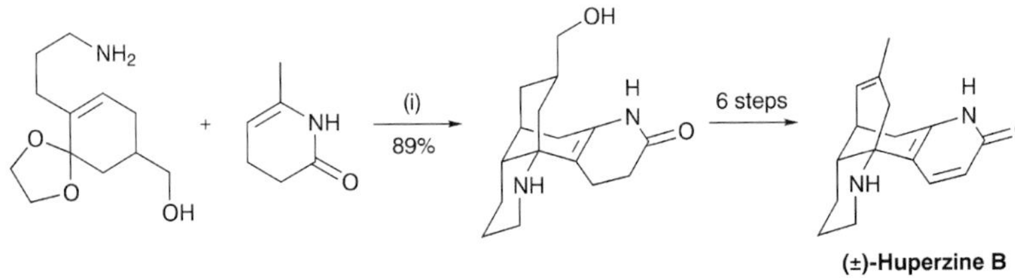

Scheme 4.21   (i) aq. HClO₄.

4.16   Cephalotaxine

Approaches to cephalotaxine, the parent alkaloid of the cytotoxic harringtonines, have been many and varied,[41] and two recent syntheses will be used as illustrations. Mori's stereospecific synthesis of the (−)-antipode[41h]—the first route to a single enantiomer—centred on electrophilic aromatic substitution by a cyclopentenyl cation to construct the seven-membered ring. Also of note in this route is an application of Mori's mild procedure for vinylcarbanion generation and cyclisation (Scheme 4.22). Tietze's efficient 'domino' sequence generated the racemate after *N*-alkylation by a π-allyl palladium(II) electrophile followed by Heck reaction under more forcing conditions to install the seven-membered ring.[41i] This constituted a formal synthesis since Kuehne had already shown that the cyclopentene ring could be elaborated to the hydroxyl and enol ether functionality found in the natural product.[41c]

4.17   Morphine

Efforts to effect the total synthesis of natural (−)-morphine—and of the unnatural enantiomer and the racemate—have been reported throughout the 1990s and almost all the total and formal syntheses contain novel

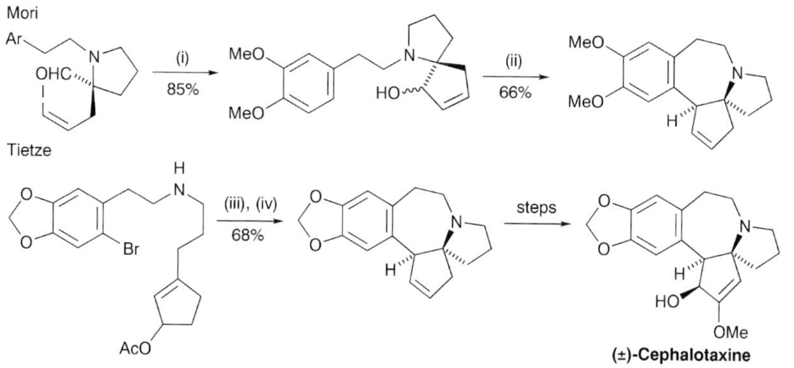

Scheme 4.22 (i) Me₃SiSnBu₃, CsF; (ii) polyphosphoric acid; (iii) Pd(PPh₃)₄, Et₃N; (iv) {o-[(o-Tol)₂P]C₆H₄CH₂PdOAc}₂, Bu₄NOAc.

elements of general interest. The approaches of Fuchs[42] and Parker[43] (Scheme 4.23) are conceptually related since powerful tandem carbon–carbon bond-forming reactions were used to assemble the B- and E-rings in a single step from precursors wherein the existing A- and C-rings were tethered by the (eventual) dihydrofuran oxygen. Fuchs's route built on his wider interest in sulfone chemistry, the sulfone group serving to activate the cyclohexene ring to conjugate addition of an intermediate aryllithium, localise the negative charge sufficiently to effect a second alkylation and, later in the synthesis, to function as an eliminable leaving group prior to closure of the piperidine ring. Parker's route to (±)-dihydrocodeinone relied on a tandem 5-*exo*-trig/6-*endo*-trig free-radical

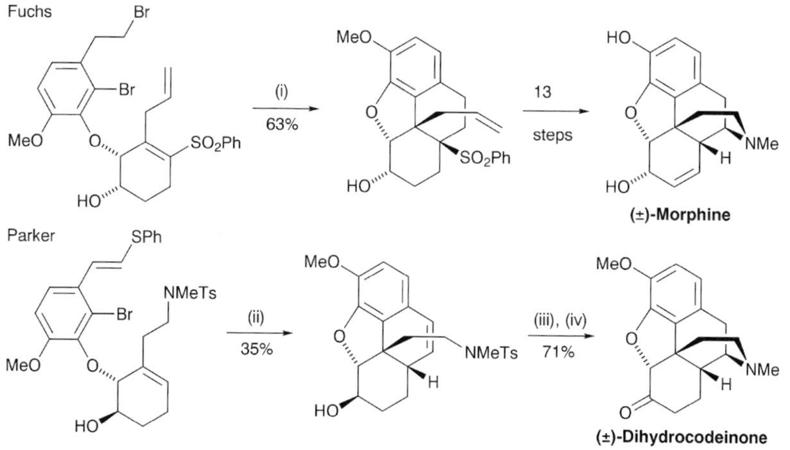

Scheme 4.23   (i) LiH then BuLi; (ii) Bu₃SnH, AIBN; (iii) Li, *t*-BuOH, NH₃; (iv) Swern.

process followed by piperidine formation that usefully accompanied reductive detosylation.

Overman's excellent synthesis[44] of both enantiomers of morphine (via dihydrocodeinone) was initiated with asymmetric reduction of 2-allyl-cyclohexenone and elaboration to an allylsilane [the enantiomer leading to (−)-morphine is shown]. Treatment of this with the A-ring precursor aldehyde and zinc iodide initiated allylsilane capture of the intermediate iminium ion with high (greater than 20:1) diastereoselectivity. Heck reaction allowed construction of the aryl-4°-carbon bond, the final di-hydrofuran ring forming after selective deprotection of the O-benzyl phenolic oxygen and epoxidation of the cyclohexene (Scheme 4.24).

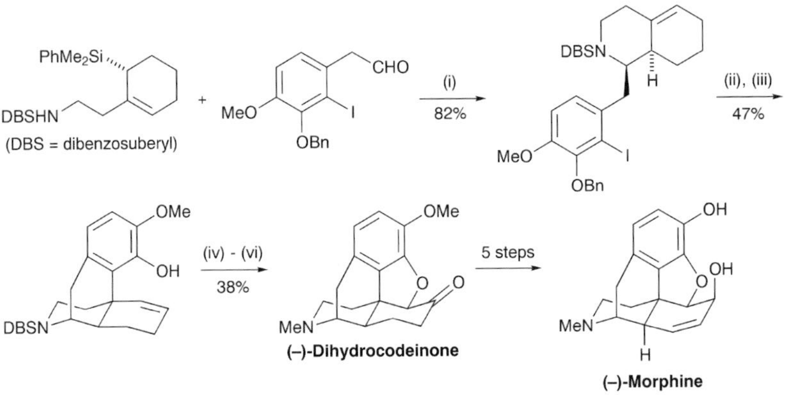

Scheme 4.24   (i) ZnI$_2$; (ii) Pd(O$_2$CCF$_3$)$_2$(PPh$_3$)$_2$, PMP; (iii) BF$_3 \cdot$ OEt$_2$, EtSH; (iv) 3,5-dinitro-peroxybenzoic acid, CSA; (v) TPAP, NMO; (vi) H$_2$, Pd(OH)$_2$, HCHO.

In Mulzer's synthesis of (−)-dihydrocodeine[45] construction of the final bond to the 4°-centre was achieved by conjugate addition of a vinyl cup-rate reagent under specifically optimised conditions, and in White's total synthesis of unnatural (+)-morphine[46] the same centre was completed by selective C—H insertion of an α-keto rhodium-carbenoid species en route to the piperidine ring (Scheme 4.25).

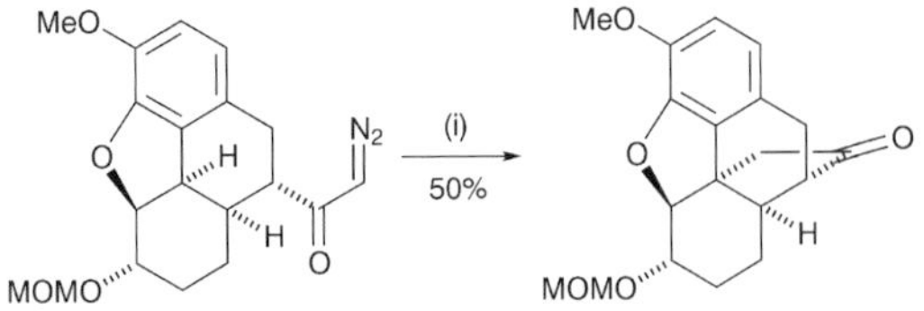

Scheme 4.25   (i) Rh$_2$(OAc)$_4$.

4.18 *Erythrina* alkaloids

The tetracyclic structure at the heart of most of the *Erythrina* alkaloids continues to be an inspiration and a test-bed for new synthetic methodology, and some efficient partial and total syntheses of these alkaloids have resulted. Attention is drawn to Wasserman's application of vicinal tricarbonyl chemistry to initiate a rapid synthesis of 3-demethoxyerythratidinone[47] and the concise enantiospecific synthesis of an erythratidinone intermediate by Lee and Park[48] in which intramolecular Wittig reaction (on an imide carbonyl group) and acyliminium ion cyclisation are key steps. Towards the start of the decade Rigby disclosed a general method for heterocycle synthesis in which the conversion is equivalent to [4+2]-cycloaddition of an alkyl isocyanide with a vinyl isocyanate.[49] N-Alkylation and electrophilic substitution led to a known intermediate for the synthesis of erysotrine (Scheme 4.26). Later reports from the same laboratory applied this reaction in the synthesis of (±)-α-lycorane.[50]

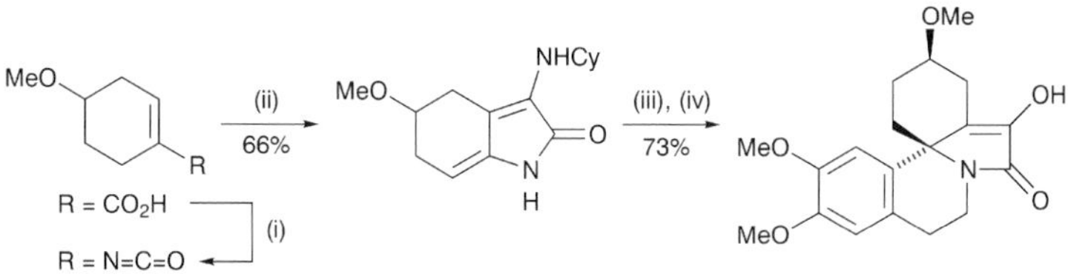

Scheme 4.26 (i) $(PhO)_2PO \cdot N_3$, heat; (ii) CyNC; (iii) NaH, (3,4-dimethoxyphenyl)ethyl mesylate; (iv) H_3PO_4 then aq. MeOH.

Highlights in this area are found in the spectacular cascade sequences engineered by Padwa.[51] For example, three rings of an advanced tetracyclic intermediate in an earlier[52] synthesis of (±)-erysotramidine were obtained in a single step.[53] Amide trapping of the first-formed α-thiocarbocation, tautomerisation to the furan, intramolecular Diels–Alder reaction, fragmentation and aromatic electrophilic substitution by the so-formed acyl iminium species were induced efficiently on exposure of the readily assembled precursor to Lewis acid. Palladium-catalysed reductive coupling of the derived enol triflate and hydrolysis of the vinylthioether completed the formal synthesis (Scheme 4.27).

4.19 Lycoridine and pancratistatin

Much elegant work has been targeted towards the synthesis of alkaloids in the lycorane series [e.g. Schultz's intramolecular Staudinger and

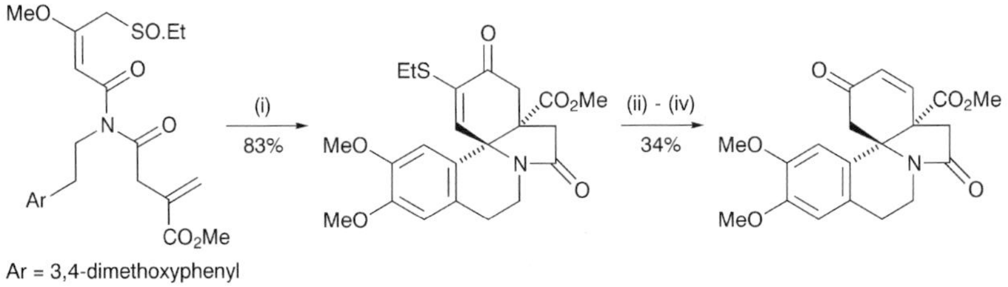

Ar = 3,4-dimethoxyphenyl

Scheme 4.27   (i) TFAA, Et_3N then $BF_3 \cdot OEt_2$; (ii) KH, $PhNTf_2$; (ii) $Pd(PPh_3)_2Cl_2$, Et_3N, HCO_2H; (iii) $TiCl_4$, aq. AcOH.

free-radical sequence towards (+)-deoxylycorine,[54] and Boeckman's convergent cyclopropyl imine rearrangement and Diels–Alder route to (±)-lycorine[55]], but it is the structurally related and more highly oxygenated alkaloids lycoricidine and pancratistatin that have stimulated the more diverse chemistry. For their lycoricidine syntheses both Martin[56] and Hudlicky[57] reported routes that established the 1,4-*cis*-aminoalcohol functionality about the cyclohexene ring by the Diels–Alder reaction of a nitroso dienophile. Hudlicky's version produced the (+)-enantiomer in only nine steps, from bromobenzene following asymmetric oxidation to the dihydroxycyclohexadiene derivative with *Pseudomonas putida*. Acyl nitroso Diels–Alder reaction followed by reductive cleavage of the nitrogen–oxygen bond gave a key intermediate that was cyclised in a Heck reaction, deprotection completing the route (Scheme 4.28). Keck's later synthesis[58] of the unnatural (−)-enantiomer was built around an efficient vinyl radical cyclisation onto an oxime radicophile. Samarium(II) iodide was used to effect reductive desulfurisation, nitrogen–oxygen

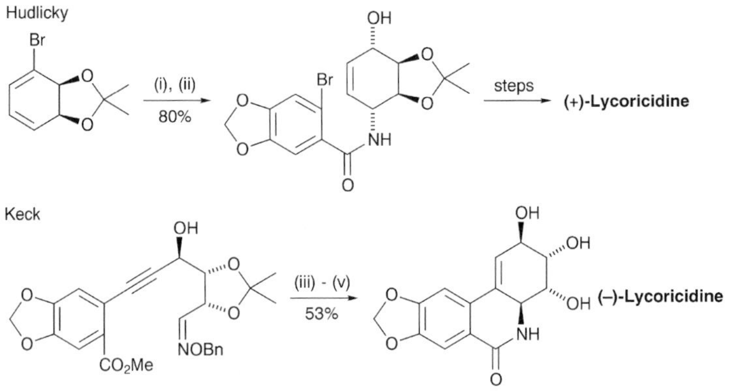

Scheme 4.28   (i) α-Bromopiperonoyl-NHOH, $Bu_4N^+IO_4^-$; (ii) Al(Hg); (iii) PhSH, $h\nu$; (iv) SmI_2; (v) TFA.

bond cleavage and lactamisation to yield (−)-lycoricidine after diol deprotection.

Moving up in oxidation level, the *Amaryllidaceae* alkaloid (+)-pancratistatin has been a favourite target for synthesis during the review period as a result of its clinical potential as an antitumour agent and the subtle problems posed by the controlled assembly of the six contiguous stereogenic centres around the aminocyclitol ring. In 1989 Danishefsky[59] published the first total synthesis (of the racemate), and Hudlicky followed with a route to the single enantiomer.[60] Haseltine[61] published a formal route that linked up with the Danishefsky approach but it is the asymmetric syntheses of Trost[62] and Magnus[63] that will be discussed here.

The Trost synthesis started with an asymmetric azidation of a *meso*-biscarbonate followed by an *anti-* stereoselective S_N2' addition of the aryl component as a cuprate to provide the coupled product in greater than 95% enantiomeric excess (ee). After functional group manipulation, conversion of the azide to an isocyanate and *in situ* metallation and cyclisation almost completed the route for which electronically controlled regioselective inversion (using Sharpless's cyclic sulfate methodology) and deprotection were also required (Scheme 4.29).

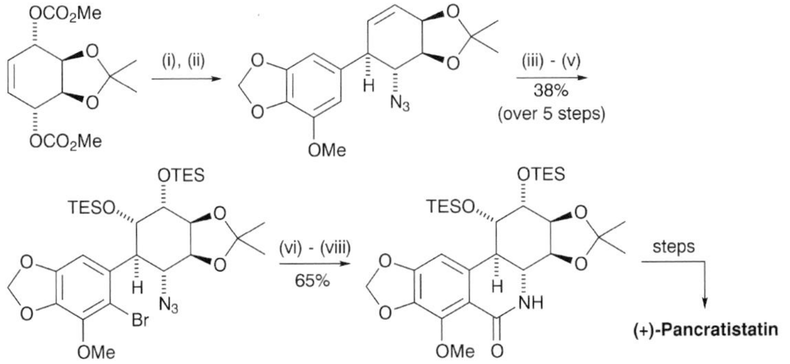

Scheme 4.29   (i) (π-C_3H_7PdCl)$_2$, Ph_2P-L^*-PPh_2, $TMSN_3$; (ii) ArMgBr, CuCN; (iii) OsO_4, NMO·H_2O; (iv) TESOTf; (v) NBS; (vi) Me_3P, aq. THF; (vii) $COCl_2$, Et_3N; (viii) *t*-BuLi.

In Magnus's enantioselective synthesis, his group's methodology for introducing azide functionality at the β position of a triisopropylsilyl enol ether (in this case formed from Koga–Simpkins asymmetric deprotonation of a prochiral 4-arylcyclohexanone) was used to provide a key intermediate in excellent overall yield. From this point the remaining functionality was introduced in a relatively straightforward fashion and relied on careful exploitation of six-membered ring steric and stereoelectronic control. The lactam ring was closed through a modified

Bischler–Napieralski reaction. Deprotection of all the hydroxyl groups then afforded (+)-pancratistatin in good yield (Scheme 4.30).

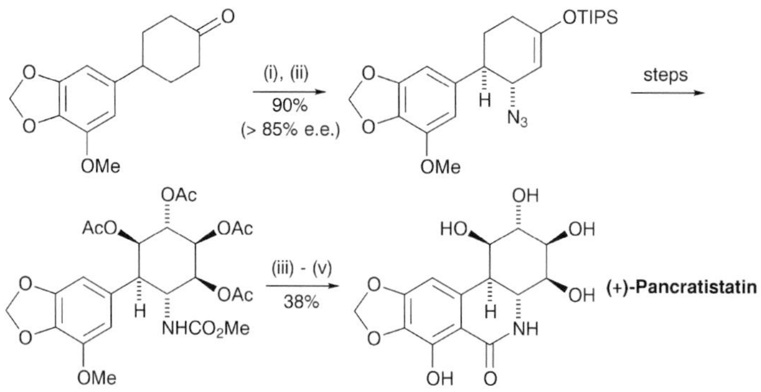

Scheme 4.30   (i) Lithium R,R-bis(α-methylbenzyl)amine, LiCl, TIPSOTf; (ii) (PhIO)$_n$, TMSN$_3$; (iii) Tf$_2$O, DMAP then H$_3$O$^+$; (iv) BBr$_3$; (v) NaOMe, MeOH.

4.20   Pyrrolizidines, indolizidines and related alkaloids

The wide structural variety embodied in the toxic pyrrolizidine alkaloids has long stimulated interest amongst synthetic chemists. Although many of these compounds occur as mono or bisesters or lactones, the vast majority of efforts have been expended on devising ways to assemble the core necine bases with control of the extent and stereochemistry of oxygenation about the ring. The highlight in this very broad area must be the ongoing contribution made by Denmark and his group[64] who have developed powerful chiral-auxiliary-controlled tandem nitronate Diels–Alder procedures for establishing most of the features of target molecules such as, *inter alia*, (−)-hastanecine, (−)-rosmarinecine, (+)-crotanecine and (−)-platynecine. The most recent example (at the time of writing), the synthesis of 7-epiaustraline,[65] typifies the approach (Scheme 4.31).

In an adjacent paper[66] White described a totally different approach to the stereoisomer (+)-australine; Grubbs's ring-closing metathesis generated an azacyclooctene which was epoxidised and induced to undergo transannular cyclisation to afford the pyrrolizidine. Transannular cyclisation was also a key feature in Vedejs's recent synthesis[67] of (±)-otonecine, an unusual pyrrolizidine betaine previously synthesised by Niwa.[68] The successful difficult oxidation—of a 2°-hydroxyl in the presence of a 3°-amine—was followed by a two-step sequence to generate the target molecule which exists in equilibrium with the eight-membered ring form (Scheme 4.32).

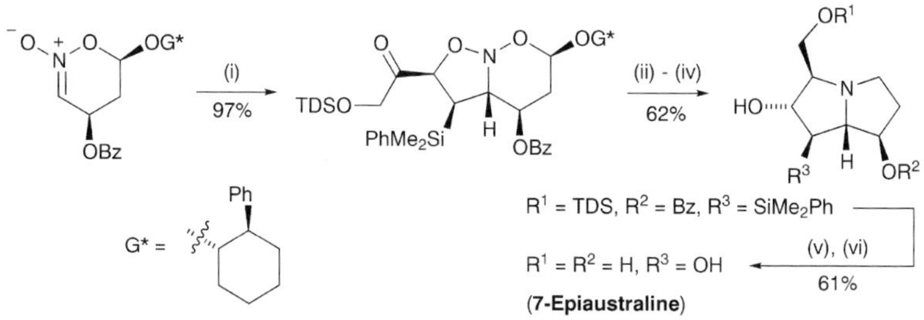

Scheme 4.31 (i) *Z*-1-(thexyldimethylsilyl)oxy-4-phenyldimethylsilylbut-3-en-2-one, PhH; (ii) L-selectride (diastereomeric ratio 14:1); (iii) Ms$_2$O, pyridine; (iv) H$_2$, Raney nickel; (v) Hg(O$_2$CCF$_3$)$_2$, TFA, AcOH, AcO$_2$H; (vi) H$_2$, Pd-C.

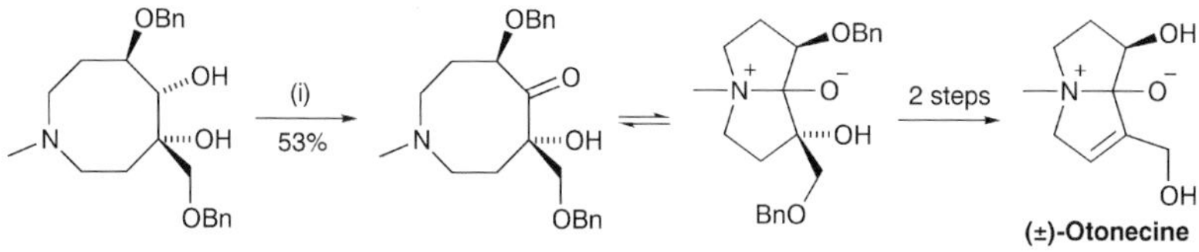

Scheme 4.32 (i) K$_3$Fe(CN)$_6$, OsO$_4$, CaO, aq. *t*-BuOH.

As with the pyrrolizidines, interest in the indolizidines has led to a wide array of new synthetic methodology being developed, but the more useful biological activity profile of certain members of this class has tended to focus efforts, and a host of syntheses of certain individual compounds have been recorded. For example there have been at least 20 syntheses of monomorine in both racemic and enantiomerically pure forms, but those by Livinghouse[69] and Mori[70] are amongst the most interesting. Livinghouse's catalytic procedure for pyrroline assembly from alkynyl amines was combined with classical reductive amination to afford the natural product in excellent overall yield from an acyclic precursor (Scheme 4.33).

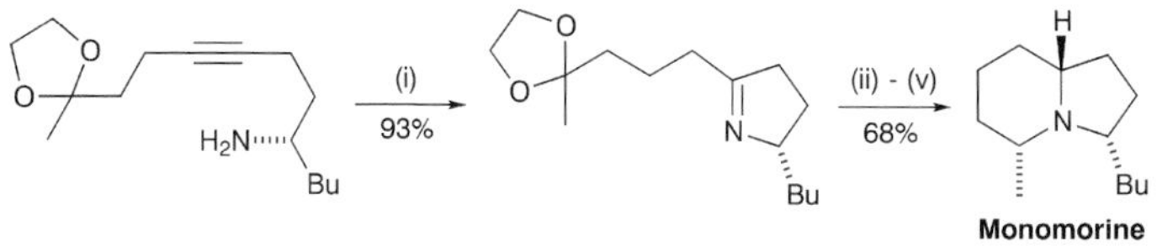

Scheme 4.33 (i) CpTiCl$_3$, Et$_3$N; (ii) DIBAL; (iii) aq. HCl; (iv) aq. K$_2$CO$_3$; (v) NaBH$_3$CN.

Mori's synthesis (Scheme 4.34) is of interest since incorporation of atmospheric nitrogen into molecules represents a potentially powerful new concept in alkaloid assembly, as attested by her group's recent synthesis of (±)-lycopodine.[71]

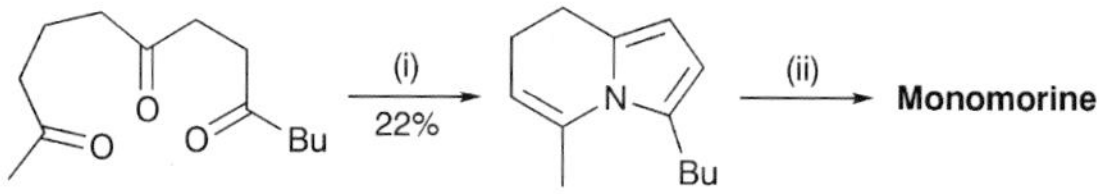

Scheme 4.34   (i) $TiCl_4$, Li, TMSCl, dry air; (ii) H_2, Rh-Al_2O_3.

Approaches to oxygenated indolizidines often take advantage of the defined stereochemical disposition of hydroxyl functionality to be found in carbohydrates, Fleet's synthesis of (−)-swainsonine[72] and Chamberlin's syntheses of both (−)-swainsonine and (+)-castanospermine[73] being good examples; most other enantiospecific routes originate with proteinogenic amino acids. Sibi's synthesis of (−)-slaframine[74] combined fragments derived from serine and hydroxyproline to generate the carbon framework, the final nitrogen–carbon bond being introduced by thermal deblocking of the N-Boc function and cyclisation with loss of carbon dioxide (Scheme 4.35).

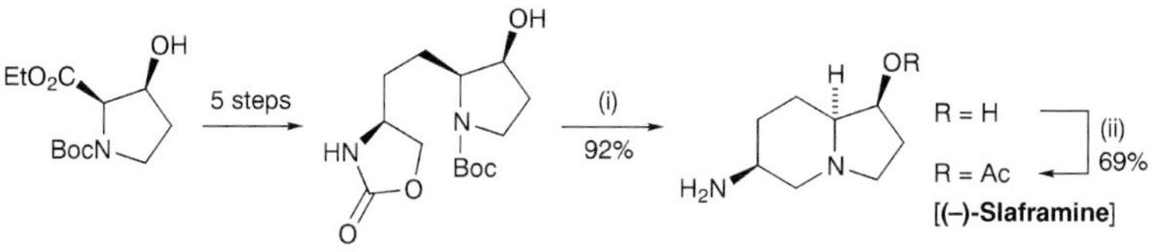

Scheme 4.35   (i) 270°C, 5 min.; (ii) HCl, HOAc.

The independent, but mutually related, methodology developed by Pearson and by Cha to address the preparation of more heavily oxygenated indolizidines provides examples of enantioselective (as opposed to enantiospecific) approaches. Both of these authors have developed related routes to single enantiomers of slaframine[75] and swainsonine,[76] with the Cha synthesis of (+)-castanospermine[77] being illustrative of a double cyclisation idea [step (vii), Scheme 4.36] that had been pioneered by Sharpless and Hashimoto during the mid-1980s.

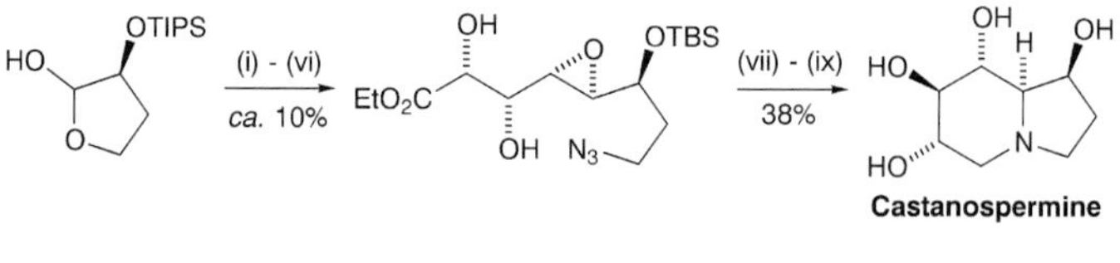

Scheme 4.36   (i) Ph_3P=$CHCO_2Et$ then TsCl, Et_3N, DMAP; (ii) TBAF; (iii) t-BuOOH, Ti(O-iPr)$_4$, (+)-DIPT; (iv) NaN_3; (v) $K_2Os(OH)_4O_2$, $K_3Fe(CN)_6$, $(DHQ)_2PHAL$, $MeSO_2NH_2$, K_2CO_3; (vi) TBSOTf; (vii) H_2, Pd-C; (viii) $BH_3 \cdot Me_2S$; (ix) TFA.

The pumiliotoxins and allopumiliotoxins—often based on the indol-izidine framework—have been popular synthetic targets, with the bulk of the chemistry summarised in Overman's recent review in which is described much of his own group's major contribution to the area.[78] Subsequent to that review Sato published[79] an interesting, concise approach to the ring system in an enantiospecific sequence (seven steps, 27% overall yield from L-proline) of allopumilitoxin 267A, a cardiotoxic *Dendrobatidae* alkaloid. The key carbon–carbon bond-forming step was achieved by acylation and stereospecific protonation of an intermediate titanacyclopropene in an extension of the earlier investigations of Kulinkovich (Scheme 4.37).[80]

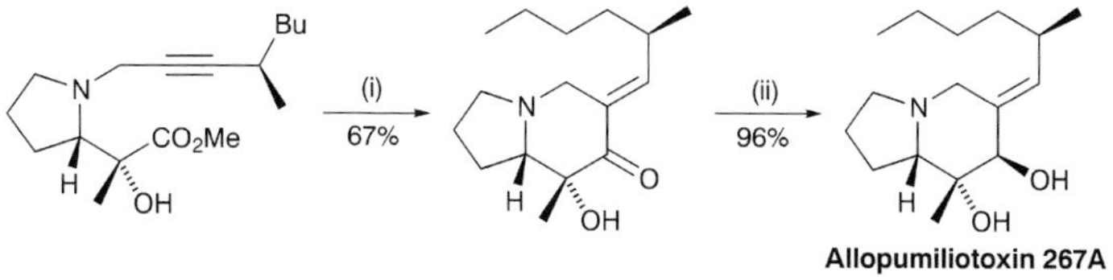

Scheme 4.37   (i) Ti(O*i*-Pr)$_4$, *i*-PrMgCl; (ii) Me$_4$NBH(OAc)$_3$.

Building on their earlier aza–Robinson annulation studies, Danishef-sky's group applied the intramolecular reaction of an α-ketocarbenoid with the carbon–sulfur double bond of a thiolactam to generate the indolizidine ring system in the unstable antibiotic (±)-indolizomycin[81] which was obtained from a mutant *Streptomyces* strain. *O*-Methylation of the vinylogous amide and reduction of the iminium ion were fol-lowed by cleavage of the ring-fusion bond to provide, after epoxidation, the nine-membered azacycle shown in Scheme 4.38. Epoxy ketone

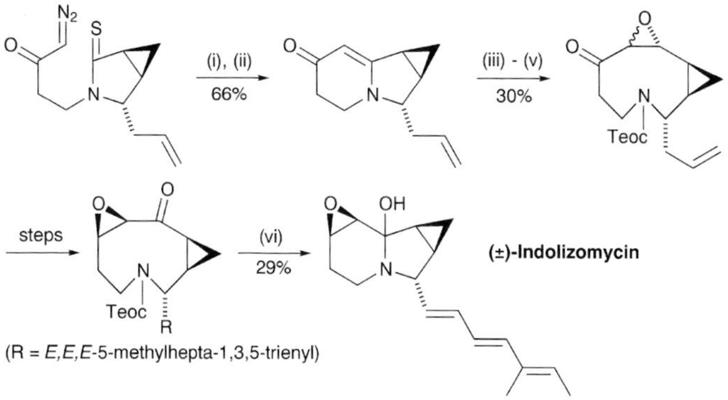

Scheme 4.38   (i) Rh$_2$(OAc)$_4$; (ii) W-2 Raney nickel; (iii) MeO$_3$BF$_4$; NaBH$_4$; (iv) Teoc-Cl; (v) H$_2$O$_2$, NaOH; (vi) TBAF.

transposition and chain elongation, followed by fluoride-mediated deprotection and cyclisation, generated the target molecule in 29% yield.

Comins has also been very active in the area of indolizidine synthesis, with most of his elegant chemistry, based on the rapid synthesis and exploitation of (enantiopure) N-acyldihydropyridones, recently summarised.[82]

4.21 Amaryllidaceae alkaloids

The reduced indole structure bearing an oxygenated aryl substituent at the bridgehead occurs widely amongst the Amaryllidaceae alkaloids; examples of particular synthetic importance include tazettine and mesembrine. Pearson has developed a general route to this ring system, based on aza–allyl anion cyclisation, to provide syntheses of (±)-crinine, (±)-epicrinine, (−)-amabiline, and (−)-augustamine,[83] and an incomplete extension of this idea to the marine alkaloid lepadiformine,[84] the augustamine synthesis providing a nice illustration of the approach (Scheme 4.39).

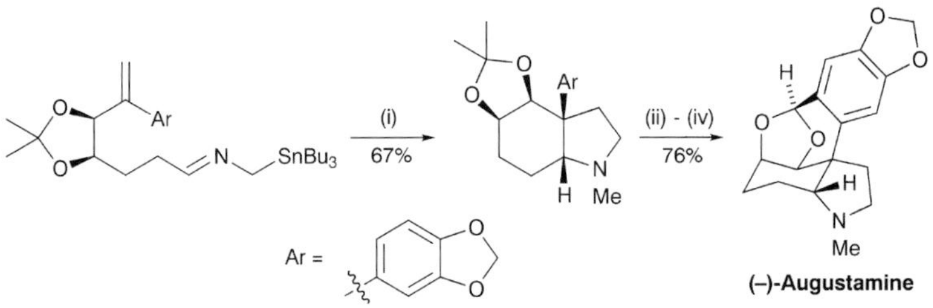

Scheme 4.39   (i) BuLi then MeI; (ii) c. HCl, MeOH; (iii) (MeO)₃CH; (iv) MeSO₃H.

Modification of the first stage, by omitting the methyl iodide quench, and successive treatment with Eschenmoser's salt and methanolic HCl led to (−)-amabiline.

From the wide variety of other syntheses of this ring system, less obvious approaches include an early application of Winkler's vinylogous amide [2+2] cycloaddition–fragmentation approach in the synthesis of mesembrine[85] and Mori's zirconocene-mediated cyclisation to establish the 4°-bridgehead centre whilst allowing the stereocontrolled introduction of oxygenation.[86] In a further exploration of the high chemical potential of vinyl isocyanates, Rigby[87] has prepared (±)-tazettine[88] based on a key step involving trapping of the Curtius rearrangement product with dimethoxycarbene (Scheme 4.40).

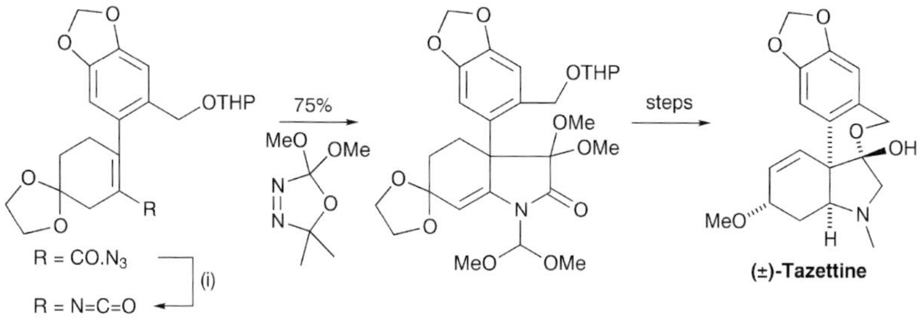

Scheme 4.40   (i) Heat.

4.22   Roseophilin

Roseophilin, a cytotoxic antibiotic obtained as a fermentation product of *Streptomyces griseoviridis*,[89] has stimulated wide interest amongst synthetic chemists, who have been attracted to solving the problems posed by assembling the unusual tricyclic framework that forms the right-hand half of the molecule. Both Fuchs[90] and Terashima[91] have reported syntheses of a complete right-hand half synthon, the latter researcher also being the first to describe a synthesis of the pyrrolo-furan left-hand half.[92] However the honour of the first total synthesis fell to Fürstner[93] in an excellent route that hinged on two applications of π-allyl palladium chemistry to effect sequential formation of the 13-membered ring and the pyrrole unit. The cyclopentanone ring was formed by intramolecular Friedel–Crafts acylation (a disconnection subsequently employed by Terashima in his route), and the isopropyl group was introduced stereoselectively under highly specific conditions in which elimination of sulfinate was followed by immediate conjugate addition to the so-formed reactive enone. The final steps to (±)-roseophilin proceeded along fairly conventional lines (Scheme 4.41).

4.23   Hispidospermidin

Very soon after the structure and phospholipase C inhibitory properties of (−)-hispidospermidin had been reported, two total syntheses were disclosed; these proceeded along conceptually similar lines to generate the tricyclic carbon framework but differed in the mode of installation of the final (oxygenated) ring.[94] For example, Overman's construction of the tricycle was effected by alkene-trapping of an incipient acylium ion with elimination of incorporated bromide. Under Evans aziridination conditions the alkene was transformed directly into an allylic amine, a

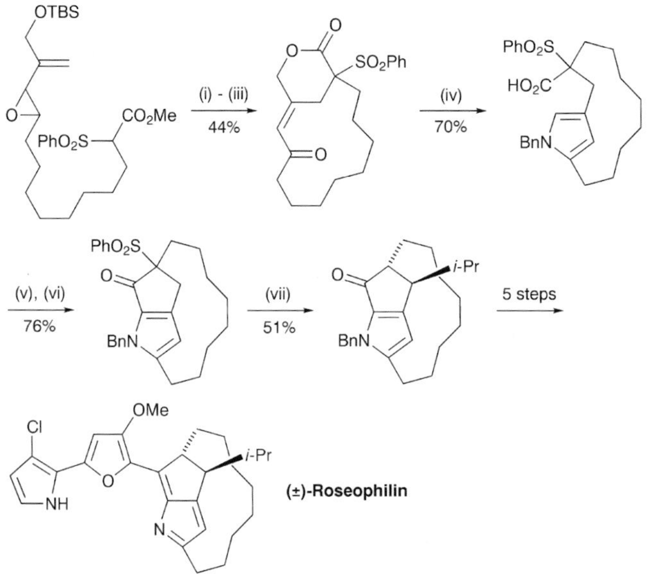

Scheme 4.41   (i) Pd(PPh$_3$)$_4$, dppe; (ii) TBAF, NH$_4$F; (iii) Dess–Martin; (iv) BnNH$_2$, Pd(PPh$_3$)$_4$; (v) Me$_2$C=C(Cl)NMe$_2$; (vi) SnCl$_4$; (vii) i-PrMe$_2$ZnMgCl, t-BuOK.

previously unrecognised reaction mode for these conditions. Epoxidation of the alkene, after *exo*-selective addition of methyllithium to the carbonyl group, then sequential cyclisation and acid-catalysed dehydration, generated the key features of the final structure (Scheme 4.42).

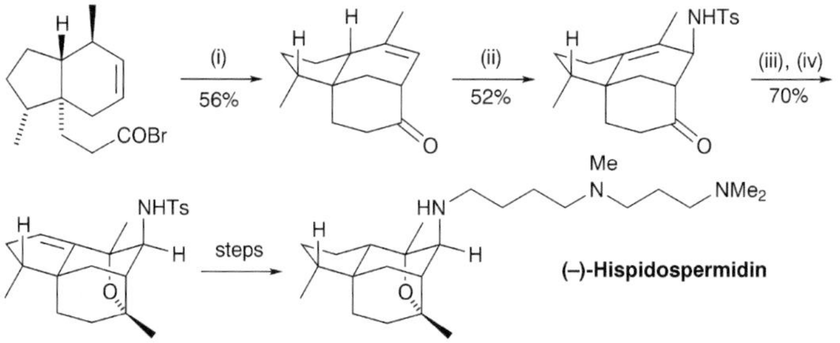

Scheme 4.42   (i) TiCl$_4$ then DBU; (ii) PhI=NTs, Cu(OTf)$_2$; (iii) MeLi; (iv) MCPBA then CSA.

4.24   Manzamine A

A now-famous proposal for the biosynthesis of the structurally diverse manzamine β-carboline alkaloids was put forward by Baldwin and

Whitehead[95] which suggested previously unrecognised links between various structural types and provided a biosynthetic context for subsequently discovered alkaloids whose structures corresponded to hypothetical intermediates in the Baldwin–Whitehead scheme. Although much synthetic work had been undertaken prior to the appearance of this proposal the paper highlighted the synthetic problem and added impetus to the field. The recent review by Langlois[96] summarised most of the reported synthetic work during the 1990s but came a few months too early to incorporate the first total synthesis of manzamine A—a potent cytotoxin and one of the most structurally complex members—by Winkler[97] in a synthesis that was anything but biomimetic. Conjugate addition of an azacyclooctene derivative[98] with Z-12-hydroxydodec-7-en-1-yn-3-one afforded (99%) the vinylogous amide precursor which was subjected to photolysis to initiate a cascade process involving intramolecular [2+2] cycloaddition, retro-Mannich reaction and aminal formation. Regeneration of the iminium ion and C-alkylation of the isomerised enol led to four of the five core rings, the fifth, 13-membered, ring being formed in

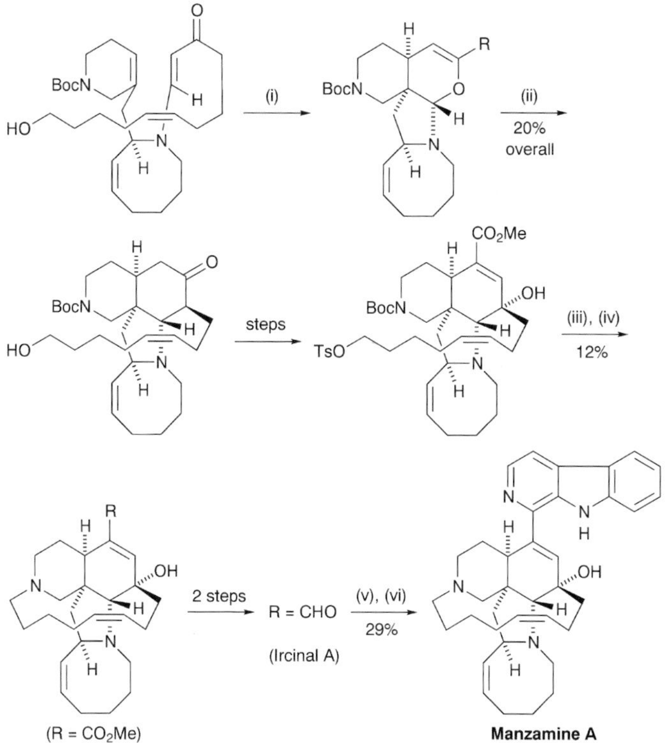

Scheme 4.43   (i) $h\nu$; (ii) pyridine, AcOH; (iii) TFA; (iv) i-Pr$_2$NEt; (v) tryptamine, TFA; (vi) DDQ.

low yield by *N*-alkylation. Functional group manipulation led, in four more steps, via ircinol A and ircinal A, to manzamine A (Scheme 4.43).

Notable amongst the many partial syntheses is that of Pandit,[99] who assembled the 13-membered ring in moderate yield by ring-closing metathesis, albeit with a full equivalent of Grubbs's ruthenium catalyst (Scheme 4.44).

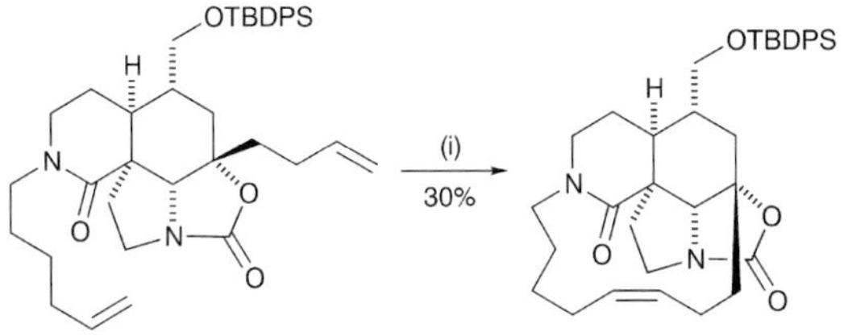

Scheme 4.44   (i) $(PCy_3)_2Cl_2Ru{=}CHCH{=}CPh_2$.

References

1. G. Stork, 1989, *Pure Appl. Chem.*, 61, 439.
2. (a) A.M. Gomez, J.C. Lopez, and B. Fraser-Reid, 1994, *J. Org. Chem.*, 59, 4048; (b) A.M. Gomez, J.C. Lopez, and B. Fraser-Reid, 1995, *J. Org. Chem.*, 60, 3859.
3. C.S. Chu, C.C. Liao, and P.D. Rao, 1996, *J. Chem. Soc., Chem. Commun.*, 1537.
4. S. Hanessian, J.W. Pan, A. Carnell, H. Bouchard, and L. Lesage, 1997, *J. Org. Chem.*, 62, 465.
5. G. Stork and K. Zhao, 1990, *J. Am. Chem. Soc.*, 112, 5875.
6. G. Stork, Y. Kobayashi, T. Suzuki, and K. Zhao, 1990, *J. Am. Chem. Soc.*, 112, 1661.
7. G. Stork and K. Zhao, 1989, *Tetrahedron Lett.*, 30, 2173.
8. D. Tanner and L. Hagberg, 1998, *Tetrahedron*, 54, 7907.
9. M. Ihara, M. Suzuki, K. Fukumoto, and C. Kabuto, 1990, *J. Am. Chem. Soc.*, 112, 1164.
10. M. Ihara, M. Suzuki, K. Fukumoto, T. Kametani, and C. Kabuto, 1988, *J. Am. Chem. Soc.*, 110, 1963.
11. (a) C.H. Heathcock, 1992, *Angew. Chem., Int. Ed. Engl.*, 31, 665; (b) C.H. Heathcock, 1996, *Proc. Natl. Acad. Sci. USA*, 93, 14323.
12. C.H. Heathcock, S. Piettre, R.B. Ruggeri, J.A. Ragan, and J.C. Kath, 1992, *J. Org. Chem.*, 57, 2554.
13. C.H. Heathcock, R.B. Ruggeri, and K.F. McClure, 1992, *J. Org. Chem.*, 57, 2585.
14. C.H. Heathcock, J.C. Kath, and R.B. Ruggeri, 1995, *J. Org. Chem.*, 60, 1120.
15. C.H. Heathcock, J.A. Stafford, and D.L. Clark, 1992, *J. Org. Chem.*, 57, 2575.
16. C.-K. Sha, R.-T. Chiu, C.-F. Yang, N.-T. Yao, W.-H. Tseng, F.-L. Liao, and S.-L. Wang, 1997, *J. Am. Chem. Soc.*, 119, 4130.
17. C.H. Lee, M. Westling, T. Livinghouse, and A.C. Williams, 1992, *J. Am. Chem. Soc.*, 114, 4089.
18. N. Uesaka, F. Saitoh, M. Mori, M. Shibasaki, K. Okamura, and T. Date, 1994, *J. Org. Chem.*, 59, 5633.
19. S.F. Martin and W. Li, 1991, *J. Org. Chem.*, 56, 642.
20. (a) P. Garner, W.B. Ho, and H. Shin, 1992, *J. Am. Chem. Soc.*, 114, 2767; (b) P. Garner, W.B. Ho, and H. Shin, 1993, *J. Am. Chem. Soc.*, 115, 10742.

21. T. Fukuyama and J.J. Nunes, 1988, *J. Am. Chem. Soc.*, 110, 5196.
22. T. Katoh and S. Terashima, 1996, *Pure Appl. Chem.*, 68, 703, and references therein.
23. G.C. Hirst, T.O. Johnson Jr and L.E. Overman, 1993, *J. Am. Chem. Soc.*, 115, 2992.
24. (a) L.A. Paquette, D. Friedrich, E. Pinard, J.P. Williams, D. St. Laurent, and B.A. Roden, 1993, *J. Am. Chem. Soc.*, 115, 4377; (b) J.P. Williams, D.R. St. Laurent, D. Friedrich, E. Pinard, B.A. Roden, and L.A. Paquette, 1994, *J. Am. Chem. Soc.*, 116, 4689.
25. (a) M.L. Bennasar, B. Vidal, and J. Bosch, 1993, *J. Am. Chem. Soc.*, 115, 5340; (b) M.L. Bennasar, B. Vidal, and J. Bosch, 1997, *J. Org. Chem.*, 62, 3597.
26. J.E. Saxton, 1992, *Nat. Prod. Rep.*, 393.
27. (a) Z. Sheikh, R. Steel, A.S. Tasker, and A.P. Johnson, 1994, *J. Chem. Soc., Chem. Commun.*, 763; (b) J.K. Dutton, R.W. Steel, A.S. Tasker, V. Popsavin, and A.P. Johnson, 1994, *J. Chem. Soc., Chem. Commun.*, 765.
28. N.J. Newcombe, F. Ya, R.J. Vijn, H. Hiemstra, and W.N. Speckamp, 1994, *J. Chem. Soc., Chem. Commun.*, 767.
29. (a) D. Kuzmich, S.C. Wu, D.-C. Ha, C.-S. Lee, S. Ramesh, S. Atarashi, J.-K. Choi, and D.J. Hart, 1994, *J. Am. Chem. Soc.*, 116, 6943; (b) S. Atarashi, J.-K. Choi, D.-C. Ha, D.J. Hart, D. Kuzmich, C.-S. Lee, S. Ramesh, and S.C. Wu, 1997, *J. Am. Chem. Soc.*, 119, 6226.
30. T. Fukuyama and G. Liu, 1996, *J. Am. Chem. Soc.*, 118, 7426.
31. B.J. Baker, P.J. Scheuer, and J.N. Shoolery, 1988, *J. Am. Chem. Soc.*, 110, 965.
32. (a) A.G.M. Barrett, M.L. Boys, and T.L. Boehm, 1994, *J. Chem. Soc., Chem. Commun.*, 1881; (b) A.G.M. Barrett, M.L. Boys, and T.L. Boehm, 1996, *J. Org. Chem.*, 61, 685; (c) R.M. Borzilleri, S.M. Weinreb and M. Parvez, 1994, *J. Am. Chem. Soc.*, 116, 9789; (d) R.M. Borzilleri, S.M. Weinreb, and M. Parvez, 1995, *J. Am. Chem. Soc.*, 117, 10905; (e) T.S. McDermott, A.A. Mortlock, and C.H. Heathcock, 1996, *J. Org. Chem.*, 61, 700.
33. E.J. Corey, D.Y. Gin, and R.S. Kania, 1996, *J. Am. Chem. Soc.*, 118, 9202.
34. D.R. Williams, D.L. Brown, and J.W. Benbow, 1989, *J. Am. Chem. Soc.*, 111, 1923.
35. S.F. Martin and K.J. Barr, 1996, *J. Am. Chem. Soc.*, 118, 3299.
36. W. Oppolzer, H. Bienaymé, and A. Genevois-Borella, 1991, *J. Am. Chem. Soc.*, 113, 9660.
37. R.B. Woodward, M.P. Cava, W.D. Ollis, A. Hunger, H.U. Daeniker, and K. Schenker, 1954, *J. Am. Chem. Soc.*, 76, 4749.
38. (a) G. Stork, 1992, reported at the Ischia Advanced School of Organic Chemistry, Ischia Porto, Italy, 21 September; (b) P. Magnus, M. Giles, R. Bonnert, C.S. Kim, L. McQuire, A. Merritt, and N. Vicker, 1992, *J. Am. Chem. Soc.*, 114, 4403; (c) P. Magnus, M. Giles, R. Bonnert, G. Johnson, L. McQuire, M. Deluca, A. Merritt, C.S. Kim, and N. Vicker, 1993, *J. Am. Chem. Soc.*, 115, 8116; (d) M.E. Kuehne and F. Xu, 1993, *J. Org. Chem.*, 58, 7490; (e) S.D. Knight, L.E. Overman, and G. Pairaudeau, 1993, *J. Am. Chem. Soc.*, 115, 9293; (f) S.D. Knight, L.E. Overman, and G. Pairaudeau, 1995, *J. Am. Chem. Soc.*, 117, 5776; (g) V.H. Rawal and S. Iwasa, 1994, *J. Org. Chem.*, 59, 2685; (h) S.F. Martin, C.W. Clark, M. Ito, and M. Mortimore, 1996, *J. Am. Chem. Soc.*, 118, 9804.
39. B. Wu and D. Bai, 1997, *J. Org. Chem.*, 62, 5978.
40. D. Schumann and A. Naumann, 1983, *Liebigs Ann. Chem.*, 220.
41. (a) T.P. Burkholder and P.L. Fuchs, 1988, *J. Am. Chem. Soc.*, 110, 2341; (b) T.P. Burkholder and P.L. Fuchs, 1990, *J. Am. Chem. Soc.*, 112, 9601; (c) M.E. Kuehne, W.G. Bornmann, W.H. Parsons, T.D. Spitzer, J.F. Blount, and J. Zubieta, 1988, *J. Org. Chem.*, 53, 3439; (d) H. Ishibashi, M. Okano, H. Tamaki, K. Maruyama, T. Yakura, and M. Ikeda, 1990, *J. Chem. Soc., Chem. Commun.*, 1436; (e) M. Okano, N. Nishimura, K. Maruyama, K. Kosaka, H. Ishibashi, and M. Ikeda, 1991, *Chem. Pharm. Bull.*, 41, 276; (f) X. Lin, R.W. Kavash, and P.S. Mariano, 1994, *J. Am. Chem. Soc.*, 116, 9791; (g) X. Lin, R.W. Kavash, and P.S. Mariano, 1996, *J. Org. Chem.*, 61, 7335; (h) N. Isono and M. Mori, 1995, *J. Org. Chem.*, 60, 115; (i) L.F. Tietze and H. Schirok, 1997, *Angew. Chem., Int. Ed. Engl.*, 36, 1124.

42. J.E. Toth, P.R. Hamann, and P.L. Fuchs, 1988, *J. Org. Chem.*, 53, 4694.

43. K.A. Parker and D. Fokas, 1992, *J. Am. Chem. Soc.*, 114, 9688.

44. C.Y. Hong, N. Kado, and L.E. Overman, 1993, *J. Am. Chem. Soc.*, 115, 11028.

45. D. Trauner, J.W. Bats, A. Werner, and J. Mulzer, 1998, *J. Org. Chem.*, 63, 5908.

46. J.D. White, P. Hrnciar, and F. Stappenbeck, 1997, *J. Org. Chem.*, 62, 5250.

47. H.H. Wasserman and R.M. Amici, 1989, *J. Org. Chem.*, 54, 5843.

48. J.Y. Lee, Y.S. Lee, B.Y. Chung, and H. Park, 1997, *Tetrahedron*, 53, 2449.

49. J.H. Rigby and M. Qabar, 1991, *J. Am. Chem. Soc.*, 113, 8975.

50. J.H. Rigby and M.E. Mateo, 1996, *Tetrahedron*, 52, 10569.

51. A. Padwa, 1998, *J. Chem. Soc., Chem. Commun.*, 1417.

52. Y. Tsuda, S. Hosoi, A. Nakai, Y. Sakai, T. Abe, Y. Ishi, F. Kiuchi, and T. Sano, 1991, *Chem. Pharm. Bull.*, 39, 1365.

53. A. Padwa, R. Hennig, C.O. Kappe, and T.S. Reger, 1998, *J. Org. Chem.*, 63, 1144.

54. A.G. Schultz, M.A. Holoboski, and M.S. Smyth, 1993, *J. Am. Chem. Soc.*, 115, 7904.

55. R.K. Boeckman, S.W. Goldstein, and M.A. Walters, 1988, *J. Am. Chem. Soc.*, 110, 8250.

56. S.F. Martin and H.H. Tso, 1993, *Heterocycles*, 35, 85.

57. (a) T. Hudlicky, H.F. Olivo, and B. McKibben, 1994, *J. Am. Chem. Soc.*, 116, 5108; (b) T. Hudlicky and H.F. Olivo, 1992, *J. Am. Chem. Soc.*, 114, 9694.

58. G.E. Keck and T.T. Wager, 1996, *J. Org. Chem.*, 61, 8366.

59. S.J. Danishefsky and J.Y. Lee, 1989, *J. Am. Chem. Soc.*, 111, 4829.

60. (a) X. Tian, T. Hudlicky, and K. Königsberger, 1995, *J. Am. Chem. Soc.*, 117, 3643; (b) X. Tian, T. Hudlicky, K. Königsberger, R. Maurya, J. Rouden, and B. Fan, 1996, *J. Am. Chem. Soc.*, 118, 10752.

61. T.J. Doyle, M. Hendrix, D. VanDerveer, S. Javanmard, and J. Haseltine, 1997, *Tetrahedron*, 53, 11153.

62. B.M. Trost and S.R. Pulley, 1995, *J. Am. Chem. Soc.*, 117, 10143.

63. P. Magnus and I.K. Sebhat, 1998, *J. Am. Chem. Soc.*, 120, 5341.

64. S.E. Denmark and A. Thorarensen, 1996, *Chem. Rev.*, 96, 137.

65. S.E. Denmark and B. Herbert, 1998, *J. Am. Chem. Soc.*, 120, 7357.

66. J.D. White, P. Hrnciar, and A.F.T. Yokochi, 1998, *J. Am. Chem. Soc.*, 120, 7359.

67. E. Vedejs, R.J. Galante, and P.G. Goekjian, 1998, *J. Am. Chem. Soc.*, 120, 3613.

68. H. Niwa, T. Sakata, and K. Yamada, 1994, *Bull. Chem. Soc. Jpn.*, 67, 2345.

69. P.L. McGrane and T. Livinghouse, 1992, *J. Org. Chem.*, 57, 1323.

70. M. Mori, M. Hori, and Y. Sato, 1998, *J. Org. Chem.*, 63, 4832.

71. M. Mori, K. Hori, M. Akashi, M. Hori, Y. Sato, and M. Nishida, 1998, *Angew. Chem., Int. Ed. Engl.*, 37, 636.

72. N.M. Carpenter, G.W.J. Fleet, I. Cenci di Bello, B. Winchester, L.E. Fellows, and R.J. Nash, 1989, *Tetrahedron Lett.*, 30, 7261.

73. S.A. Miller and A.R. Chamberlin, 1990, *J. Am. Chem. Soc.*, 112, 8100.

74. M.P. Sibi, J.W. Christensen, B. Li, and P.A. Renhowe, 1992, *J. Org. Chem.*, 57, 4329.

75. (a) J.-R. Choi, S. Han, and J.K. Cha, 1991, *Tetrahedron Lett.*, 32, 6469; (b) W.H. Pearson and S.C. Bergmeier, 1991, *J. Org. Chem.*, 56, 1976; (c) W.H. Pearson, S.C. Bergmeier, and J.P. Williams, 1992, *J. Org. Chem.*, 57, 3977.

76. (a) R.B. Bennett III, J.-R. Choi, W.D. Montgomery, and J.K. Cha, 1989, *J. Am. Chem. Soc.*, 111, 2580; (b) W.H. Pearson and K.-C. Lin, 1990, *Tetrahedron Lett.*, 31, 7571.

77. N.-S. Kim, J.-R. Choi, and J.K. Cha, 1993, *J. Org. Chem.*, 58, 7096.

78. A.S. Franklin and L.E. Overman, 1996, *Chem. Rev.*, 96, 505.

79. S. Okamoto, M. Iwakubo, K. Kobayashi, and F. Sato, 1997, *J. Am. Chem. Soc.*, 119, 6984.

80. O.G. Kulinkovich, S.V. Sviridov, and D.A. Vasilevski, 1991, *Synthesis*, 234.

81. (a) G. Kim, M.Y. Chu-Moyer, and S.J. Danishefsky, 1990, *J. Am. Chem. Soc.*, 112, 2003; (b) G. Kim, M.Y. Chu-Moyer, S.J. Danishefsky, and G.K. Schulte, 1993, *J. Am. Chem. Soc.*, 115, 30.

82. D.L. Comins, D.H. LaMunyon, and X. Chen, 1997, *J. Org. Chem.*, 62, 8182.

83. W.H. Pearson and F.E. Lovering, 1998, *J. Org. Chem.*, 63, 3607.

84. W.H. Pearson, N.S. Barta, and J.W. Kampf, 1997, *Tetrahedron Lett.*, 38, 3369.

85. J.D. Winkler, C.L. Muller, and R.D. Scott, 1988, *J. Am. Chem. Soc.*, 110, 4831.

86. M. Mori, S. Kuroda, C.-S. Zhang, and Y. Sato, 1997, *J. Org. Chem.*, 62, 3263.

87. J.H. Rigby, A. Cavezza, and M.J. Heeg, 1998, *J. Am. Chem. Soc.*, 120, 3664.

88. M.W. Abelman, L.E. Overman, and V.D. Tran, 1990, *J. Am. Chem. Soc.*, 112, 6959.

89. Y. Hayakawa, K. Kawakami, H. Seto, and K. Furihata, 1992, *Tetrahedron Lett.*, 33, 2701.

90. S.H. Kim, I. Figueroa, and P.L. Fuchs, 1997, *Tetrahedron Lett.*, 38, 2601.

91. T. Mochizuki, E. Itoh, N. Shibata, S. Nakatani, T. Katoh, and S. Terashima, 1998, *Tetrahedron Lett.*, 39, 6911.

92. S. Nakatani, M. Kirihara, K. Yamada, and S. Terashima, 1995, *Tetrahedron Lett.*, 36, 8461.

93. (a) A. Fürstner and H. Weintritt, 1997, *J. Am. Chem. Soc.*, 119, 2944; (b) A. Fürstner and H. Weintritt, 1998, *J. Am. Chem. Soc.*, 120, 2817.

94. (a) L.E. Overman and A.L. Tomasi, 1998, *J. Am. Chem. Soc.*, 120, 4039; (b) A.J. Frontier, S. Raghavan, and S.J. Danishefsky, 1997, *J. Am. Chem. Soc.*, 119, 6686.

95. J.E. Baldwin and R.C. Whitehead, 1992, *Tetrahedron Lett.*, 33, 2059.

96. E. Magnier and Y. Langlois, 1998, *Tetrahedron*, 54, 6201.

97. J.D. Winkler and J.M. Axten, 1998, *J. Am. Chem. Soc.*, 120, 6425.

98. J.D. Winkler, J. Axten, A.H. Hammach, Y.-S. Kwak, U. Lengweiler, M.J. Lucero, and K.N. Houk, 1998, *Tetrahedron*, 54, 7045.

99. B.C. Borer, S. Deerenberg, H. Bieräugel, and U.K. Pandit, 1994, *Tetrahedron Lett.*, 35, 3191.

5 Recent developments in the chemical synthesis of naturally occurring aromatic heterocycles

G.A. Sulikowski and M.M. Sulikowski

5.1 Introduction

This chapter highlights total syntheses of various aromatic hetereocycles reported between the years 1988 and 1998. Obviously, because of space limitations only a minor representation of the outstanding work done in this area has been covered. Part of our selection criteria has been the illustration of new tactics, strategies and methods. As a result, the total syntheses described here are among the most concise reported within the past 10 years.

The chapter is broadly organised into two sections, which deal with the synthesis of five- and six-membered aromatic heterocycles, respectively. Furthermore, total syntheses of many natural aromatic heterocycles fall within the headings of other chapters of this book.

5.2 Five-membered heterocycles

5.2.1 Duocarmycin A

In 1996 Boger and co-workers described an enantioselective total synthesis of (+)-duocarmycin A (**9**) and related congeners.[1,2] The synthetic sequence started with the regiospecific addition of allyltributyl-stannane to *p*-quinodiimide **1** to give **2** (Scheme 5.1). Sharpless asymmetric dihydroxylation of **2** using (DHQD)$_2$-PHAL proceeded in 78% enantiomeric excess, which was later improved by semipreparative chromatographic resolution of piperidine **4**. Exchange of *N*-benzoyl protecting groups in **4** for BOC groups set the stage for production of oxazolidone **6** and subsequent Dieckmann cyclisation for introduction of the C(6) stereocentre (cf. **7**). Dieckmann cyclisation of **6** was accomplished under thermodynamic conditions (6 equivalents of LDA, inverse addition) with greater than 10:1 diastereoselectivity providing imine **7**, after methoxide assisted removal of the oxazolidone auxiliary. Conversion to amide **8** was completed following imine hydrolysis, acid-catalysed selective BOC deprotection and EDCI mediated coupling with 5,6,7-trimethoxyindole-2-carboxylic acid. Hydrogenolysis of the remaining benzyl group followed by transannular spirocyclisation under Mitsunobu conditions gave (+)-duocarmycin A (**9**).

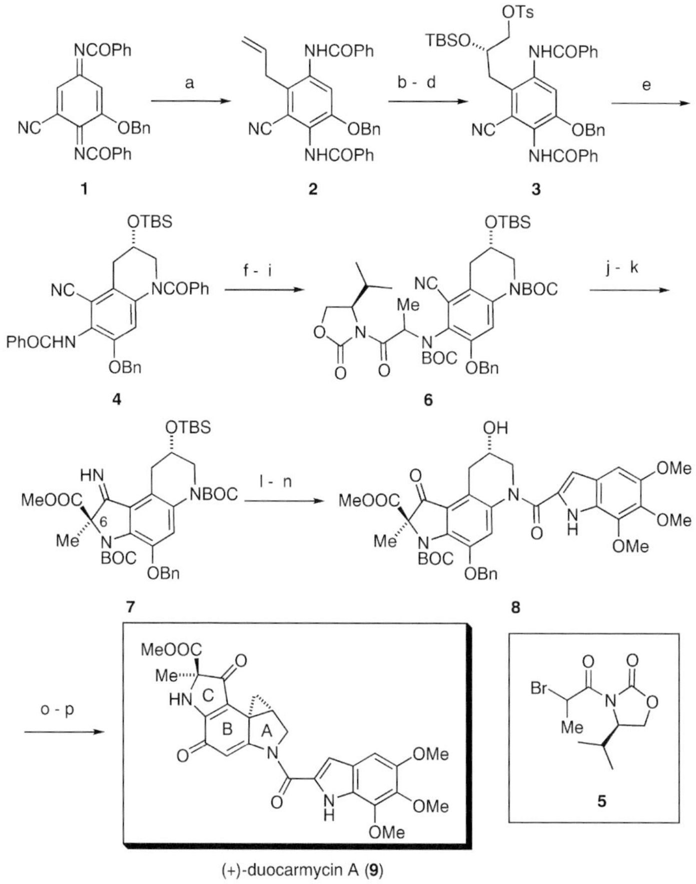

Scheme 5.1 (a) Allyltributylstannane, $BF_3 \cdot OEt_2$, CH_2Cl_2, $-20°C$ (83%–89%); (b) $(DHQD)_2$-PHAL, OsO_4, $K_3Fe(CN)_6$, K_2CO_3, $CH_3SO_2NH_2$, THF (aq) (92%, 78% ee); (c) Bu_3SnO, PhMe-THF, reflux, then p-TsCl, Et_3N (89%); (d) TBSOTf, 2,6-lutidine, CH_2Cl_2, $0°C$ (75%); (e) NaH, THF, $0°C$ (92%–97%); (f) NH_2NH_2, EtOH, $140°C$ (58%–65%); (g) BOC_2O, DMAP, THF, reflux, TFA (95%); (h) oxazolidinone **5**, NaH, DMF, $0°C$ (93%); (i) LDA (6 equiv., inverse addition), THF, $-78 \rightarrow 50°C$ (78%); (j) LiOMe, THF-MeOH, $0°C$ (78%–84%); (k) p-TSA, THF (aq), $0°C$ (76%–80%); (l) 4 M HCl-MeOH, $0°C$; (m) 5,6,7-trimethoxyinodole-2-carboxylic acid, EDCl, DMF (95%–98%, two steps); (n) H_2, 10% Pd-C, MeOH (95%–98%); (o) ADDP, Bu_3P, PhH, $50°C$ (99%).

5.2.2 Indolocarbazoles

The first total synthesis of staurosporine was reported in 1995 by Danishefsky and co-workers (Scheme 5.2).[3] Indole **10** was first glycosylated with 1,2-anhydro-sugar **11** using reaction conditions established in related investigations.[4] Next, the C(2′) hydroxyl group of the coupled

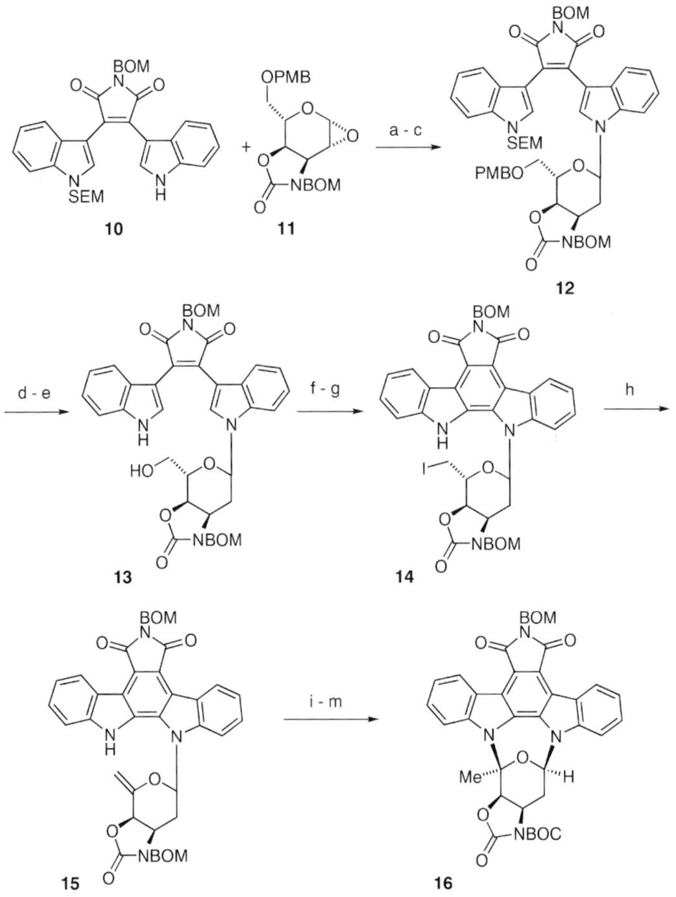

Scheme 5.2 (a) **10**, NaH, THF, RT then **11**, RT to reflux (47%); (b) thiophosgene, DMAP, pyr, CH$_2$Cl$_2$, reflux, then C$_6$H$_5$OH (79%); (c) n-Bu$_3$SnH, AIBN, PhH, reflux (74%); (d) DDQ, CH$_2$Cl$_2$, H$_2$O, 0°C to RT (97%); (e) TBAF, THF, reflux (91%); (f) $h\nu$, cat. I$_2$, air, PhH (73%); (g) I$_2$, PPh$_3$, imidazole, CH$_2$Cl$_2$, 0°C to RT (84%); (h) THF, DBU, RT (89%); (i) t-BuOK, I$_2$, THF, MeOH, RT (65%); (j) n-Bu$_3$SnH, AIBN, PhH, reflux (99%); (k) H$_2$, Pd(OH)$_2$, EtOAc, MeOH, RT, then NaOMe, MeOH (92%); (l) (BOC)$_2$O, THF, cat. DMAP (81%); (m) NaH, DMF then BOMCl (82%).

product was removed by using the Barton deoxygenation protocol. Subsequent removal of the O(6′) and indole nitrogen protecting groups led to **13**. Following oxidative photocyclisation of **13**, attention was turned towards establishing the second glycoside linkage. To this end, an exocyclic glycal (**15**) was introduced following the usual dehydroiodination procedure. Treatment of **15** with potassium *tert*-butoxide and iodine led to an intramolecular iodinative coupling. Finally, the iodo group was reduced to give the core structure of staurosporine (**16**).

Subsequent to Danishefsky's work, Wood and co-workers developed a concise unified strategy for the synthesis of various indolocarbazole alkaloids.[5] This research eventually led to the total synthesis of K252A (**21**), MLR-52, TAN-1030a as well as staurosporine. K252A (**21**) was the first indolocarbazole to be synthesised in this series. A critical step in the Wood approach was the rhodium(II)-catalysed condensation of 2,2'-biindole (**17**) with diazo lactam (**18**) to afford N-protected aglycone (**19**) (Scheme 5.3). A single-step acid-promoted coupling of **19** and furanose **20** (either enantiomer of **20** is available) provided a 2:1 mixture of N-protected K252A and the corresponding isomer. Acid-catalysed removal of the dimethoxybenzyl nitrogen protecting groups finally delivered (+)-K252A (**21**) and **22**.

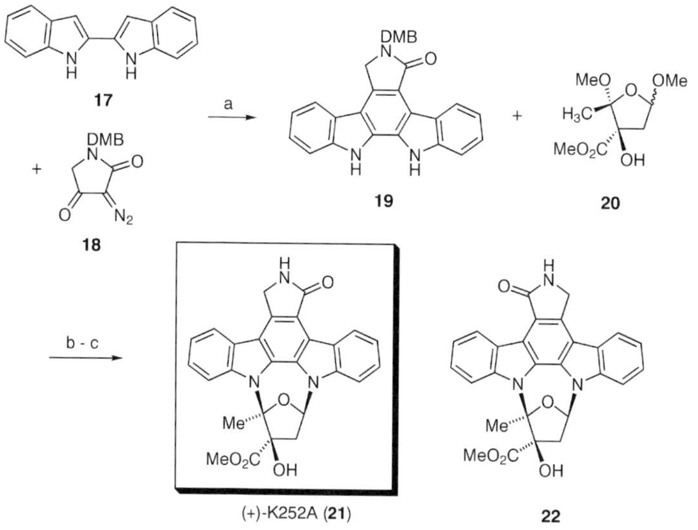

Scheme 5.3 (a) Rh$_2$(OAc)$_4$, 120°C, pinacolone, sealed tube (62%); (b) CSA, 48 h, reflux, C$_2$H$_4$Cl$_2$ (80%); (c) TFA, CH$_2$Cl$_2$, thioanisole, 6 h (83%).

The indolocarbazole alkaloids may be divided into two groups in which the aglycone is either furanosylated (cf. K252A **21**) or pyranosylated (cf. **16**). After establishing a concise route to the furanosylated congener **23**, Wood and co-workers developed an entry into the pyranosylated indolocarbazoles through Lewis-acid-catalysed ring expansion of aldehyde **24** to ketone **25**.[6] Aldehyde **24** was derived from **23** in two steps (Scheme 5.4). Ketone **25** served as a common synthetic intermediate en route to RK-286c (**26**), MLR-52, TAN-103a and staurosporine.

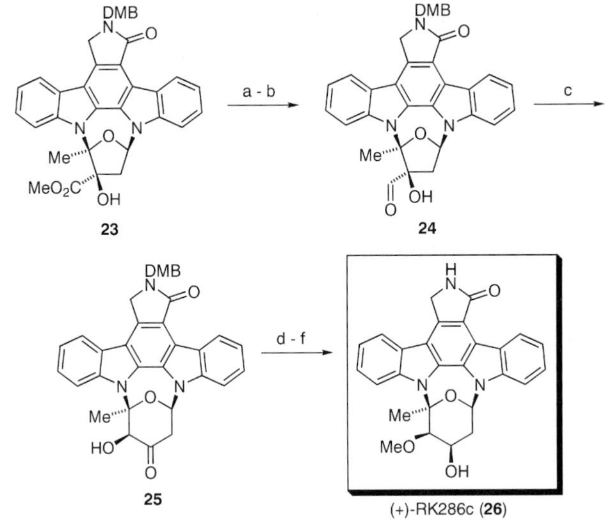

Scheme 5.4   (a) LiBH$_4$; (b) Moffatt oxidation (63%, two steps); (c) BF$_3 \cdot$ OEt$_2$, Et$_2$O; (d) NaBH$_4$ (95%); (e) NaH, MeI (80%); (f) TFA, anisole (75%).

A subgroup of dissymetric indolocarbazoles possessing only one glycosidic linkage is represented by tjipanazole F2 (**30**). In 1996 Van Vranken and Gilbert described an efficient synthesis of **30** starting with the high-yielding intramolecular Mannich dimerisation of bis-indole **27**. This underwent a regioselective bromination within the indoline ring to give **28** (Scheme 5.5).[7] In a remarkably efficient reaction, **28** was glycosylated with D-xylose to provide a 1:1 mixture of diastereomers which

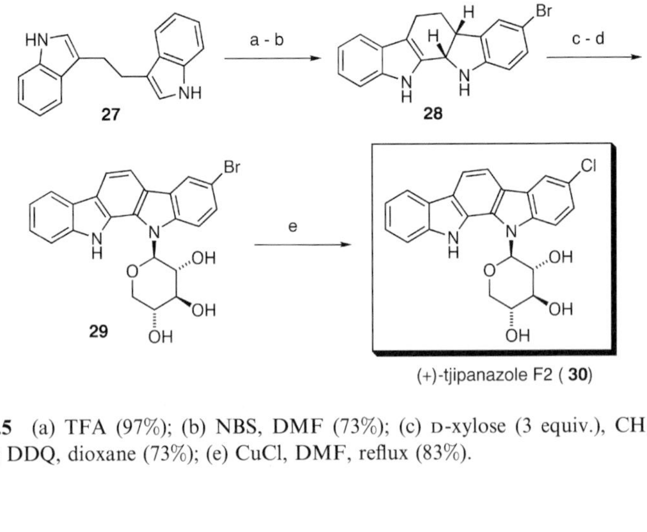

Scheme 5.5   (a) TFA (97%); (b) NBS, DMF (73%); (c) D-xylose (3 equiv.), CH$_3$OH, reflux (82%); (d) DDQ, dioxane (73%); (e) CuCl, DMF, reflux (83%).

converged to a single product (**29**) upon DDQ oxidation. Finally, halogen exchange with copper(I) chloride afforded tjipanazole F2 (**30**).

5.2.3   Mitomycin K and FR-900482

The first total synthesis of mitomycin K (**37**) was reported by Danishefsky and co-workers.[8] The parent ring system was established early by the photolysis of nitro arene **31** (Scheme 5.6). In this photochemical reaction, an initial internal redox results in oxidation of the benzylic alcohol to a ketone and reduction of the nitro group to a nitroso group. Intramolecular [4 + 2]-cycloaddition between the nitroso group and the proximal diene unit produces an intermediate oxazine (**32**) which undergoes a second internal redox reaction to give **33**. Oxidation of aminal **33** produced an unsaturated amide which on addition of (phenylthio)methyl azide gave triazole **34**. L-Selectride reduction of **34** created a hydroxyl group that was removed by Barton–McCombie deoxygenation. Aziridine **35** was formed following a photochemically assisted loss of nitrogen. Raney nickel desulfurisation and (trimethylsilyl)methyl lithium addition gave methyl aziridine **36**. Finally, oxidation of **36** and acid-promoted collapse of the β silanol provided mitomycin K (**37**).

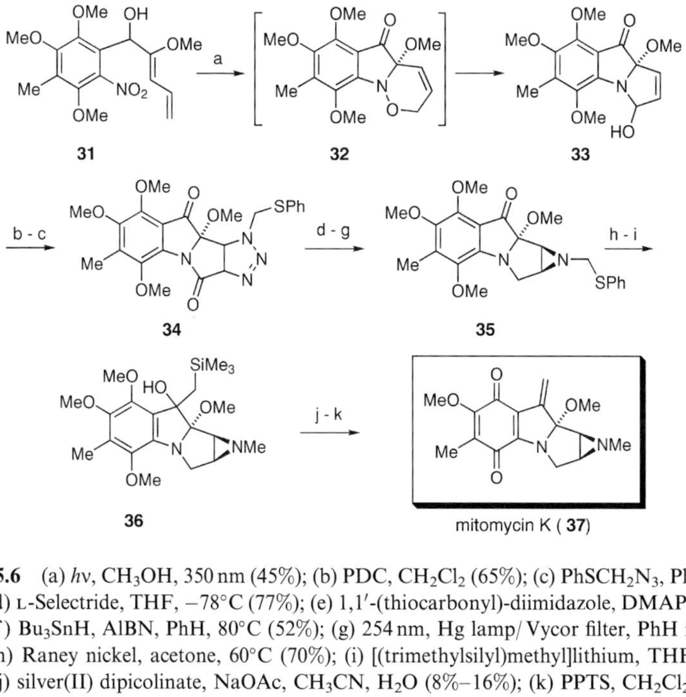

Scheme 5.6   (a) *hv*, CH$_3$OH, 350 nm (45%); (b) PDC, CH$_2$Cl$_2$ (65%); (c) PhSCH$_2$N$_3$, PhH, 80°C (90%); (d) L-Selectride, THF, −78°C (77%); (e) 1,1′-(thiocarbonyl)-diimidazole, DMAP, CH$_2$Cl$_2$ (65%); (f) Bu$_3$SnH, AlBN, PhH, 80°C (52%); (g) 254 nm, Hg lamp/ Vycor filter, PhH in quartz (48%); (h) Raney nickel, acetone, 60°C (70%); (i) [(trimethylsilyl)methyl]lithium, THF, −10°C (90%); (j) silver(II) dipicolinate, NaOAc, CH$_3$CN, H$_2$O (8%–16%); (k) PPTS, CH$_2$Cl$_2$ (81%).

In 1996 a second total synthesis of mitomycin K (**37**) was reported by Jimenez and Wang.[9, 10] The key transformation was the reaction of indole **38** with dimethylvinylsulfonium iodide under basic conditions (Scheme 5.7). This effected a two-carbon annulation and simultaneous introduction of a 2,3-epoxide.[11] The epoxide was not isolated but was reacted directly with sodium azide to give a *trans* azido alcohol which was subsequently converted to mesylate **39**. Oxodiperoxymolybdenum(pyridine) (hexamethylphosphoric triamide) (MoOPH) oxidation of indole **39** in methanol gave ketone **40**. Reduction of the azide group triggered aziridine formation which, after *N*-methylation, provided **41**. As with the Danishefsky synthesis of mitomycin K, Peterson methylenation and oxidation afforded mitomycin K (**37**).

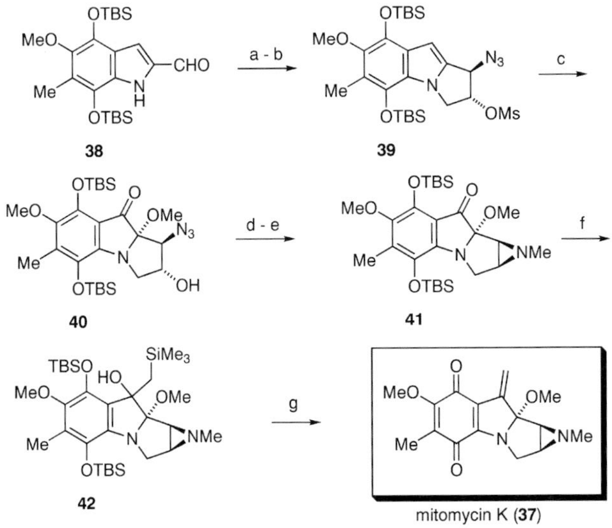

Scheme 5.7 (a) Dimethylvinylsulfonium iodide, NaH, THF, 0°C, then NaN$_3$, acetone (aq) (65%); (b) MsCl, Et$_3$N, CH$_2$Cl$_2$ (90%); (c) MoO$_5$·HMPA (2 equiv.), MeOH, 5–10°C, 3 d (71%); (d) PPh$_3$ (1.5 equiv.), Et$_3$N (2 equiv.), 10:1 THF:H$_2$O, RT, 10 h (70%); (e) CH$_3$OTf (30 equiv.), pyr (30 equiv.), CH$_2$Cl$_2$, 0°C (78%); (f) 1.0 M (CH$_3$)$_3$SiCH$_2$Li, THF, −10°C, 10 min (76%); (g) PCC (3 equiv.), CH$_2$Cl$_2$, 0°C, 10 min (63%).

The first total synthesis of FR-900482 was described by Fukuyama and Goto in 1992.[12] Danishefsky and Schkeryantz described a second total synthesis in 1995 which is partially outlined in Scheme 5.8.[13, 14] One of the highlights of Danishefsky's synthesis included the hetero-Diels–Alder reaction between nitroso arene **43** and diene **44** to afford dihydrobenzoxazine **45**. Following introduction of the aziridine group (**45** to **47**), primary alcohol **47** was manipulated to alkene **48** in anticipation of

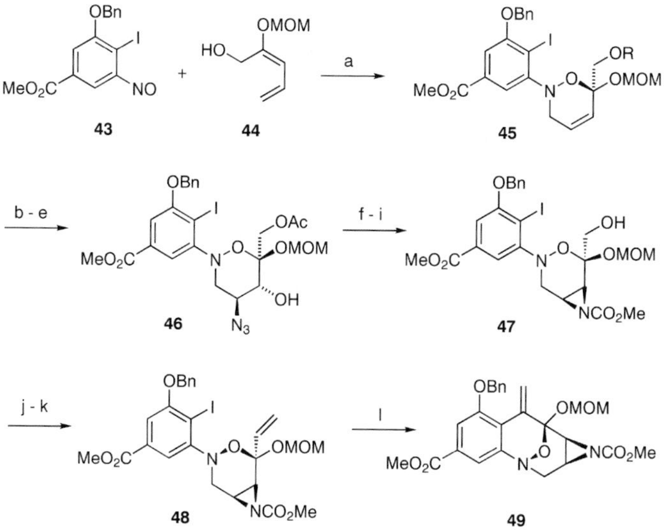

Scheme 5.8 (a) PhH, 80°C (80%); (b) Ac$_2$O, pyr, CH$_2$Cl$_2$ (92%); (c) OsO$_4$, Me$_3$NO·H$_2$O, CH$_2$Cl$_2$, PhH (71%); (d) triflic anhydride, pyr, CH$_2$Cl$_2$, 0°C (71%); (e) Bu$_4$NN$_3$, DMF, RT (74%); (f) triflic anhydride, pyr, CH$_2$Cl$_2$, 0°C; (g) Ph$_3$P, THF then NH$_4$OH; (h) methyl chloroformate (1.5 equiv.), pyr, CH$_2$Cl$_2$, 0°C (72%, three steps); (i) K$_2$CO$_3$, CH$_3$OH, RT (100%); (j) DMSO, (COCl)$_2$, CH$_2$Cl$_2$, −78°C, Et$_3$N; (k) Ph$_3$PCH$_3$Br, NaHMDS, THF, −20°C (75%, two steps); (l) (Ph$_3$P)$_4$Pd, Et$_3$N, CH$_3$CN, 90°C, 18 h (93%).

an intramolecular Heck arylation (**48** to **49**) which established a fully functionalised FR-900482 core structure.

5.2.4 Halenaquinone

In 1996 Shibasaki and co-workers described the asymmetric synthesis of halenaquinone (**58**) and halenaquinol (**59**).[15] A cascade reaction consisting of a Suzuki cross-coupling followed by an asymmetric Heck reaction (**50** + **51** → **52**) served to install the quaternary benzylic carbon and effect a four-carbon annulation (Scheme 5.9). Although this reaction proceeded in modest chemical yield (20%) the level of asymmetric induction was very good (85% ee). Desilylation and reduction followed by triflation afforded **53** which was coupled with acyl anion equivalent **54** to give **55** after ketalisation. A series of oxidations followed by deketalisation gave ketone **56**. Palladium(0) promoted cyclisation of **56** completed the pentacyclic ring structure common to halenaquinone and halenoquinol. Oxidation of **57** gave halenoquinone (**58**) which was reduced to halenoquinol (**59**).

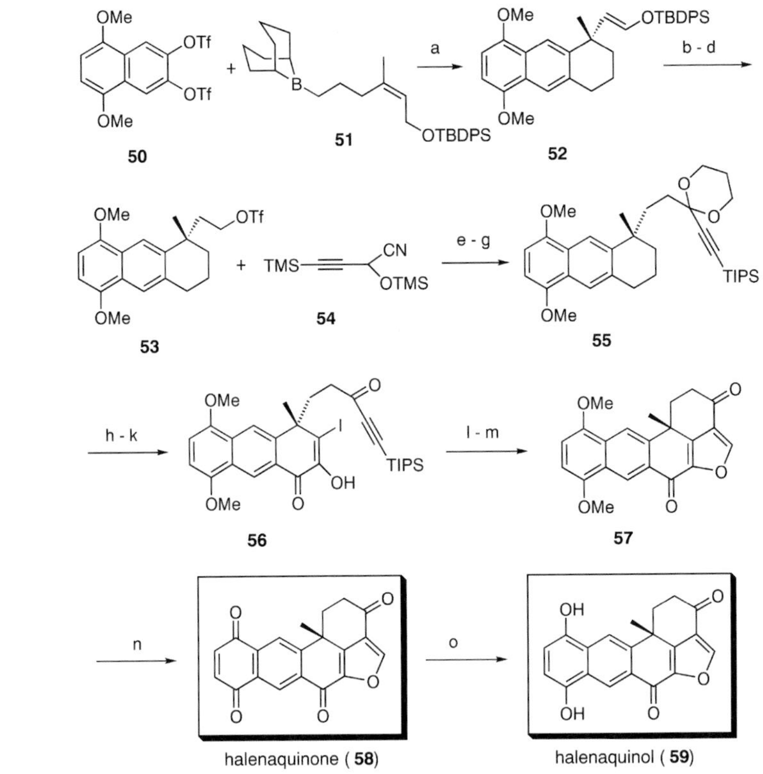

Scheme 5.9   (a) 20 mol% Pd(OAc)$_2$, 40 mol% (S)-BINAP, K$_2$CO$_3$, THF, 60°C, (20%) 85% ee; (b) n-Bu$_4$NF (2 equiv.), AcOH (3 equiv.), THF, 0°C to RT; (c) NaBH$_4$ (5 equiv.), CH$_3$OH, 0°C to RT (93%, two steps); (d) Tf$_2$O (1.2 equiv.), pyr, CH$_2$Cl$_2$, −78°C to RT; (e) LDA/**54** 1.5 equiv.), THF, −78°C, then 2% NaF (68%, two steps); (f) HO(CH$_2$)$_3$OH (10 equiv.), TsOH·H$_2$O, PhH, reflux (98%); (g) n-BuLi (2 equiv.), THF, −78 to 50°C, then TIPSCl (2 equiv.), −78°C to RT (98%); (h) DDQ (3 equiv.), CH$_2$Cl$_2$, H$_2$O, RT (6%); (i) O$_2$ (1 atm), t-BuOK (5 equiv.), t-BuOH, 35°C (79%); (j) NaI (10 equiv.), CuSO$_4$·5H$_2$O (10 equiv.), CH$_3$OH, H$_2$O, RT (97%); (k) TsOH·H$_2$O, acetone, H$_2$O, 60°C (98%); (l) Pd(dba)$_3$, CHCl$_3$ (0.28 equiv.), K$_2$CO$_3$, DMF, RT (72%); (m) n-Bu$_4$NF (16 equiv.), AcOH (24 equiv.), THF, CH$_3$CN, 60°C (83%); (n) CAN (5 equiv.), CH$_3$OH, H$_2$O, 0°C (45%); (o) Na$_2$S$_2$O$_4$ (20 equiv.), acetone, H$_2$O, 0°C (100%).

5.3   Six-membered heterocycles

5.3.1   Camptothecin

In the early 1990s a resurgent interest in the development of an efficient asymmetric synthesis of (+)-camptothecin (**67**) occurred. This was triggered by the preclinical development of various camptothecin derivatives. Comins was among the first researchers to describe a concise (10-step) synthesis of (+)-camptothecin (Scheme 5.10).[16] His synthetic sequence

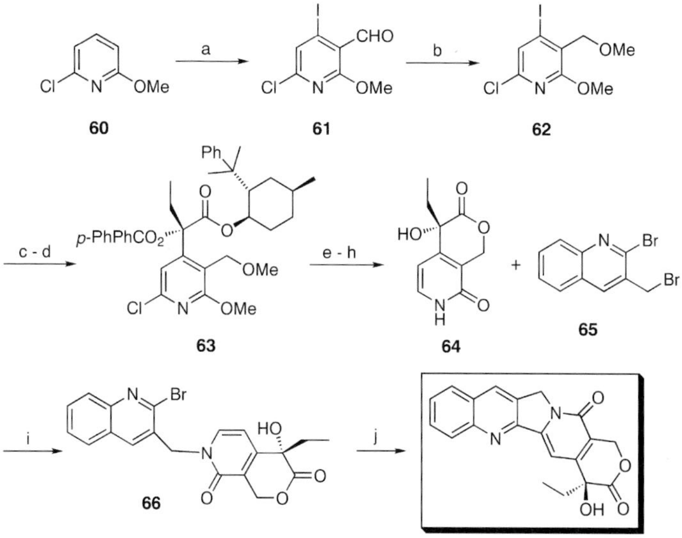

Scheme 5.10 (a) i. *t*-BuLi, THF, −78°C, 1 h; ii. (CH$_3$)$_2$NCH$_2$CH$_2$N(CH$_3$)CHO; iii. *n*-BuLi, −23°C, 2 h; iv. I$_2$ (78%); (b) CH$_3$OH, EtSiH, TFA (98%); (c) *n*-BuLi, THF, −78°C then α-keto-8-phenylmenthyl butyrate; (d) *p*-PhPhCOCl (60% from **62**); (e) 2 N NaOH, EtOH (76%); (f) TMSCl, NaI, CH$_3$CN, Dabco, reflux; (g) 6 N HCl, 100°C (77%, two steps); (h) H$_2$, Pd(C); (i) *t*-BuOK, DME, reflux (87%); (j) Pd(OAc)$_2$, Bu$_4$N$^+$Br$^-$, KOAc, DMF, 90°C, 3 h (59%).

began with a two-step, one-pot, reaction. Commercially available 2-chloro-6-methoxy pyridine (**60**) was subjected to alkoxy-directed lithiation and the intermediate aryl lithium was trapped with a formamide. Immediate addition of an equivalent of *n*-BuLi generated an anion which upon quenching with iodine gave aldehyde **61**. Reduction of **61** with Et$_3$SiH/TFA in methanol afforded methyl ether **62**. Lithium halogen exchange followed by the addition of α-keto-8-phenylmenthyl butyrate produced an intermediate alkoxide that reacted with 4-phenylbenzoyl chloride to give ester **63** in 87% diastereomeric excess. Recrystallisation from petroleum ether afforded diastereomerically pure ester **63**. Exhaustive saponification provided an intermediate hydroxy acid which without purification was subjected to TMSCl/NaI/Dabco and acid-promoted lactonisation. Subsequent hydrogenolysis effected removal of the C(6) chloro group to give lactone **64**. Pyridone **64** coupled to quinoline dibromide **65** by simple bromide displacement to give **66**, the substrate for an ensuing intramolecular Heck reaction, which afforded (+)-camptothecin **67**. Following this initial report Comins reported an improved synthesis of pyridones **64** and **66** as well as a radical-mediated cyclisation of **66**.[17]

Curran has used an elegant radical cascade reaction to assemble camptothecin (Scheme 5.11).[18] In a second generation synthesis, the optically enriched pyridone **73** was assembled starting from allyl ether **68** (prepared from 2-bromo-6-methoxypyridine in a reaction sequence analogous to Comin's procedure for the preparation of methyl ether **62**).[19] Allyl ether **68** was subjected to an intramolecular Heck reaction to give enol ether **69**. Sharpless asymmetric dihydroxylation of enol ether **69** followed by oxidation of the intermediate α-hydroxy lactol furnished **70** in 94% enantiomeric excess. Following exchange of the TMS group for an iodo group, demethylation of **71** gave pyridone **72** which was *N*-propargylated to afford **73**. Heating a solution of **73**, phenyl isonitrile **74** and hexamethyldistannane in benzene triggered a radical cascade reaction to give (+)-camptothecin in 63% yield.

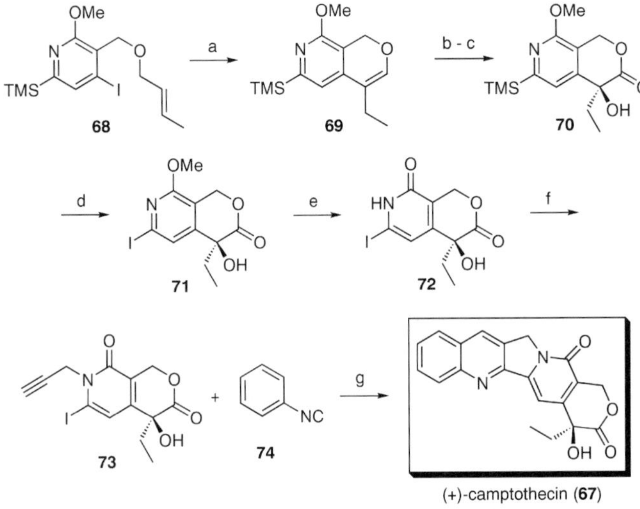

Scheme 5.11   (a) Pd(OAc)$_2$, K$_2$CO$_3$, Bu$_4$NBr (69%); (b) OsO$_4$, (DHQD)$_2$-pyr; (c) I$_2$, CaCO$_3$ (85%, two steps, 94% ee); (d) ICl (47%); (e) aq. HI (72%); (f) 3-bromopropyne, NaH, LiBr, DMF, DME (88%); (g) Me$_3$SnSnMe$_3$, PhH, 70°C, *hv* (63%).

To date, Ciufolini and Roschangar have developed the most concise synthesis of (+)-camptothecin (**67**) starting from chloroquinoline **75**, which underwent a palladium-mediated carbomethoxylation to **76**.[20] Benzylic bromination of **76** followed by methanolysis gave methyl ether **77** (Scheme 5.12). Addition of the anion derived from dimethyl methylphosphonate afforded keto phosphonate **78** which was condensed with aldehyde **79** (greater than 98% ee, derived from enzymatic desymmetrisation of the corresponding dimethylmalonate ester). Michael addition of the potassium enolate derived from **81** to enone **80** provided amide **82**.

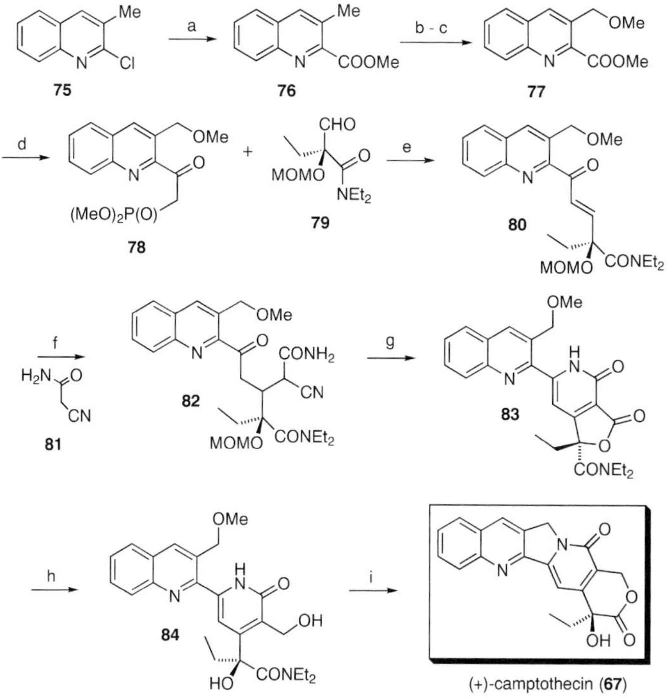

Scheme 5.12 (a) [Pd(dppp)$_2$Cl$_2$] 5 mol%, NaOAc, CH$_3$OH, DMF, 105 atm CO, 140°C (98%); (b) NBS, CCl$_4$, (BzO)$_2$, hv, (c) CH$_3$OH, 5% H$_2$SO$_4$, reflux (55%, two steps); (d) (CH$_3$O)$_2$P(O)-CH$_2$Li (2 equiv.), THF, −78°C (100%); (e) t-BuOK, DME, 50°C, 12 h (80%); (f) KOBu-t, DMSO, 81; (g) 5% SeO$_2$·silica gel (0.2 equiv.), t-BuOOH, AcOH, 110°C, then 10% H$_2$SO$_4$ (aq) (68%, two steps); (h) NaBH$_4$ (10 equiv.), CeCl$_3$ (2.5 equiv.), EtOH, 0°C to 45°C; (i) 60% H$_2$SO$_4$ in EtOH, 115°C (94%, two steps).

Oxidation of **82** followed by acidification gave pyridone **83**. Finally, Luche reduction of **83** afforded diol **84** which upon heating in ethanolic H$_2$SO$_4$ gave synthetic (+)-camptothecin (**67**).

5.3.2 Pyridoacridines

The pyridoacridine diplamine (**94**) has been synthesised by Sczepankie-wicz and Heathcock starting from quinolone **85**, prepared by a Knorr quinoline synthesis from 4-methoxyphenol.[21] Quinoline **85** was reacted with POCl$_3$ to afford chloroquinoline **86** which, upon oxidation, gave quinone **87** (Scheme 5.13). Unexpectedly, when **87** was treated with POCl$_3$ furan **88** was isolated rather than the expected primary chloride. Fortunately, the newly installed chloride later facilitated the introduction

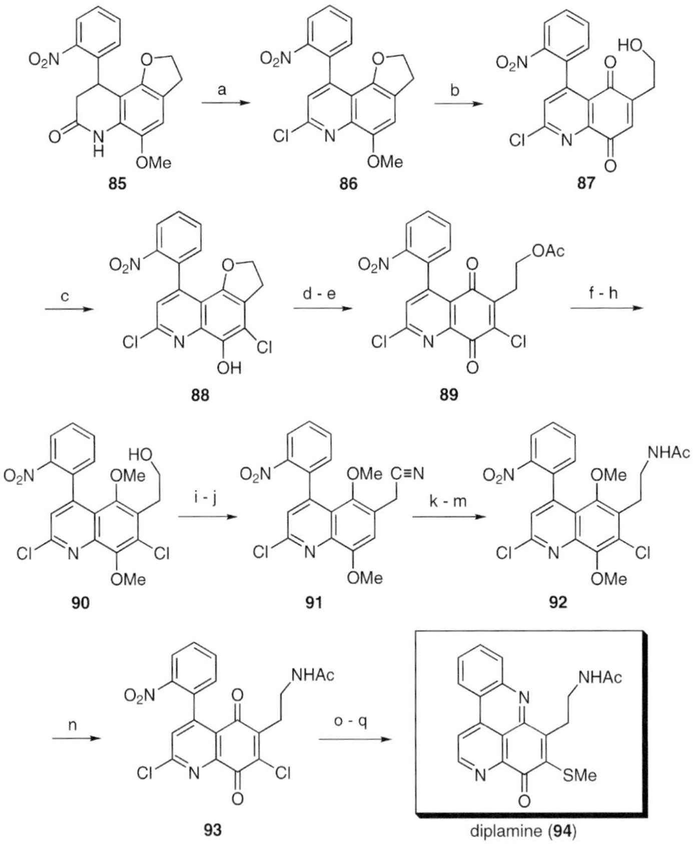

Scheme 5.13   (a) POCl$_3$, 50°C (94%); (b) HNO$_3$, AcOH (83%); (c) POCl$_3$, CHCl$_3$ (80%); (d) HNO$_3$, AcOH (79%); (e) AcCl, pyr, CH$_2$Cl$_2$ (96%); (f) 47% HI, acetone; (g) CH$_2$N$_2$, Et$_2$O, EtOAc; (h) K$_2$CO$_3$, THF/MeOH/H$_2$O (82%, three steps); (i) SO$_3 \cdot$pyr, DMSO (81%); (j) H$_2$NSO$_3$H, EtOH/pyr. (83%); (k) MeOBH$_2$, THF; (l) 1 M HCl; (m) Ac$_2$O (75%, three steps); (n) CAN, aq. CH$_3$CN (73%); (o) NaSCH$_3$, CHCl$_3$, Et$_3$N (51%); (p) Zn, 50% AcOH; (q) aq. CAN (37%, two steps).

of the methylthio group of diplamine (**94**). Oxidation of **88** to the corresponding quinone followed by acetylation gave quinone **89** in which the primary alcohol was protected. Selective quinone reduction, methylation and deacetylation provided **90**. Primary alcohol **90** was then oxidised to the aldehyde and this condensed with hydroxylamine-*O*-sulfonic acid. Following loss of sulfuric acid, nitrile **91** was obtained. Reduction and acetylation afforded acetamide **92** which underwent oxidation to quinone **93**. At this point the chloride was substituted for a methyl sulfide. Zinc reduction and CAN oxidation finally provided an amino quinone which spontaneously cyclised to diplamine (**94**).

Ciufolini and co-workers have developed a unified synthetic strategy for assembling several classes of pyridoacrididine alkaloids including the dercitins, kuanoniamine (**101**), shermilamine B (**102**) and cystodytin J (**103**).[22] Hetero Diels–Alder reaction of ethyl vinyl ether and readily available cyclohexadienone **95** provided pyran **96** (Scheme 5.14). Condensation of **96** (a masked 1,5-dicarbonyl system) with hydroxylamine hydrochloride (which completed a modified Knoevenagel–Stobbe pyridine synthesis) and ozonolysis gave ketone **97**. α-Bromination afforded **98** which served as a common intermediate for kuanoniamine (**101**) and shermilamine B (**102**). En route to kuanoniamine (**101**), bromo ketone **98** was reacted with thiourea to provide aminothiazole **99**. Deamination of **99** was followed by introduction of the acetamide sidechain, a Gabriel reaction and amine acetylation. Finally, azide **100** was subject to a thermophotolysis reaction under Meth–Cohn's conditions to obtain kuanoniamine (**101**).

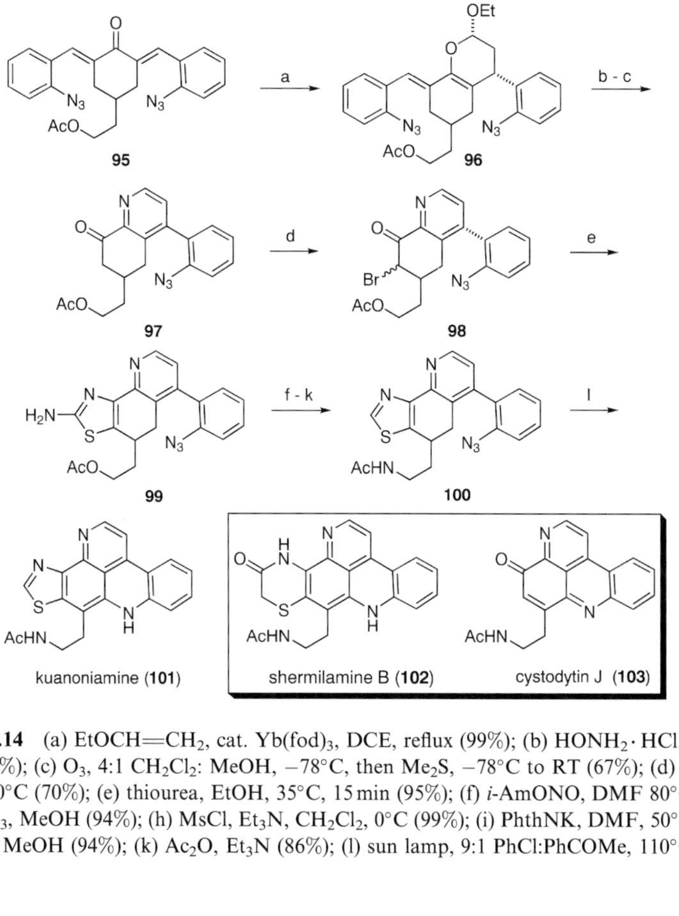

Scheme 5.14   (a) EtOCH=CH$_2$, cat. Yb(fod)$_3$, DCE, reflux (99%); (b) HONH$_2$·HCl, MeCN, reflux (62%); (c) O$_3$, 4:1 CH$_2$Cl$_2$: MeOH, −78°C, then Me$_2$S, −78°C to RT (67%); (d) PyHBr$_3$, AcOH, 50°C (70%); (e) thiourea, EtOH, 35°C, 15 min (95%); (f) *i*-AmONO, DMF 80°C (80%); (g) K$_2$CO$_3$, MeOH (94%); (h) MsCl, Et$_3$N, CH$_2$Cl$_2$, 0°C (99%); (i) PhthNK, DMF, 50°C (84%); (j) N$_2$H$_4$, MeOH (94%); (k) Ac$_2$O, Et$_3$N (86%); (l) sun lamp, 9:1 PhCl:PhCOMe, 110°C (62%).

5.3.3   Tropoloisoquinolines

The tropoloisoquinolines are represented by grandirubrine (**114**), imerubrine (**115**), isoimerubruine and pareirubruines A and B. Boger and Takahashi have described the total synthesis of grandirubrine and imerubrine starting from isoquinoline **104**.[23] Peterson olefination of **104** with **105** gave ketene thioacetal **106** which on mercury(II)-mediated methanolysis afforded methyl ester **107**. Successive treatment of isoquinoline **107** with p-TsCl and KCN in a two-phase system (CH_2Cl_2–H_2O) then provided cyano-1,2-dihydroisoquinoline **108**.[24] After Dieckmann-type cyclisation and acid-catalysed decarboxylation, ketone **109** was obtained, and this converted to α-pyrone **111** via a five-step reaction sequence. The key Diels–Alder reaction of α-pyrone **111** with cyclopropenone ketal

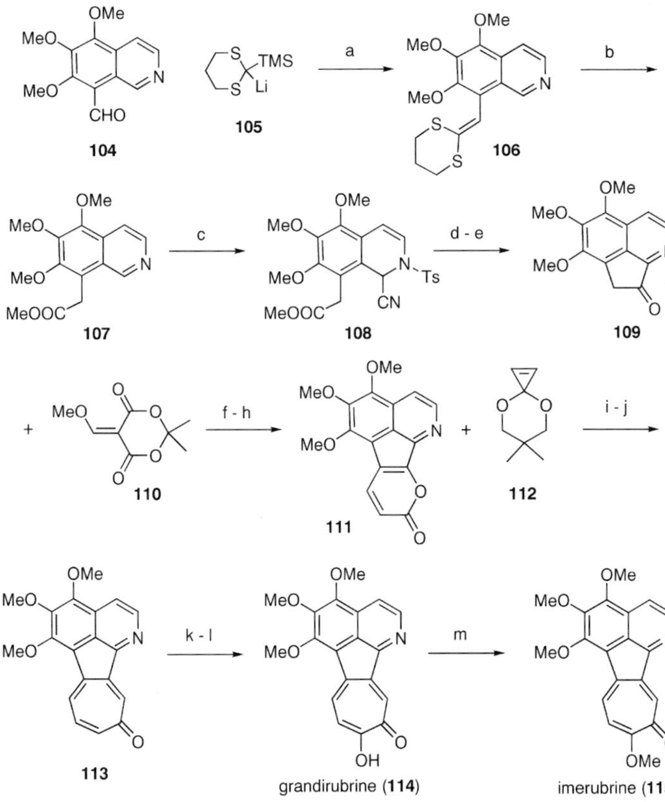

Scheme 5.15   (a) **104** + **105** (77%); (b) $HgCl_2$, CH_3OH, H_2O; (c) TsCl·KCN (91%, two steps); (d) t-BuOK (98%); (e) HCl (aq) then KOH (aq) (89%); (f) t-BuOK, THF, −78°C, **110**; (g) TFA (neat); LiOH (aq) then HCl (aq); (h) Ac_2O (52%); (i) 13 kbar, $CHCl_3$, pyr; (j) HCl, EtOAc (10%–35%, two steps); (k) NH_2NH_2 (78%); (l) KOH, CH_3OH (70%); (m) TMSCHN₂ (76%).

112 was accomplished under high-pressure reaction conditions. Subsequent loss of carbon dioxide by retro-Diels–Alder reaction, and norcaradiene rearrangement, led to an intermediate cycloheptatrienone ketal which was hydrolysed to tropone **112**. Treatment of tropone **112** with hydrazine provided an amine which on saponification afforded grandirubrine (**114**). Treatment of **114** with TMSCHN$_2$ provided imerubrine (**115**).

References

1. D.L. Boger, K. Machiya, D. Hertzog, P. Kitos and O. Holmes, 1993, *J. Am. Chem. Soc.*, 115, 9025-9036.
2. D.L. Boger, J.A. McKie, T. Nishi and T. Ogiku, 1997, *J. Am. Chem. Soc.*, 119, 311-325.
3. J.T. Link, S. Ragharan and S.J. Danishevsky, 1995, *J. Am. Chem. Soc.*, 117, 552-553.
4. M. Galant, J.T. Link and S.J. Danishevsky, 1993, *J. Org. Chem.*, 58, 343-349.
5. J.L. Wood, B.M. Stolz, H.J. Dietnich, D.A. Pflum and D.T. Petsch, 1997, *J. Am. Chem. Soc.*, 119, 9641-9651.
6. J.L. Wood, B.M. Stolz, S.N. Goodman and K. Onwoeme, 1997, *J. Am. Chem. Soc.*, 119, 9652-9661.
7. E.J. Gilbert and D.L. Van Vranken, 1996, *J. Am. Chem. Soc.*, 118, 5500-5501.
8. J.W. Benbow, K.F. McClune and S.J. Danishevsky, 1993, *J. Am. Chem. Soc.*, 115, 12305-12314
9. Z.Wang and L.S. Jimenez, 1996, *Tetrahedron Lett.*, 37, 6049-6052.
10. Z. Wang and L.S. Jimenez, 1996, *J. Org. Chem.*, 61, 816-818.
11. Z. Wang and L.S. Jimenez, 1994, *J. Am. Chem. Soc.*, 116, 4977-4978.
12. T. Fukuyama, L. Xu and S. Goro, 1992, *J. Am. Chem. Soc.*, 114, 383-385.
13. K.F. McClure and S.J. Danishefsky, 1993, *J. Am. Chem. Soc.*, 115, 6094-6100.
14. J.M. Schkeryantz and S.J. Danishefsky, 1995, *J. Am. Chem. Soc.*, 117, 4722-4723.
15. A. Kojima, T. Takemoto, M. Sodeoka and M. Shibasaki, 1996, *J. Org. Chem.*, 61, 4876-4877.
16. D.L. Comins, M.F. Baevsky and H. Hong, 1992, *J. Am. Chem. Soc.*, 114, 10971-10972.
17. D.L. Comins, H. Hong and G. Jianhua, 1994, *Tetrahedron Lett.*, 35, 5331-5334.
18. D.P. Curran and J. Liu, 1992, *J. Am. Chem. Soc.*, 114, 5863-5864.
19. D.P. Curran, S.B. Ko and H. Josein, 1995, *Angew. Chem. Int., Ed. Engl.*, 34, 2683-2684.
20. M.A. Ciufolini and F. Roschangar, 1996, *Angew. Chem., Int. Ed. Engl.*, 35, 1692-1694.
21. B.G. Szczepankiewicz and C.H. Heathcock, 1994, *J. Org. Chem.*, 59, 3521-3523.
22. M.A. Ciufolini, Y.-C. Shen and M.J. Bishop, 1995, *J. Am. Chem. Soc.*, 117, 12460-12469.
23. D.L. Boger and K. Takahashi, 1995, *J. Am. Chem. Soc.*, 117, 12452-12459.
24. D.L. Boger, C.E. Brotherton, J.S. Panek and D. Yohannes, 1984, *J. Org. Chem.*, 49, 4056-4058.

6 Carboaromatic compounds

R.S. Coleman and M.L. Madaras

6.1 Introduction

This chapter presents literature highlights on the synthesis of carbocyclic aromatic natural products.[1] The literature is covered from mid-1992 through late-1998. The coverage is intended to be neither comprehensive nor restricted to entirely carbocyclic systems. Rather, we have attempted to present the most notable achievements and interesting structures where the focus is on the synthesis of the carboaromatic systems.

6.2 Natural products with axial chirality

6.2.1 Korupensamines and michellamines

Michellamines A and B (1) are dimeric atropisomers isolated from *Ancistrocladus korupensis*[2,3] that show antiviral activity against human immunodeficiency virus strains HIV-1 and HIV-2, including some resistant variants. A third atropisomer, michellamine C was also isolated, but was later found be an artifact formed from overly harsh isolation conditions.[4]

The first synthesis of michellamines was achieved by Bringmann et al.,[5] following the biosynthetic pathway that involves the dimerisation of korupensamines (2). These are monomeric 'subunits' of michellamines that were also isolated from *A. korupensis*. They do not exhibit anti-HIV activity, but instead are antimalarial.[6] Korupensamine A and B were synthesised en route to michellamines by a Stille coupling reaction between bromide 6 and stannane 7 (Scheme 6.1),[7] and were homo- or cross-coupled under oxidative conditions using Ag_2O to afford the corresponding michellamines. Hoye[8] and Dawson[9] have reported similar syntheses that follow the same bond-forming strategy, the central axis being made by oxidative dimerisation using Ag_2O[8] or by a Suzuki coupling.[9] More recently, Bringmann et al.,[10] reported a convergent total synthesis of the michellamines that first constructed the inner biaryl axis by an oxidative dimerisation with Ag_2O and subsequently formed the other two stereogenic axes via a double Suzuki coupling.

Gossypol (8) was first isolated from cotton seeds over 100 years ago,[11] and was later shown to have an antispermatogenic effect.[12] In addition,

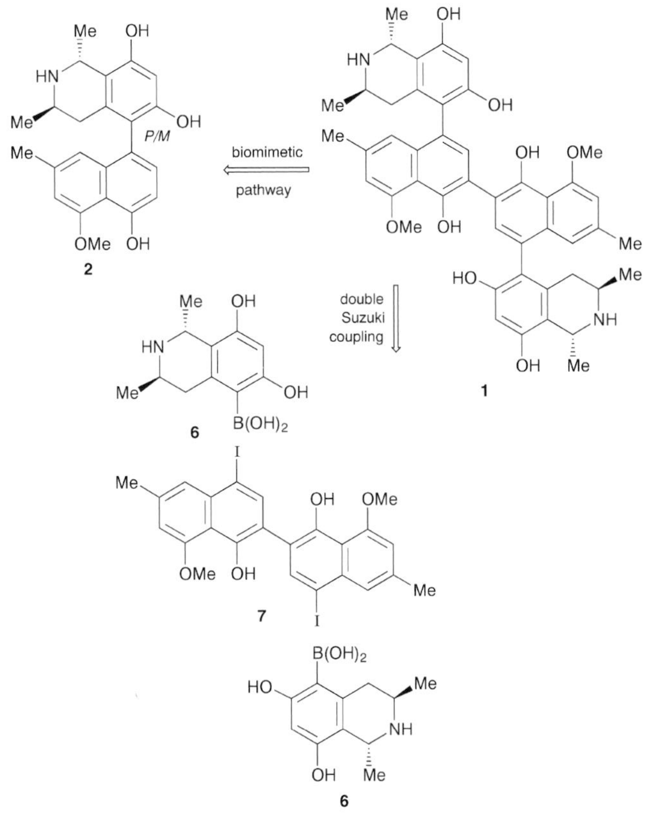

2 ⇌ biomimetic pathway

double Suzuki coupling

6

7

6

1

(R)-gossypol is effective as an anticancer[13] and anti-HIV-1[14] agent. The (S)-enantiomer, by comparison, exhibits activity against herpes simplex[15] and influenza viruses.[16]

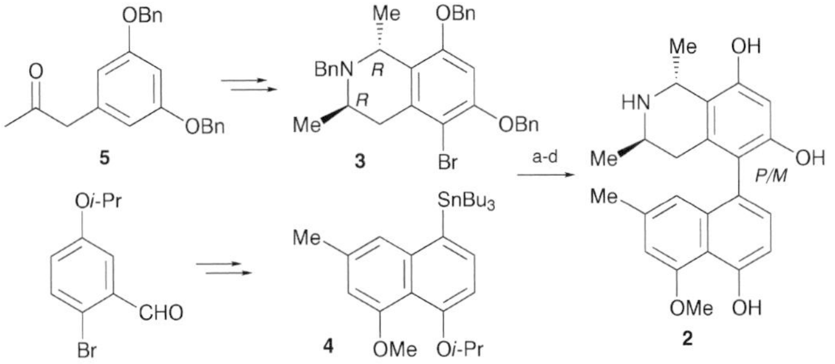

Scheme 6.1 Reaction conditions: (a) PdCl$_2$(PPh$_3$)$_2$, LiCl, CuBr, DMF, 40 h, 135°C; (b) BCl$_3$, CH$_2$Cl$_2$, 10 min, 25°C; (c) H$_2$, Pd/C, MeOH; (d) separation of atropisomers on amino-bonded phase column.

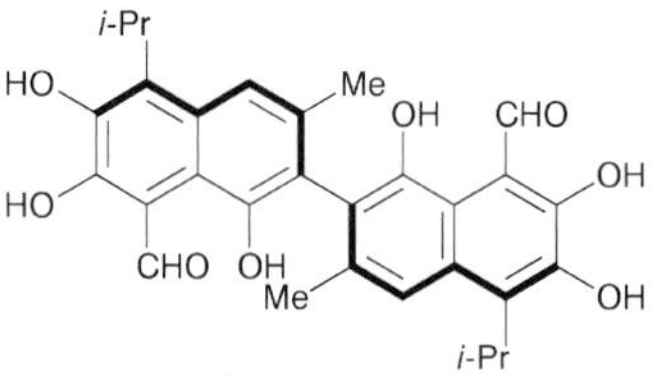

8 (S)-gossypol

Meyers and Willemsen[17] reported the synthesis of (S)-(+)-gossypol via an asymmetric Ullmann reaction using a chiral oxazoline as a naphthyl substituent. The sequence was initiated by transforming 2,3,4-trimethoxy-benzene into the oxazoline 14. This oxazoline underwent a series of transformations including 2-methoxy substitution with an isopropyl group, formylation/reduction, Stobbe condensation and cyclisation to build the naphthoic acid 15 (Scheme 6.2). Conversion of 15 to the chiral oxazoline was achieved with (S)-(+)-tert-leucinol, which was brominated to the bromonaphthalene 16. This compound was subjected to Ullmann coupling conditions. The diastereomeric ratio of the resulting dimer 12 was 17:1, and after a series of transformations, (S)-(+)-gossypol was obtained in 9% overall yield.

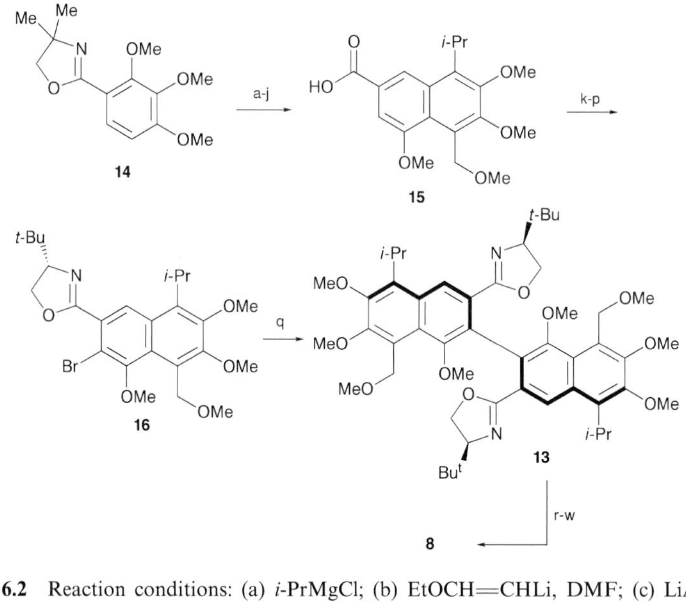

Scheme 6.2   Reaction conditions: (a) i-PrMgCl; (b) EtOCH=CHLi, DMF; (c) LiAlH₄; (d) NaH, MeI; (e) MeOSO₂CF₃; (f) NaBH₄; (g) H₃O⁺; (h) NaH, dimethyl succinate; (i) AcOH, Ac₂O; (j) KOH; (k) K₂CO₃, Me₂SO₄; (l) KOH; (m) (COCl)₂; (n) (S)-tert-leucinol; (o) SOCl₂; (p) Br₂, AcOH; (q) Cu⁽⁰⁾, DMF reflux; (r)TFA; (s) Ac₂O; (t) LiAlH₄; (u) H₂, Pd/C; (v) BBr₃; (w) DMSO, (COCl)₂.

Bioxanthracene (−)-ES-242-4 (9) was isolated from the culture of a *Verticillium* sp. and was shown to be an antagonist for the *N*-methyl-

9 ES-242-4

D-aspartate receptor,[18] making it of potential therapeutic interest for the treatment of neurodegenerative diseases. The synthesis of (−)-ES-242-4 was accomplished by Tatsuta and co-workers[19] using an oxidative dimerisation of a naphthopyran that was derived from an α,β-unsaturated lactone and methyl 2,4-dimethoxy-6-methylbenzoate through tandem Michael–Dieckmann reactions.

In 1971, Büchi and co-workers[20] isolated the fungal metabolites kotanin (**14**) and demethylkotanin (**15**) from *Aspergillus clavatus*. Several years later, Cutler and co-workers[21] isolated another optically active 8,8′-bicoumarin from *Aspergillus niger* that they named orlandin (**16**). Wheat

14	Kotanin	$R^1 = R^2 = Me$
15	Demethylkotanin	$R^1 = H, R^2 = Me$
16	Orlandin	$R^1 = R^2 = H$

coleoptile and chick bioassays indicated that the biological activity of orlandin in plants, and kotanin in chicks, is closely related to the 7,7′-hydroxyl groups. The absolute configuration of (+)-kotanin has been determined by asymmetric synthesis and comparison of the CD Cotton effects with a stereochemically known compound.[22]

6.2.2 *Perylenequinones*

The family of antibiotic natural products known as the 4,9-dihydroxy-3,10-perylenequinones was first established in 1957 with the isolation of cercosporin (**17**).[23] Other members have been added to this family, including phleichrome (**18**) from *Cladosporium phlei*[24] and the calphostins (**19–23**) from *Cladosporium cladosporioides*.[25] The calphostins have

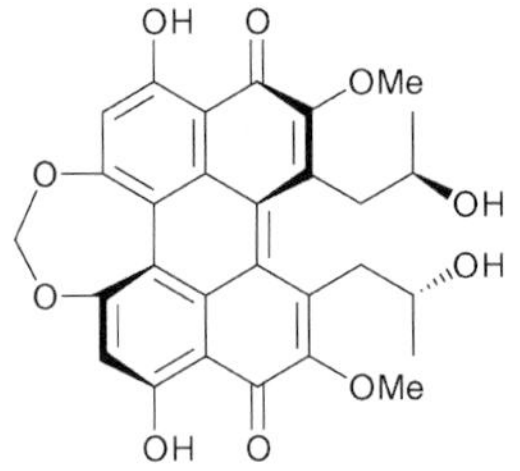

17 Cercosporin

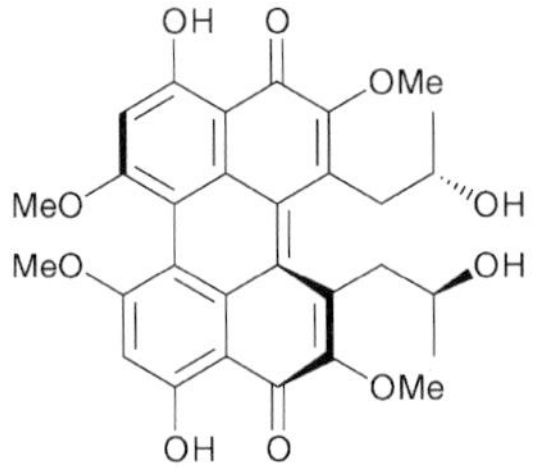

18 Phleichrome

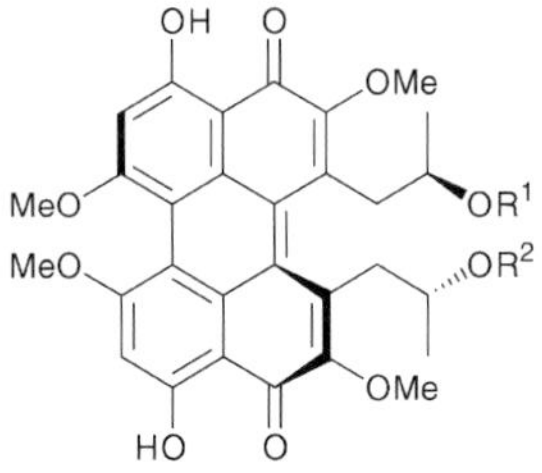

19 Calphostin A $R^1 = R^2 = COPh$
20 Calphostin B $R^1 = COPh$, $R^2 = H$
21 Calphostin C $R^1 = COPh$, $R^2 = CO_2C_6H_4\text{-}pC$
22 Calphostin D $R^1 = R^2 = H$
23 Calphostin I $R^1 = R^2 = CO_2C_6H_4\text{-}pOH$

been shown to inhibit protein kinase C (PKC)[26] by binding in the regulatory site of the enzyme. For some cancers, PKC overexpression has been correlated with the extent of malignancy. Inhibiting PKC enzymes could offer a method for controlling cancer, making calphostins potential candidates for the treatment of this disease. The inhibition of PKC by calphostin C is light-dependent and occurs in the presence or absence of oxygen and radical scavengers.[27] The pentacyclic aromatic dione core of the perylenequinones generates singlet oxygen upon irradiation in the presence of oxygen. Perylenequinones are among the most efficient $^1O_2$ generators in nature, with quantum yields as high as 81%.[28] Coleman and Grant[29, 30] reported a 13-step synthesis for phleichrome and a 14-step synthesis for calphostin A; their syntheses were both enantioselective and atropdiastereoselective.

These syntheses started with the *de novo* construction of a regio-specifically oxygenated and selectively protected naphthalene subunit using a novel Diels–Alder cycloaddition[31] between Brassard's diene **24** and cyclohexadienone **25**, which was followed by a carefully controlled series of elimination reactions to effect aromatisation to afford naph-thalene **26** (Scheme 6.3). Protection of the remaining free phenol and

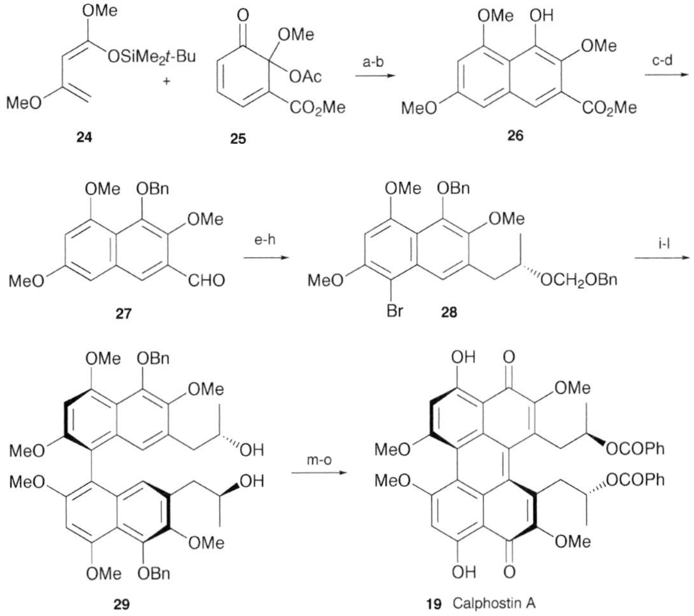

Scheme 6.3 Reaction conditions: (a) 23°C, 2–3 days; (b) SiO$_2$; (c) NaH, BnBr; (d) *i*-Bu$_2$AlH, toluene, −78°C; (e) (*S*)-Me-CHLi-OCH$_2$OBn, THF, −78°C; (f) (imid)$_2$C=S, THF, reflux; (g) *n*-Bu$_3$SnH, toluene, reflux; (h) NBS, THF, 0°C; (i) *n*-BuLi, THF, −78°C; (j) CuCN·TMEDA, −78°C; (k) O$_2$, −78°C; (l) AcCl, MeOH/THF, 23°C; (m) *i*-PrO$_2$CN=NCO$_2$*i*-Pr, Ph$_3$P, PhCO$_2$H, toluene, THF; (n) MnO$_2$, Et$_2$O, 23°C; (o) Br$_2$/acetone, CHCl$_3$, 23°C.

reduction of the carboxylate to the corresponding aldehyde afforded **27**. The stereogenic side-chains were attached by the addition of a chiral, enantiomerically pure lithium reagent to the aldehyde, and the naph-thalene subunits were connected via a highly atropdiastereoselective, low-temperature Cu(I)-promoted biaryl coupling.[32] In the case of calphostin, use of the (*S*)-configuration at the stereogenic centre in the naphthalene side-chains (which is opposite to that in the natural product) induced the formation of the desired (*S*$_a$)-configuration of the biaryl linkage in **29**.

The configuration of the side-chains was subsequently inverted through a low-temperature variant of the Mitsunobu esterification reaction[33] which occurred concomitantly with introduction of the ester appendages of the calphostin A. Oxidative cyclisation to the perylenequinone was followed by a facile, regioselective demethylation of the methyl ether proximal to the quinone carbonyl group to afford calphostin A.

6.2.3   Pigments

The fresh water protozoan cilliate *Stentor coeruleus* is blue-green and exhibits a characteristic red fluorescence. The compound responsible for this coloration is stentorin (**30**), a pigment localised in vesicles close to the cell membrane that is believed to function as the primary photoreceptor in the light-induced responses of this organism.[34] Hypericin (**31**), a similar pigment, exhibits highly specific activity against HIV and related retro-viruses.[35] It has been proposed that the virucidal activity of hypericin is due to an increased acidity in the photoexcited state.[36] More recently, the structures of blepharismin 1–5 (**32–36**), five new photoreceptor molecules mediating the photobehaviour of the unicellular organism *Blepharisma japonicum*, were determined spectroscopically.[37]

30   Stentorin

31   Hypericin

Blepharismins

32   BL-1 $R^1 = R^2 = Et, R^3 = H$
33   BL-2 $R^1 = Et, R^2 = i\text{-}Pr, R^3 = H$
34   BL-3 $R^1 = R^2 = i\text{-}Pr, R^3 = H$
35   BL-4 $R^1 = Et, R^2 = i\text{-}Pr, R^3 = Me$
36   BL-5 $R^1 = R^2 = i\text{-}Pr, R^3 = Me$

The chemical structure of stentorin remained unclear until recently, when Cameron and Riches[38, 39] achieved its synthesis and compared synthetic material with an authentic sample. The synthesis involved the controlled oxidative couplings of anthrone **38**, available via two consecutive Diels–Alder reactions of 2,6-dichlorobenzoquinone. The selective placement of the isopropyl groups in the helianthrone **40** was achieved by a selective oxidation of the resorcinol rings of bianthrone **39** with oxygen in the absence of light (Scheme 6.4). Further intramolecular

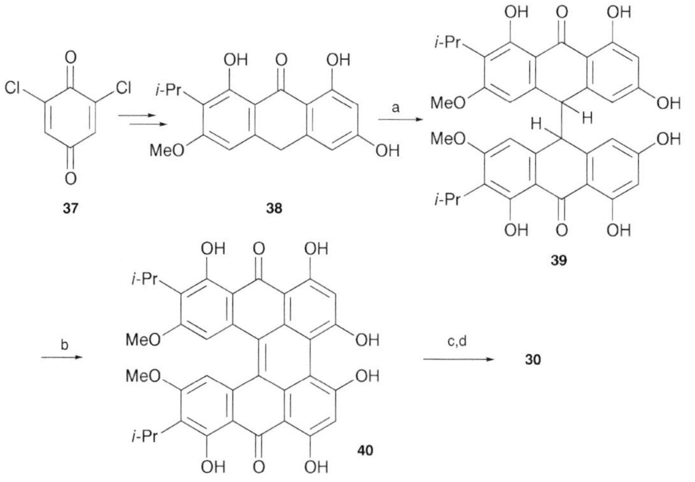

Scheme 6.4 Reaction conditions: (a) K$_3$FeCN$_6$, EtOH/pH 4.5 buffer; (b) EtOH/NH$_4$OH, Δ, O$_2$, dark; (c) $h\nu$ (Hg lamp), DMSO, O$_2$; (d) HI, AcOH.

coupling of helianthrone **40** occurred readily upon irradiation of its dimethylsulfoxide solution in oxygen with a mercury vapour lamp to give a fluorescent naphthodianthrone which could be demethylated to afford a compound identical to natural stentorin.

6.3 Dibenzocyclooctadiene lignans

The fruits of *Schizandra chinensis* Bail. (Schizandraceae) are used in Asia as antitussive, astringent and tonic remedies. More than three dozen dibenzocyclooctadiene (DBCO) lignans have been isolated from this plant since 1961. These lignans are classified into three types, namely, compounds with a *cis* configuration for the C(6) and C(7) methyl groups, such as schizandrin (**42**) and gomisin A (**43**); compounds with *trans* configuration of the C(6) and C(7) methyl groups such as isoschizandrin (**46**); and deoxy-type compounds such as deoxyschizandrin (**41**) and wuweizisu C (**47**). Schiarisanrins A–D (**50**–**53**) were recently isolated from the ethanolic extracts of *Schizandra arisanensis* by a bioassay-directed

41 (+)-Deoxyschizandrin	R^1, R^2, R^3, R^4, R^5 = Me, R^6 = H
42 (+)-Schizandrin	R^1, R^2, R^3, R^4, R^5 = Me, R^6 = OH
43 (+)-Gomisin	R^1, R^2 = CH$_2$, R^3, R^4, R^5 = Me, R^6 = OH
44 γ-Schizandrin (racemic)	R^1, R^2 = CH$_2$, R^3, R^4, R^5 = Me, R^6 = H
45 Gomisin M$_1$ (racemic)	R^1, R^2 = Me, R^3, R^6 = H, R^4, R^5 = CH$_2$

46 Isoschizandrin	R^1, R^2, R^3, R^4, R^5 = Me, R^6 = OH

47 (−)-Wuweizisu C	R^1 = R^2 = CH$_2$, R^3 = R^4 = CH$_2$
48 (−)-Gomisin N	R^1 = R^2 = Me, R^3 = R^4 = CH$_2$
49 (−)-Gomisin J	R^1 = R^4 = H, R^2 = R^3 = Me

50 Schiarisanrin A	R = OCOs-Bu
51 Schiarisanrin B	R = OCOMe
52 Schiarisanrin C	R = OCOPh
53 Schiarisanrin D	R = OCOCH=CHPh

fractionation.[40] These unique lignans, with a 5,4′-butano-2,4-cyclohexa-dienone-6-spiro-3′-(2′,3′-dihydrobenzo[*b*]furan) skeleton, exhibit cyto-toxicity against human nasopharynx, epidermal and colon carcinomas as well as hepatoma and cervical tumour cell lines.

Tobinaga *et al.*,[41] addressed the issue of a general synthesis for DBCO lignans using a common intermediate to access a series of these compounds. The synthetic plan used spirodienone (*erythro*-**56 IE-56**) or *threo*-**56** (**T-56**) as the key intermediate, prepared from the correspond-ing bisaryl butanol **E-55** or **T-55** by the oxidation with Weitz's aminium salt [tris(4-bromophenyl)aminium hexachloroantimonate; BAHA]. Hydrogenolysis of the spirodienone afforded the monodeprotected phenol which, after silylation, underwent a second BAHA oxidation to form an orthoquinone (Scheme 6.5). Upon reduction with sodium boro-hydride the corresponding catechol was alkylated. The aryl–aryl bond was constructed in the final stages of the synthesis by an iron(III)

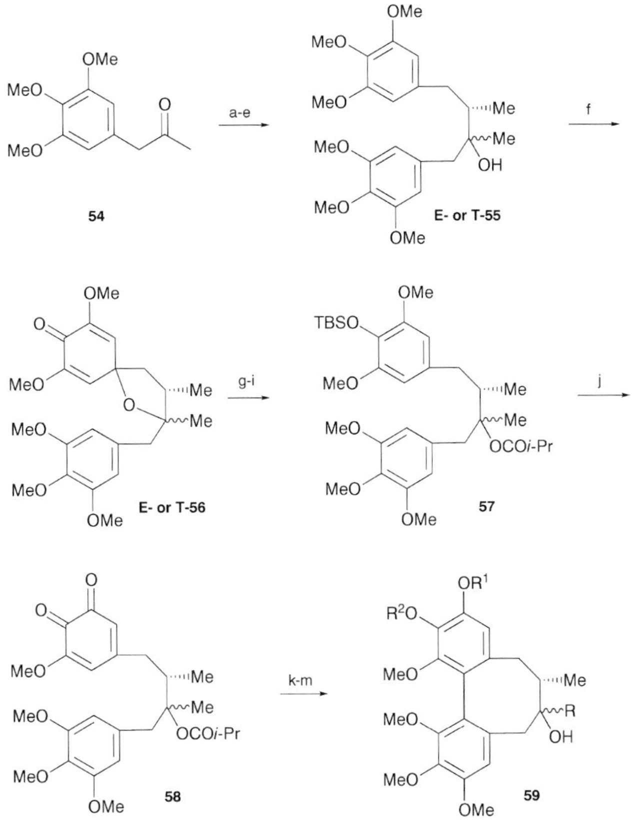

Scheme 6.5 Reaction conditions: (a) TiCl$_4$-Zn; (b) CH(OEt)$_3$, PPTS, CH$_2$Cl$_2$; (c) Ac$_2$O; (d) BH$_3\cdot$THF; (e) NaOH, H$_2$O$_2$; (f) BAHA, Na$_2$CO$_3$, THF, 0°C; (g) H$_2$, 10% Pd/C, EtOH; (h) t-BuMe$_2$SiCl, DBU, benzene; (i) (i-PrCO)$_2$O, TsOH; (j) BAHA, Na$_2$CO$_3$, THF, −20°C; (k) NaBH$_4$; MeOH; (l) MeI, K$_2$CO$_3$ or BrCH$_2$Cl, Cs$_2$CO$_3$, DMF; (m) Fe(ClO$_4$)$_3\cdot$9H$_2$O, CH$_2$Cl$_2$/MeCN.

perchlorate oxidation. A mixture of atropisomers was formed in which one isomer predominated.

Using a stereoselective approach for the construction of the cyclo-octadiene ring, Tanaka and co-workers[42] accomplished a synthesis of deoxyschizandrin, wuweizisu C, gomisin J, gomisin N and γ-schizandrin. The strategy, exemplified in Scheme 6.6 for wuweizisu C, employed an enantioselective catalytic hydrogenation on unsaturated ester **60** to introduce the first stereogenic centre. The second aryl group was installed via an aldol/elimination tactic on lactone **61**. Oxidative coupling using iron(III) perchlorate, followed by a hydrogenation and reduction reaction sequence, afforded wuweizisu C (**47**). Thermal stability studies

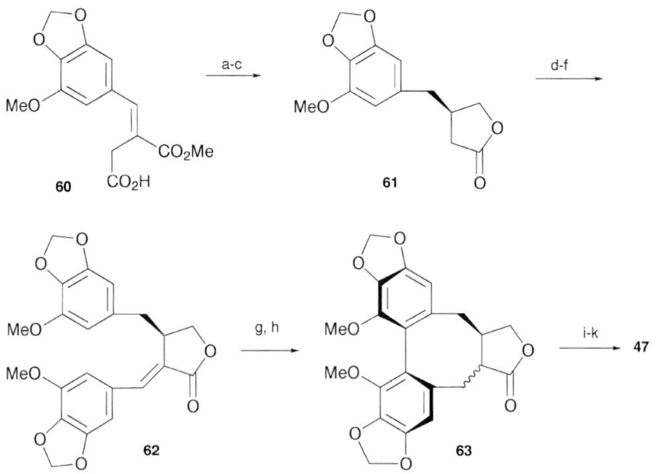

Scheme 6.6 Reaction conditions: (a) (S,S)-MOD-DIOP, H_2, $Rh(COD)_2BF_4$; (b) $Ca(BH_4)_2$, KOH; (c) HCl; (d) $LiN(i\text{-}Pr)_2$, 3-methoxy-4,5-methylenedioxybenzaldehyde; (e) Ac_2O, NEt_3, DMAP; (f) DBU; (g) $Fe(ClO_4)_3$, CF_3CO_2H, CH_2Cl_2; (h) H_2, Pd/C; (i) $i\text{-}Bu_2AlH$; (j) $MeSO_2Cl$, NEt_3; (k) $LiBHEt_3$.

performed on the aryl–aryl bond in DBCO lignans suggest that rotation does not occur up to 200°C. Consequently, the occurrence of racemic DBCO lignans in nature must be ascribed to a different mechanism, presumably low selectivity in the aryl–aryl oxidative coupling, a likely step in the biogenetic path.

6.4   Podophyllotoxins

Podophyllotoxins are naturally occurring lignans found in plants, particularly in the genus *Podophyllum*. The class includes several closely related chemical structures such as podophyllotoxin (**64**), deoxypodo-phyllotoxin (**65**) and the peltanins (**68** and **69**). A recently published review[43] on the discovery of podophyllotoxins traces the isolation and

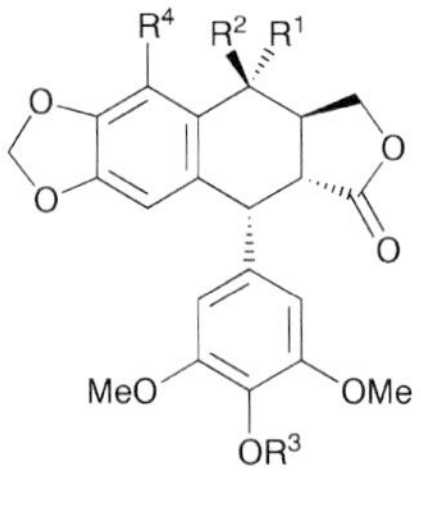

64 Podophyllotoxin $R^1 = OH$, R^2, $R^4 = H$, $R^3 = Me$
65 Deoxypodophyllotoxin R^1, R^2, $R^4 = H$, $R^3 = Me$
66 4′-Demethylpodophyllotoxin $R^1 = OH$, R^2, R^3, $R^4 = H$
67 4′-Demethylepipodophyllotoxin R^1, R^3, $R^4 = H$, $R^2 = OH$
68 α-Peltatin R^1, R^2, $R^3 = H$, $R^4 = OH$
69 β-Peltatin R^1, $R^2 = H$, $R^3 = H$, $R^4 = OH$

purification of this class of compounds back to the first half of the nineteenth century. Since then, a multitude of reports on the synthesis of the podophyllotoxins have emerged, motivated mainly by their significant anticancer effects.[44] Etoposide and teniposide are two examples of anticancer drugs used in the treatment of lung and bladder cancer that were discovered as a result of the efforts made in the area of podophyllotoxin chemistry.

Despite the abundance of reports on podophyllotoxin and analogues, few reports have been concerned with the asymmetric aspect of their synthesis. In 1996, Bush and Jones[45] reported an efficient asymmetric synthesis of (−)-podophyllotoxin in 8 steps and 15% overall yield. The key step in the reaction sequence was an enantioselective Diels–Alder reaction involving Feringa's dienophile (71) and the *ortho*-quinonoid pyrone 70, which proceeded with high regio- and stereoselectivity (Scheme 6.7). The correct stereochemistry of all the substituents was set in the hydrogenolysis step of the unsaturated lactone, but after a lead tetraacetate oxidation a mixture of epimers at C(4) was formed. These could be separated and both epimers used in the synthesis to afford podophyllotoxin of 98% enantiomeric purity.

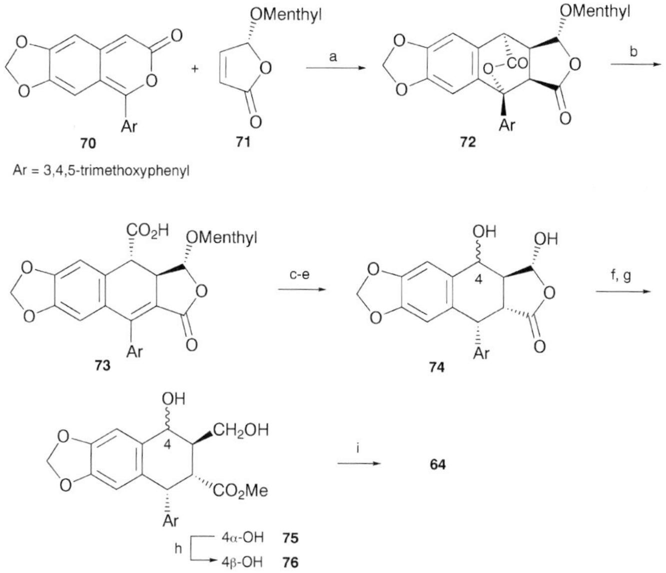

Scheme 6.7 Reaction conditions: (a) 50°C, MeCN, base-washed glassware; (b) AcOH, 49°C; (c) H$_2$, 10% Pd/C, EtOAc; (d) Pb(OAc)$_4$, 1:5 AcOH/THF; (e) HCl, dioxane; (f) CH$_2$N$_2$, Et$_2$O/MeOH; (g) LiEt$_3$BH, THF, −78°C; (h) HCl, THF; (i) ZnCl$_2$, THF, molecular sieves.

6.5 Quinones and hydroquinones

Hongconin (**82**) was isolated from the rhizome of *Eleutherine americana*, a plant found in southern China. It exhibits cardioprotective activity in angina pectoris by increasing coronary blood flow.[46] Swenton *et al.*[47] reported the synthesis of (−)-hongconin by the annulation reaction of levoglucosenone (**78**), available from the pyrolysis of paper, with cyanophthalide **77** (Scheme 6.8). Reductive ring-opening with Zn/Cu couple and subsequent methylation of the pyranone provided a facile entry into the naphthopyran ring system. A similar approach, based on annulation with cyanophthalide **77**, was used by Baker and co-workers[48] to synthesise both enantiomers of hongconin.[49]

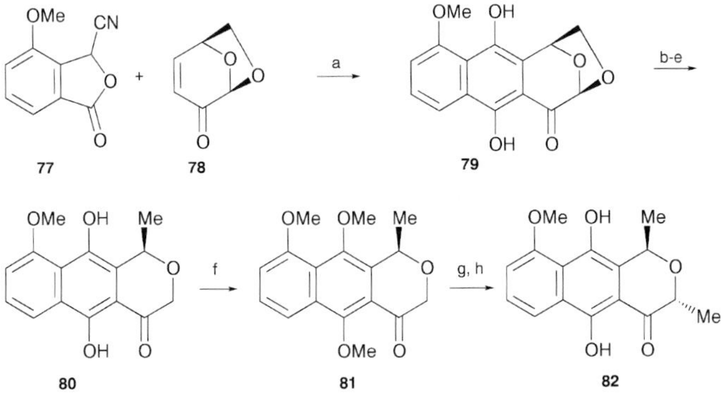

Scheme 6.8 Reaction conditions: (a) dimsyl lithium, THF, 0°C; (b) Zn/Cu, THF/AcOH, reflux; (c) $MeSO_2Cl$, pyr, CH_2Cl_2; (d) NaI, acetone; (e) n-Bu_3SnH, AIBN, benzene; (f) LiTMP, DMPU, MeI; (g) Me_2SO_4, K_2CO_3; (h) AgO then $Na_2S_2O_4$.

Frenolicin B (**83**) is a metabolite of *Streptomyces roseofulvus* which belongs to the family of naturally occurring naphthopyranquinones. Other members of the same family are kalafungin (**84**), nanaomycin D (**85**), medermycin (**86**) and deoxyfrenolicin (**87**). These compounds show antibiotic, antimicotic and in some cases antineoplastic activity. The broad spectrum of biological activities exhibited by naphthopyranquinones has led to numerous reports on their synthesis.[50–53a]

(+)-Favelanone (**96**) is an interesting tetracyclic dione isolated from the Brazilian plant Favela (*Cnidoscolus phyllacantus*) that shows antileukemic activity *in vitro*.[53b] Ng and Wege (1996)[53c] synthesised favelanone by using a Diels–Alder reaction between polarised cyclopropene derivative **91** and the electron-rich push–pull isobenzofuran **94**. 1,7,7-Tribromo-5,5-dimethylbicyclo[4.1.0]heptane **90**, the precursor to the cyclopropene dienophile, was prepared from α-bromoketone **88** in three steps: reduction

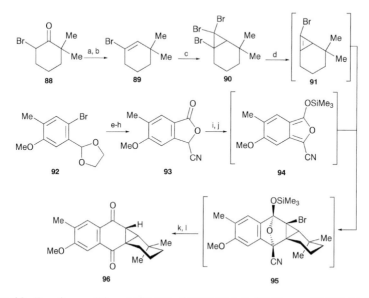

83 Frenolicin B R¹ = Pr, R² = H
84 Kalafungin R¹ = Me, R² = H

85 Nanaomycin D R¹ = H, R² = Me

86 Medermycin

87 Deoxyfrenolicin R¹ = Pr, R² = H

of the ketone to alcohol, elimination with the formation of cyclohexene **89** and addition of dibromocarbene (Scheme 6.9). The cyclopropene **91** was generated *in situ* by treatment of **90** with *n*-butyllithium at $-100°C$.

Scheme 6.9 Reaction conditions: (a) $NaBH_4$; (b) $(CF_3SO_2)_2O$, CH_2Cl_2, pyr, DMAP; (c) $CHBr_3$, NaOH; (d) *n*-BuLi, THF, $-100°C$, (e) *n*-BuLi, CO_2; (f) H_3O^+; (g) $(COCl)_2$, DMF, pyr; (h) HCN; (i) $LiN(i$-$Pr)_2$, THF, $-78°C$; (j) Me_3SiCl; (k) H_3O^+; (l) *n*-Bu_3SnH, AIBN, benzene.

The isobenzofuran **94** was available from protected aldehyde **92** through the intermediacy of cyanophthalide **93**.

Palmarumycins (**97, 98**),[54] diepoxins (**99**)[55] and preussomerins (**100**)[56] are natural products isolated from *Coniothyrium palmarum* that possess as a unique feature a spiroketal entity formally derived from naphthalene-1,8-diol and 1,4-naphthoquinone. In addition to general antitumour and antibiotic activities, selective inhibition of Ras-farnesyl transferase with $IC_{50} \geq 1.2\,\mu M$ was demonstrated for these compounds.

97
Palmarumycin CP$_1$

98
Palmarumycin CP$_3$

99
Diepoxin α

100
Preussomerin E

101
Deoxypreussomerin A

Recently, Wipf and Jung[57, 58] reported a total synthesis of palmarumycin CP$_1$ (**97**) and (±)-deoxypreussomerin A (**101**) that involved the formation of the biaryl ether **103** through a Cu$_2$O-catalysed reaction (Scheme 6.10). Subsequent oxidation of the demethylated hydrobenzoquinone ring of **103** afforded a quinone that underwent immediate acetalisation to form spiroketal **104**, a common intermediate for the synthesis of both **97** and **101**. For the preparation of palmarumycin CP$_1$, spiroketal **104** was oxidised with Dess–Martin reagent and the resulting decalindione aromatised with activated MnO$_2$. The same protocol was applied for the synthesis of (±)-deoxypreussomerin A, after an epoxidation of spiroketal **103** by cumene hydroperoxide anion.

Furaquinocin D (**105**)[59] is a representative of the furaquinocin class of cytocidal antibiotics that was isolated form the culture of *Streptomyces*

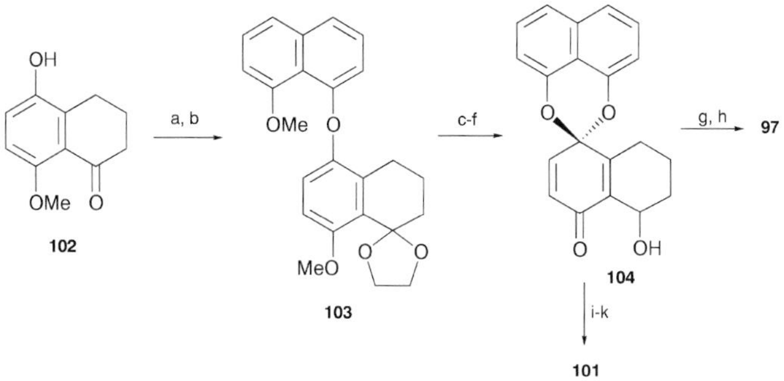

Scheme 6.10 Reaction conditions: (a) $(CH_2OH)_2$, PPTS, benzene; (b) 8-iodo-1-methoxy-naphthalene, Cu_2O, K_2CO_3, pyr; (c) TsOH, acetone/H_2O; (d) BBr_3, CH_2Cl_2; (e) $LiAlH_4$, Et_2O; (f) $PhI(OAc)_2$, CF_3CH_2OH; (g) Dess–Martin periodinane, CH_2Cl_2; (h) MnO_2, CH_2Cl_2; (i) cumene hydroperoxide, NaH, THF; (j) Dess–Martin periodinane, CH_2Cl_2; (k) MnO_2, CH_2Cl_2.

sp. KO-3988.[60] Their structure contains two fragments that have different biosynthetic origins, the polyketide-derived naphthoquinone ring and the isoprenoid side-chain. In 1992, Seto *et al.*[61] reported the isolation of stealthin A (**106**) and B (**107**) from *Streptomyces viridochromogenes*.

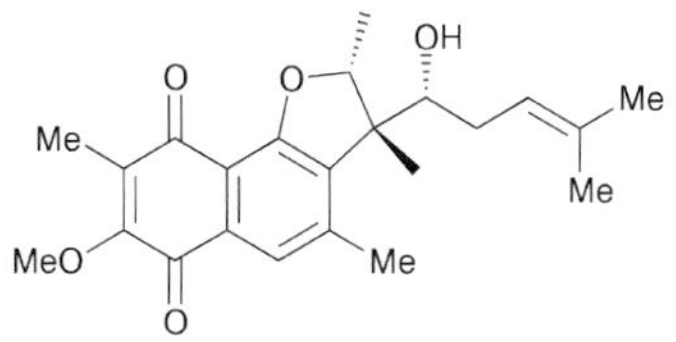

105
Furaquinocin D

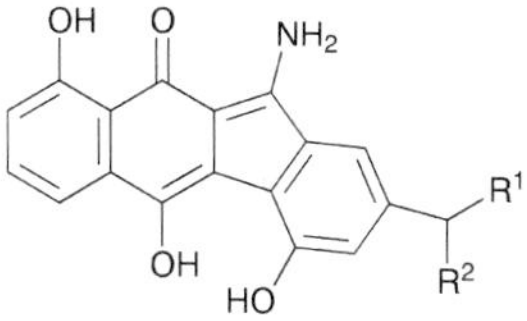

106 Stealthin A R^1 = OH, R^2 = H
107 Stealthin B R^1, R^2 = O
108 Stealthin C R^1 = R^2 = H

These workers showed that stealthins are potent radical scavengers with activity 20–30 times higher than vitamin E. The interest generated by these compounds and their biological activity prompted Koyama and Kamikawa[62] to undertake a synthesis of O^4,O^9-dimethylstealthins A and C, which featured a Suzuki coupling followed by a Friedel–Crafts cyclisation as key steps. Gould *et al.*[63] synthesised stealthin C (**108**) and proved that it was an intermediate in kinamycin biosynthesis.

The bioassay-guided fractionation of an organic extract from the endemic Australian shrub *Conospermum* sp. (Proteaceae) resulted in the isolation of conocurvone (**109**), a trimeric naphthoquinone derivative with anti-HIV activity.[64] The structure and absolute stereochemistry of conocurvone was elucidated by X-ray crystallography, and its

109
Conocurvone

semisynthesis was performed by using teretifolione B, the hydroxylated monomer unit, also a component of the *Conospermum* extract.

In 1989, Konishi and co-workers[65] elucidated the structure of dynemicin A (**117**), a metabolite of *Micromonospora chersina* which showed important antitumour activity *in vitro* against certain cancer cell lines. Its unique structure incorporates an anthraquinone unit, an epoxide functionality and an unusual enediyne handle which seem to have an important role in the interaction of dynemicin with DNA strands.[66] Danishefsky *et al.*[67] elaborated a strategy for the synthesis of dynemicin A which allows the generation of similar but simpler compounds for clinical evaluation. The synthesis commenced with commercially available aldehyde **110**, which was transformed in a series of eight reactions, including a stereoselective intramolecular Diels–Alder reaction, into the quinoline **111** (Scheme 6.11). This was dihydroxylated and the resulting diol suitably protected in order to render the β face of the ring system more accessible for the ethynylation reaction. The second ethynyl group was introduced into **113** by operating on the silyloxymethylene group in compound **112**. This prepared the ground for the formation of the enediyne bridge. First, the allyloxycarbonyl (Alloc) protecting group was removed with Pd(PPh₃)₄ in combination with morpholine, in order not to interfere with the palladium-mediated coupling. A trimethylsilylethoxy-carbonyl (TEOC) group was then installed. Stereoselective epoxidation was performed to obtain a bridged *cis*-epoxide which brought the alkyne moieties into alignment for the simultaneous coupling to an ethylene unit. The enediyne bridge was built through a Stille coupling between bis-iodo-alkyne **114** and *cis*-1,2-distannylethylene. The resulting compound was carboxylated and oxidised to the quinone imine **115**, in preparation for

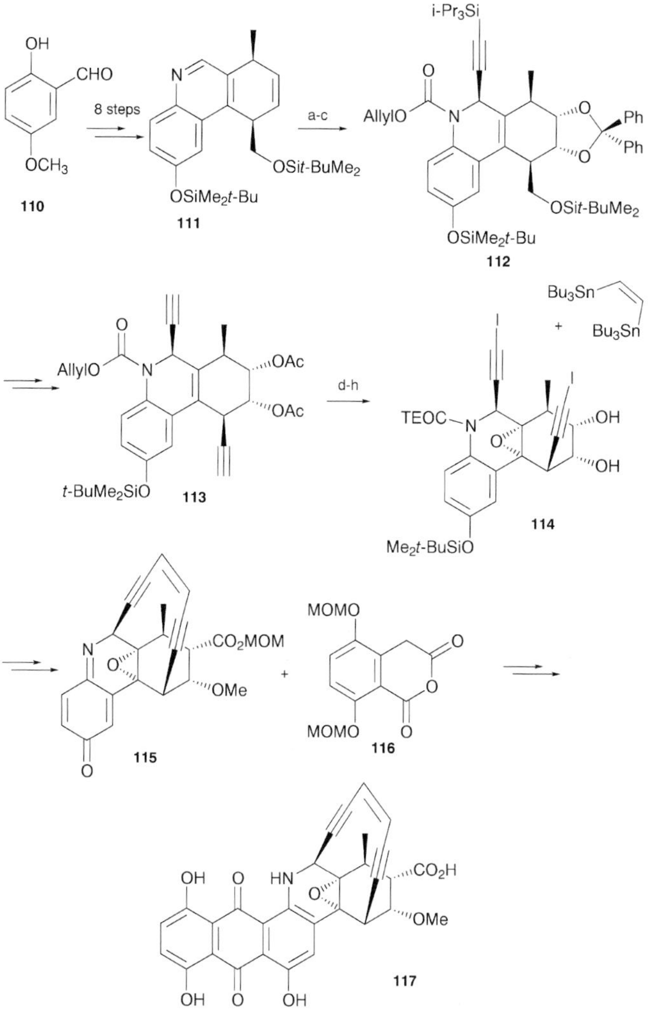

Scheme 6.11 Reaction conditions: (a) OsO₄, *N*-methylmorpholine-*N*-oxide; (b) Ph₂C(OMe)₂, H₂SO₄, CH₂Cl₂, 40°C then *t*-BuMe₂SiCl, imidazole; (c) (triisopropylsilyl)ethynyl magnesium bromide, ClCO₂allyl, THF, −20°C; (d) Pd(PPh₃)₄, morpholine, THF, 0°C; (e) TEOC–Cl, NaH, DMAP, THF, 0°C–25°C; (f) NH₃, MeOH; (g) *m*-CPBA, CH₂Cl₂; (h) AgNO₃ (cat.), *N*-iodosuccinimide, THF.

the attachment of two more rings through the reaction with homo-phthalic anhydride **116**. Although the stability of some intermediates proved to be troublesome, the synthesis of dynemicin A was eventually accomplished and provided a vehicle for the synthesis of more accessible analogues.

6.6   Chromans, chromenes and chromanols

The dipyranocoumarins isolated from plants of the genus *Calophyllum* have received considerable attention in recent years as a result of their potency in the inhibition of HIV-1 reverse transcriptase.[68] Although several mutants that are resistant to these compounds have been identified, the agents seem to induce a different pattern of aminoacid changes in the development of resistance than other nonnucleoside drugs.[69] This feature could aid in identifying drugs or drug combinations less prone to induce viral resistance. Calanolide A (**124**) and B (**123**) were isolated from *Calophyllum* more than 40 years ago[70] but the reports about their syntheses are concentrated mainly in the past decade.[71] Trost and Toste[72] reported an enantioselective synthesis of both **123** and **124** that involves an asymmetric regioselective *O*-alkylation of a phenol (Scheme 6.12). The synthesis begins with a palladium-catalysed coupling between the phloroglucinol-derived chroman **118** and alkyne **119** to form the coumarin ring of **120** at the less sterically hindered hydroxyl group (4:1 ratio of regioisomers). The asymmetric allylic alkylation of **120** with

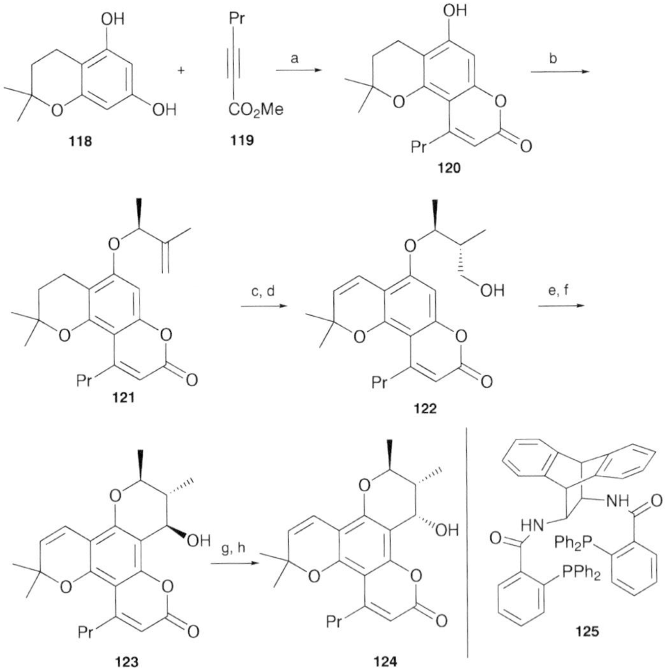

Scheme 6.12   Reaction conditions: (a) 2.5% Pd$_2$(dba)$_3$, HCO$_2$H; (b) tiglyl methyl carbonate, 2.5% Pd$_2$(dba)$_3$, 7.5% ligand **125**; (c) DDQ, dioxane; (d) 9-BBN, THF then NaOH, H$_2$O$_2$; (e) Dess–Martin periodinane; (f) ZnCl$_2$, Et$_2$O; (g) Me$_3$P, EtO$_2$CN=NCO$_2$Et, ClCH$_2$CO$_2$H; (h) K$_2$CO$_3$, MeOH.

tiglyl methyl carbonate in the presence of Pd and chiral ligand **125** afforded the secondary ether **121** with excellent regio- (92:8) and enantioselectivity [98% enantiomeric excess (ee)]. Further oxidation of ring C with DDQ, hydroboration of the alkene side-chain and Dess–Martin oxidation of the resulting alcohol afforded an aldehyde that was cyclised with zinc chloride to produce the thermodynamically less stable *ent*-calanolide B (**123**). This was converted to *ent*-calanolide A via a Mitsunobu inversion.

Thielocin A1β (**131**) is an unusual pentameric compound, isolated from the culture of ascomycetes *Thielavia terricola* RF-143,[73] which exhibits inhibitory activity toward group II phospholipase (PLA$_2$).[74] The first total synthesis of (±)-thielocin A1β was reported by Young *et al.*[75] It is based upon the efficient chemo- and stereoselective condensation of hydroxydienone **126** with the quinone methide **128** formed *in situ* from **127** after treatment with fluoride (Scheme 6.13). The resulting tricyclic

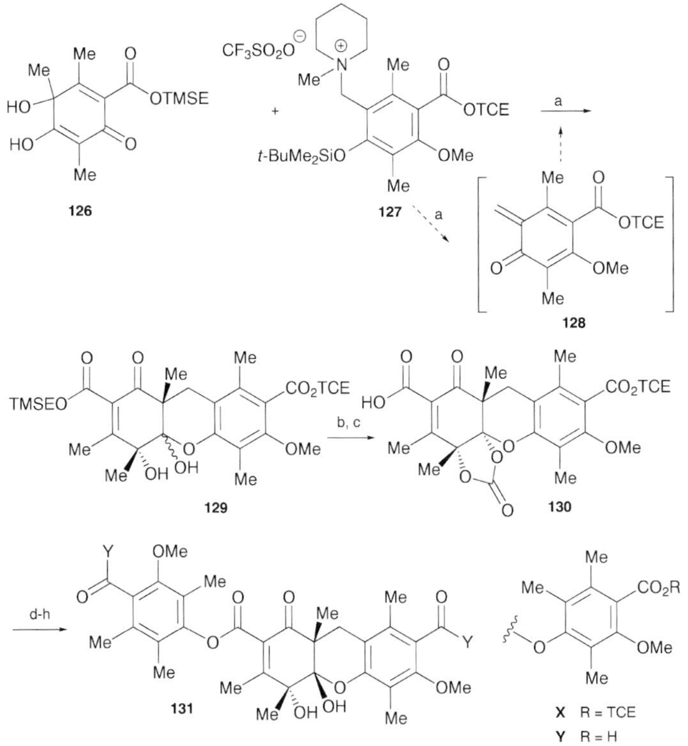

Scheme 6.13 Reaction conditions: (a) *n*-Bu$_4$NF, CH$_2$Cl$_2$; (b) carbonyldiimidazole, NEt$_3$, CH$_2$Cl$_2$; (c) *n*-Bu$_4$NF, DMF; (d) Cl$_2$CHOCH$_3$, CH$_2$Cl$_2$, reflux, then **XH**, NEt$_3$, 25°C; (e) 50 equiv. Cd^0, DMF/AcOH (1:1); (f) 5 equiv. **XH**, 50 equiv. (CF$_3$CO)$_2$O, benzene; (g) 50 equiv. Cd^0, DMF/AcOH (1:1); (h) 0.5 N NaOH/dioxane (1:1), −10°C.

compound **129** was treated with carbonyldiimidazole to obtain a single diastereomer of carbonate **130**. Attachment of three ester fragments **Y**, and hydrolysis of carbonate moiety brought about inversion of the hemiacetal chiral centre to yield thielocin A1β.

The Thai plant *Puraria mirifica* contains the estrogenic phenol miro-estrol (**137**), a molecule that was isolated more than 50 years ago,[76] but which eluded synthetic efforts until Corey and Wu reported its synthesis in 1993.[77] Their strategy involves a convergent Stille coupling between **133** and **134**, followed by a Lewis-acid catalysed transannular double cation–olefin cyclisation to build up the condensed cyclic framework (Scheme 6.14). After conversion of the α,β-enone **136** to the more stable β,γ-enone by using triethylamine further functional group manipulation led to a compound identical to natural miroestrol.

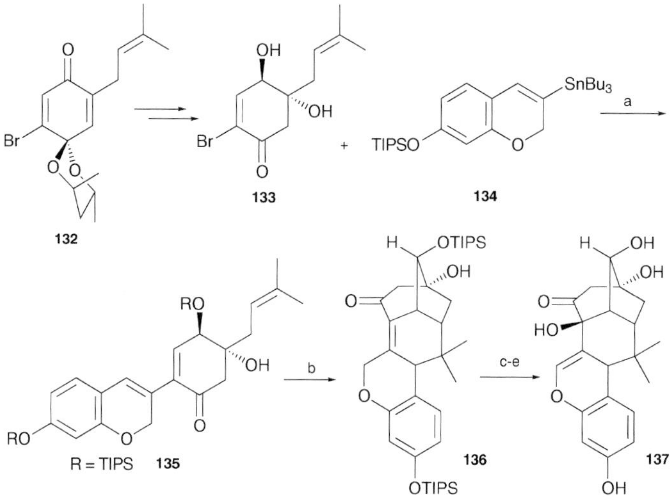

Scheme 6.14   Reaction conditions: (a) Pd(PPh₃)₄, toluene, 90°C; (b) *i*-Bu₂AlCl, CH₂Cl₂, 0°C; (c) NEt₃, CH₂Cl₂, reflux; (d) SeO₂, *t*-BuOOH, CH₂Cl₂; (e) *n*-Bu₄NF, THF, 0°C.

6.7   Benzofurans

(−)-Aplysin (**146**) is a halogenated sesquiterpene isolated from *Aplysia kurodai*,[78] a sea hare of the Pacific ocean. It can also be obtained from opistobranchs,[79] which are found along the North American coasts. This compound exhibits antifeedant properties and is used by the host molluscs to protect themselves from raptor enemies. Aplysin was dis-covered along with its dehalogenated form, (−)-debromoaplysin (**145**), their co-occurrence suggesting that (−)-debromoaplysin might function as an antioxidant and a reactive halogen scavenger. Fukumoto *et al.*[80, 81] described an enantioselective synthesis of (−)-aplysin and its debrominated counterpart by employing a tandem asymmetric epoxidation and enantiospecific ring expansion of cyclopropylidene alcohol (**139**). The resulting cyclobutanone **140**, obtained in 95% ee was transformed into triethylsilyl ether **141**, and this subjected to the palladium-mediated ring expansion reaction to give the unsaturated cyclopentanone **142** (Scheme 6.15). After forming the furan ring and after other functional group manipulations, (−)-debromoaplysin (**145**) and (−)-aplysin (**146**) were obtained in high enantiomeric excess.

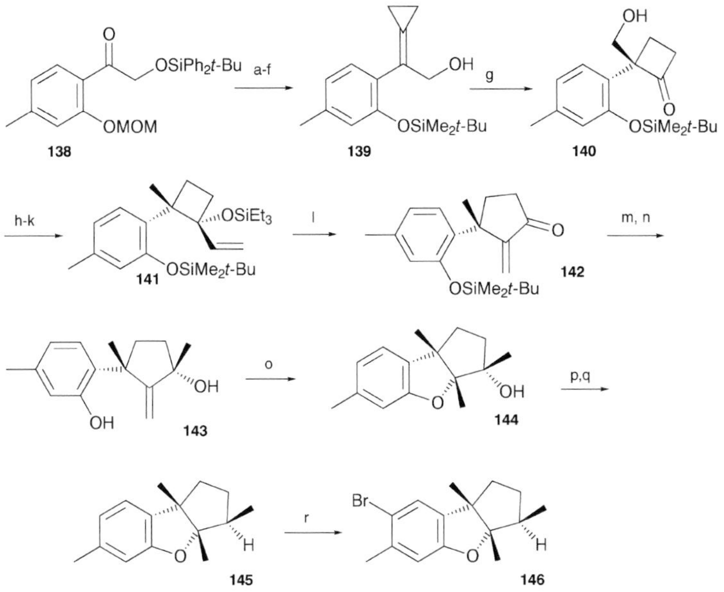

Scheme 6.15 Reaction conditions: (a) cyclopropyltriphenylphosphonium bromide, NaH, THF, reflux; (b) EtSH, BF₃·OEt₂, CH₂Cl₂, −78°C to −23°C; (c) *n*-Bu₄NF, THF; (d) pivaloyl chloride, pyr; (e) *t*-BuMe₂SiCl, DMAP, imidazole, DMF, 0°C; (f) *i*-Bu₂AlH, CH₂Cl₂, −23°C; (g) (+)-diisopropyl tartrate, Ti(O*i*-Pr)₄, *t*-BuOOH, CH₂Cl₂, −50°C; (h) PhSSPh, *n*-Bu₃P, THF, reflux; (i) RaNi (W2), acetone; (j) vinylmagnesium bromide, CeCl₃, THF, −78°C; (k) Et₃SiOTf, 2,6-lutidine, CH₂Cl₂, 0°C; (l) PdCl₂(MeCN)₂, THF, reflux or Pd(OAc)₂, AsPh₃, CH₂Cl₂; (m) MeLi, CeCl₃, Et₂O, −78°C; (n) *n*-Bu₄NF, THF; (o) Hg(OCOCF₃)₂, THF then 10% NaOH, NaBH₄; (p) POCl₃, pyr; (q) H₂, PtO₂, EtOH; (r) Br₂, NaHCO₃, CHCl₃.

Another interesting compound with a dihydrobenzofuran ring is rocaglamide (**156**), isolated from *Aglaia elliptifolia*. It shows good activity against P388 lymphocytic leukemia in mice and human epidermoid carcinoma cells of the nasopharynx *in vitro*.[82] Trost and co-workers[83, 84] reported the first enantioselective synthesis of (−)-rocaglamide and as a consequence elucidated the absolute stereochemistry of this natural product. His synthetic strategy exploits an oxidative cyclisation of the regioisomeric olefin mixture **151**, obtained from asymmetric palladium-catalysed cycloaddition of precursors **147** and **148** (Scheme 6.16). Since the stereochemistry of the oxidative cyclisation produced the unexpected and undesired diastereoisomer (the aryl and phenyl groups in a *trans* configuration), the route ultimately had to involve an inversion of the stereochemistry of the phenyl-bearing carbon.

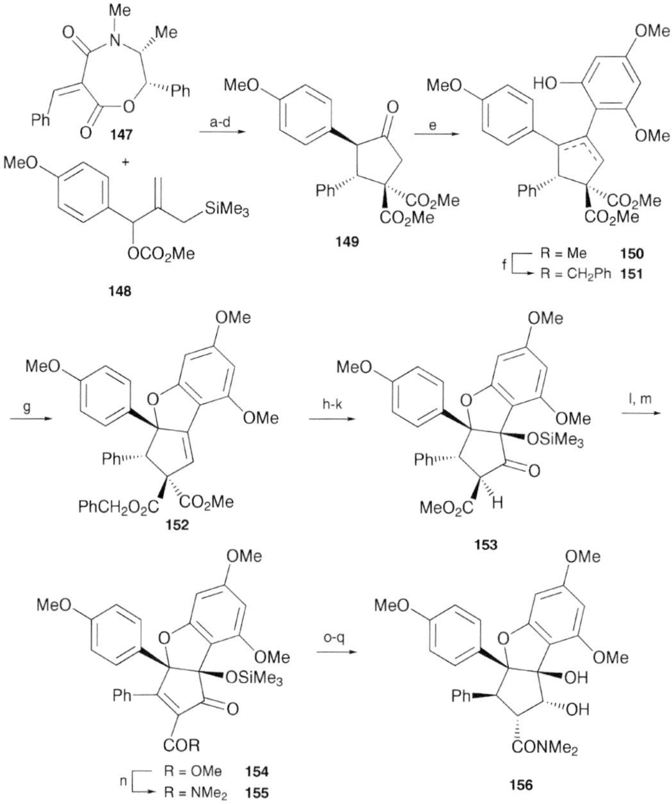

Scheme 6.16 Reaction conditions: (a) 5% Pd(OAc)$_2$, 30% (*i*-PrO)$_3$P, toluene, reflux, (b) NaOH, EtOH, reflux; (c) CH$_2$N$_2$, EtOAc; (d) O$_3$, MeOH/CH$_2$Cl$_2$ (2:1); (e) dimethylphloroglucinol, BF$_3$·MeOH, CH$_2$Cl$_2$; (f) Ti(OCH$_2$Ph)$_4$, PhCH$_2$OH, 100°C; (g) DDQ, THF, reflux; (h) 4% OsO$_4$, *N*-methylmorpholine-*N*-oxide, DABCO, THF/H$_2$O (5:1); (i) pyr·SO$_3$, DMSO, NEt$_3$; (j) Me$_3$SiOTf, *i*-Pr$_2$NEt, benzene; (k) H$_2$, 10% Pd/C, EtOH; (l) NaH, PhSCl, THF; (m) *m*-CPBA, NaHCO$_3$, CH$_2$Cl$_2$, −20°C; (n) Me$_2$NH$_2$Cl, Me$_3$Al, benzene, 45°C; (o) H$_2$, 20% Pd(OH)$_2$, EtOH; (p) KF, MeOH, 40°C; (q) Me$_4$NBH(OAc)$_3$, MeCN, AcOH.

6.8 Diterpene quinones

The extract obtained from the root of Chinese sage *Salvia miltiorrhiza*, or Dan Shen, has been used in Chinese traditional medicine for the treatment of heart disease, menstrual disorders, miscarriage, hypertension and viral hepatitis, among other ailments.[85] The extract contains the interesting abietane diterpenoid quinones danshexinkun A (**157**), B (**158**) and C (**159**), dyhydrotanshinone I (**160**), tanshinone I (**161**) and IIA (**162**), cryptotanshinone (**163**), neocryptotanshinone (**164**) and royleanone (**165**).[86] Unfortunately, none of the individual compounds match the therapeutic properties of the crude extract.

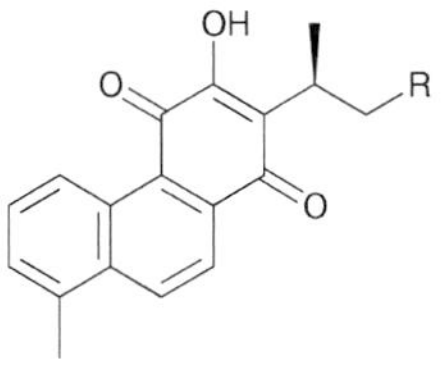

157 Danshexinkun A R = OH
158 Danshexinkun B R = H

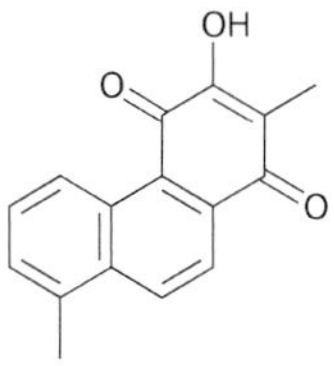

159 Danshexinkun C

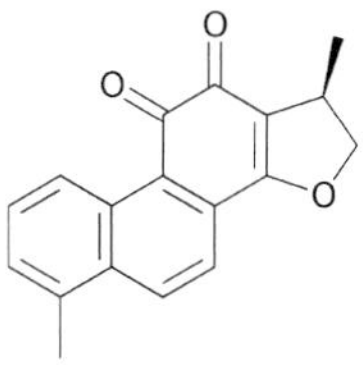

160 Dihydrotanshinone I

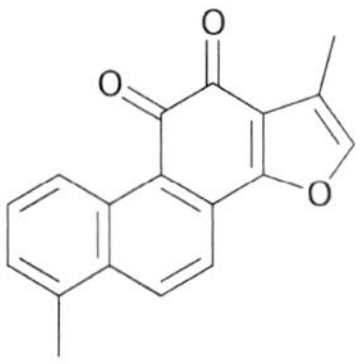

161 Tanshinone I

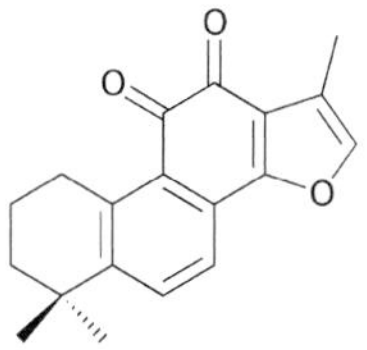

162 Tanshinone IIA

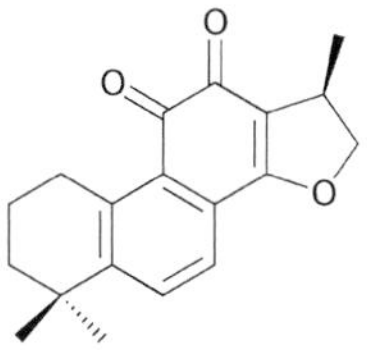

163 Cryptotanshinone

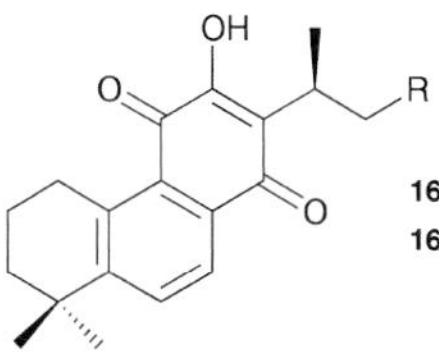

164 Neocryptotanshinone R = OH
165 Royleanone   R = H

Danheiser and co-workers[87] have developed a photochemical aromatic annulation strategy that provides efficient routes to all the aforementioned compounds. The methodology is illustrated by the synthesis of danshexinkun A (**157**), dihydrotanshinone I (**160**) and tanshinone I (**161**). The key step takes place upon irradiation of diazoketone **166**, whereupon a photochemical Wolff rearrangement is triggered to produce arylketene **169** (Scheme 6.17). This undergoes a regiospecific [2 + 2] cycloaddition with alkyne **167**. Continued irradiation leads to a 4π electrocyclic opening

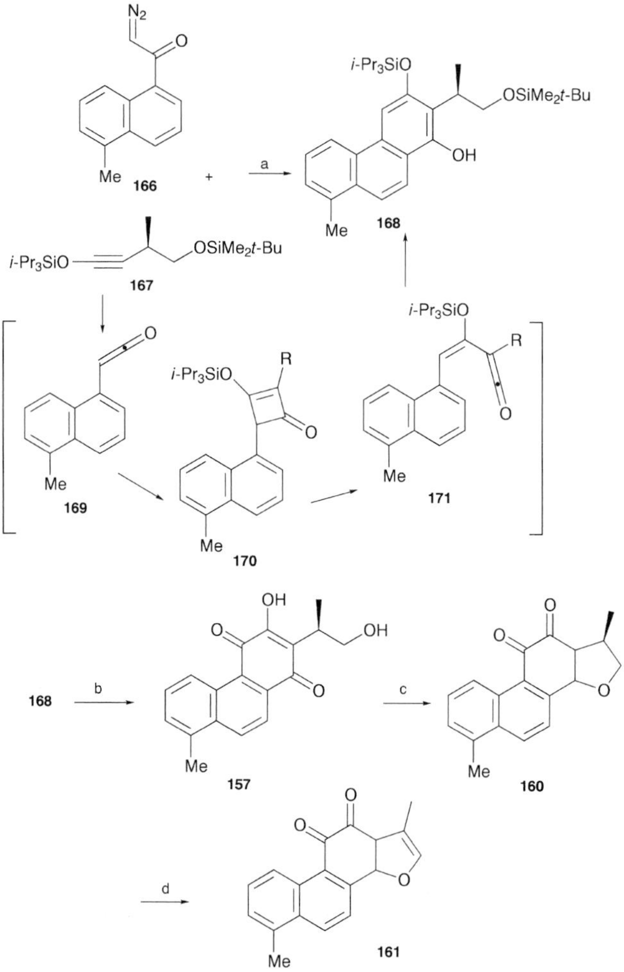

Scheme 6.17 Reaction conditions: (a) *h*v (Vycor filter), ClCH₂CH₂Cl; (b) *n*-Bu₄NF, then O₂, −78°C to 25°C; (c) H₂SO₄, EtOH; (d) DDQ, benzene.

of cyclobutenone **170** to provide vinylketene **171**, which undergoes 6π electrocyclisation to afford the tricyclic product **168**.

In 1996, Matsumoto *et al.*,[88] reported an interesting rearrangement of an angular diterpene to a linear abietane skeleton to produce pygmaeo-cine E (**178**), a compound isolated from the roots of *Pygmaeopremna herbaceae*. The synthesis started with ketone **172**, available from (+)-dehydroabietic acid. Compound **172** was reduced with sodium boro-hydride in the presence of cerium chloride heptahydrate to give alcohol **173** (Scheme 6.18). In the presence of boron trifluoride etherate, this

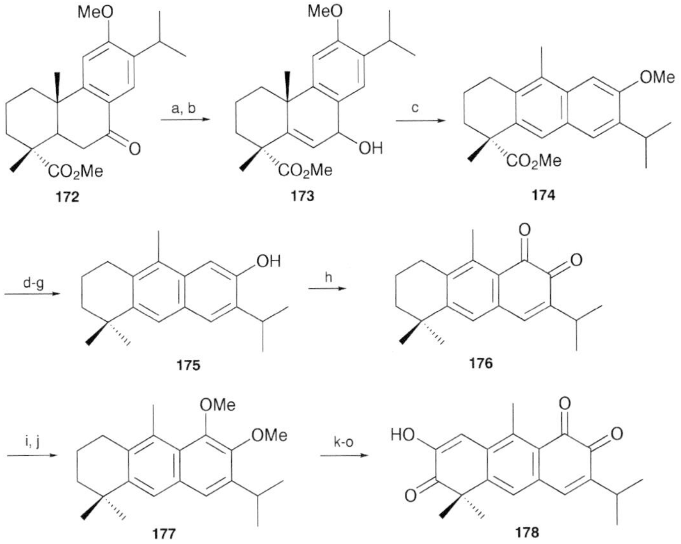

Scheme 6.18 Reaction conditions: (a) SeO$_2$, AcOH/H$_2$O (2:1), reflux; (b) NaBH$_4$, CeCl$_3$·H$_2$O, MeOH/THF (1:1), 0°C; (c) BF$_3$·OEt$_2$, CH$_2$Cl$_2$; (d) LiAlH$_4$, THF, 0°C; (e) pyridinium chlorochromate, CH$_2$Cl$_2$; (f) N$_2$H$_4$·H$_2$O, diethyleneglycol then NaOH, 185°C; (g) EtSH, AlCl$_3$, CH$_2$Cl$_2$, 0°C; (h) Fremy's salt (potassium nitrosodisulfonate), KH$_2$PO$_4$, DMF/H$_2$O (1:1.5); (i) H$_2$, PtO$_2$, MeOH; (j) MeI, K$_2$CO$_3$, methyl ethyl ketone, reflux; (k) DDQ, 1,4-dioxane, reflux; (l) H$_2$, PtO$_2$; (m) Jones reagent, acetone, 0°C; (n) O$_2$, t-BuOK, t-BuOH; (o) EtSH, AlCl$_3$, CH$_2$Cl$_2$, 0°C.

alcohol rearranged to the optically active ester **174** possessing the hydroxyanthracene skeleton. The next steps involved the transformation of the methoxycarbonyl into a methyl group and additional oxidation of rings A and C to afford pygmaeocine E.

6.9 Benzonaphthopyranones

Gilvocarcins are *C*-glycosides in which a 6*H*-benzo[*d*]-naphtho[1,2-*b*] pyran-6-one nucleus is attached to a fucose residue. They have been isolated from various strains of *Streptomyces* spp.[89] and they exhibit significant antibiotic and antitumour activity.[90] Until recently, their synthesis remained a challenge, mainly because of the difficulties encountered in the formation of the *C*-glycosidic bond. Suzuki *et al.*[91] reported the first total synthesis of gilvocarcin M (**184**) in 1992 and thereby established unequivocally the absolute configuration of this natural product. This synthesis fashioned the *C*-glycosidic bond at an early stage by employing the hafnocene dichloride/silver perchlorate system as a promoter. Initially an *O*-glycoside was formed which subsequently rearranged thermally to the *ortho C*-glycoside in a contrasteric manner (8.2:1 α:β anomer). The annulation of ring B was performed regioselectively through a benzyne-furan [4 + 2] cycloaddition. The connection between rings B and D was established through a palladium-mediated reaction. The nucleus of gilvocarcin V (**186**) was built in an analogous fashion, and installation of the vinyl group was achieved by employing organoselenium chemistry (Scheme 6.19).

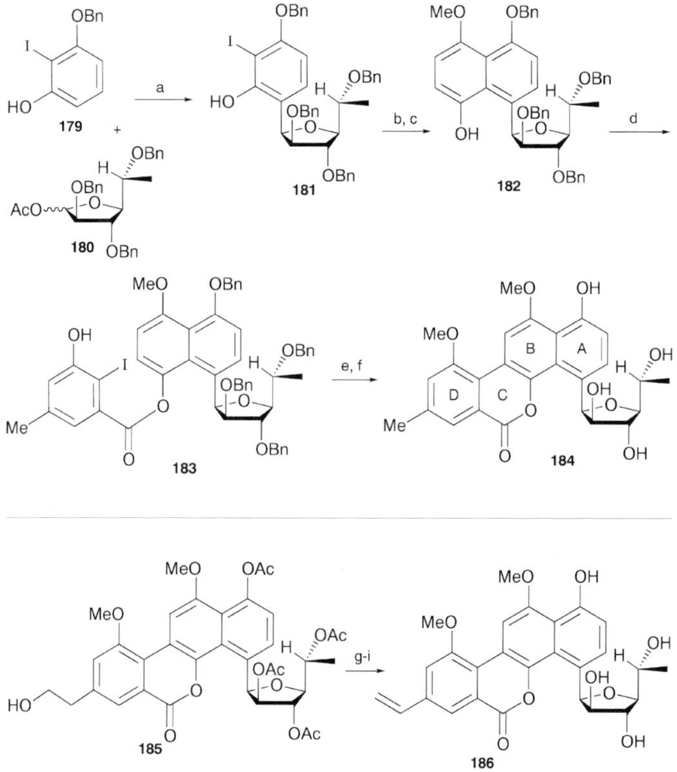

Scheme 6.19 Reaction conditions: (a) Cp$_2$HfCl$_2$-AgClO$_4$, CH$_2$Cl$_2$, −78°C to −20°C; (b) (CF$_3$SO$_2$)$_2$O, *i*-Pr$_2$NEt, CH$_2$Cl$_2$, −78°C; (c) *n*-BuLi, 2-methoxyfuran, THF, −78°C; (d) 2-iodo-3-methoxy-5-methylbenzoyl chloride, *i*-Pr$_2$NEt, DMAP, THF; (e) 26 mol% (Ph$_3$P)$_2$PdCl$_2$, NaOAc, *N,N*-dimethylacetamide, 125°C; (f) H$_2$, 10% Pd/C, MeOH/THF; (g) *o*-nitrophenyl seleno-cyanate, *n*-Bu$_3$P, THF; (h) 35% aqueous H$_2$O$_2$, 0°C–25°C; (i) NaOMe, MeOH.

6.10 Angucycline antibiotics

Angucycline antibiotics are compounds, isolated from the *Actinomycete* group of microorganisms, that have as a characteristic structural feature an angular tetracyclic benz[a]anthracene framework.[92] They display a wide variety of biological activities, ranging from antitumour properties, blood platelet aggregation inhibition and enzyme inhibition.[93] The main structural difference among the members of the angucycline family resides in the aromatic/hydroaromatic nature of rings A and B. The majority of synthetic efforts have concentrated on the angucyclines with aromatic A and/or B rings. However, owing to the challenging nature of angucyclines with hydroaromatic A and B rings, as a consequence of the stereogenic centres found in these compounds, reports on the synthesis of such compounds have been scarce.[94] Emyein A (187), SF 2315A (188) and B (189) possess a structure that belongs to the hydroaromatic subclass of angucycline antibiotics, with angular hydroxyl groups that have a strong tendency to undergo dehydration under acidic conditions.

187 (+)-Emycin A   188 SF 2315A   189 SF 2315B

Sulikowski and co-workers[95] reported the assemblage of epoxyquinol 196, a close analogue of SF 2315B, by using an interesting oxidation of the naphthoquinone precursor 192, obtained via a Diels–Alder reaction (Scheme 6.20). The oxidation process occurs in the presence of oxygen and tetrabutylammonium fluoride and generates a mixture of α-193 and α-194 with a composition dependent on the silyloxy group substituent, presumably through the intermediacy of a quinone methide. The epoxy-quinol 196 is obtained after the reduction of the C(5)=C(6) double bond and a stereodirected reduction of the C(12) keto group.

6.11 Benzenoids

Differanisole A (201) is a differentiation-inducing factor active against mouse leukemia cells that was isolated from the culture broth of a

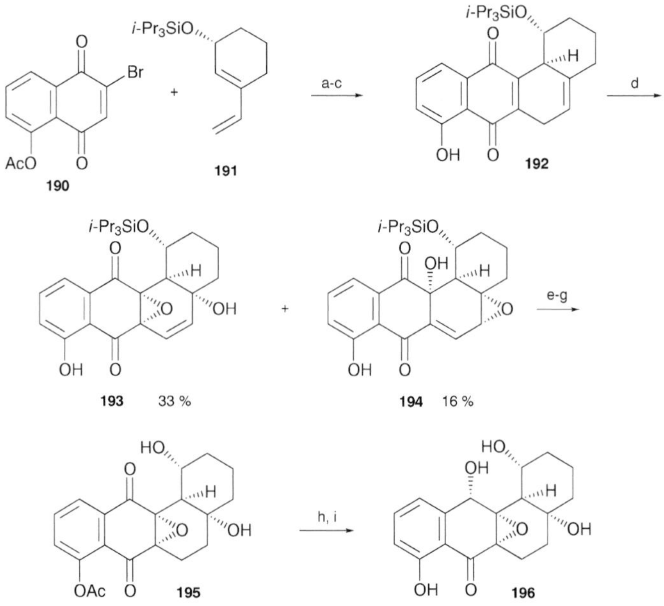

Scheme 6.20 Reaction conditions: (a) toluene, reflux; (b) DBU, benzene, 0°C; (c) n-Bu$_4$NOH, THF; (d) O$_2$, n-Bu$_4$NF, THF, −78°C to 25°C; (e) H$_2$, PtO$_2$, EtOAc, 0°C; (f) Ac$_2$O, pyridine, CH$_2$Cl$_2$; (g) HF·pyridine, MeCN; (h) Me$_4$NBH(OAc)$_3$, AcOH/MeCN, −10°C; (i) n-Bu$_4$NOH, THF, 0°C.

Chaetomium strain.[96] The first synthesis of this hexasubstituted benzene derivative was that reported by Mori *et al.*[97] It proceeded in an overall yield of 2.5% and relied on a sensitive ring-forming reaction. Recently, Green[98] described two similar routes for the synthesis of differanisole A that start with a mixture of *E*- and *Z*-**197**, available from 3,5-dimethoxybenzaldehyde. After a Vilsmeier formylation, hydrogenation of the propenyl side-chain and a selective demethylation, tetrasubstituted benzene derivative **200** was obtained (Scheme 6.21). Chlorination and oxidation afforded differanisole A (**201**). The second route provided the same final product by a variation of the reaction sequence.

Snieckus and co-workers[99] developed the synthesis of naturally occur-ring fluorenone dengibsinin (**205**) by using a remote aromatic met-allation–cross-coupling sequence (Scheme 6.22). This process is the equivalent of an anionic Friedel–Crafts, with the advantage of having the regioselectivity dictated by the availability of the directed metallation group, in this case the *N,N*-diethylamido group.

Macrocyclic bis(benzyls) such as perrottetin (**207**) and marchantin (**208**) are constituents of liverwort species and are derived from a single

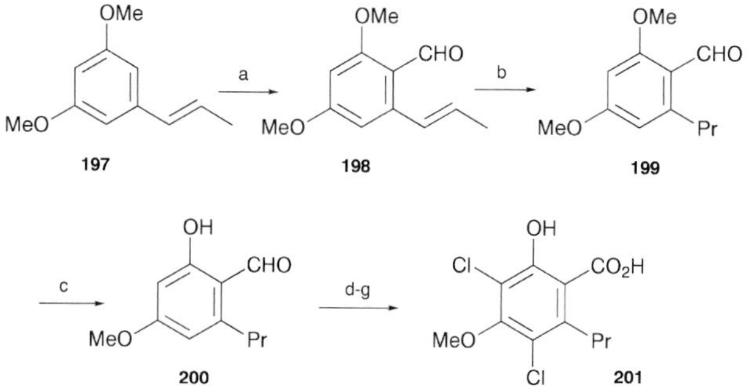

Scheme 6.21 Reaction conditions: (a) POCl$_3$, DMF, CH$_2$Cl$_2$; (b) H$_2$, Pd/C, EtOAc; (c) BCl$_3$, CH$_2$Cl$_2$; (d) SOCl$_2$, Et$_2$O; (e) Ac$_2$O, pyr; (f) NaClO$_2$, NH$_3$SO$_3$H; (g) NaOH, EtOH, H$_2$O.

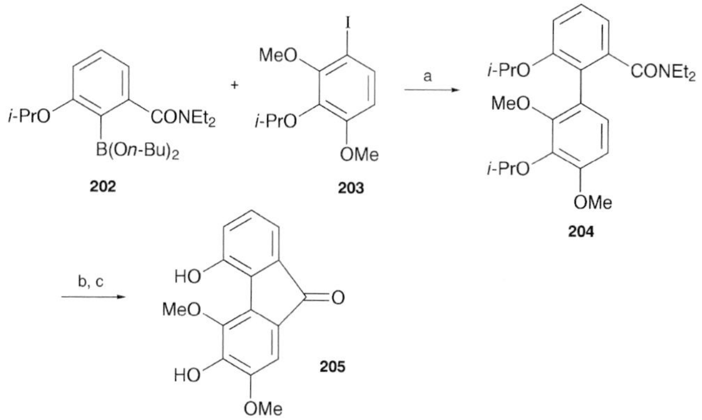

Scheme 6.22 Reaction conditions: (a) PdCl$_2$(dppf), K$_3$PO$_4$, DMF; (b) i-Pr$_2$NLi, 0°C, (c) BCl$_3$, CH$_2$Cl$_2$, 0°C.

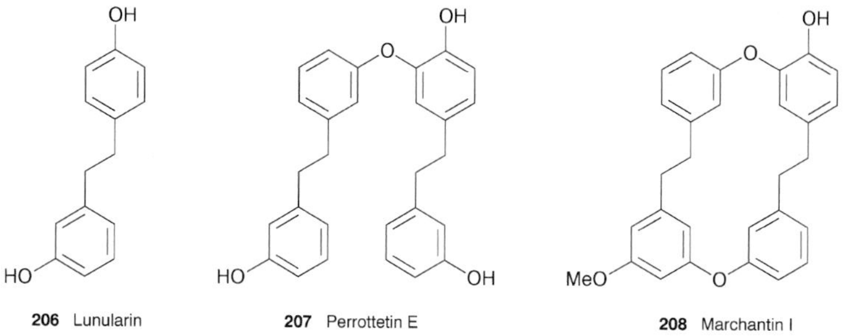

hypothetical precursor, lunularin (**206**), by oxidative C–C or C–O couplings at different positions. A large number of cyclic bis(benzyl) systems such as isoplagiochin (**209**) and riccardin (**210**) have been

209 Isoplagiochin C

210 Riccardin C

211 (+)-Cavicularin

synthesised, the strategies mostly relying upon Ullmann and Suzuki protocols for formation of the dibenzylether and dibenzyl linkages.[100] One of the most interesting structures belonging to this class is (+)-clavicularin (**211**).[101] The X-ray crystallographic analysis revealed that its cyclic dibenzyl-dihydrophenantrene skeleton has a highly strained structure in which the benzene ring A is twisted out of the plane by as much as 15°. Owing to restricted rotation and the highly strained geometry, cavicularin exhibits axial and planar chirality.

References

1. For a review of the literature from 1985 through the middle of 1992, see: M. Gill, 1993, in *The Chemistry of Natural Products* (R.H. Thompson, ed.), Blackie, Glasgow, p. 60.
2. K.P. Manfredi, J.W. Blunt, J.H. Cardellina II, J.B. McMahon, L.L. Panell, M.C. Gordon, and M.R. Boyd, 1991, *J. Med. Chem.*, 34, 3402.
3. M.R. Boyd, Y.F. Hallock, J.H. Cardellina II, K.P. Manfredi, J.W. Blunt, J.B. McMahon, R.W. Buckheit, G. Bringmann, M. Schaffer, G.M. Cragg, D.W. Thomas, and J.G. Jato, 1994, *J. Med. Chem.*, 37, 1740.
4. G. Bringmann, S. Harmsen, J. Holenz, T. Geuder, R. Gotz, P.A. Keller, R. Walter, Y.F. Hallock, J.H. Cardellina II, and M.R. Boyd, 1994, *Tetrahedron*, 50, 9643.
5. G. Bringmann, R. Gotz, S. Harmsen, J. Holenz, and R. Walter,1996, *Liebigs Ann. Chem.*, 2045.
6. Y.F. Hallock, K.P. Manfredi, J.W. Blunt, J.H. Cardellina II, M. Schaffer, K.-P. Gulden, G. Bringmann, A.Y. Lee, J. Clardy, G. Francois, and M.R. Boyd, 1994, *J. Org. Chem.*, 59, 6349.

7. G. Bringmann, R. Gotz, P.A. Keller, R. Walter, P. Henschel, M. Schaffer, M. Stablein, T.R. Kelly, and M.R. Boyd, 1994, *Heterocycles*, 39, 503.

8. T.R. Hoye, M. Chen, L. Mi, and O.P. Priest, 1994, *Tetrahedron Lett.*, 47, 8747.

9. P.D. Hobbs, V. Upender, and M.I. Dawson, 1997, *Synlett*, 965.

10. G. Bringmann, R. Gotz, P.A. Keller, R. Walter, M.R. Boyd, F. Lang, A. Garcia, J.J. Walsh, I. Tellitu, K.V. Bhaskar, and T.R. Kelly, 1998, *J. Org. Chem.*, 63, 1090.

11. J. Longmore, 1886, *Indian J. Chem.*, 5, 200.

12. National Coordinating Group on Male Fertility Agents, 1978, *China's Med.*, 6, 417.

13. A.E.A. Joseph, S.A. Martin, and P. Knox, 1986, *Br. J. Cancer*, 54, 511.

14. T.S. Lin, R. Schinazi, B.P. Griffith, E.M. August, B.H. Ericksson, D.K. Zheng, and W.H. Prusoff, 1989, *Antimicrob. Agents Chemother.*, 33, 2149.

15. (a) P.H. Dorsett and E.E. Kerstine, 1975, *J. Pharm. Sci.*, 64, 1073; (b) R.J. Radloff, L.M. Deck, R.E. Royer, and P.L. Vanderjagt, 1986, *Pharmacol. Res. Commun.*, 18, 1063.

16. L.V. Goryunova and S.A. Vichkanova, 1969, *Farmakol. Toksikol.*, 18, 615.

17. (a) A.I. Meyers and J.J. Willemsen, 1997, *J. Chem. Soc., Chem. Commun.*, 1573; (b) A.I. Meyers and J.J. Willemsen, 1998, *Tetrahedron*, 54, 10493.

18. (a) S. Toki, K. Ando, M. Yoshida, I. Kawamoto, H. Sano, and Y. Matsuda, 1992, *J. Antibiot.*, 45, 88; (b) S. Toki, K. Ando, I. Kawamoto, H. Sano, M. Yoshida, and Y. Matsuda, 1992, *J. Antibiot.*, 45, 1047; (c) S. Toki, E. Tsukuda, M. Nazawa, H. Nonaka, M. Yoshida, and Y. Matsuda, 1992, *J. Biol. Chem.*, 267, 14884.

19. K. Tatsuta, T. Yamazaki, M. Takanobu, and T. Yoshimoto, 1998, *Tetrahedron Lett.*, 39, 1771.

20. G. Büchi, D.H. Klaubert, R.C. Shank, S.M. Weinreb, and G.N. Wogan, 1971, *J. Org. Chem.*, 36, 1143.

21. H.G. Cutler, F.G. Crumley, R.H. Cox, O. Hernandez, R.J. Cole, and J.W. Dorner, 1979, *J. Agric. Food Chem.*, 27, 592.

22. G.-Q. Lin and M. Zhong, 1997, *Tetrahedron: Asymm.*, 8, 1369.

23. S. Kuyama and T. Tamura, 1957, *J. Am. Chem. Soc.*, 81, 5725 and 5726.

24. (a) A. Arnone, L. Camarda, G. Nasini, and L. Merlini, 1985, *J. Chem. Soc., Perkin Trans. 1*, 1387; (b) T. Yoshihara, T. Shimanuki, T. Araki, and S. Sakamura, 1975, *Agric. Biol. Chem.*, 39, 1683.

25. (a) T. Iida, E. Kobayashi, K. Ando, M. Yoshida, and H. Sano, 1989, *J. Antibiot.*, 42, 1475; (b) E. Kobayashi, K. Ando, H. Nakano, T. Iida, H. Ohno, M. Morimoto, and T. Tamaoki, 1989, *J. Antibiot.*, 42, 1470; (c) E. Kobayashi, K. Ando, H. Nakano, and T. Tamaoki, 1989, *J. Antibiot.*, 42, 153.

26. E. Kobayashi, H. Nakano, M. Morimoto, and T. Tamaoki, 1989, *Biochem. Biophys. Res. Comm.*, 159, 548.

27. R.F. Bruns, F.D. Miller, R.L. Meriman, J.J. Howbert, W.F. Heath, E. Kobayashi, I. Takahashi, T. Tamaoki, and H. Nakano, 1991, *Biochem. Biophys. Res. Comm.*, 176, 288.

28. T. Aminian-Saghafi, G. Nasini, T. Carrona, A.M. Braun, and E. Oliveros, 1992, *Helv. Chim. Acta*, 75, 531.

29. R.S. Coleman and E.B. Grant, 1994, *J. Am. Chem. Soc.*, 116, 8795.

30. R.S. Coleman and E.B. Grant, 1995, *J. Am. Chem. Soc.*, 117, 10889.

31. R.S. Coleman and E.B. Grant, 1991, *J. Org. Chem.*, 56, 1357.

32. R.S. Coleman and E.B. Grant, 1993, *Tetrahedron Lett.*, 34, 2225.

33. R.S. Coleman and E.B. Grant, 1994, *Tetrahedron Lett.*, 35, 8341.

34. (a) P.-S. Song, D.-P. Hader, and K.L. Poff, 1980, *Photochem. Photobiol.*, 32, 781; (b) P.-S. Song, D.-P. Hader, and K. L. Poff, 1980, *Arch. Microbiol.*, 126, 181.

35. (a) G. Lavie, F. Valentine, B. Levin, Y. Mazur, G. Gallo, D. Lavie, D. Weiner, and D. Meruelo, 1989, *Proc. Natl. Acad. Sci. U.S.A.*, 86, 5963; (b) D. Meruelo, G. Lavie, and D. Lavie, 1988, *Proc. Natl. Acad. Sci. U.S.A.*, 85, 5230.

36. M.J. Fehr, S.L. Carpenter, and J.W. Petrich, 1994, *Biomed. Chem. Lett.*, 4, 1339.

37. M. Maeda, H. Naoki, T. Matsuoka, Y. Kato, H. Kotsuki, K. Utsumi, and T. Tanaka, 1997, *Tetrahedron Lett.*, 38, 7411.

38. (a) D.W. Cameron and A.G. Riches, 1995, *Tetrahedron Lett.*, 36, 2331; (b) D.W. Cameron and A.G. Riches, 1997, *Aust. J. Chem.*, 50, 409.

39. For other reports on the synthesis of stentorin, see: (a) H. Iio, K. Zenfuku, and T. Tokoyama, 1995, *Tetrahedron Lett.*, 36, 5921; (b) H. Falk and E. Mayr, 1995, *Monatsch. Chem.*, 126, 1311.

40. Y.-H. Kuo, L.-M.Y. Kuo, and C.-F. Chen, 1997, *J. Org. Chem.*, 62, 3242.

41. T. Takeya, S. Yamaki, T. Itoh, H. Hosogai, and S. Tobinaga, 1996, *Chem. Pharm. Bull.*, 44, 909.

42. (a) M. Tanaka, H. Mitsuhashi, M. Maruno, and T. Wakamatsu, 1994, *Tetrahedron Lett.*, 35, 3733; (b) M. Tanaka, T. Ohshima, H. Mitsuhashi, M. Maruno, and T. Wakamatsu, 1995, *Tetrahedron*, 51, 11693.

43. T.F. Imbert, 1998, *Biochimie*, 80, 207.

44. For more recent syntheses not discussed here, see: (a) J.W. Bode, M.P. Doyle, M.N. Protopopova, and Q.-L. Zhou, 1996, *J. Org. Chem.*, 61, 9146; (b) D.M. Coltart and J.L. Charlton, 1996, *Can. J. Chem.*, 74, 88; (c) T. Kuroda, M. Takahashi, K. Kondo, and T. Iwasaki, 1996, *J. Org. Chem.*, 61, 9560; (d) M. Medarde, A.C. Ramos, E. Caballero, J.L. Lopez, R. Pelaez-Lamamie de Clairac, and A.C. Feliciano, 1996, *Tetrahedron Lett.*, 37, 2663; (e) S.B. Hadimani, R.P. Tanpure, and S.V. Bhat, 1996, *Tetrahedron Lett.*, 37, 4791; (f) S. Hanessian and S. Ninkovic, 1996, *Can. J. Chem.*, 74, 1880; (g) D.E. Bogucki and J.L. Charlton, 1995, *J. Org. Chem.*, 60, 588; (h) Y. Ara, T. Takeya, and S. Tobinaga, 1995, *Chem. Pharm. Bull.*, 43, 1977.

45. E.J. Bush and D.W. Jones, 1996, *J. Chem. Soc., Perkin Trans. 1*, 151.

46. C. Zhengxiong, H. Huizhu, W. Chengrui, L. Yuhui, D. Jianmi, U. Sankawa, H. Noguchi, and Y. Iitaka, 1986, *Chem. Pharm. Bull.*, 14, 2743.

47. J.S. Swenton, J.N. Freskos, P. Dalidowicz, and M.L. Kerns, 1996, *J. Org. Chem.*, 61, 459.

48. P.P. Deshpande, K.N. Price, and D.C. Baker, 1996, *J. Org. Chem.*, 61, 455.

49. For reports on the synthesis of racemic hongconin, see: (a) G.A. Kraus, J. Li, M. Gordon, and J.H. Hensen, 1994, *J. Org. Chem.*, 59, 2219; (b) I.R. Green, 1996, *Synth. Commun.*, 26, 867.

50. Frenolicin B: T. Masquelin, U. Hengartner, and J. Streith, 1997, *Helv. Chim. Acta*, 80, 43.

51. Frenolicin B and kalafungin: G.A. Kraus, J. Li, M.S. Gordon, and J.H. Jensen, 1995, *J. Org. Chem.*, 60, 1154.

52. Nanaomycin D: M.P. Winters, M. Stranberg, and H.W. Moore, 1994, *J. Org. Chem.*, 59, 7572.

53. (a) Medermycin: K. Tatsuta, H. Ozeki, M. Yamaguchi, M. Tanaka, and T. Okui, 1990, *Tetrahedron Lett.*, 31, 5495; (b) Y. Endo, T. Ohta, and S. Nozoe, 1991, *Tetrahedron Lett.*, 34, 5555; (c) W. Ng and D. Wege, 1996, *Tetrahedron Lett.*, 37, 6797.

54. K. Krohn, A. Michel, U. Florke, H.-J. Aust, S. Draeger, and B. Schulz, 1994, *Liebigs Ann. Chem.*, 1093 and 1099.

55. (a) G. Schlingmann, R.R. West, L. Milne, C.J. Pearce, and G.T. Carter, 1993, *Tetrahedron Lett.*, 34, 7225; (b) F. Petersen, T. Moerker, F. Vanzanella, and H.H. Peter, 1994, *J. Antibiot.*, 47, 1098; (c) R. Thiergardt, G. Rihs, P. Hug, and H.H. Peter, 1995, *Tetrahedron*, 51, 733.

56. (a) H.A. Weber, N.C. Baezinger, and J. B. Gloer, 1990, *J. Am. Chem. Soc.*, 112, 6718; (b) H.A. Weber and J.B. Gloer, 1991, *J. Org. Chem.*, 56, 4355.

57. P. Wipf and J.-K. Jung, 1998, *J. Org. Chem.*, 63, 3530.

58. For another report on the synthesis of palmarumycins, see: A.G.M. Barrett, H. Hamprecht, and T. Meyer, 1998, *J. Chem. Soc., Chem. Commun.*, 809.

59. For a recent synthesis, see: T. Saito, M. Morimoto, C. Akiyama, T. Matsumoto, and K. Suzuki, 1995, *J. Am. Chem. Soc.*, 117, 10757.

60. S. Funayama, M. Ishibashi, Y. Anraku, K. Komiyama, and S. Omura, 1989, *Tetrahedron Lett.*, 30, 7427.

61. K. Shin-ya, K. Furihata, Y. Teshima, Y. Hayakawa, and H. Seto, 1992, *Tetrahedron Lett.*, 33, 7025.

62. (a) H. Koyama and T. Kamikawa, 1997, *Tetrahedron Lett.*, 38, 3973; (b) H. Koyama and T. Kamikawa, 1998, *J. Chem. Soc., Perkin Trans. 1*, 203.

63. (a) S.J. Gould, J. Chen, M.C. Cone, M.P. Gore, C.R. Melville, and N.J. Tamaya, 1996, *J. Org. Chem.*, 61, 5720; (b) S.J. Gould, C.R. Melville, M.C. Cone, J. Chen, and J.R. Carney, 1997, *J. Org. Chem.*, 62, 320.

64. L.A. Decosterd, I.C. Parsons, K.R. Gustafson, J.H. Cardellina II, J.B. McMahon, G.M. Cragg, Y. Murata, L.K. Pannell, J.R. Steiner, J. Clardy, and M.R. Boyd, 1993, *J. Am. Chem. Soc.*, 115, 6673.

65. (a) M. Konishi, H. Ohkuma, K. Matsumoto, T. Tsuno, H. Kamei, T. Mivaaki, T. Oki, H. Kawaguchi, G.D. VanDuyne, and J. Clardy, 1989, *J. Antibiot.*, 42, 1449; (b) M. Konishi, H. Ohkuma, T. Tsuno, T. Oki, G.D. VanDuyne, and J. Clardy, 1990, *J. Am. Chem. Soc.*, 112, 3715.

66. M.F. Semmelhack, J. Gallagher, and D. Cohen, 1990, *Tetrahedron Lett.*, 31, 1521.

67. M.D. Shair, T.Y. Yoon, K.K. Mosny, T.C. Chou, and S.J. Danishefsky, 1996, *J. Am. Chem. Soc.*, 118, 9509.

68. Y. Kashman, K.R. Gustafson, R.W. Fuller, J.H. Cardellina, J.B. McMahon, M.J. Currens, R.W. Buckheit Jr, S.H. Hughes, G.M. Cragg, and M.R. Boyd, 1992, *J. Med. Chem.*, 35, 2735.

69. P.L. Boyer, M.J. Currens, J.B. McMahon, M.R. Boyd, and S.H. Hughes, 1993, *J. Virol.*, 67, 2412.

70. (a) J. Polonsky, 1956, *Bull. Soc. Chim. Fr.*, 914; (b) J. Polonsky and Z. Baskevitch, 1958, *Bull. Soc. Chim. Fr.*, 929.

71. For reports on the synthesis of calanolides not discussed here, see: (a) A. Khilevich, A. Mar, M.T. Flavin, J.D. Rizzo, L. Lin, S. Dzekhtser, D. Brankovic, H. Zhang, W. Chen, S. Liao, D.E. Zembower, and Z.-Q. Xu, 1996, *Tetrahedron:Asymm.*, 7, 3315; (b) K.S. Rehder and J.A. Kepler, 1996, *Synth. Commun.*, 26, 4005; (c) P.P. Deshpande, F. Tagliaferri, S.F. Victory, S. Yan, and D.C. Baker, 1995, *J. Org. Chem.*, 60, 2964; (d) A. Kucherenko, M.T. Flavin, W.A. Boulanger, A. Khilevich, R.L. Shone, J.D. Rizzo, A.K. Sheinkman, and Z.-Q. Xu, 1995, *Tetrahedron Lett.*, 36, 5475; (e) C.J. Palmer and J.L. Josephs, 1995, *J. Chem. Soc., Perkin Trans. I*, 3135; (f) C.J. Palmer and J.L. Josephs, 1995, *Tetrahedron Lett.*, 35, 5363; (g) B. Chenera, M.L. West, J.A. Finkelstein, and G.B. Dreyer, 1993, *J. Org. Chem.*, 58, 5605.

72. B.M. Trost and F.D. Toste, 1998, *J. Am. Chem. Soc.*, 120, 9074.

73. K. Inoue, S. Matsutani, and Y. Kawamura, EP-395,418, 31 October 1990.

74. (a) T. Yoshida, S. Nakamoto, R. Sakazaki, K. Matsumoto, Y. Terui, T. Sato, H. Arita, S. Matsutani, K. Inoue, and I. Kudo, 1991, *J. Antibiot.*, 44, 1467; (b) K. Tanaka, S. Matsutani, K. Matsumoto, and T. Yoshida, 1992, *J. Antibiot.*, 45, 1071.

75. Y. Genisson, P.C. Tyler, and R.N. Young, 1994, *J. Am. Chem. Soc.*, 116, 759.

76. W. Schoeller, M. Dohrn, and W. Hohlweg, 1940, *Naturwiesenschaften*, 28, 532.

77. E.J. Corey and L.I. Wu, 1993, *J. Am. Chem. Soc.*, 115, 9327.

78. S. Yamamura and Y. Hirata, 1963, *Tetrahedron*, 19, 1485.

79. P.J. Scheuer, 1973, *Chemistry of Marine Natural Products*, Academic Press, New York, p. 10.

80. H. Nemoto, M. Nagamochi, H. Ishibashi, and K. Fukumoto, 1994, *J. Org. Chem.*, 59, 74. For other recent reports, see: A. Nath, A. Gosh, and R.V. Venkateswaran, 1992, *J. Org. Chem.*, 57, 1467.

81. For another synthesis, see: A. Nath, A. Ghosh, and R.V. Venkateswaran, 1992, *J. Org. Chem.*, 57, 1467.

82. (a) A.T. McPhail, M.L. King, C.C. Chiang, H.C. Ling, E. Fujita, and M. Ochiai, 1982, *J. Chem. Soc., Chem. Commun.*, 1150; (b) M.L. King, H.C. Ling, C.B. Wang, and S.C. Leu, 1975, *Med. Sci.*, 1, 11.

83. B.M. Trost, P.D. Greenspan, B.V. Yang, and M.G. Saulnier, 1990, *J. Am. Chem. Soc.*, 112, 9022. For other synthetic reports, see: (a) A.E. Davey, M. Schaeffer, and R.J.K. Taylor, 1991, *J. Chem. Soc., Chem. Commun.*, 1137; (b) A.E. Davey, M. Schaeffer, and R.J.K. Taylor, 1992, *J. Chem. Soc., Perkin Trans. 1*, 2657.

84. For another report on the synthesis of rocaglamide, see: A.E. Davey, M.J. Schaeffer, and R.J.K. Taylor, 1992, *J. Chem. Soc., Perkin Trans. I*, 2657.

85. *Pharmacology and Applications of Chinese Materia Medica* (H.M. Chang and P.P.H. But, eds), World Scientific Publishing, Singapore, 1986, vol. 1, pp. 255-268.

86. (a) M. Nakao and T. Fukushima, 1934, *J. Pharm. Soc. Jpn.*, 54, 154; (b) A.-R. Lee, W.-L. Wu, W.-L. Chang, H.-C. Lin, and M.-L. King, 1987, *J. Nat. Prod.*, 50, 157; For a review on the synthesis of tanshinones and related diterpenoid quinones reported until 1991, see: (c) D. Goldsmith, 1992, in *The Total Synthesis of Natural Products*, (J. ApSimon ed.), Wiley, New York, vol. 8, p. 1; (d) R.H. Thompson, 1992, in *The Total Synthesis of Natural Products*, (J. ApSimon ed.), Wiley, New York , vol. 8, p. 311.

87. (a) R.L. Danheiser, D.S. Casbeier, and J.L. Loebach, 1992, *Tetrahedron Lett.*, 33, 1149; (b) R.L. Danheiser, D.S. Casbeier, and F. Firoonzia, 1995, *J. Org. Chem.*, 60, 8341.

88. T. Matsumoto, Y. Takeda, K. Soh, H. Gotoh, and S. Imai, 1996, *Chem. Pharm. Bull.*, 44, 1318.

89. (a) K. Hatano, E. Higashide, M. Shibata, Y. Kameda, S. Horii, and K. Mizuno, 1980, *J. Agric. Biol. Chem.*, 44, 1157; (b) S. Horii, H. Fukase, E. Mizuta, K. Hatano, and K. Mizuno, 1980, *Chem. Pharm. Bull.*, 28, 3601.

90. (a) O. Kikuchi, T. Eguchi, K. Kakinuma, Y. Koezuma, K. Shindo, and N. Otake, 1993, *J. Antibiot.*, 46, 985; (b) R.M. Knobler, F.B. Radlwimmer, and M.J. Lane, 1992, *Nucl. Acids Res.*, 20, 4553; (c) Y.Yamashita and H. Nakano, 1988, *Nucl. Acids Res. Symp. Ser.*, 20, 60; (d) M. Greenstein, T. Monji, R. Yeung, W.M. Maiese, and R.J. White, 1986, *Antimicrob. Agents Chemother.*, 29, 861 and references therein.

91. (a) T. Matsumoto, T. Hosoya, and K. Suzuki, 1992, *J. Am. Chem. Soc.*, 114, 3568; (b) T. Hosoya, E. Takashiro, T. Matsumoto, and K. Suzuki, 1994, *J. Am. Chem. Soc.*, 116, 1004.

92. Reviews: (a) R.H. Thomson, 1996, *Naturally Occurring Quinones IV*, 4th edn, Blackie, London; (b) J. Rohr and R. Thiericke, 1992, *Nat. Prod. Rep.*, 9, 103.

93. (a) S. Kondo, S. Gomi, D. Ikeda, M. Hamada, T. Takeuchi, H. Iwai, J. Seki, and H. Hoshino, 1991, *J. Antibiot.*, 44, 1228; (b) M. Ogasawara, M. Hasegawa, Y. Hamagishi, H. Kamei, and T. Oki, 1992, *J. Antibiot.*, 45, 129.

94. For reports on the synthesis of hydroaromatic angucyclines not discussed here, see: (a) K. Krohn and J. Micheel, 1998, *Tetrahedron*, 54, 4827; (b) K. Krohn, N. Boker, U. Florke, and C. Freund, 1997, *J. Org. Chem.*, 62, 2350; (c) D.S. Larsen, M.D. O'Shea, and S. Brooker, 1996, *J. Chem. Soc., Chem. Commun.*, 203; (d) D. Larsen and M.D. O'Shea, 1996, *J. Org. Chem.*, 61, 5681; (e) D. Larsen and M.D. O'Shea, 1995, *J. Chem. Soc., Perkin Trans. I*, 1019; (f) G. Matsuo, Y. Miki, M. Nakata, S. Matsumura, and K. Toshima, 1996, *J. Chem. Soc., Chem. Commun.*, 225; (g) V.A. Boyd and G.A. Sulikowski, 1996, *J. Am. Chem. Soc.*, 117, 8472; (h) K. Krohn, K. Khanbabaee, and P.G. Jones, 1995, *Liebigs Ann.*, 1981; (i) K. Kim and G. Sulikowski, 1995, *Angew. Chem., Int. Ed. Engl.*, 34, 2396.

95. K. Kim, Y. Guo, and G.A. Sulikowski, 1995, *J. Org. Chem.*, 60, 6866.

96. Isolation: (a) H. Oka, K. Asahi, H. Morishita, M. Senda, K. Shiratori, Y. Iimura, T. Sakurai, I. Uzawa, S. Iwadare, and N. Takahashi, 1985, *J. Antibiot.*, 38, 1100; structure elucidation: (b) Y. Iimura, T. Sakurai, K. Asahi, N. Takahashi, and H. Oka, 1984, *Acta Crystallogr., Sect. C*, 40, 2058.

97. K. Mori, A. Kamada, and H. Mori, 1989, *Liebigs Ann. Chem.*, 303.

98. R.H. Green, 1997, *Tetrahedron Lett.*, 38, 4697.
99. J.-M. Fu, B.-P. Zhao, M.J. Sharp, and V. Snieckus, 1994, *Can. J. Chem.*, 72, 227.
100. (a) J. Gerencser, G.M. Keseru, I. Macsari, M. Nogradi, M. Kajtar-Peredy, and A. Szollosy, 1997, *J. Org. Chem.*, **62**, 3666, and references therein; (b) T. Eicher, S. Fey, W. Puhl, E. Buchel, and A. Spiecher, 1998, *Eur. J. Chem.*, 877, and references therein.
101. M. Toyota, T. Yoshida, Y. Kan, S. Takaoka, and Y. Asakawa, 1996, *Tetrahedron Lett.*, 37, 4745.

7 Highlights in the chemical synthesis of terpenes and terpenoids

E. Tyrrell

7.1 Introduction

The complex structures of many terpenes continue to challenge the modern-day synthetic organic chemist. Frequently, their very potent biological activity or rarity makes a total synthesis desirable. This chapter will therefore reflect these factors. It will describe those molecules that possess unusual topologies whose use is limited by their scarcity or whose synthesis has employed novel methodologies or reagents.

7.2 Sesquiterpenes

7.2.1 Seychellene

The tricyclic sesquiterpene seychellene (4) has aroused considerable interest because of its unique topography, and as a result, several syntheses have been reported.[1] Rao and Vijaya Bhaskar[2] reported a total synthesis that involved treatment of 1 with Bu_3SnH to effect the key cyclisation step. This gave 2 via a vinyl-radical-mediated Michael addition reaction (Scheme 7.1). Methylene oxidation, dehydration and catalytic hydrogenation gave 3 with the desired equatorial methyl group. This was then converted to seychellene (4) in four highly efficient transformations. A strategy that provided access to compound 4, as well as other members of this class, was reported by Cory et al.[3] Their concise approach to a variety of isolates from patchouli oil commenced from the same carvone derivative (5) (Scheme 7.2). Reductive cleavage and alkylation of intermediate 5 gave the bicyclooctanone (6). Olefinic oxidation, alkylation of the resulting keto-aldehyde and oxidation gave the dicarbonyl (7) as a single stereoisomer. Ring closure, to afford 3-oxopatchouli alcohol (8) via an intramolecular aldol reaction was effected by treatment of 7 with LDA. By using the same intermediate, 5, the authors were able to synthesis the norcycloseychellene derivative 10 by an analogous methodology via the diketone 9.

Hagiwara et al.[4] employed an elegant threefold Michael addition of the trimethylsilyl enol ether (11) to divinyl ketone to obtain the dione derivative (12) (Scheme 7.3). In this process, the authors succeeded in

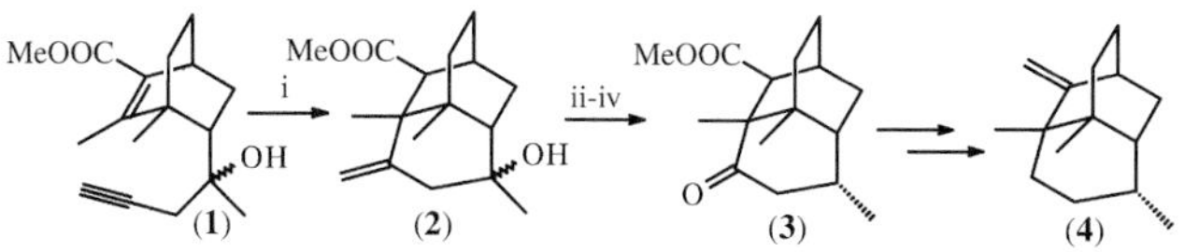

Scheme 7.1 Reagents: (i) Tributyl tin hydride; (ii) ruthenium trichloride–sodium metaperiodate; (iii) pTSA/silica; (iv) hydrogenation 10% Pd-C.

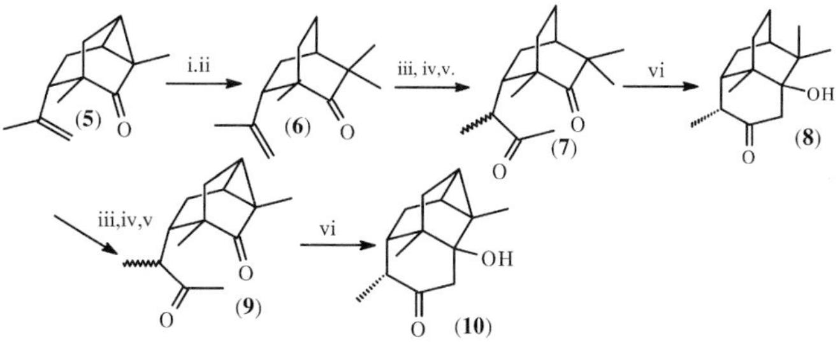

Scheme 7.2 Reagents: (i) Li, ammonia; (ii) MeI; (iii) chromyl chloride; (iv) MeMgBr; (v) PCC; (vi) LDA.

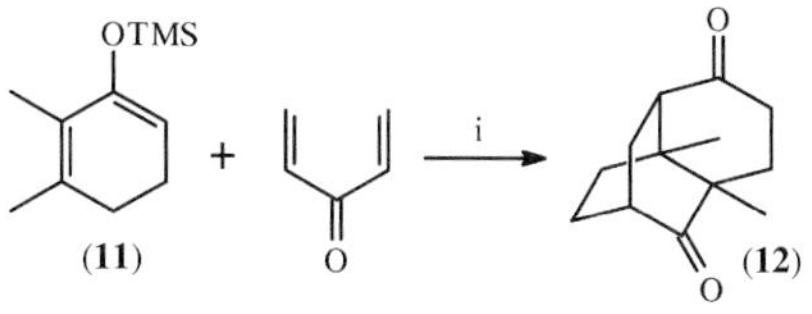

Scheme 7.3 Reagents: (i) excess diethyl aluminium chloride.

forming three new bonds and two new six-membered rings in a one-pot formal synthesis, as compound **12** had been transformed into (±)-seychellene (**4**) previously.[5]

7.2.2 Longifolene

The plant-growth promotor longifolene (**20**) is a bridged-ring sesquiterpene that occupies a particularly important niche in synthetic organic chemistry because of its unusual architecture. It is not surprising, therefore, that the first total syntheses reported[6] spurred the development of new strategies for the attainment of this and related compounds. More recently, the Diels–Alder reaction has featured in several syntheses of longifolene (**20**). In an intramolecular approach, Fallis and Lei[7] obtained the cyclisation precursor **18** in the following way (Scheme 7.4). Treatment of the known fulvene (**13**)[8] with methyllithium gave the cyclopentadienyl

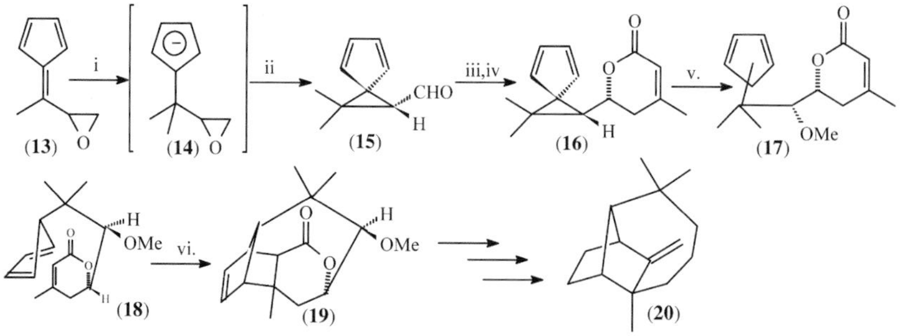

Scheme 7.4 Reagents (i) MeLi; (ii) manganese dioxide; (iii) LDA; (iv) methyl 3-methylcrotonate; (v) Lewis acid; (vi) sealed tube.

anion **14** which cyclised, in an *exo*-tet mode, onto the epoxide. The concomitant alcohol was resolved[9] and the *R*-(+) isomer oxidised to aldehyde **15**. Condensation with the anion derived from methyl 3-methylcrotonate led to attack at the less-hindered *re* face of the aldehyde. The product then cyclised to the lactone **16**. Heat treatment of triene **17**, obtained by reaction of **16** with Lewis acid, gave a single Diels–Alder adduct **19** (97%) from *exo* transition state **18**. Selective functional group transformations then converted the adduct, **19**, to longifolene (**20**).

An alternative approach for the synthesis of longifolene (**20**), reported by Ho *et al.*,[10] involved an intermolecular Diels–Alder reaction followed by ring expansion. This gave the seven-membered ring of longifolene (**20**) (Scheme 7.5). The intermolecular Diels–Alder reaction between fulvene and maleic anhydride provided the desired *exo* adduct (**21**) which was readily transformed to the *endo*-ester **22** in five steps. Cyclodehydration[11] gave the predominant enone, **23**, which was methylated to afford **24**

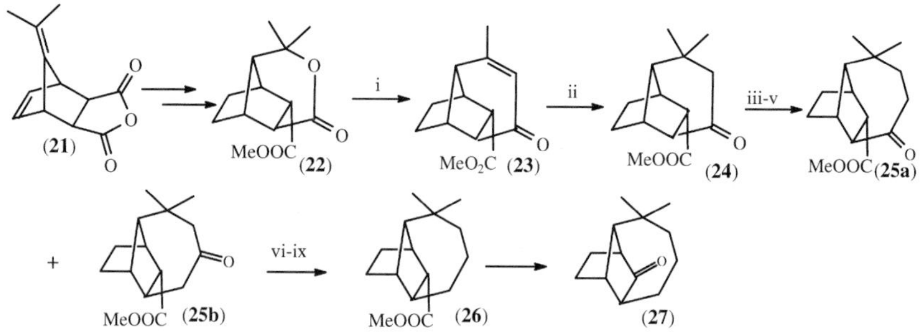

Scheme 7.5 Reagents: (i) phosphorus pentoxide, methanesulfonic acid; (ii) lithium dimethylcuprate; (iii) ethyl diazoacetate; (iv) LiI, collidine; (v) potassium carbonate, MeI; (vi) sodium borohydride; (vii) mesyl chloride; (viii) NaI, Zn; (ix) hydrogenation.

quantitatively. Construction of the longifolene skeleton was achieved by a ring expansion reaction. The major regioisomers were then decarbethoxylated and re-esterified to afford the inseparable keto-esters **25a** and **25b** in a ratio of 2.5:1. Deoxygenation of the ketone moiety by sequential reduction, mesylation, elimination and hydrogenation with Adams catalyst gave the ester **26**. This was then converted to the known ketone **27** in a formal synthesis[12] of longifolene (**20**).

An alternative strategy for the synthesis of this framework was reported by Kuo and Money (Scheme 7.6).[13] Treatment of the trimethylsilyl enol ether **28**, derived from (+)-8-cyanocamphor ethylene

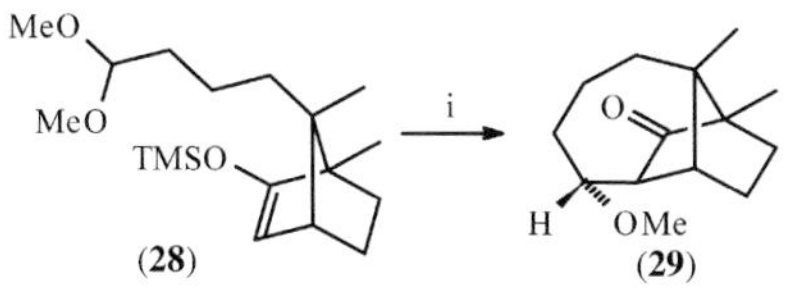

Scheme 7.6   Reagents: (i) titanium tetrachloride.

ketal, with Lewis acid, effected an intramolecular Mukaiyama reaction to afford the tricyclic ketone **29** as the major isomer. This process provided a key intermediate for an enantiospecific synthesis of longifolene (**20**).

7.2.3   α-Cedrene

The tricyclic sesquiterpene α-cedrene (**30**), isolated from cedar wood oil,[14] contains a synthetically challenging tricyclo[5.3.1.0^{1,5}]undecane skeleton

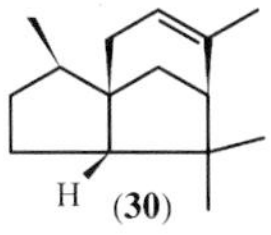

as well as four asymmetric centres and an isolated double bond in one of the rings. Several syntheses have been reported,[15] the most recent of which is that of Mukherjee *et al.*[16] who used an intramolecular anionic cyclisation reaction in their first stereocontrolled total synthesis of (±)-Δ^2-cedrene (**31**) (Scheme 7.7). The cyclisation precursor **33**, obtained from

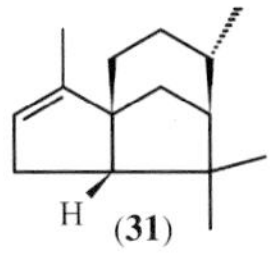

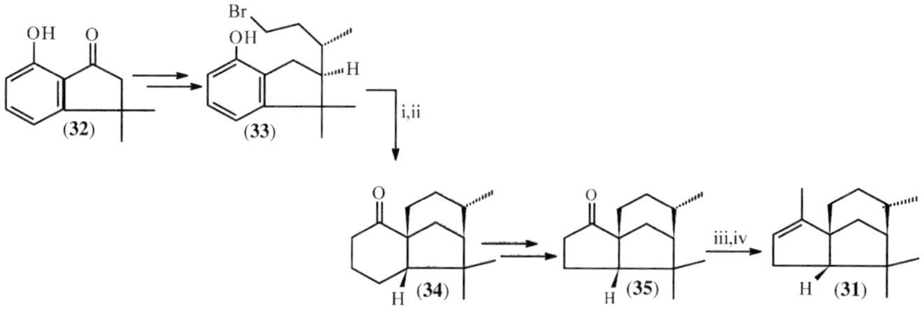

Scheme 7.7   Reagents: (i) *t*-BuOK; (ii) hydrogenation; (iii) MeMgBr; (iv) DMSO, heat.

the hydroxyindanone **32**,[17] was treated with *t*-BuOK and then hydrogenated to furnish the A/B *cis*-fused ketone **34**. Oxidative cleavage of the cyclohexanone ring of **34**, and Dieckmann cyclisation, provided the ring-contracted product **35** which was transformed into (±)-Δ^2-cedrene (**31**) by methylation and dehydration.

A recent synthesis of β-cedrene (**40**), by Rigby and Kirova-Snover,[18] utilised the tricyclic compound **37** as a starting material. This was synthesised as a 1:1 mixture of C(7) epimers by an *endo*-selective chromium(0)-mediated intramolecular $[6\pi + 2\pi]$ cycloaddition reaction of the cycloheptatriene complex **36** (Scheme 7.8). Routine modification of C(4)

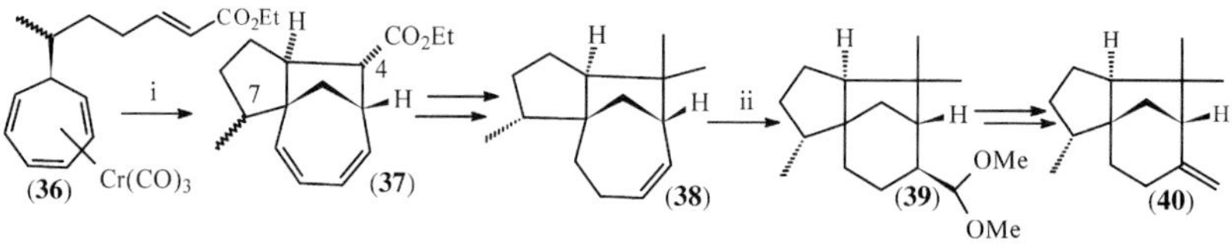

Scheme 7.8   Reagents: (i) dioxane; (ii) Tl(III) nitrate, MeOH.

provided the *gem*-dimethyl intermediate which was regioselectively reduced, and the desired C(7) epimer separated, to provide the alkene **38**. Using the Taylor–McKillop protocol[19] for effecting the key oxidative rearrangement/ring-contraction step, the tricycle **39** was obtained as a single regio- and stereoisomer. This compound was readily transformed into β-cedrene (**40**) in a further three steps.

A formal synthesis of (±)-α-cedrene (**30**), by Ghosh and Patra,[20] involved irradiation of diene **41** to access the cyclobutane **42** (Scheme 7.9). Acid treatment of **42** effected a 1,5-bond pinacol-style rearrangement reaction to afford the spiropentanone **43**. This was then converted, by using standard transformations, to the known ketone **44**[21] to effect a formal synthesis of α-cedrene (**30**).

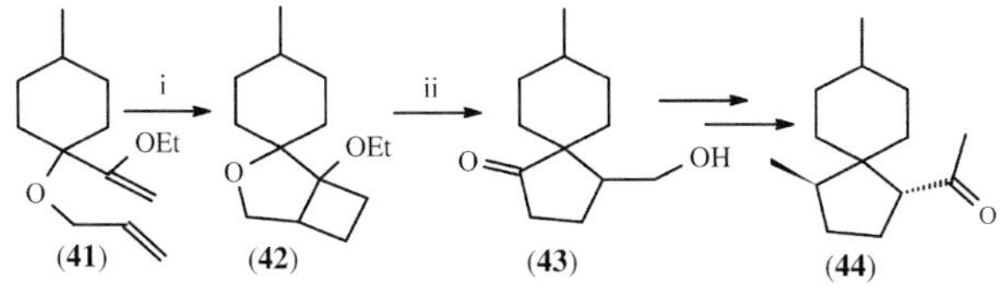

Scheme 7.9   Reagents: (i) *hv* (ii) TfOH.

7.2.4   Quadrone

The tetracyclic polyquinane quadrone **49** has aroused considerable interest ever since its isolation from *Aspergillus terreus*[22] and its cytotoxic activity was subsequently discovered.[23] As a result many total syntheses have been reported.[24] A synthesis by Liu and Llinas-Brunet[25] utilised α-campholenic acid (**45**) to establish the B-ring and part of the C-ring of quadrone as well as some of the appropriate chirality (Scheme 7.10).

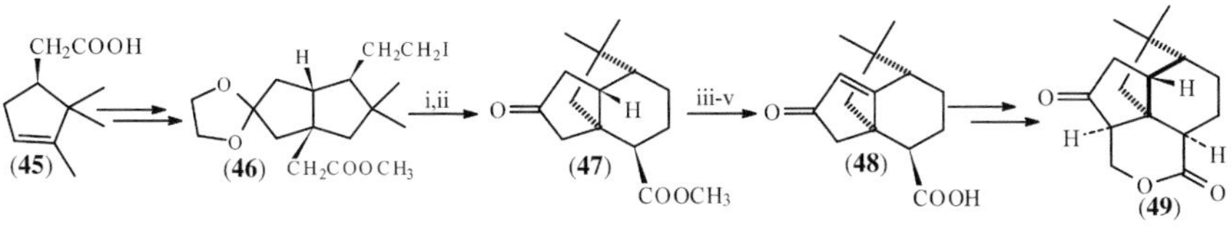

Scheme 7.10   Reagents: (i) lithium hexamethyldisilazide; (ii) dilute acid; (iii) NaOH; (iv) PhSeCl; (v) hydrogen peroxide.

Compound **45** was converted to the key cyclisation precursor **46** by a multistep synthesis. Treatment of the iodo-ketal **46** with lithium hexamethyldisilazide then effected the key cyclisation reaction to form a tricyclic compound, which could be deprotected to afford the (−)- keto-ester **47**. A racemic form of **47** had previously been transformed to (±)-quadrone.[26] The conversion of **47** into enone **48** required a reduction, a selenylation and an oxidative work-up. This enone was then converted, in three steps, to (−)-quadrone (**49**).

A short route to Danishefsky's intermediate, **53**, by Parsons and Neary[27] exploited a radical cyclisation on alkene **50** to access **51** in 80% yield (Scheme 7.11). Oxidative cleavage of **51** gave the desired axial acid **52** which cyclised, under basic conditions, to **53**, which had previously[28] been transformed into (±)-quadrone.

A stereoselective electroreductive cyclisation reaction (Scheme 7.12) of the aldehyde **54** was used by Little *et al.*[29] to build the first ring system of quadrone **49** in their formal total synthesis. Compound **55** was then

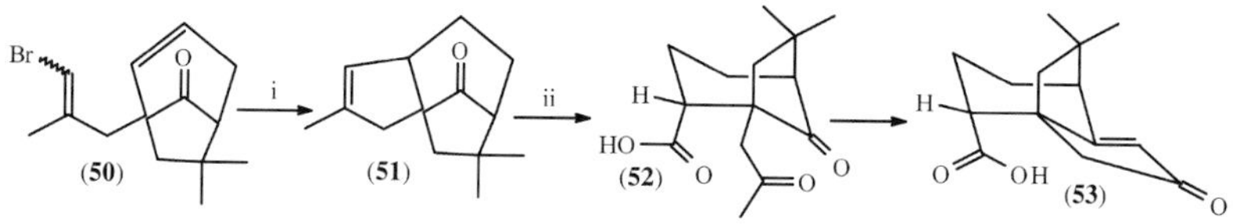

Scheme 7.11 Reagents: (i) tributyltin hydride, AlBN; (ii) ruthenium (IV) oxide, sodium periodate, MeCN.

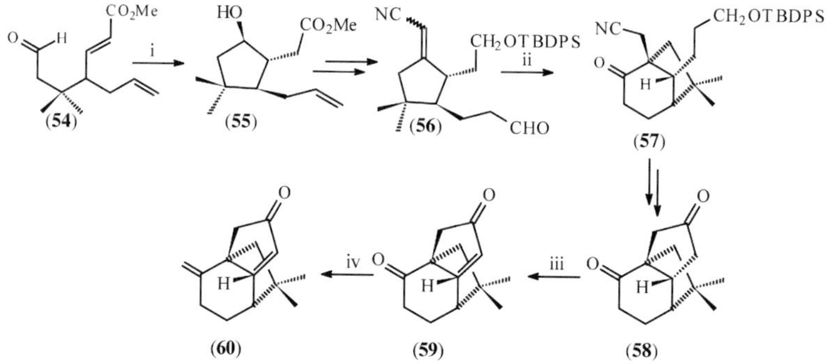

Scheme 7.12 Reagents: (i) Hg-SCE; (ii) −2.4 V, Hg-SCE; (iii) LTMP, THF, −78°C then PhSecl, THF, −78°C then hydrogen peroxide, DCM, 0°C; (iv) Ph₃P=CH₂, THF/hexanes, 0°C.

transformed into the nitrile-aldehyde **56** and this also cyclised electro-chemically. Subsequent oxidation of the diastereoisomeric alcohols gave the [3:2:1] framework **57** as a single diastereoisomer. Cleavage of the silyl ether and oxidation of the primary alcohol to the corresponding car-boxylic acid set the scene for a novel oxidative decarboxylation reaction[30] followed by a radical closure onto the nitrile. The resulting imine was hydrolysed to the ketone **58** and this was transformed to the Kende enone[31] **60** via the enone **59**.

The cornerstone of a racemic and parallel chiral synthesis of quadrone **49**, by Smith III et al.,[32] involved a rearrangement reaction of the [4:3:2] propellane **62** to generate the tricarbocyclic quadrone skeleton in **63** (Scheme 7.13). Conversion of the resolved ester **61**[33] to **62** was followed by the pivotal rearrangement reaction to obtain **63** as a single compound in 85% yield. This was then transformed, in five steps, to the nonracemic

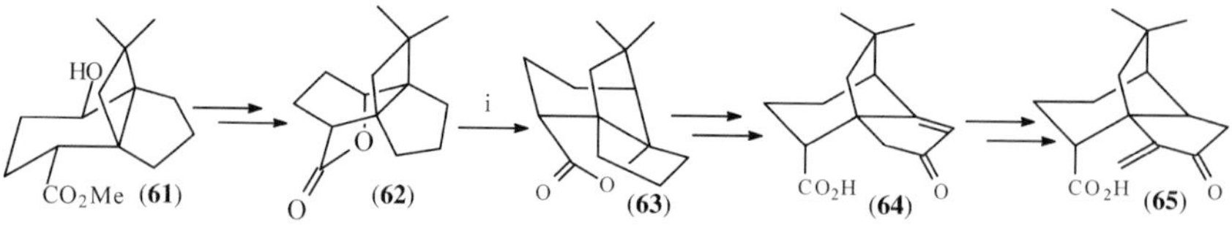

Scheme 7.13 Reagents: (i) 40% dilute sulfuric acid.

enone **64** which had previously been converted to terrecyclic acid **65**[24] and thence to quadrone (**49**).

7.2.5   Triquinanes

Members of the triquinane family of sesquiterpenes exhibit a diverse range of biological activities ranging from coral defence[34] to antibiotic activity.[35] Over the past decade interesting methodologies have been developed to access this framework.[36] However, the use of an intra-molecular Pauson–Khand (IMPK) [2 + 2 + 1] cycloaddition reaction[37] provides a method of accessing a bicyclopentanoid skeleton in one step. In an approach for the synthesis of linearly fused triquinanes[38] the dicobalt hexacarbonyl complexed derivative of the propynyl acetate **66** (Scheme 7.14) underwent an IMPK reaction to afford the tricyclopenta-noids **67a** and **67b**, with the best stereoselectivity for this reaction being a 15:1 ratio of **67a:67b**.

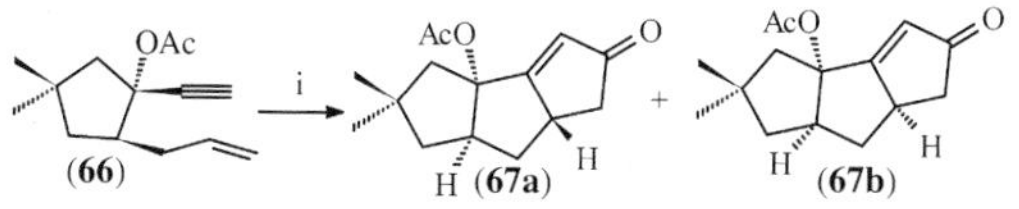

Scheme 7.14   Reagents: (i) dicobalt octacarbonyl/DCM.

A further example of this chemistry being used for triquinane synthesis was reported by Schore and Rowley.[39] Heating of the dicobalt hexa-carbonyl complex **68** (Scheme 7.15) in a sealed tube brought about an IMPK reaction to produce **69a** and **69b** in a ratio of 1:8. The major isomer **69b** was then transformed into (±)-pentalenene (**71**) by enone reduction, Wittig olefination and isomerisation.

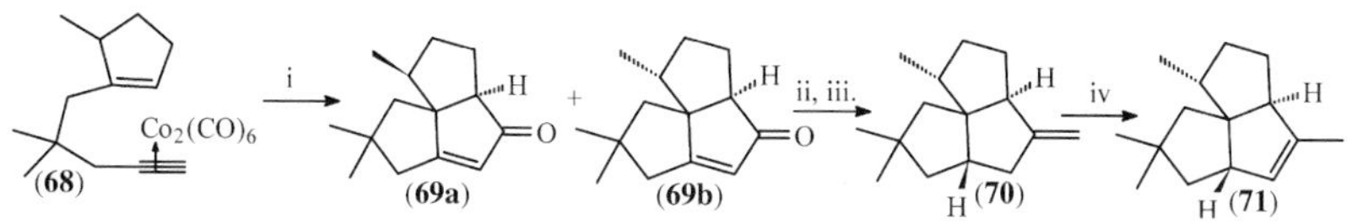

Scheme 7.15   Reagents: (i) heat; (ii) sodium/liquid ammonia; (iii)$Ph_3P{=}CH_2$; (iv) TsOH.

In a complementary formal synthesis of (±)-pentalenic acid (**75**) (Scheme 7.16) the siloxy enynes **72a** and **72b** were treated with dicobalt hexacarbonyl and heated in a sealed tube to afford a mixture of enones. The 9-*exo*-adduct **73** was reduced to provide **74**, a compound which had previously been converted to **75** by Hudlicky *et al.*[40] This represents a formal synthesis of this interesting molecule.

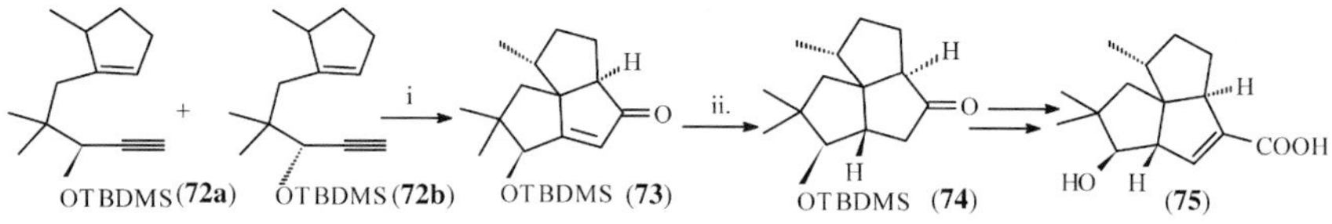

OTBDMS (72a)   OTBDMS (72b)   OTBDMS (73)   OTBDMS (74)   HO H (75)

Scheme 7.16   Reagents: (i) dicobalt octacarbonyl, heat; (ii) sodium/liquid ammonia.

7.3   Diterpenes

7.3.1   Clerodane

Clerodanes comprise a large class of diterpenes, with over 800 examples having been described to date.[41] Although biological testing has yet to be comprehensively carried out on all members of this family of natural products, many of those that have been evaluated have been found to exhibit a wide range of activities, including insect antifeedant properties[42] as well as antiviral, antibiotic, and antitumour effects.[43] Synthetic approaches to the clerodanes have been reviewed.[44] During the past decade a resurgence of interest in these natural products (possibly associated with their biological activity) has led to several recent syntheses.[45] As a model, in a synthesis of the *trans*-decalin **80**,[46] the pendant alkene of ester **76** was oxidised by sequential dihydroxylation and oxidative cleavage. The resulting aldehyde underwent spontaneous intramolecular aldol condensation to produce the *trans*-decalin **77** as the sole product after alcohol protection (Scheme 7.17). Reduction and acetylation provided the diacetate **78** which was then deprotected, oxidised and converted by a Wittig reaction to the *exo*-methylene derivative **79**. This compound was then transformed in a further three steps to epoxide **80**, which itself displayed interesting antifeedant activity as well as representing a useful model for clerodin **81** itself.

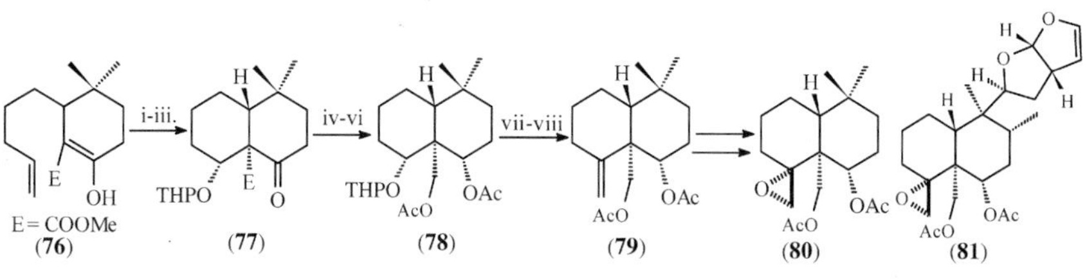

E = COOMe
(76)   (77)   (78)   (79)   (80)   (81)

Scheme 7.17   Reagents: (i) osmium tetroxide, NMO; (ii) sodium periodate; (iii) DHP cat.H⁺; (iv) sodium borohydride; (v) lithium aluminium hydride; (vi) acetic anhydride; (vii) PCC; (viii) methyltriphenylphosphonium halide.

Ley *et al.*[47] have reported a synthesis of the polycyclic epoxy diacetate **87** (Scheme 7.18) which contains many of the features found in the natural

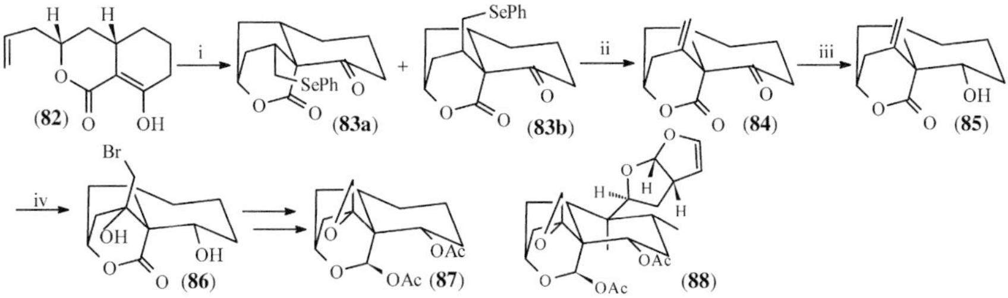

Scheme 7.18 Reagents: (i) NPSP, zinc iodide; (ii) TBHP, cat. titanium(IV) isopropoxide; (iii) sodium borohydride; (iv) bromine in MeCN, sat. aq. ammonium chloride.

product jodrellin A (**88**). Significantly, this compound also displays insect antifeedant properties. The key reaction used to access this unusual polycyclic framework was an enol/phenylepiselenonium ion trapping. The cyclisation precursor **82** was treated with *N*-phenylselenophthalimide (NPSP) and zinc iodide to give the isomeric selenides **83a** and **83b**. This mixture underwent efficient oxidative-elimination to produce alkene **84** as a single product. The transformation of alkene **84** to diacetate **87** exploited a stereoselective reduction on alkene **84** to give lactone **85** and a bromohydration to obtain **86**. Exposure of **86** to Amberlite resin gave an epoxide which underwent reduction to a lactol and sequential di-acetylation to afford **87**, a model for jodrellin A (**88**).

In a recent Diels–Alder approach[48] to (±)-2-oxo-5α,8,-13,14,15,16-tetranorclerod-3-en-12-oic acid (**94**), the dienone ester **89** was treated with *trans*-piperylene to give the *cis*-decalin **90** as the major product isomer (Scheme 7.19). This adduct contained four of the contiguous stereogenic centres that are found in many of the naturally occurring *cis*-clerodanes. Conversion of adduct **90** to the decalone **91** required a further four chemical transformations. Wolff–Kishner decarbonylation gave the two isomeric decalenes **92a** and **92b** in a ratio of 3:1. Irradiation of this

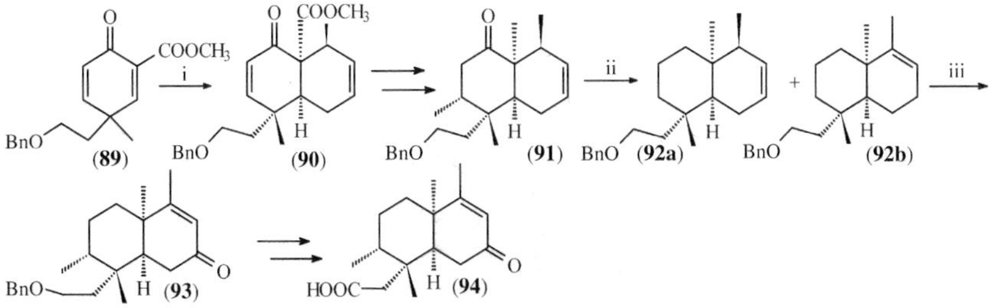

Scheme 7.19 Reagents: (i) *trans*-piperylene, zinc chloride; (ii) hydrazine, KOH, ethylene glycol; (iii) *hv*, oxygen.

inseparable mixture gave the enone **93** which was debenzylated and oxidised to obtain the desired clerodenoic acid **94**.

Liao and Liu[49] have also employed Diels–Alder chemistry in their recent approach to the *cis*-clerodane **99**. The Diels–Alder adduct **95**, obtained from the reaction of methyl tiglate with a masked bromo-stabilising *o*-benzoquinone (Scheme 7.20), was debrominated and the car-

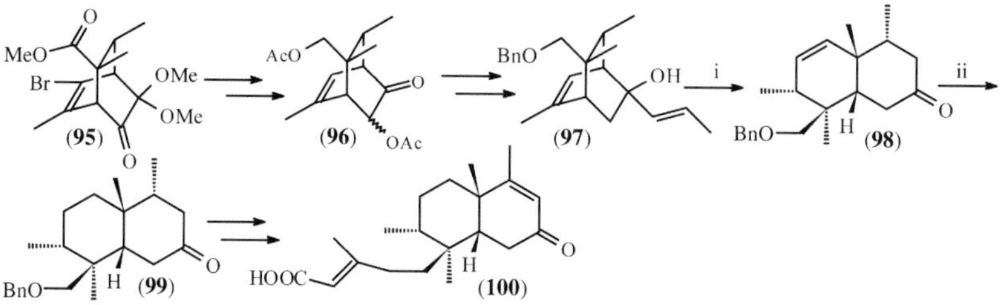

Scheme 7.20   Reagents: (i) KH, 18-crown-6; (ii) hydrogenation, Pd/C.

bonyl group α-transposed to obtain the ketone **96**. This compound was then transformed in six steps to the crucial rearrangement precursor **97**. Exposure of **97** to potassium hydride and 18-crown-6 effected an anionic oxy-Cope rearrangement reaction to give **98** (84%). Hydrogenation of this compound to decalone **99** was achieved with total stereocontrol. The generality of this chemistry was further established with their subsequent synthesis of the *cis*-clerodane diterpenic acid (**100**).

Kato *et al.*[50] have recently reported a stereocontrolled synthesis of (+)-*trans*-decalone (**107**) using (−)-verbenone (**102**),[51] derived from (1*R*,5*S*)-(+)-nopinone (**101**), to provide four of the chiral centres present in the *neo-trans*-clerodane **108** (Scheme 7.21). Thus, **102** was efficiently converted to the keto-acetate **103** by standard transformations. Regio- and

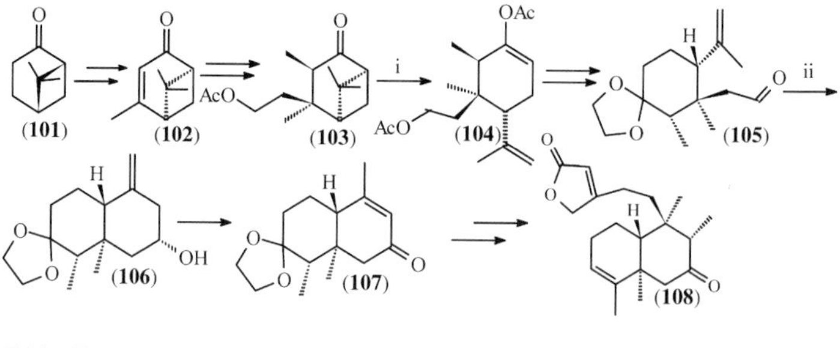

Scheme 7.21   Reagents: (i) acetic anhydride/zinc acetate/boron trifluoride diethyl etherate; (ii) diethylaluminium chloride.

stereoselective cleavage of **103** to obtain enol acetate (**104**) was effected by methodologies developed in their laboratories.[52] Hydrolysis of **104** was accompanied by concomitant epimerisation of the secondary methyl group. This was followed by a protection and oxidation procedure to furnish the aldehyde **105**. Stereoselective ene reaction occurred efficiently to provide the *trans*-decalol **106** which was then oxidised and isomerised to the conjugated enone **107**, a potential intermediate for the synthesis of **108**.

7.3.2   Labdane diterpenoids

The highly oxygenated labdane forskolin **109** was first discovered as a result of a screening programme aimed at identifying new medicinal

compounds from Indian plants used in traditional medicine.[53] The methanolic extract from the roots of one particular plant, *Coleus forskohlii*, exhibited marked hypotensive and antispasmolytic activity associated with its activation of the enzyme adenylate cyclase.[54] Owing to its interesting topology and biological activity, forskolin has been regarded as a prime candidate for synthesis by many organic chemistry groups. A number of total syntheses,[55] and syntheses of advanced intermediates,[56] have been reported and reviewed[57] and the Ziegler intermediate **110**[58] has been the target of several formal approaches.

Kanematsu and Nagashima[59] reported a pathway to compound **116** which exploited the intramolecular cycloaddition reaction of propynyl ether **111** to obtain the tricyclic compound **113** via the allenyl ether intermediate **112** (Scheme 7.22). This compound was then transformed in six steps to enone **114** which, upon conjugate methylation, selenation and selenoxide elimination, produced enone **115**. The latter was then

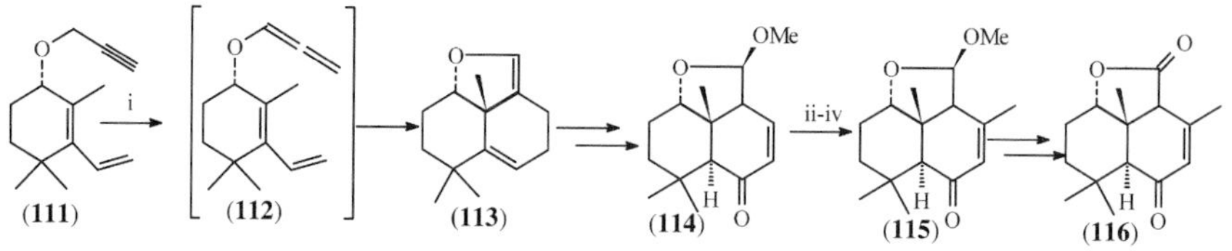

Scheme 7.22   Reagents: (i) *tert*-butanol, *tert*-potassium butoxide; (ii) dimethylcopper lithium; (iii) PhSeCl; (iv) *m*CPBA.

converted, using the Grieco protocol,[60] to lactone **116**, which had previously been used as an intermediate in Ziegler's synthesis.[58]

Leclaire *et al.*[61] reported the transformation of lactone **117**, synthesised previously,[62] into the Ziegler intermediate **110** (Scheme 7.23). Initially, compound **117** was converted to **118** via a sequence of two new rearrangement reactions. *O*-Methylation and reduction then afforded compound **119**. This compound was then treated with *m*CPBA, to obtain the ketodiol, and this reduced and protected to obtain lactol **120**, an intermediate in which epimerisation of the C(11) OMe occurred. Oxidation of this compound followed by dehydration finally produced the desired intermediate **110** in 15% overall yield from **117**.

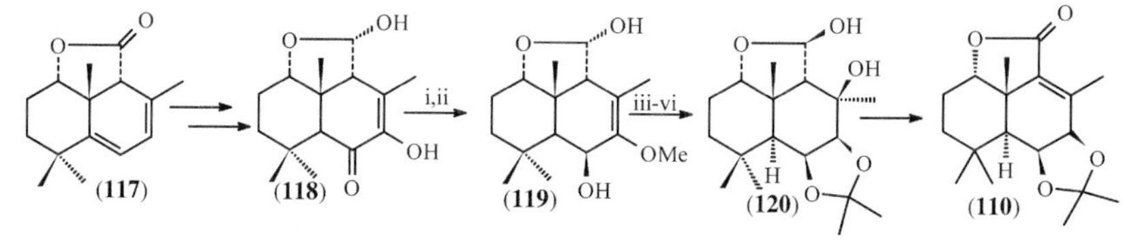

Scheme 7.23   Reagents: (i) MeI, NaH, THF; (ii) DIBAL-H, THF; (iii) *m*CPBA; (iv) sodium borohydride; (v) *p*TSA, 2,2-dimethoxypropane; (vi) thionyl chloride, pyridine.

Lett *et al.*[63] have published a trilogy of papers on the total synthesis of forskolin (**109**). Their aim was to establish a significantly different approach to those total syntheses that were already known,[55] and this was achieved by their synthesis[63a] of the novel advanced intermediate **121** (Scheme 7.24). Thus, reduction of enone **121** and stereodirected epoxidation gave the α-epoxide **122**. Ring opening of the epoxide gave the carbonate **123** which was desilylated and oxidised to aldehyde **124**. Propargylation of aldehyde **124** was somewhat problematic, it affording retroaldol products. However, condensation with an excess of propynyl-lithium (prepared *in situ*) gave the propynyl alcohol **125** as a mixture of two diastereoisomers. Collins oxidation of this mixture and conversion to the dihydro-γ-pyrone **126** was accomplished in low yield by using caesium carbonate. Stereoselective conjugate addition gave **127** which underwent

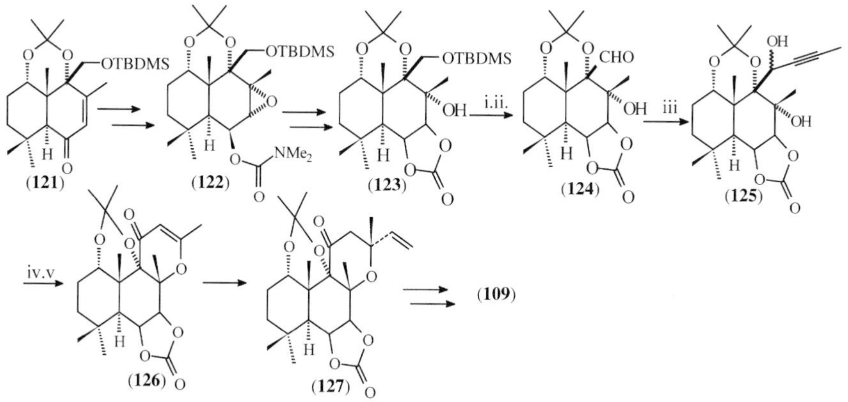

Scheme 7.24 Reagents: (i) TBAF, THF; (ii) Swern oxidation; (iii) propynyllithium; (iv) Collins oxidation; (v) caesium carbonate.

hydrolysis and acetylation to afford forskolin (109) in 73% yield, from compound 127.

7.3.3 Vinigrol

The diterpenoid vinigrol (134), first isolated from the fungus *Virgaria nigra*,[64] was found to exhibit antihypertensive and platelet aggregation-inhibiting activity.[65] Particularly topical, however, was finding that vinigrol arrests the progression of AIDS-related complex (ARC) to full-blown AIDS (acquired immune deficiency syndrome).[66] The structure of vinigrol (134) was determined in 1987.[67] The molecule possesses a unique tricyclic skeleton incorporating a synthetically challenging bridged eight-membered ring. To date only a few synthetic approaches have been recorded to this molecule, with the first synthesis of the decahydro-1,5-butanonaphthalene skeleton 133 not being reported until 1993 by Hanna *et al.*[68] This synthesis was predicated on the intermolecular Diels–Alder reaction between cyclohexadiene (128) and 1,4-benzoquinone (Scheme 7.25), which furnished an adduct that was immediately reduced to the cyclic ether 129. Hydroxyl group protection and acid hydrolysis produced the hydroxy ketone 130a and its corresponding hemiketal 130b. Dehydration and selective hydrogenation of the less-hindered double bond, by using Wilkinson's catalyst, followed by deprotection, gave the ketone 131. Treatment of this compound with excess vinyl Grignard reagent gave the *endo*-adduct 132 exclusively, as a result of chelation control. Exposure of this adduct to KH, in the presence of 18-crown-6, effected an anionic oxy-Cope rearrangement reaction to produce 133 an

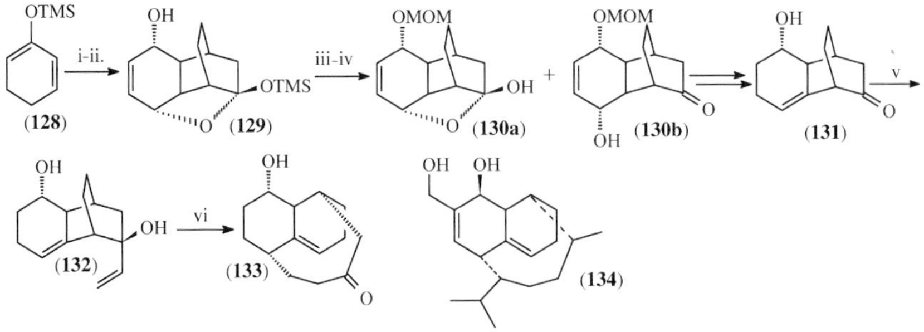

Scheme 7.25 Reagents: (i) 1,4-benzoquinone; (ii) sodium borohydride/cerium(III) chloride; (iii) MOMCl; (iv) $BF_3 \cdot Et_2O$, THF, $-78°C$, 15 min; (v) vinyl-magnesium chloride; (vi) KH, THF, 18-crown-6.

intermediate for vinigrol (**134**). Elaboration of compound **133** to vinigrol (**134**) will require reductive decarbonylation of C(10), incorporation of a tertiary alcohol at C(8) and functionalisation of ring A.

Two diterpenes that have captured the imagination of many synthetic chemists in recent years have been taxol[70] and the gibberellins.[71] As the synthesis of these natural products has been extensively reviewed elsewhere, this chapter will conclude by looking at other tetracyclic diterpenoids.

7.3.4   Kaurane and other tetracyclic diterpenes

In common with taxol and the gibberellins, the bridged bicyclic ring system is a feature found in several tetracyclic diterpenes such as phyllocladane (**144**) and kaurane (**145**). Malacria et al.[72] reported a sequence of three cycloaddition reactions to access either framework (Scheme 7.26). A noteworthy feature of this approach was that the stereochemical outcome of the B/C-ring junction could be readily manipulated by varying the C(12) substituent. Treatment of the diester **135** with the acetate **136** in the presence of palladium(II) acetate and triisopropyl phosphite effected a Trost [3 + 2] cycloaddition reaction to afford the adduct **137** as a mixture of diastereoisomers. This mixture was protodesilylated and both the acetal **138** and the keto derivative **139** were exposed to the cobalt-catalysed cyclotrimerisation reaction conditions described by Vollhardt[73] to afford **140** and **141** in high yield. Thermolysis of compound **140** provided the precursor for phyllocladane **142**. By way of contrast, thermolysis of **141** provided a separable mixture of both **142** and **143** in a ratio of 58:42.

Aphidicolin (**151**) and its structural relative stemodin have attracted considerable interest[74] because of their potent biological activity.[75] Aphidicolin (**151**), for instance, has selective action on DNA polymerase

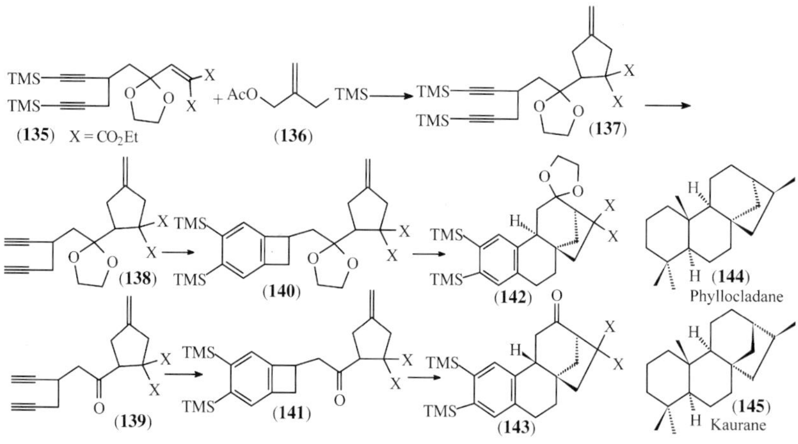

Scheme 7.26.

and shows potential as an antiviral[76] and an antitumour agent.[77] Iwata *et al.*[78] have reported a facile formal synthesis of ($\pm$)-aphidicolin (**151**) via a Lewis-acid-mediated stereoselective spiroannulation reaction. Treatment of the bisbenzyl acetal **146** with TMSOTf gave the spiroannulated compounds **147a** and **147b** in 88% combined yield as a 2:1 mixture (Scheme 7.27). The tricyclic ketone **148** was obtained by base treatment of the mixture followed by chromatographic separation of the major isomer. Hydrogenation of enone **148**, ketalisation, and PCC oxidation gave the ketal-enone **149** containing the B/C/D-ring of **151**. Stereoselective construction of the A-ring was then accomplished by a synthetic sequence that had been previously reported by the authors[79] for the formation of **150**. This constitutes a formal total synthesis of aphidocolin **151**.

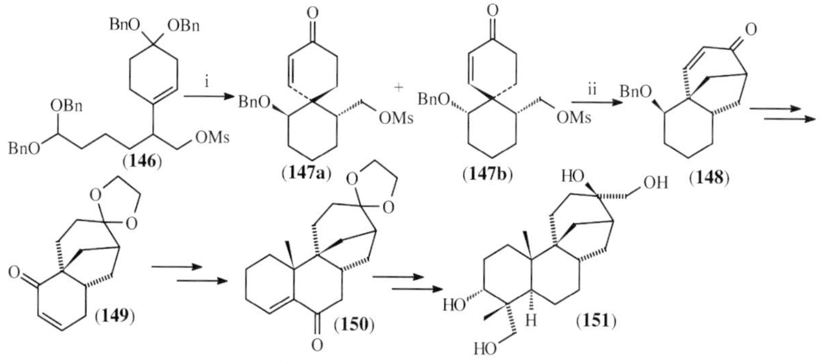

Scheme 7.27   Reagents: (i) TMSOTf; (ii) *t*-BuOK, then separate diastereoisomers.

References

1. B.B. Snider and R.B. Beal, 1988, *J. Org. Chem.*, 53, 4508-4515 and references cited therein. Also, see: (a) D. Spitzner and I. Klein, 1990, *Liebigs Ann. Chem.*, 63-68; (b) H. Hagiwara, 1992, *J. Synth. Org. Chem. (Japan)*, 50, 713-725.
2. (a) G.S.R. Subba Rao and K. Vijaya Bhaskar, 1993, *J. Chem. Soc., Perkin Trans.1.*, 2813-2816; (b) G.S.R. Subba Rao and K. Vijaya Bhaskar, 1989, *Tetrahedron Lett.*, 30, 225-228.
3. R.M. Cory *et al.*, 1990, *Tetrahedron Lett.*, 31, 6839-6842.
4. H. Hagiwara *et al.*, 1990, *J. Chem. Soc., Perkin Trans.1.*, 2109-2113.
5. M.E. Jung *et al.*, 1981, *J. Am. Chem. Soc.*, 103, 6677.
6. E.J. Corey *et al.*, 1964, *J. Am. Chem. Soc.*, 86, 478.
7. (a) A.G. Fallis and B. Lei, 1990, *J. Am. Chem. Soc.*, 112, 4609-4610; (b) A.G. Fallis and B. Lei, 1993, *J. Org. Chem.*, 58, 2186-2191.
8. A.G. Fallis *et al.*, 1984, *Can. J. Chem.*, 62, 2451.
9. J.W. Westley and B. Halpern, 1968, *J. Org. Chem.*, 33, 3978.
10. T.-L. Ho *et al.*, 1992, *Can. J. Chem.*, 70, 1375-1384.
11. P.E. Eaton *et al.*, 1973, *J. Org. Chem.*, 35, 4071.
12. W.S. Johnson *et al.*, 1972, *J. Am. Chem. Soc.*, 97, 4777.
13. D.L. Kuo and T. Money, 1988, *Can. J. Chem.*, 66, 1794-1804.
14. G. Stork and R. Breslow, 1953, *J. Am. Chem. Soc.*, 75, 3291.
15. M. Horton and G. Pattenden, 1983, *Tetrahedron Lett.*, 24, 2125-2128 and references cited therein.
16. D. Mukherjee *et al.*, 1996, *Tetrahedron Lett.*, 37, 4421-4422.
17. N.F. Hayes and R.H. Thomson, 1956, *J. Chem. Soc.*, 1585-1589.
18. J.H. Rigby and M. Kirova-Snover, 1997, *Tetrahedron Lett.*, 38, 8153-8156.
19. A. McKillop *et al.*, 1973, *J. Am. Chem. Soc.*, 95, 3635.
20. S. Ghosh and D. Patra, 1995, *J. Chem. Soc., Perkin. Trans. 1*, 2635-2641; for a formal synthesis see: Y.-J. Chen *et al.*, 1993, *Tetrahedron Lett.*, 34, 2961-2962 and references cited therein.
21. P.T. Lansbury *et al.*, 1974, *J. Am. Chem. Soc.*, 96, 896.
22. R.L. Ranieri and G.J. Calton, 1978, *Tetrahedron Lett.*, 19, 499-502.
23. G.J. Calton *et al.*, 1978, *J. Antibiot.*, 31, 38.
24. R.L. Funk and M.M. Abelman, 1986, *J. Org. Chem.*, 51, 3248 and references cited therein.
25. H.J. Liu and M. Llinas-Brunet, 1988, *Can. J. Chem.*, 66, 528-530.
26. S. Danishefsky *et al.*, 1980, *J. Am. Chem. Soc.*, 102, 4262.
27. P.J. Parsons and A.P. Neary, 1989, *J. Chem. Soc., Chem. Commun.*, 1090.
28. S. Danishefsky *et al.*, 1981, *J. Am. Chem. Soc.*, 193, 4136.
29. R.D. Little *et al.*, 1990, *Tetrahedron Lett.*, 31, 485-488.
30. Y.N. Ogibin *et al.*, 1975, *Izvestiya, Acadeii Nauk SSSR, Senya Khimi*, 1461.
31. A.S. Kende *et al.*, 1982, *J. Am. Chem. Soc.*, 104, 5808; also, see: (a) M.S. Ermolenko and M. Pipelier, 1997, *Tetrahedron Lett.*, 38, 5975-5976; (b) T. Imanishi *et al.*, 1992, *Chem. Pharm. Bull.*, 40, 2691-2693; (c) M. Nakagawa *et al.*, 1991, *J. Chem. Soc., Chem. Commun.*, 1598-1599.
32. A.B. Smith III *et al.*, 1991, *J. Am. Chem. Soc.*, 113, 3533-3542.
33. J.K. Whitesell and D. Reynolds, 1983, *J. Org. Chem.*, 48, 3548-3551.
34. R.P. Burkholder and L.M. Burkholder, 1958, *Science*, 127, 1174.
35. B.K. Koe and W.D. Celmer, 1956, *Antibio. Annu.*, 672.
36. (a) B.B. Snider and R.B. Beal, 1988, *J. Org. Chem.*, 53, 4508-4515; (b) G. Pattenden and S.J. Teague, 1988, *J. Chem. Soc., Perkin Trans. 1*, 1077-1083; (c) H.R. Sonawane *et al.*, 1991, *Tetrahedron*, 47, 8259-8276; (d) H.R. Sonawane *et al.*, 1993, *Synlett*, 875-883; (e) A.

Gambacorta *et al.*, 1992, *Tetrahedron*, 48, 4459-4464; (f) M. Lautens and J. Blackwell, 1998, *Synthesis*, 537; (g) S. Kim *et al.*, 1998, *Synlett*, 981-982; (h) K.C. Nicolaou and E.J. Sorenson, 1996, *Classics in Total Synthesis*, VCH, 221-225; (i) M. Channon *et al.*, 1998, *Synthesis*, 1559-1583.

37. See: N.E. Shore, 1990, *Organic Reactions*, 40, 1.
38. D.O. Koltun *et al.*, 1995, *Tetrahedron Lett.*, 36, 4651-4654.
39. N.E. Schore and E.G. Rowley, 1992, *J. Org. Chem.*, 57, 6853-6861.
40. T. Hudlicky *et al.*, 1987, *J. Org. Chem.*, 52, 4641.
41. (a) A.T. Merritt and S.V. Ley, 1992, *Nat. Prod. Reports*, 9, 243-287; (b) T. Tokoroyama, 1993, *J. Synth. Org. Chem. (Japan)*, 51, 1164-1177.
42. S.V. Ley *et al.*, 1979, *J. Chem. Soc., Chem. Commun.*, 97-98.
43. (a) T.A. Van Beek and A. De Groot, 1986, *Recl. Trav. Pays-Bas.*, 105, 513; (b) ibid., 1987, 106, 1; (c) H. Nakamura *et al.*, 1984, *Tetrahedron Lett.*, 25, 2989; (d) H. Wu *et al.*, 1986, *Bull. Chem. Soc. Jpn.*, 59, 2495.
44. (a) T. Tokoroyama *et al.*, 1988, *Tetrahedron*, 44, 6607; (b) T. Tokoroyama *et al.*, 1983, *J. Chem. Soc., Chem. Commun.*, 1516.
45. (a) T. Tokoroyama *et al.*, 1990, *Bull. Chem. Soc. Jpn.*, 63, 1720-1728; (b) E. Piers and J.Y. Roberge, 1991, *Tetrahedron Lett.*, 32, 5219-5222; (c) E. Piers and J.Y. Roberge, 1992, *Tetrahedron Lett.*, 33, 6923-6926; (d) E. Piers *et al.*, 1995, *J. Chem. Soc., Perkin. Trans. 1*, 963; (e) H.-J. Liu *et al.*, 1995, *Synlett*, 545-546.
46. J.Y. Lallemand and H. Bouchard, 1990, *Tetrahedron Lett.*, 31, 5151-5152; also, see: J.Y. Lallemand and J. Lejeune, 1992, *Tetrahedron Lett.*, 33, 2977-2980.
47. S.V. Ley *et al.*, 1996, *J. Chem. Soc., Perkin Trans. 1*, 611-620.
48. H.-J. Liu *et al.*, 1997, *Canadian Journal of Chemistry*, 75, 646-656.
49. C.-C. Liao and W.-C. Liu, 1998, *Synlett*, 912-913.
50. M. Kato *et al.*, 1997, *Tetrahedron Lett.*, 39, 6845-6848.
51. M. Kato *et al.*, 1993, *J. Org. Chem.*, 58, 3923-3927.
52. M. Kato *et al.*, 1989, *J. Org. Chem.*, 54, 1536-1538.
53. S.V. Bhat *et al.*, 1977, *Tetrahedron Lett.*, 18, 1669-1672.
54. J.W. Daly and K.B. Seamon, 1986, *Adv. Cyclic Nucleotide Res.*, 20, 1-150.
55. (a) F.E. Ziegler *et al.*, 1987, *J. Am. Chem. Soc.*, 109, 8115; (b) S. Ikegami *et al.*, 1988, *J. Am. Chem. Soc.*, 110, 3670-3672; (c) E.J. Corey *et al.*, 1988, *J. Am. Chem. Soc.*, 110, 3672-3673.
56. (a) K.C. Nicolaou *et al.*, 1989, *J. Chem. Soc., Chem. Commun.*, 512-513; (b) S. Benetti *et al.*, 1991, *Il Farmaco*, 46, 1281-1295; (c) J.Y. Lallemand *et al.*, 1995, *Tetrahedron Lett.*, 36, 2075-2078; (d) E.A. Ruveda and M.D. Preite, 1994, *Synthetic Communications*, 24, 2809-2825.
57. (a) E.A. Ruveda *et al.*, 1992, *Tetrahedron*, 48, 963-1037 and references cited therein; (b) S.V. Bhat, 1993, *Progress in the Chemistry of Organic Natural Products*, Springer, New York, 3-74.
58. F.E. Ziegler *et al.*, 1985, *Tetrahedron Lett.*, 26, 3307.
59. K. Kanematsu and S. Nagashima, 1989, *J. Chem. Soc., Chem. Commun.*, 1028-1029.
60. P.A. Grieco *et al.*, 1978, *Tetrahedron Lett.*, 19, 419.
61. M. Leclaire *et al.*, 1995, *J. Chem. Soc., Chem. Commun.*, 1333-1334.
62. (a) J.-Y. Lallemand *et al.*, 1993, *Synth. Commun.*, 23, 1923; (b) J.-Y. Lallemand *et al.*, 1993, *Tetrahedron Asymmetry*, 4, 1775.
63. Synthesis of compound **118**: (a) R. Lett *et al.*, 1996, *Tetrahedron Lett.*, 37, 1015-1018; (b) Total synthesis: ibid., 1019-1022; (c) Asymmetric approach via a 'Ziegler'-type intermediate: ibid., 1023-1024.
64. I. Uchida *et al.*, 1987, *J. Org. Chem.*, 52, 5292-5293.
65. T. Ando *et al.*, 1988, *J. Antibiot.*, 41, 25-30.
66. D.B. Norris *et al.*, 1991, *Chem. Abstr.*, 115, 64776h.
67. M. Hashimoto *et al.*, 1987, *J. Org. Chem.*, 52, 5293-5294.

68. I. Hanna *et al.*, 1993, *J. Org. Chem.*, 58, 2349-2350.
69. I. Hanna *et al.*, 1997, *J. Org. Chem.*, 62, 5063-5068.
70. (a) K.C. Nicolaou, 1996, *Classics in Total Synthesis* (K.C. Nicolaou and E.J. Sorenson, eds), VCH Publishers, New York, pp. 655-671 and references cited therein; (b) A.N. Boa *et al.*, 1994, *Contemporary Organic Synthesis*, 1, 47-75; (c) J.D. Winkler, 1992, *Tetrahedron*, 48, 6953-7056.
71. (a) L.N. Mander, 1992, *Chem. Rev.*, 92, 573-611; (b) L.N. Mander, 1988, *Natural Product Reports*, 541-579.
72. M. Malacria *et al.*, 1993, *J. Org. Chem.*, 58, 4298-4305.
73. K.P.C. Vollhardt, 1984, *Angew. Chem., Int. Ed. Engl.*, 23, 539.
74. (a) K.P.C. Vollhardt *et al.*, 1991, *J. Am. Chem. Soc.*, 113, 4006 and references cited therein; (b) J. Mann and P. Hegarty, 1993, *Synlett*, 553-554 [approach to stemodin (**151**)]; (c) A. Abad *et al.*, 1994, *J. Chem. Soc., Perkin Trans. 1*, 2987-2991 [approaches to beyerane diterpenoids].
75. (a) J.R. Hanson, 1988, *Nat. Prod. Rep.*, 3, 211-227; (b) R.A. Bucknall *et al.*, 1973, *Antimicrob. Agents Chemother.*, 4, 294.
76. G. Pedrali-Noy and S. Spadari, 1980, *Virology*, 36, 457.
77. J. Douros and M. Suffness, 1980, *New Anticancer Drugs* (S.K. Carter and Y. Sakurai, eds), Springer Verlag, Berlin, p. 29.
78. C. Iwata *et al.*, 1995, *Chem. Pharm. Bull.*, 43, 1407-1411.
79. T. Tanaka *et al.*, 1995, *Chem. Pharm. Bull.*, 43, 193.

8 Strategies for building and modifying naturally occurring steroids

C.M. Marson

8.1 General chemistry

In such an all-encompassing field, this overview cannot be comprehensive and thus reflects the author's interests. However, a fascinating account of steroid research during the past 60 years[1] forms a fitting prologue.

8.1.1 Alcohols and their derivatives

Inversion of steroidal alcohols can be effected by treatment of their chloromethanesulfonates with caesium acetate in the presence of 18-crown-6.[2] Acylation of cholestan-3α- and -3β-alcohols to give the epimeric esters proceeds more satisfactorily using N,N,N',N'-tetramethyl-azodicarboxamide and tributylphosphine rather than the diethylazo-dicarboxylate-triphenylphosphine reagent of conventional Mitsunobu reactions.[3] Steroidal alcohols can be rapidly acylated by using acid anhydrides in the presence of trimethylsilyl trifluoromethanesulfonate as a catalyst.[4]

Some steroidal primary and secondary alcohols can be efficiently oxidised to the corresponding aldehydes and ketones using o-iodobenzoic acid.[5] Variations on the Oppenauer oxidation include the use of *tert*-butyl hydroperoxide with zirconium *tert*-butoxide[6] or with chloral as the hydride acceptor.[7] 3β-Hydroxy-Δ^5-steroids can be efficiently oxidised to the corresponding 4-en-3-ones by using tris(triphenylphosphine)ruthenium dichloride.[8] Tetrapropylammonium perruthenate with N-methylmorpholine N-oxide is a useful oxidising reagent for sensitive steroidal alcohols;[9] it oxidises cholesterol to the Δ^4-3,6-dione.[10] In contrast, cholesterol is oxidised to the 3β-hydroxy-4,5-en-6-one using silver chromate and iodine.[11]

6-Oxygenated estradiols can be obtained by the metallation of protected estradiols (1) at position -6 by using lithium diisopropylamide: potassium 2-methyl-2-butoxide, quenching with trimethylborate and oxidation with hydrogen peroxide to give the alcohol (2) (9:1 6α:6β-OH) in 88% yield.[12]

A double fragmentation involving seven atoms leads to cleavage of both the A and B rings of the 3-mesylate derivative of 3,7-dihydroxy-cholestane when treated with sodium *tert*-amylate.[13] 3-Ene steroids have

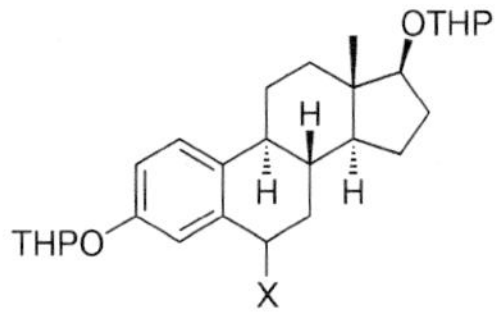

(1) X = H

(2) X = OH

been prepared by reacting 4-en-3-ols with *o*-nitrobenzenesulfonylhydrazine, warming to generate the allylic diazene, followed by sigmatropic elimination of nitrogen.[14]

In a search for metabolic blocking agents of 20-hydroxyecdysone, fluoroponasterone A (**3**) has been prepared by selective replacement of the 20-hydroxy group of 20-hydroxyecdysone 2,3:20,22-diacetonide by fluorine, using diethylaminosulfur trifluoride, followed by hydrolysis.[15]

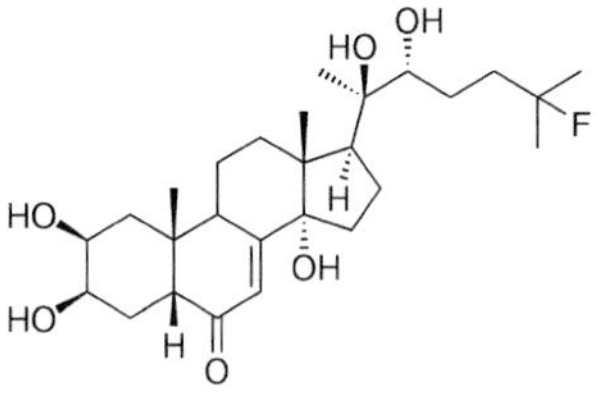

(3)

8.1.2   *Epoxide ring-opening reactions*

Neighbouring-group participation is involved in the oxidation of certain steroidal 4,5-epoxides by chromium trioxide; whereas the 4,5-α-epoxide (**5**) affords the 3-keto steroid (**4**), the 4,5-β-diastereoisomer (**6**) gives the 4-keto product (**7**) (Scheme 8.1).[16] Neighbouring-group participation can also arise in the acid-catalysed cleavage of 6β-hydroxy-4α,5α-epoxides; whereas epoxide **9** undergoes the usual transdiaxial hydrolytic

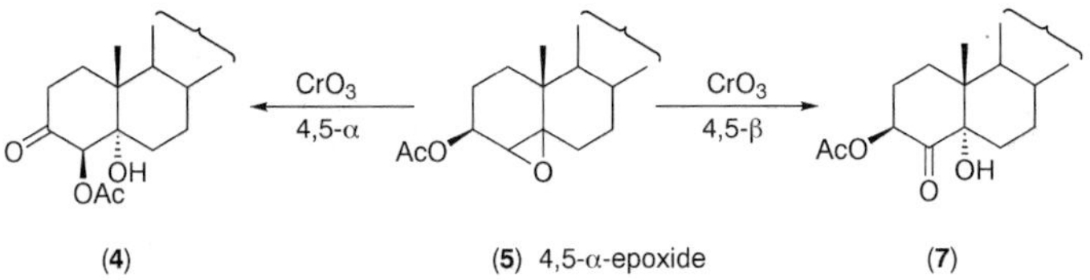

(4)   (5) 4,5-α-epoxide   (7)

(6) 4,5-β-epoxide

Scheme 8.1.

ring-opening to give **11** and its unprotected triol, the inductive effect of the methoxy group in the derivative **10** inhibits such attack, and instead the adjacent 6β-hydroxy group participates in a transdiaxial ring-opening to give an epoxide which itself undergoes transdiaxial ring-opening, but at the 5 position, thereby leading to the protected triol **8**, epimeric at C(4) (Scheme 8.2).[17]

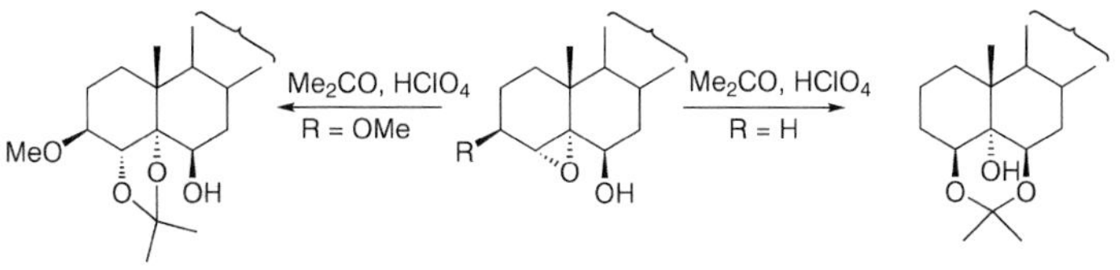

Scheme 8.2.

The isomerisation of oxirane (**12**) to allylic alcohol (**13**) has been used to synthesise 16-dehydro-20-oxopregnenes and other steroids (Scheme 8.3).[18]

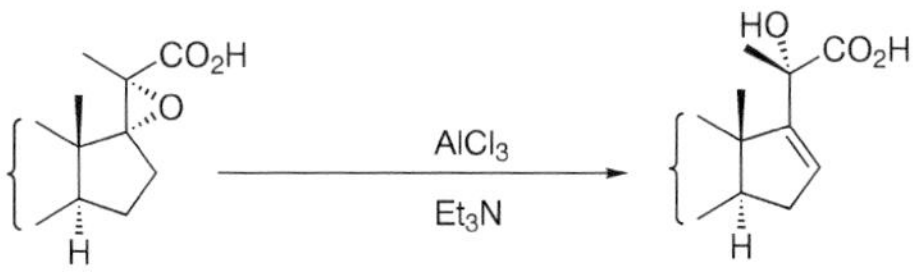

Scheme 8.3.

α, β-Epoxyketones undergo sonochemical reductive cleavage by aluminium amalgam giving 16α-hydroxy-20-ketones.[19]

A library of 3β-amido-3α-hydroxy-5α-androstane-17-ones was prepared by a sequence involving treatment of a spiro-3(R)-oxirane with an amine, followed by N-acylation.[20]

8.1.3   Unsaturated steroids

Introduction of C8(9)-unsaturation in a total synthesis of zymosterol was achieved via a regioselective hydroboration followed by a Barton-type deoxygenation.[21] Dehydration of 4,5-epoxyestra-3,17-diones affords 4-hydroxyestrone.[22]

11β-Fluoro-5α-dihydrotestosterone has been prepared by bromofluorination of the corresponding 9(11)alkene followed by reductive

debromination using tributyltin hydride. 11β-Fluoro-5α-dihydrotestost-
erone and its 19-nor analogue, if labelled with ^{18}F, could be used as
androgen-receptor-based imaging agents for prostate cancer.[23]

Generally, the stereochemistry of the C(10) methyl group of an
androst-4-ene controls which face of the steroid undergoes hydroboration
and results in attack of the α face. However, that directing feature is
overridden by a 3α-hydroxy group, so that hydroboration of such a 4-
en-3-ol affords the 3α,4β-alcohol.[24]

Catalytic oxidation of 5-ene steroids by manganese metalloporphy-
rin–molecular oxygen systems affords the 5α-hydroxy derivatives.[25]
Pyridinium chlorochromate is convenient for the oxidation of tetrahy-
dropyranyl ethers (14) to the corresponding diones (15) (Scheme 8.4);[26]

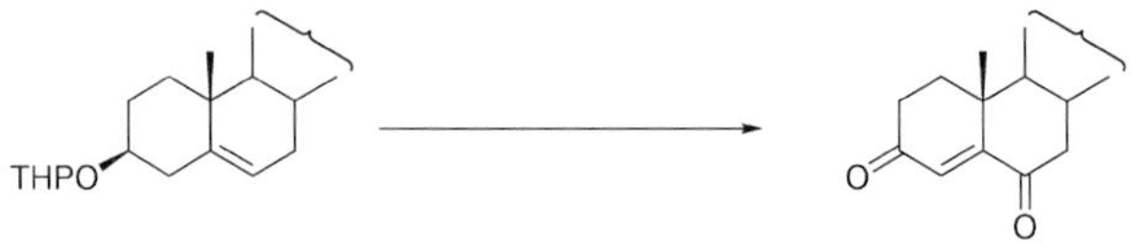

Scheme 8.4.

however, chromyl diacetate in dichloromethane effects epoxidation.[27] In
contrast to peracids that usually epoxidise Δ[5]-steroids from the less-
hindered α face, β face epoxidation has been achieved by using potassium
permanganate with various metal sulfates.[28] Oxidative dearomatisation
of estrone methyl ether by using antimony(V) chloride has been reported
to effect *ipso* chlorination, giving the ring-A chlorocyclohexadienone.[29]

Ene reactions of alkene **16** with TBDMSOCH$_2$CHO in the presence
of dimethylaluminium chloride at −78°C afforded solely diastereoisomer
17 in 90% yield, via a nonchelated *endo* transition state. However, an
analogue of alkene **16** reacted with PhCH$_2$OCH$_2$CHO in the presence of
SnCl$_4$ at −78°C giving the alcohol **18**; chelation control leading to an *exo*
transition state accounts for this reversal of diastereoselection.[30]

An intramolecular Pauson–Khand reaction has been used to construct
rings D and E of a pentacyclic precursor related to the histamine release
inhibitor xestobergsterol-A.[31]

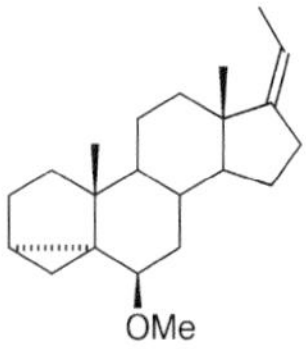

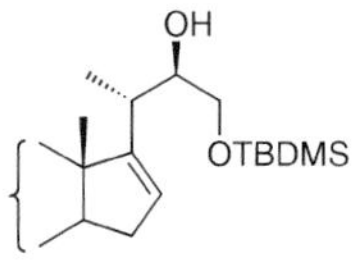

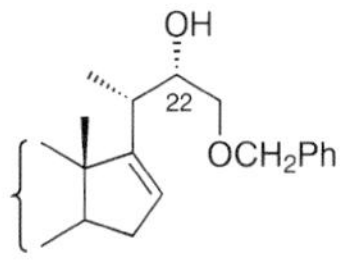

17-Iodo-Δ^{16}-steroids undergo dehalodimerisation in the presence of palladium catalysts.[32]

8.1.4   Carbonyl compounds

Factors influencing the reduction of 1-oxo-steroids have been reviewed.[33] A modification of the Meerwein–Ponndorf–Verley reduction using zirconium tetra-*tert*-butoxide and 1-(4-dimethylaminophenyl)ethanol as the hydrogen donor has been applied to steroidal ketones.[34] Deoxygenation of 3-oxosteroids can be effected by using zinc dust in acetic acid activated by ultrasound.[35] Conversely, 7-oxosteroids have been prepared by the copper-catalysed allylic oxidation of Δ^5-steroids by *tert*-butyl hydroperoxide.[36]

3β-Hydroxyandrost-5-ene-7,17-dione is a metabolite of dehydroepiandrosterone, from which it has been prepared by an improved oxidative procedure.[37]

Transition-metal-catalysed conjugate addition of alkylaluminiums and of methyltitanium ate complexes to androsta-1,4-diene-3,17-dione (**19**)

(19)

can be regioselectively controlled, catalytic amounts of copper(I) salts favouring addition to C(1) over C(5) (97:3), whereas nickel acetylacetonate leads to addition in favour of C(5) (11:89).[38] The use of antimony chemistry, particularly of tributylstibine in Knoevenagel condensations, facilitates the preparation of alkylidene malonates from 3-keto steroids.[39]

1-Fluoro-2-bromo steroids have been prepared by reaction of the 1-phenylsulfide with *N*-bromosuccinimide and either HF or diethylaminosulfur trifluoride.[40] γ-Fluorination of α,β-unsaturated ketones such as 4-cholest-4-en-3-one can be achieved by reacting the potassium dienoxyborates (obtained by treatment of the potassium enolates with 2-phenyl-1,3,2-benzodioxaborole) with *N*-fluorobenzenesulfonimide.[41]

Hydrolysis of ring-A methoxy enol ethers by using oxalic acid in the presence of silica gel provides a route to the 5(10)-en-3-ones (**20**).[42]

(20)

8.1.5   CH activation, including remote functionalisation

Remote intramolecular functionalisation by radicals has been reviewed[43] and continues to be an important strategy in steroid chemistry, with many developments since the early 1960s and the work on photolysis of nitrite esters conducted in Barton's group. In a radical relay approach, a covalently attached catalytic template can direct chlorination efficiently to either C(9) or C(17) in 3α-cholestanyl esters.[44] Photolysis of the hypoiodite of 5α-cholestan-7α-yl 4-(α-hydroxyphenylmethyl)phenylacetate leads to alkoxy radicals that abstract hydrogen from C(25) of the side-chain of cholestane, resulting in the macrocyclic lactone (21).[45] Subsequent reductive cleavage of the benzyl ether and hydrolysis gave 5α-cholestane-7α,25-diol.

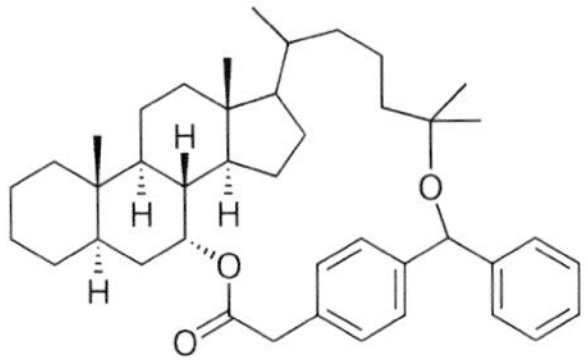

(21)

In the cholesterol derivative (22), the tether is converted into an alkoxy radical which effects a dual functionalisation of ring C, giving the allylic alcohol (23) (Scheme 8.5).[46] By varying the length of a C(3) tether that contains an oxometalloporphinate, oxygen functions can be introduced at the 12-, 14- and 17-positions.[47]

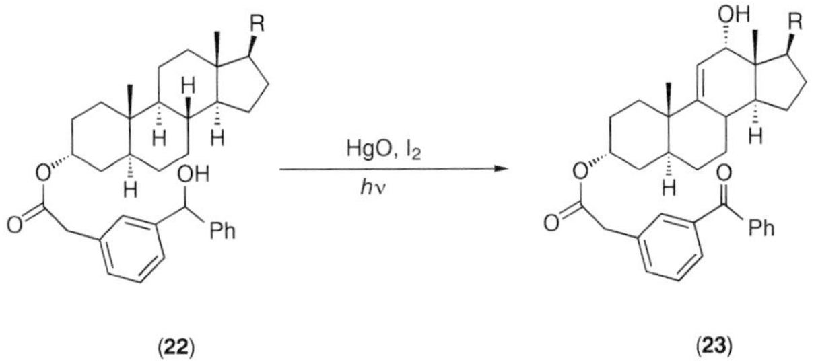

Scheme 8.5.

Oxidation of progesterone with aqueous hydrogen peroxide in the presence of iron(III) chloride and picolinic acid in pyridine-acetic acid affords products including Δ^4-pregnene-3,6,20-trione and Δ^4-pregnene-3,12,20-trione.[48] The biomimetic oxidation of vitamin D_3 with oxygen in the presence of an iron–sulfur cluster gave (5E)-10-oxo-10-norvitamin D$_3$, mimicking biological oxidation.[49] Oxyfunctionalisation of 5β-steroids by dimethyldioxirane or methyltrifluoromethyldioxirane gives the 5β-hydroxy steroids;[50] likewise, perfluorodialkyloxaziridines have been used to obtain 5β-hydroxy steroids,[51] but some of those reagents can also effect C(25) hydroxylation.[52] A cytostatic cholestane pentol found in an Antarctic starfish has been synthesised from diosgenin, a key step being the introduction of the 15α-hydroxy group by the oxidation of an enol silane with dimethyldioxirane.[53]

Functionalisation of the 18-methyl group in Δ^5-pregnene-3β-acetoxy-20β-ol was achieved by irradiation in the presence of iodine, lead tetraacetate and calcium carbonate, and led to the 18-iodo derivative, a key intermediate in a total synthesis of mirasorvone.[54]

8.2   Rearrangements

8.2.1   Carbocation rearrangements

Multiple Wagner–Meerwein rearrangements of steroids have long been studied. The backbone rearrangement of cholesta-6,8(14)-dienes such as **24** in the presence of p-toluenesulfonic acid and acetic acid was examined in the context of the fate of steroids in fossils. These reactions generally give spirosteradienes such as **25** (Scheme 8.6).[55]

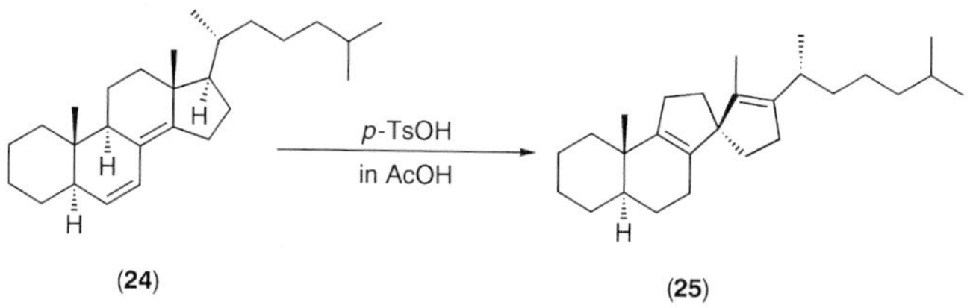

Scheme 8.6.

4,4-Dimethylcholest-5-ene (**26**) undergoes a backbone rearrangement catalysed by K-10 montmorillonite to give the Δ$^{13(17)}$-steroid **27** (Scheme 8.7).[56]

8.2.2   Miscellaneous rearrangements

The synthesis of the first analogue of the 7(Z)-isomer of vitamin D and its thermal rearrangement have been reported.[57]

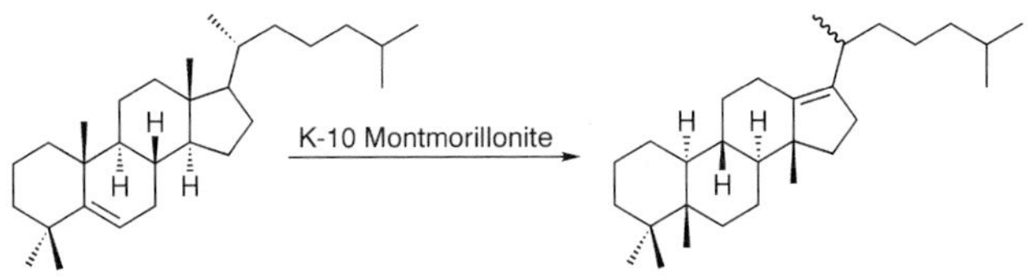

(26)   (27)

Scheme 8.7.

Migration of a silyl group from a phenolic oxygen atom to an *ortho*-carbon atom has been observed in the aromatic ring A of steroids.[58] A crossover experiment showed the migration to be intramolecular.

Mercury(II) or thallium(III) salts induce stereoselective cleavage, with retention of configuration, of cyclopropyl alcohols (**28**) to give an isolable organomercurial (**29**)[59] which with molybdenum(v) pentachloride affords cholesteryl chloride (**30**) (Scheme 8.8).[60] Transmetallation of the organomercurial (**29**) led to the fused alcohol (**31**).[61] Other molybdenum reagents act on organomercurial (**29**) inducing Grob-type fragmentation.[62]

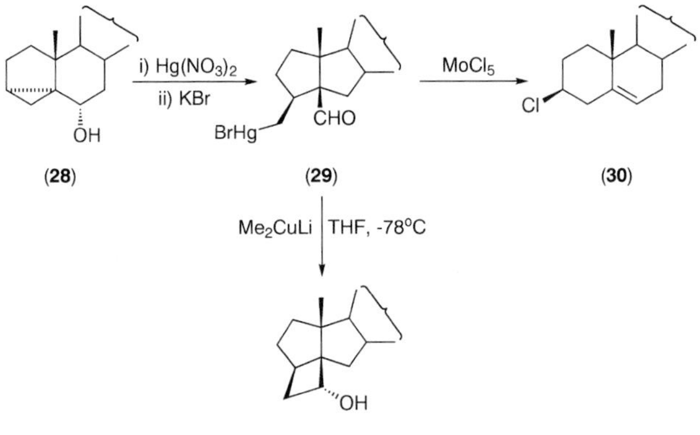

(28)   (29)   (30)

Me₂CuLi | THF, -78°C

(31)

Scheme 8.8.

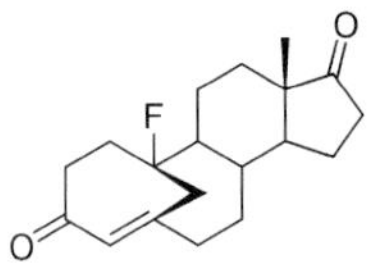

(32)

Reaction of the 19-hydroxycholest-4-en-3-one with diethylaminosulfur trifluoride affords the bridged fluoroenone **32** in 60% yield.[63]

8.2.3 Photochemical rearrangements

1α-Hydroxyprovitamin D undergoes a photochemical isomerisation initiated by cleavage of the C(1)–C(10) bond; that pathway predominates over the normal electrocyclic B ring opening if a 1β-methyl group is present.[64]

Photochemically induced oxidation of cholest-5-en-3α-ol using mercury(II) oxide and iodine affords a bridged 3,6-ether, spiroketals, and also the 5,6-epoxide.[65]

The photochemical Beckmann rearrangement of a 4,5-dehydro-6-oxime gives an amide that is a constitutional isomer of that obtained from a ground-state Beckmann rearrangement.[66]

Oxidative photochemical fragmentation of 5α-hydroxycholestan-2-one cleaves the AB ring junction giving α,β-unsaturated enones and β,γ-unsaturated ones.[67]

A double-ring contraction of 4α-homo-5α-cholest-3-en-1-one is observed under photochemical conditions, resulting in a fused cyclobutane tetracycle as ring A.[68]

8.3 Novel total and partial syntheses

8.3.1 Cationic cyclisations including biomimetic polyene cyclisations

For more than 40 years, mechanistic details of enzymic cyclisations occuring in the biosynthesis of steroids and terpenoids have been of continued interest.[69] The realisation that cyclisation of epoxide intermediates were implicated in plant and mammalian terpenoid biosynthesis led to *in vitro* studies of the cyclisation of model compounds, pioneering work being reported by Goldsmith and co-workers. Early work from the groups of van Tamelen and Corey demonstrated the intermediacy of 2,3-oxidosqualene (**33**) in the biosynthesis of lanosterol (**35**) (Scheme 8.9).

2,3-Oxidosqualene (**33**)

2,3-Oxidosqualene was also cyclised in high yield using pig or rat liver enzymes to give lanosterol. The Stork–Eschenmoser hypothesis that poly-alkenes could react in a defined conformation which, in combination

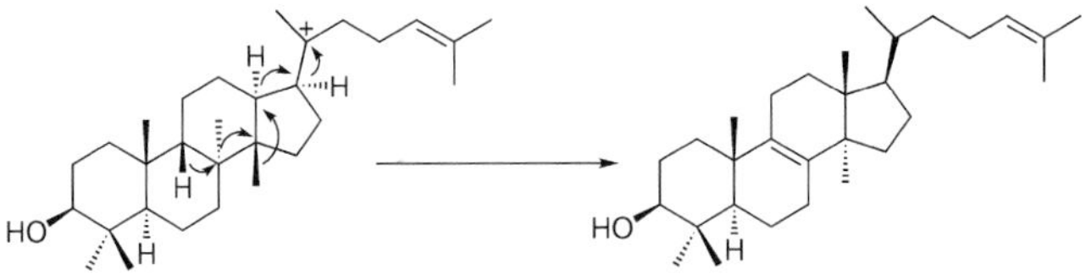

(34)   Lanosterol (35)

Scheme 8.9.

with antiperiplanar addition to double bonds, accounts for the relative stereochemistry of the cyclisation products. The chair–boat–chair conformation of 2,3-oxidosqualene secures the stereocontrol during cyclisation; recently, good evidence has been presented concerning ring closures under *enzymic* conditions that implicate cations and products predicted by Markovnikov's rule.[70] Thus, the nonenzymic formation of ring C of cation **36** is closely analogous to the corresponding step in the

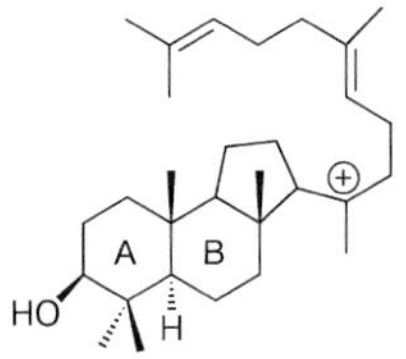

(36)

biosynthesis of lanosterol. Further steps, including a 1,2-ring expansion of ring C, lead to the protosterol cation **34**, and thence to lanosterol (**35**) by a series of 1,2-migrations. Thus, the cyclisation of (S)-2,3-oxidosqualene to the protosterol cation is probably *not* concerted, but occurs in stages involving discrete carbocationic intermediates, though under tight conformational control.[70]

The outcome of enzymic cyclisation for C(1) hydroxylated surrogate squalenoids can depend upon the configuration of the initial epoxy alcohol.[71] The complete tetracyclic framework of **39** (40% yield) was enantioselectively formed from the epoxy alcohol **37**, but the diastereo-isomer **38** led only to a bicyclic product; interactions between the enzyme and the β face of the chair–boat–chair conformer of the substrate determine the product. The efficient epoxide cyclisations of 2,3-oxidosqualene contrast to those of squalene induced by squalene-hop-22(29)-ene cyclase, from which hop-22(29)-ene, a neohopene, and five tetracyclic steroidal alkenes have been isolated.[72]

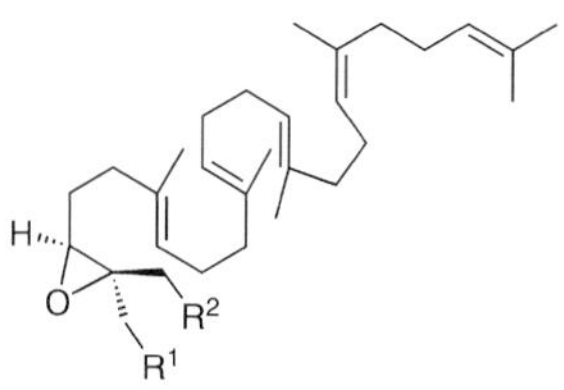

(37) R^1 = OH, R^2 = H
(38) R^1 = H, R^2 = OH

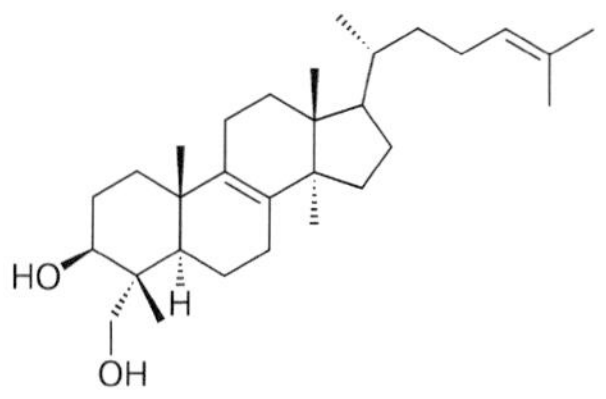

(39)

W.S. Johnson and co-workers have shown that their earlier biomimetic cyclisations can be extended by locating an auxiliary group such as isobutenyl at a position that develops positive charge in the transition state, as in polyene (41). Thus, cyclisation of 41 appeared to be complete after just one minute, and the epimers 43 were isolated (80%) after a reaction period of one hour (Scheme 8.10). In contrast, only 1%–2% of 42

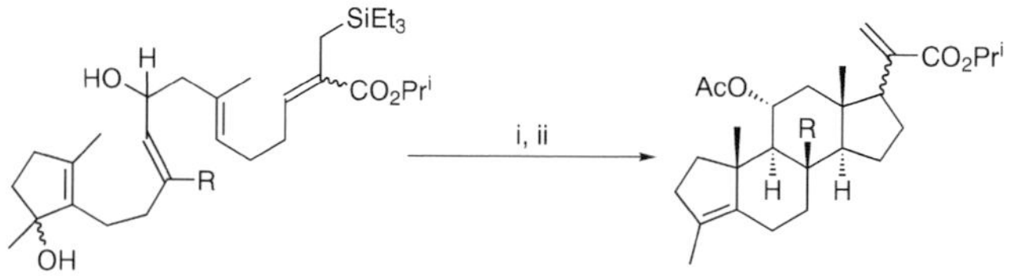

(40) R = H
(41) R = CH=C(CH$_3$)$_2$

(42) R = H
(43) R = CH=C(CH$_3$)$_2$

Scheme 8.10   Reagents: (i) CF$_3$CO$_2$H; (ii) Ac$_2$O, DMAP, Et$_3$N.

had been formed from 40 after one hour.[73] However, the isobutenyl group proved difficult to remove, and the use of fluorine to stabilise the incipient cation was then investigated.[74] Such a strategy was deployed in a synthesis of (±)-β-amyrin in which four rings incorporating seven stereocentres were formed in a single reaction, 44 giving 45, in 70% yield (Scheme 8.11).[75] Building on the efficiency of cyclisations using an isobutenyl group at *pro*-C(8), and the stereocontrol provided by a

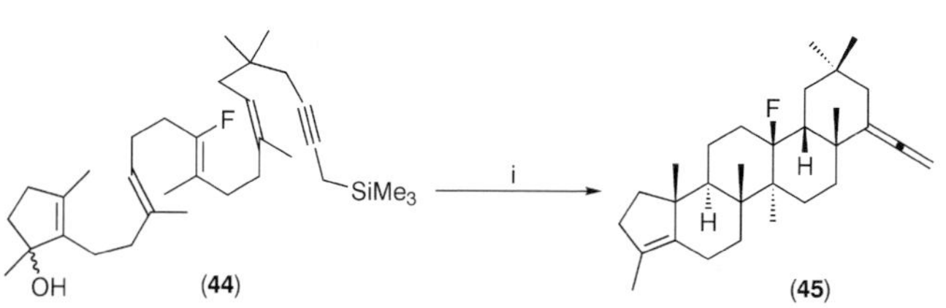

(44)

(45)

Scheme 8.11   Reagent: (i) CF$_3$CO$_2$H.

pentanediol acetal as in **46**, it was shown that a fluorine atom also enabled cyclisation of (±)-acetal (**47**) to the (±)-system (**49**) bearing the natural backbone configuration of the steroids (Scheme 8.12). Cyclisation

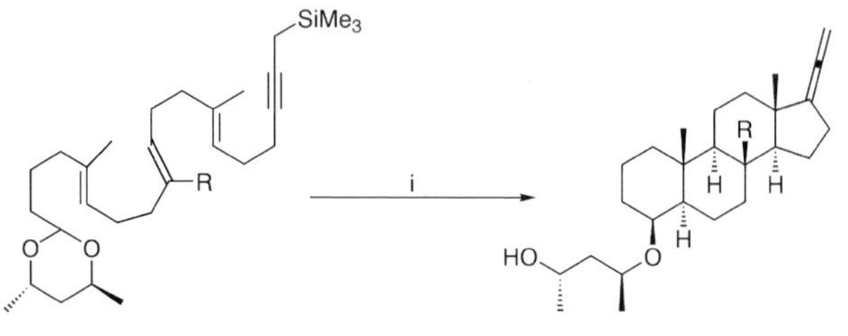

(**46**) R = CH=C(CH₃)₂   (**48**) R = CH=C(CH₃)₂
(**47**) R = F   (**49**) R = F

Scheme 8.12   Reagent: (i) SnCl₄.

of enantiopure acetal (**47**) gave allene (**49**) (38%) in 93% enantiomeric excess (ee).[76] The fluorine atom was replaced by hydrogen stereoselectively, with retention of configuration, by reaction of enantiopure **49** with potassium and dicyclohexyl-18-crown-6; the defluorinated product was subsequently converted into the known steroid 4β-hydroxyandrostan-17-one.

An enantioselective route to the CD steroidal ring system involves Friedel–Crafts-type cycliacylation of **50**, prepared from (*R*)-(+)-camphor.[77] After protection of the resulting keto group, the desired side-chain was introduced by Wicha alkylation; the route offers advantages

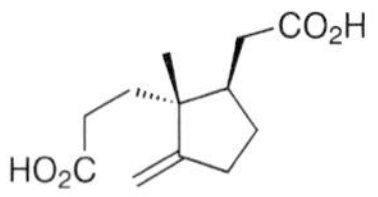

(**50**)

over earlier sterol syntheses in which the whole of the side-chain has to be attached to a steroidal 17-ketone. In a complementary approach to the CD ring system, whereas intramolecular cycliacylation of γ,δ-unsaturated carboxylate derivatives has not proven generally feasible, a corresponding cyclialkylation of ester **51** to ketone **52** was induced by *tert*-butyllithium (Scheme 8.13).[78]

3-Thioxoandrosta-1,4-dien-17-one has been reported to undergo acid-catalysed rearrangement of the dienone-phenol type to give 1-(acetyl-thio)-4-methylestra-1,3,5(10)-trien-17-one and its dithio dimer.[79]

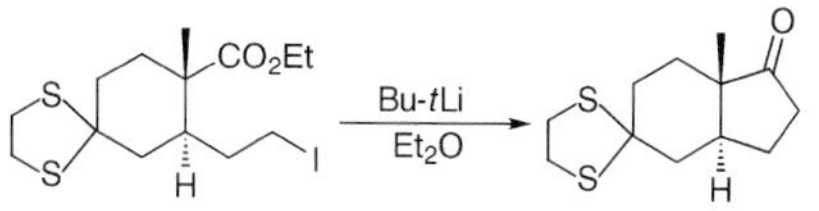

(51)   (52)

Scheme 8.13.

8.3.2   Radical   cyclisations

The application of radical processes to form steroid rings is relatively new. Of several methods studied, one may note the manganese(III)-acetate-promoted oxidative formation of keto ester radicals[80] that induce cyclisation to **54** (Scheme 8.14), subsequently converted into (±)-14-

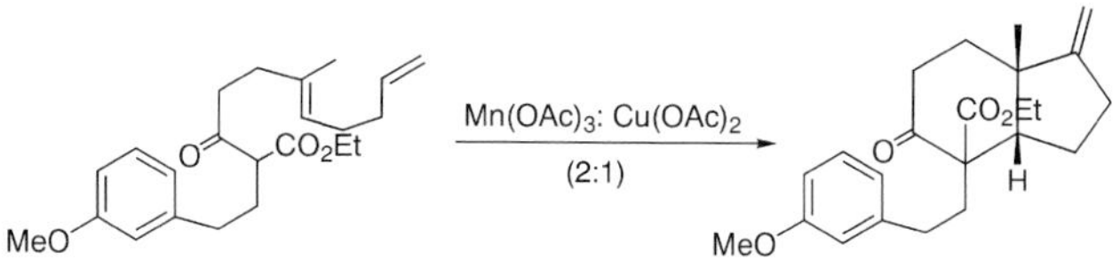

(53)   (54)

Scheme 8.14.

epi-estrone.[81] In radical processes, the size of the D ring formed governs the stereochemistry at the ring junction, a five-membered ring being *cis*-fused to ring C. Conversely, when a six-membered ring is formed, the C-ring system is *trans*-fused, a key feature that was exploited in a synthesis of the D-homo-5α-androstane-3-one (**56**) (Scheme 8.15).[82]

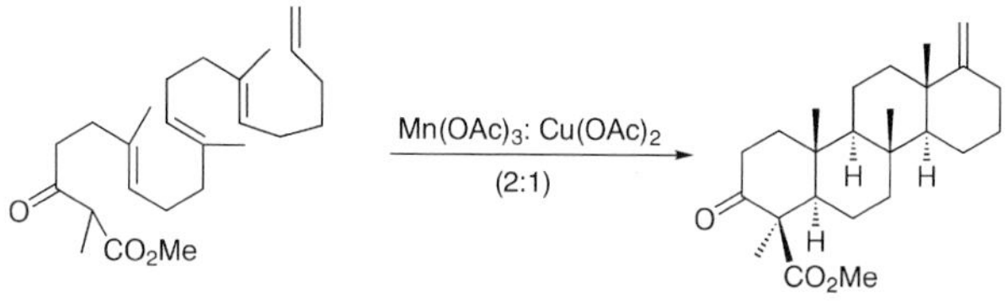

(55)   (56)

Scheme 8.15.

An interesting radical cascade reaction has been used to generate the tetracyclic steroidal ring system, although with the unusual *cis, anti, cis, anti, cis*-stereochemistry.[83] A tandem radical cyclisation on an acyclic iodide delivered in one reaction the CD steroidal ring system with the desired *trans, anti, trans*-stereochemistry.[84]

The iminyl radical generated by reacting a 17-acetoximinosteroid with nickel powder undergoes ring opening to give a C(13) radical which then reforms the D ring, but with a *cis*-ring junction; this provides a simple and powerful new route to 13-epi-17-ketosteroids.[85]

The reductive cyclisation of a 5-alkynyl iodide gave only the (*E*)-alkenyl ether attached to the newly formed D ring; however, the 5-alkynyl dithiocarbonate afforded a 7:3 *E:Z* mixture, probably owing to inter-molecular hydrogen atom abstraction.[86]

Transannular cyclisations induced by alkoxy radicals have been examined; photolysis of a steroidal 6α-substituted cyclodecenone in the presence of iodine and mercury(II) oxide gave a bicyclo[5.3.0]decane.[87]

Digitoxigenin has been synthesised by using the addition of a C(17) radical of an androstanediol to maleic anydride.[88] Ring A of 1α,25-dihydroxyvitamin D_3 has been constructed by radical cyclisation of a vinyl selenide induced by tributyltin hydride, and a subsequent elimination.[89]

8.3.3 Electrocyclic reactions

Diels–Alder reactions continue to provide elegant routes in steroid synthesis. In a very elegant formal total synthesis of (±)-batrachotoxinin A (57), an intramolecular Diels–Alder reaction involving a furan de-livered the ABCD ring assembly with substituents that smoothly led to

(57)

the tetracyclic bridge across the CD ring junction.[90] Another intra-molecular Diels–Alder strategy has afforded the *cis, anti, trans* ABC ring system of the brassinosteroids.[91]

A novel approach to steroids involves the transannular intramolecu-lar Diels–Alder cycloaddition of a macrocyclic triene such as (58) (Scheme 8.16), in which three rings are formed in one reaction.[92] Of the two chair–boat–chair conformations subjected to MM2 calculations, the one that leads to the tetracycle (59) of desired stereochemistry has less destabilisation arising from van der Waals and bending energy functions.

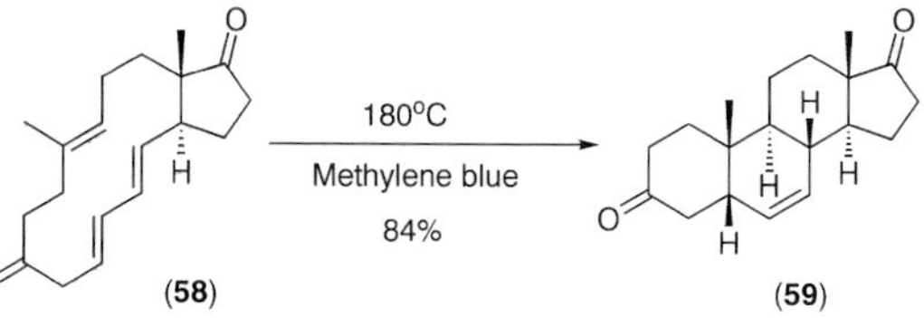

Scheme 8.16.

The importance of *o*-quinodimethanes in steroid synthesis is high-lighted in a recent review.[93] A total synthesis of the 25-hydroxy Windaus–Grundmann ketone **(60)** via an *o*-quinodimethane approach

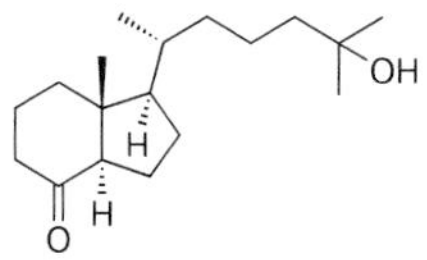

(60)

has been reported,[94] and hence a formal total synthesis of vitamin D_3. Fukumoto, Kametani and co-workers have used *o*-quinodimethane strategies in total syntheses of (±)-estrone and (±)-adrenosterone,[95] as well as (±)-19-norcanrenone[96] and a novel route to the des-*A* corticosteroids.[97] Fukumoto's group has also achieved a total enantio-selective synthesis of (+)-cortisone, the naturally occurring enantiomer.[98] The principal steps involve the thermal conversion of cyclobutane **(61)** into the tricyclic system **(62)**, proceeding via the *o*-quinodimethane (Scheme 8.17). The heterocyclic chiral auxiliary is used to introduce the desired absolute configuration at the *pro*-17 position in the cyclisation of the cyclobutane **(61)**. Later steps included the stereoselective formation of ring A, as pioneered by Stork.

Recent advances in Lewis-acid catalysis of Diels–Alder reactions have rekindled interest in intermolecular approaches to steroids, an example

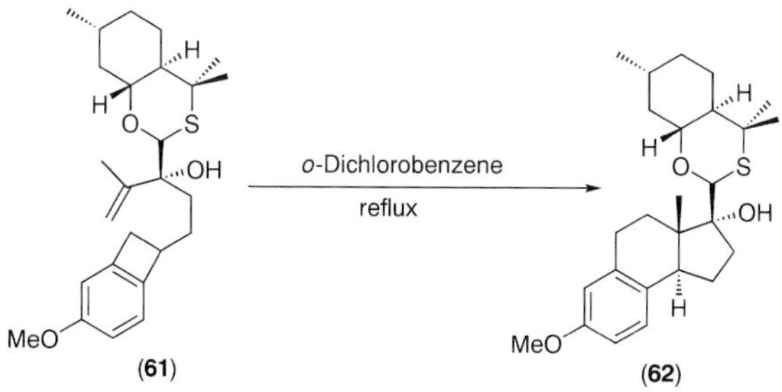

Scheme 8.17.

being the formation of **65** in 73% ee (Scheme 8.18) using a chiral complex of TiCl$_2$(O^{i}Pr)$_2$.[99] Ketol (**65**) can be converted into Torgov's pentaenone (**66**) which provides access to estrogens and progestogens.

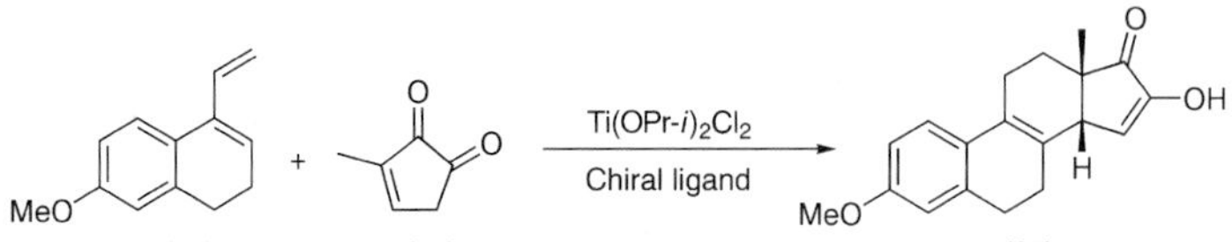

Scheme 8.18.

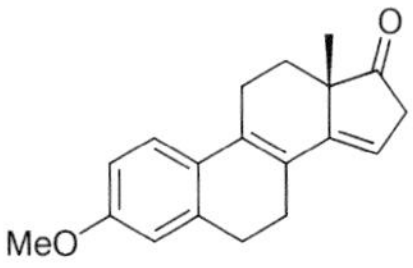

(**66**)

17-Iodoandrost-16-enes and vinyl stannanes undergo palladium-catalysed coupling to give the expected dienes which participate in Diels–Alder reactions with dienophiles to give pentacyclic steroids.[100]

Routes to aromatic steroids based upon thermal electrocyclic ring closures of trienes such as **67** have been developed; the latter were prepared by palladium-catalysed coupling, forming the BD bond.[101]

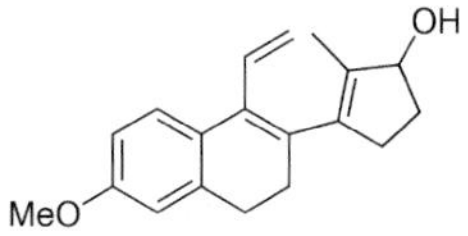

(**67**)

The first synthesis of (+)-withanolide E involved cycloaddition of benzyl nitrosoformate to a ring-D diene, which led to a 14α-hydroxy-androstane whose 17-keto functionality was manipulated to incorporate the required δ-lactone ring.[102]

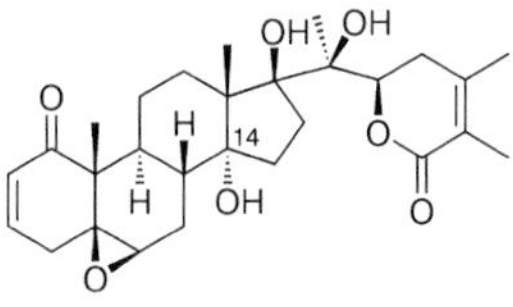

(+)-Withanolide E

A total synthesis of the naturally occurring enantiomer of 9(11)-dehydroestrone methyl ether involves a tandem Claisen-ene sequence[103] that establishes three stereocentres in ring C.[104] The (8S,6Z)-alcohol (**68**) was prepared from (R)-isopropylideneglyceraldehyde (obtainable from D-mannitol). Condensation of **68** with the enol ether **69** gave the ether that underwent the Claisen rearrangement to give enone **70**, which then cyclised in an ene reaction to give the unsaturated ester **71** (Scheme 8.19). Further conversions, and lastly a modified intramolecular McMurry coupling reaction, established ring C and its double bond and the desired estrone derivative.

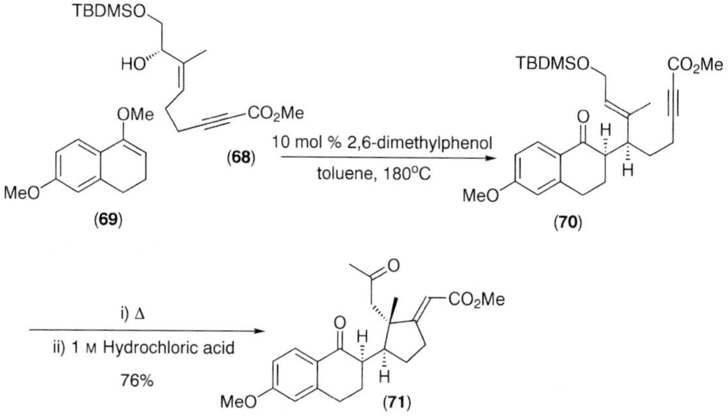

Scheme 8.19.

8.3.4   Novel structures and side-chain syntheses

Alkylation of a C(17) unsaturated ester with methyl iodide provides an efficient preparation of a Δ^{16}-(20S) ester side-chain.[105]

Samarium iodide promotes the radical addition of tetrahydrofuran to a 20-keto steroid; subsequent oxidation to the lactone, addition of methylmagnesium iodide and ring closure afforded an elegant route to (20R, 22R)-5α-cholestane-3β,20,22,25-tetraol.[106]

A stereoselective synthesis of the cholestenetriol **72**, which contains the four stereocentres of the brassinolide side-chain, has been reported.[107]

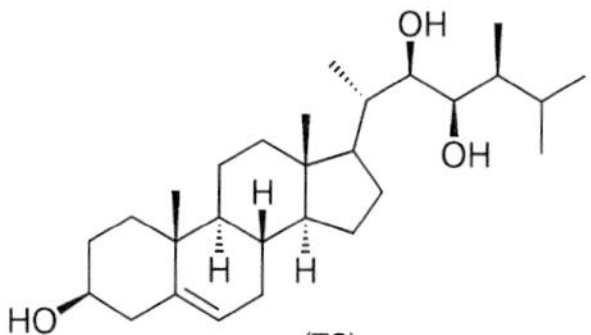

A route to 6-deoxo brassinosteroids proceeds via Grignard addition to (20S)-20-formyl-6β-methoxy-3α,5-cyclo-5α-pregnane, and subsequent orthoester Claisen rearrangement.[108] A stereocontrolled synthesis of the side-chain of withanolide D from 20-oxopregnane employs a stereo-selective hydrogenation with accompanying ring contraction to a γ-lactone.[109]

The silyl ether **73** derived from the corresponding cyanohydrin of a 4-en-3-one undergoes a migration to carbon with lithium diisopropyl-amide affording the chloroketone **74** (Scheme 8.20) which upon reductive dechlorination gives a corticosteroid side-chain.[110] That is a useful protocol, since 17-keto steroids are often available via microbial degradation of naturally occurring sterols.

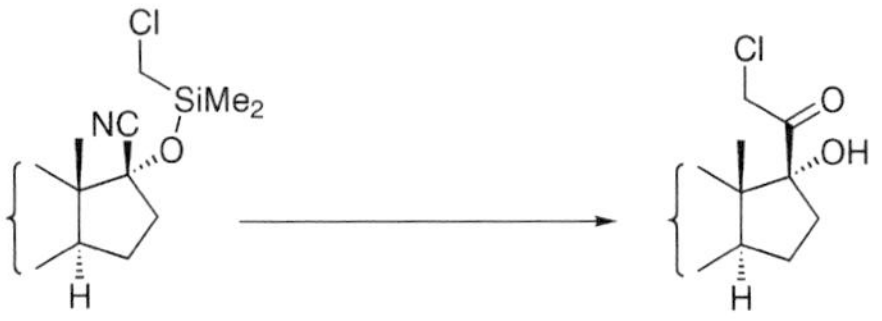

(73)   (74)

Scheme 8.20.

Steroids with a functionalised side-chain have been generated by oxidation of sapogenins with dimethyldioxirane; thus, tigogenin acetate was selectively oxidised to give the corresponding C(16) hemiketal (95%) which was treated with acetic acid to give the 16,22-dioxo-27-acetoxycholestane.[111]

Synthetic work in the area of vitamin D has been reviewed,[112] and includes extensive variations in the side-chain such as alkyl and fluoro groups. The isocalysterols possess a cyclopropene ring in the side-chain and two have been synthesised.[113] A concise synthesis of the side-chain and CD rings of 25-hydroxyvitamin D_3 involves two tandem Mukaiyama–Michael additions.[114] As part of studies that achieved a total synthesis of calcitriol, a cyano group used to assemble the side-chain was removed using potassium and *tert*-butanol.[115]

An hydrindenone with a side-chain as in vitamin D_3 has been introduced via a conjugate addition to 2-methylcyclopent-2-en-1-one.[116] That strategy has also provided a synthesis of the northern portion of 25-hydroxy vitamin D_3.[117]

8.3.5 Heterocyclic steroids

The dimeric steroid–pyrazine marine alkaloids have been reviewed,[118] and include the cephalostatins and ritterazines, trisdecacyclic pyrazines of extraordinarily potent anticancer properties that have been isolated from the marine tube worm *Cephalodiscus gilchristi* and the tunicate *Ritterella tokioka*. A key feature of those compounds is the pyrazine core which Pettit hypothesised[119] is assembled via the dimerisation and oxidation of steroidal α-amino ketones, a well-known laboratory reaction.

The timing of the dimerisation step in the biosynthesis was a question considered by Fuchs when undertaking a biomimetic total synthesis of (+)-cephalostatin 7 (**75**).[120] The strategy involved transformation of

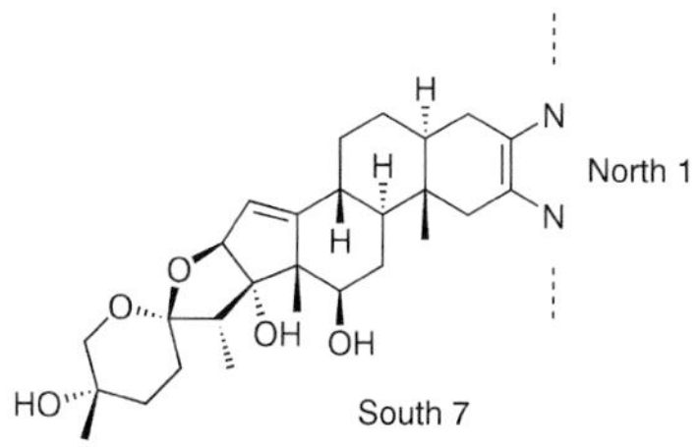
North 1
South 7
Cephalostatin 7 (**75**)

commercially available hecogenin acetate into aldehyde (**76**)[121] which served as the intermediate for the synthesis of both the 'north' segment [122] and the 'south' segment[123] of the target pyrazines. Reaction of north ketone (**77**) with phenyltrimethylammonium bromide afforded the bromoketone (**78**) which when treated with tetramethylguanidinium azide in nitromethane afforded quantitatively the desired azido ketone (**79**) (Scheme 8.21). For the southern segment, ketone **80** was treated with phenyltrimethylammonium bromide to give the bromoketone (**81**)

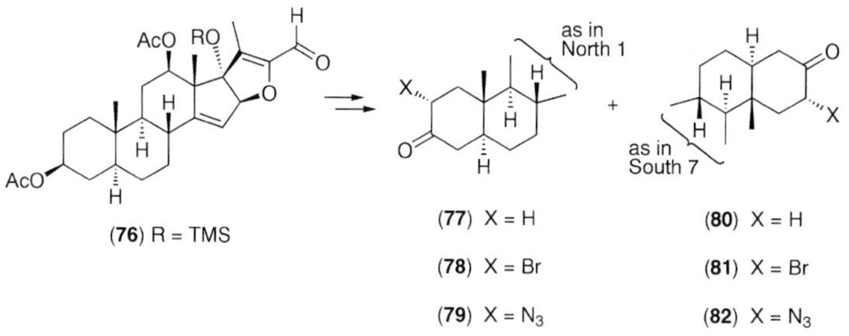
as in North 1
as in South 7
(**76**) R = TMS
(**77**) X = H
(**78**) X = Br
(**79**) X = N_3
(**80**) X = H
(**81**) X = Br
(**82**) X = N_3

Scheme 8.21.

[during which deprotection of the C(25) moiety occurred] which was converted into the azido ketone (**82**) as described above for ketone (**79**). Pyrazine formation was achieved by treating a 1:1 mixture of azido ketones (**79**) and (**83**) with excess NaHTe in ethanol at 25°C for 1 h. Chromatographic separation followed by deprotection afforded the first synthetic samples of cephalostatin 7 (**75**), cephalostatin 12 (**83**) and ritterazine K (**84**).

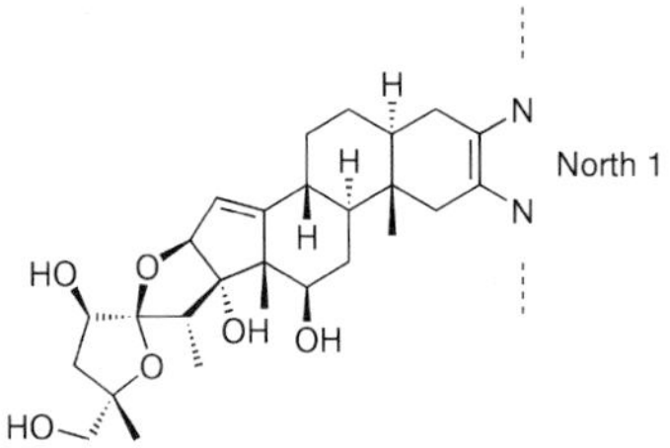

Cephalostatin 12 (**83**)

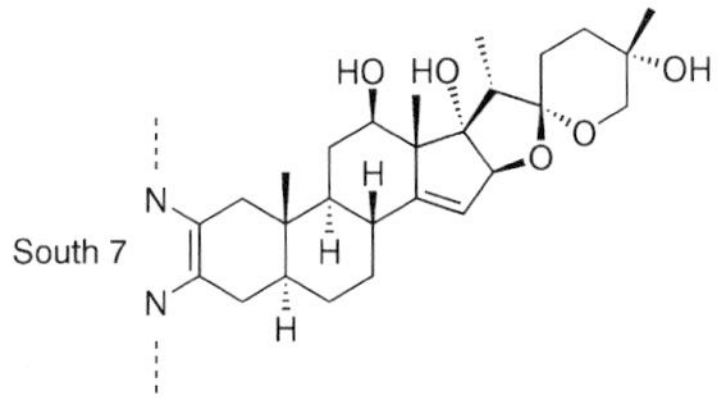

Ritterazine K (**84**)

Cephalostatin 1 (**85**)[119] is one of the most powerful anticancer agents known; however, clinical trials have been hampered by the poor accessibility of the natural source and the low abundance of the natural

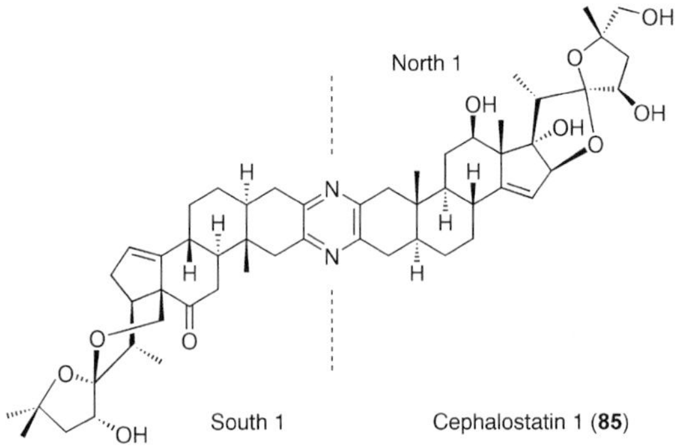

Cephalostatin 1 (**85**)

product (100 mg from 450 g of marine tube worm, found in white-shark-infested waters off the coast of East Africa). Some ritterazines exhibit cytotoxicities approaching those of the most active cephalostatins. The ritterazines are much less oxygenated, and their relative simplicity should afford greater accessibility through synthesis. The effect of structure on bioactivity has been enlarged by syntheses of dihydrocephalostatin 1,[124, 125] derivatives of ritterazine[126] and pyrazines more closely related to hecogenin.[127] As part of a programme aimed at securing novel steroidal pyrazines and testing their biological activity, Fuchs and co-workers set about devising methodology that would allow combinations of north and south units to be synthesised and condensed. This led to the first synthesis of the north segment of ritterazine G (**86**)[128] which involved an isomerisation of the spiroketal unit of hecogenin acetate, and extension of a photolysis/Prins sequence to install the Δ^{14} functionality. Such a

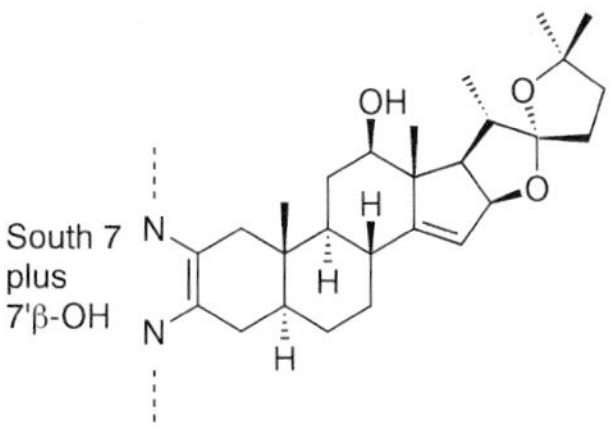

Ritterazine G (**86**)

strategy was also used in the total synthesis of cephalostatin 1 (**85**) by Fuchs's group,[128] a landmark achievement in this new area.

Related pyrazines had been prepared by the seminal procedure of Heathcock and Smith,[129] in which an α-aminomethoxime (obtained via Staudinger reduction) is reacted with an α-acetoxy ketone typically giving yields of 30%–45%. Fuchs and his group found that replacement of the α-acetoxy ketone by the corresponding azido ketone gave yields in the range 60%–90%, an improvement which can in part be accounted for by the production of a basic medium through the loss of methoxylamine (in contrast to the acidic one resulting from Heathcock's method). The condensation pathway is outlined in Scheme 8.22.

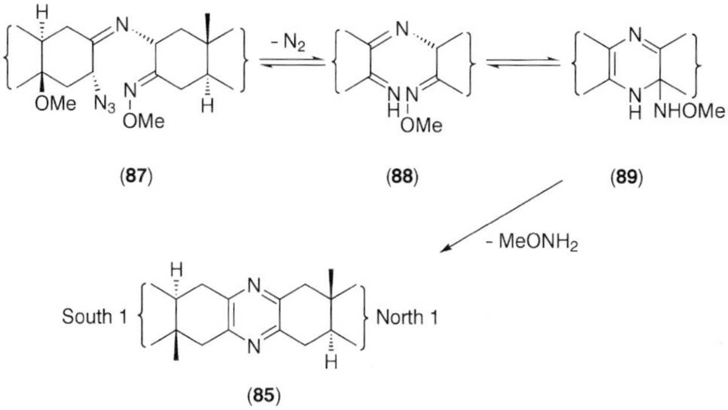

Scheme 8.22.

The hybrid analogues ritterostatin G_N1_N and ritterostatin G_N1_S were also synthesised.[128] Ritterostatin G_N1_N contains the north unit of cephalostatin 1 (**85**) fused across the pyrazine ring to the north unit of ritterazine G (**86**); accordingly, in the pyrazine formation step an azido ketone of the latter was condensed with an aminomethoxime related to the north unit of cephalostatin 1. In a closely similar sequence, the azidoketone of south cephalostatin 1 was condensed with the amino-methoxime of north ritterazine G to give ritterostatin G_N1_S. Ritterostatin

G_N1_N was found to be more potent than cephalostatin 1 (**85**) for the leukaemic K-562 cell line, and equipotent for certain ovarian, renal and breast cell lines. Ritterostatin G_N1_N showed a mean tumour inhibition activity approaching that of taxol, and superior to that of cephalostatin 7 (**75**) in all cancer cell lines tested. Ritterostatin G_N1_N also proved more effective in cell lines than several common therapeutic agents, including 5-fluorouracil and cyclophosphamide, and thus shows the importance of hybrid steroidal pyrazines. In contrast, the lack of a 17α-hydroxy group in ritterostatin G_N1_S contributes to its lower activity, as do its spiroketals which are present in low-energy isomeric forms; that is consistent with the most active pyrazines containing at least one spiroketal which calculations show to be a less stable isomer.

The steroidal saponin OSW-1 (**90**) is one of a family of cholestane glycosides that possess the 3β,16β,17α-trihydroxycholest-5-en-22-one

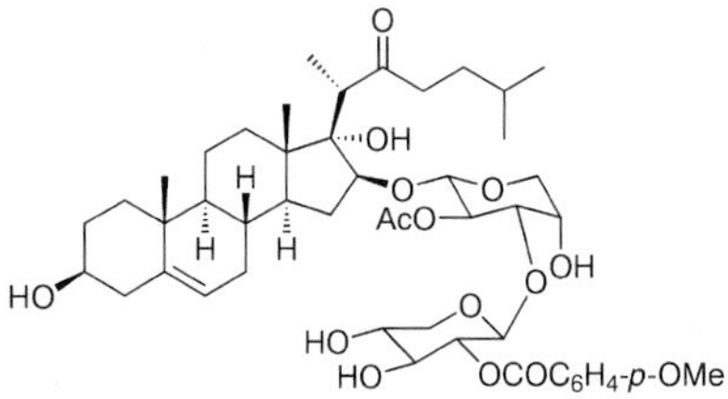

OSW-1 (**90**)

unit. Although OSW-1 (**90**) does not incorporate a fused heterocyclic ring, for example as found in the cephalostatins, it has been proposed[130] that the possibility of forming a cyclic oxonium ring after cleavage of the 16β-sugar attachment is crucial to its activity. Moreover, the very similar cytotoxicity profile of OSW-1 (**90**) compared with the cephalostatins suggests that there may be a common mechanism of action, and it has been proposed[130] that if that is so, the active intermediate of the cephalostatins could be the oxonium ion **92** (Scheme 8.23) resulting from spiroketal cleavage, an oxonium ion closely analogous to **93**, one that

Cephalostatins   (**91**)   (**92**)
(North 1 unit)

Scheme 8.23.

could be formed from the 16- and 17-substituents on ring D of OSW-1 (90) (Scheme 8.24). Inspired by the above considerations, and by the potent (subnanomolar) cytotoxicities of such cholestane glycosides, Guo

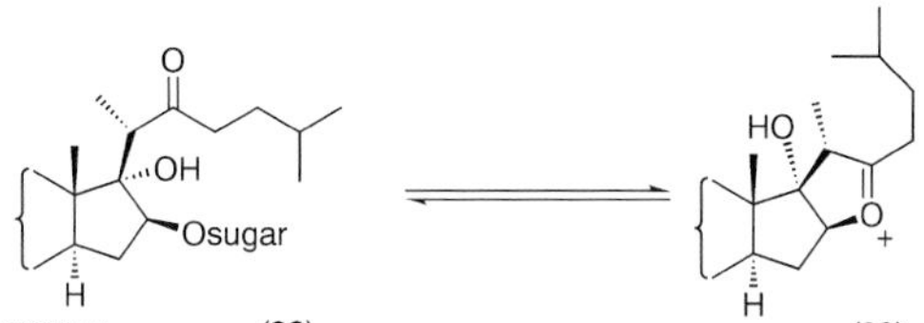

Scheme 8.24.

and Fuchs accomplished a synthesis of the aglycone of OSW-1 (90) from 5-androsten-3β-ol-17-one in nine steps (55% overall yield).[130] The side-chain was installed via an ene reaction; ring D was functionalised by a regioselective and stereoselective dihydroxylation; subsequent oxidation of the 16α-ol group to a ketone was followed by a stereoselective Luche reduction (CeCl$_3$/NaBH$_4$) which afforded exclusively the *trans*-alcohol of 16β- and 17α-stereochemistry.[130]

In a total synthesis of a trisaccharide saponin isolated from an ophiopogonis plant, the 1-hydroxy steroid acceptor was first activated with a catalytic quantity of TMSOTf (Schmidt's inverse procedure) prior to addition of the glycosyl donor.[131]

Digitoxigenin has been synthesised from 3β-acetoxyandrost-5-en-17-one via a presumed allene oxide intermediate;[132] a formal total synthesis has also been described.[133] A pentacyclic precursor of the cardenolide (–)-ouabain has been synthesised by using an intramolecular Heck reaction involving an α-sulfenyl enol trifluoromethanesulfonate in order to join the A and C rings.[134]

Thiocorticosteroid analogues have been prepared via the introduction of a disulfide bridge by reaction 3-keto-1α,5α-epidithio steroids with sulfuryl chloride.[135]

The first chemical synthesis of wortmannin, a specific inhibitor of PI 3-kinases, has been achieved, starting from cortisone.[136]

In a new asymmetric synthesis of 5β-ranol, the side-chain derived from (S)-(–)butane-1,2,4-triol was introduced by Wittig olefination.[137]

8.3.6 Strategies, total syntheses and additional biologically active steroids

The indenone 94 and its derivatives are useful intermediates in the total synthesis of steroids; their elaboration has provided key precursors of

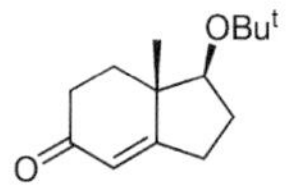

(94)

6α-methylprednisolone[138] and a total synthesis of testosterone.[139] Base-catalysed condensations provide a route to (±)-A-nor-estr-3(5)-ene-2,17-dione which contains the skeleton of dinordrin.[140]

In a new synthesis of the potent 5α-reductase and 17,20-lyase inhibitor 4-amino-20(S)-hydroxymethylpregn-4-ene-3-one, the amino group was introduced via C-nitration of a steroidal dienolate with isopropyl nitrate, followed by reduction.[141]

Allylic α-face cyclopropanes **95**, prepared by cyclopropanation directed by the 17α-hydroxy group, undergo facile isomerisation over molecular sieves to the β-face isomers.[142]

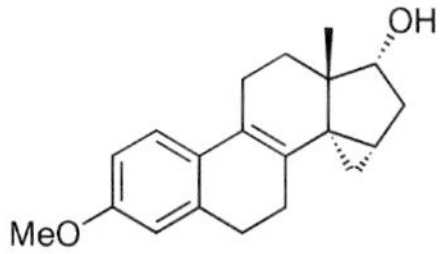

(95)

10β-Fluoro-3-oxo-1,4-estradienes have been obtained in high yields by reaction of 1,3,5(10)-estratrien-3-ols with N-fluorobis[(trifluoromethyl)-sulfonyl]imide.[143]

4-Cyanoprogesterone (**96**), is a potent inhibitor of 5α-reductase, and has been efficiently prepared by addition of the anion of acetonitrile

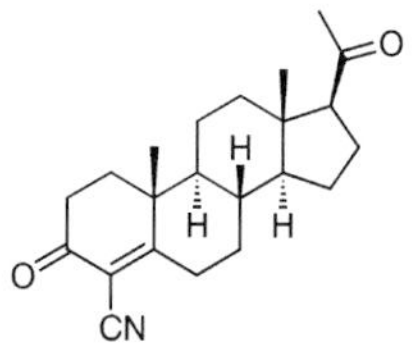

(96)

to 20-acetoxy-4-oxa-3,5-secopregna-5,17-dien-3-one.[144] Irradiation of Δ⁵-pregnene-3β-acetoxy-20β-ol in the presence of iodine, lead tetraacetate and calcium carbonate led to the 18-iodo derivative, enabling subsequent ring closure and a total synthesis of mirasorvone.[54] A total synthesis of the withanolide minabeolide 3 involved the oxidative ring expansion of a furylcarbinol to a lactol.[145] Estrone has been synthesised by aromatisation of ring A by using a thallium(III)-mediated fragmentation of a 19-hydroxyandrost-5-ene precursor.[146] An synthesis of enantiopure estrone employed a double Heck reaction.[147]

The 6-oxocholestane **97**, isolated from the defensive secretion of *Chrysomela varians*, has been synthesised from diosgenin.[148] A total

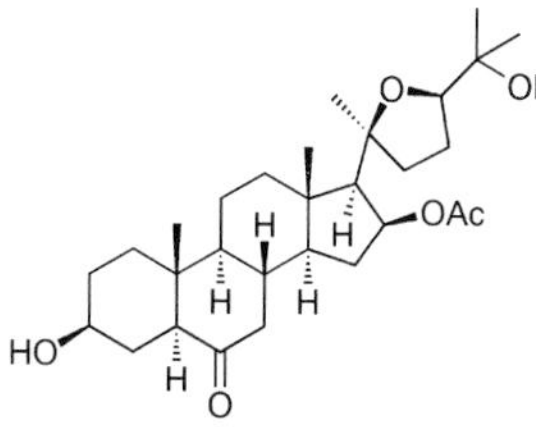

(97)

synthesis of mosesin-4, a naturally occurring shark repellant, began with cholic acid and used trimethylsilyl trifluoromethanesulfonate as a promoter for the hindered 7α-glycosylation.[149] A new strategy for cardenolide synthesis involves the palladium-catalysed cyclisation of a CD ring enol trifluoromethanesulfonate.[150]

In a synthesis of blattellastanoside A (**98**), the aggregation pheromone of the German cockroach, β-glucosidation of a β-epoxide was achieved in the presence of mercury(II) cyanide.[151] Steroids of the form **99** have been

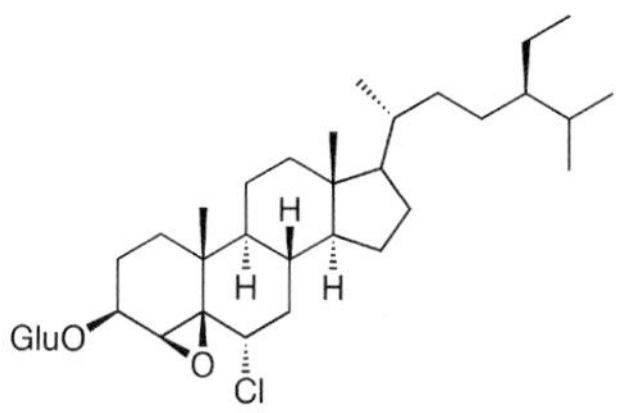

(98)

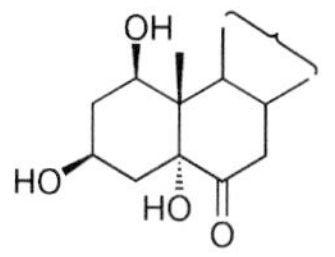

(99)

synthesised; their A and B rings were designed to overlay the carbonyl and triol units on the five- and seven-membered rings, respectively, of phorbol. Steroids **99** did exhibit weak phorbol-like activity.[152]

Syntheses of the plant growth promoting brassinosteroids have been reviewed.[118b, 153] The most efficient route to brassinolide uses the anion of phenyl *n*-propyl selenide with a subsequent selenoxide elimination.[154] The jujubosides are new dammarane oligoglycosides, some of which inhibit histamine release.[155] Recently, a synthesis of squalamine in 15 steps from stigmasterol has been completed.[156]

References

1. J. Kalvoda, 1992, *Helv. Chim. Acta*, 75, 2341.
2. T. Shimizu, S. Hiranuma, and T. Nakata, 1996, *Tetrahedron Lett.*, 37, 6145.
3. T. Tsunoda, Y. Yamamiya, Y. Kawamura, and S. Itô, 1995, *Tetrahedron Lett.*, 36, 2529.
4. P.A. Procopiou, S.P.D. Baugh, S.S. Flack, and G.G.A. Inglis, 1996, *J. Chem. Soc., Chem. Commun.*, 2625.
5. M. Frigerio and M. Santagostino, 1994, *Tetrahedron Lett.*, 35, 8019.
6. K. Krohn, I. Vinke, and H. Adam, 1996, *J. Org. Chem.*, 61, 1467.
7. K. Krohn, B. Knauer, J. Küpke, D. Seebach, A.K. Beck, and M. Hayakawa, 1996, *Synthesis*, 1341.
8. M.L.S. Almeida, P. Kocovsky, and J.-E. Bäckvall, 1996, *J. Org. Chem.*, 61, 6587.
9. C.K. Acosta, P.N. Rao, and H. K. Kim, 1993, *Steroids*, 58, 205.
10. Shamsuzzaman, S. Ahmad, B.Z. Khan, and Shafiullah, 1991, *J. Org. Chem.*, 56, 1936.
11. M.J.S.M. Moreno, M.L. Sá e Melo, and A.S. Campos Neves, 1991, *Tetrahedron Lett.*, 32, 3201.
12. R. Tedesco, R. Fiaschi, and E. Napolitano, 1995, *Synthesis*, 1493.
13. P.M.F.M. Bastiaansen, L. Kohout, M.A. Posthumus, J.B.P.A. Wijnberg, and A. de Groot, 1996, *J. Org. Chem.*, 61, 859.
14. A.G. Myers and B. Zheng, 1996, *Tetrahedron Lett.*, 37, 4841.
15. J. Tomás, F. Camps, J. Coll, E. Melé, and N. Pascual, 1992, *Tetrahedron*, 48, 9809.
16. N. Flaih, J.R. Hanson, and P.B. Hitchcock, 1991, *J. Chem. Soc., Perkin Trans. 1*, 1085.
17. G.A. Morrison and J.B. Wilkinson, 1990, *J. Chem. Soc., Perkin Trans. 1*, 345.
18. A. Toró, I. Pallagi, and G. Ambrus, 1994, *Tetrahedron Lett.*, 35, 7651.
19. M.J.S.M. Moreno, M.L. Sá e Melo, and A.S. Campos Neves, 1993, *Tetrahedron Lett.*, 34, 353.
20. R. Maltais and D. Poirier, 1998, *Tetrahedron Lett.*, 39, 4151.
21. R.E. Dolle, S.J. Schmidt, and L.I. Kruse, 1988, *J. Chem. Soc., Chem. Commun.*, 19.
22. H. Majgier-Baranowska, J.N. Bridson, K. Marat, and J.F. Templeton, 1998, *J. Chem. Soc., Perkin Trans. 1*, 1967.
23. Y.S. Choe, P.J. Lidstroem, D.Y. Chi, T.A. Bonasera, M.J. Welch, and J.A. Katzenellenbogen, 1995, *J. Med. Chem.*, 38, 816.
24. J.R. Hanson, P.B. Hitchcock, M.D. Liman, and S. Nagaratnam, 1995, *J. Chem. Soc., Perkin Trans. 1*, 2183.
25. A.B. Solovieva, V.V. Borovkov, E.A. Lukashova, and G.S. Grinenko, 1995, *Chem. Lett.*, 41.
26. E.J. Parish, S.A. Kizito, and R.W. Heidepriem, 1993, *Synth. Commun.*, 23, 223.
27. L.R. Galagovsky and E.G. Gros, 1993, *J. Chem. Res. (S)*, 137.
28. J.R. Hanson, S. Nagaratnam, and J. Stevens, 1996, *J. Chem. Res. (S)*, 102.
29. B. Ferron, J.-C. Jacquesy, M.-P. Jouannetaud, O. Karam, and J.-M. Coustard, 1993, *Tetrahedron Lett.*, 34, 2949.
30. K. Mikami, H. Kishino, and T.-P. Loh, 1994, *J. Chem. Soc., Chem. Commun.*, 495.
31. M.E. Krafft, O.A. Dasse, and B. Shao, 1998, *Tetrahedron*, 54, 7033.
32. R. Skoda-Földes, Z. Csákai, L. Kollár, G. Szalontai, J. Horváth, and Z. Tuba, 1995, *Steroids*, 60, 786.
33. M. Weissenberg and J. Levisalles, 1995, *Tetrahedron*, 51, 5711.
34. B. Knauer and K. Krohn, 1995, *Liebigs Ann.*, 677.
35. J.A.R. Salvador, M.L. Sá e Melo, and A.S. Campos Neves, 1993, *Tetrahedron Lett.*, 34, 361.
36. J.A.R. Salvador, M.L. Sá e Melo, and A.S. Campos Neves, 1997, *Tetrahedron Lett.*, 38, 119.
37. H. Lardy, N. Kneer, Y. Wei, B. Partridge, and P. Marwah, 1998, *Steroids*, 63, 158.
38. J. Westermann, H. Neh, and K. Nickisch, 1996, *Chem. Ber.*, 129, 963.

39. A.P. Davis and K.M. Bhattarai, 1995, *Tetrahedron*, 51, 8033.
40. R. Bohlmann, 1994, *Tetrahedron Lett.*, 35, 85.
41. A.J. Poss and G.A. Shia, 1995, *Tetrahedron Lett.*, 36, 4721.
42. L.G. Liu, T. Zhang, and Z.S. Li, 1996, *Synth. Commun.*, 26, 2999.
43. G. Majetich and K. Wheless, 1995, *Tetrahedron*, 51, 7095.
44. R. Batra and R. Breslow, 1989, *Tetrahedron Lett.*, 40, 535.
45. K. Orito, S. Sato, and H. Suginome, 1995, *J. Chem. Soc., Perkin Trans. 1*, 63.
46. K. Orito, M. Ohto, N. Sugawara, and H. Suginome, 1990, *Tetrahedron Lett.*, 31, 5921.
47. P.A. Grieco and T.L. Stuk, 1990, *J. Am. Chem. Soc.*, 112, 7799.
48. D.H.R. Barton and D. Doller, 1991, *Collect. Czech. Chem. Commun.*, 56, 984.
49. K. Yamamoto, Y. Imae, and S. Yamada, 1990, *Tetrahedron Lett.*, 31, 4903.
50. P. Bovicelli, A. Gambacorta, P. Lupattelli, and E. Mincione, 1992, *Tetrahedron Lett.*, 33, 7411.
51. A. Arnone, M. Cavicchiolo, V. Montanari, and G. Resnati, 1994, *J. Org. Chem.*, 59, 5511.
52. P. Bovicelli, P. Lupatelli, E. Mincione, T. Prencipe, and R. Curci, 1992, *J. Org. Chem.*, 57, 5052.
53. I. Izzo, F. De Riccardis, and G. Sodano, 1998, *J. Org. Chem.*, 63, 4438.
54. Z.-C. Yang and J. Meinwald, 1998, *Tetrahedron Lett.*, 39, 3425.
55. R.-M. Liu, X.F.D. Chillier, P. Kamalprija, U. Burger, and F.O. Gülaçar, 1996, *Helv. Chim. Acta*, 79, 989.
56. T. Li, Y. Yang, and Y. Li, 1993, *J. Chem. Res. (S)*, 28.
57. M.A. Maestro, F.J. Sardina, L. Castedo, and A. Mourino, 1991, *J. Org. Chem.*, 56, 3582.
58. H.-M. He, P.E. Fanwick, K. Wood, and M. Cushman, 1995, *J. Org. Chem.*, 60, 5905.
59. P. Kočovský, J. Strogl, M. Pour, and A. Gogoll, 1994, *J. Am. Chem. Soc.*, 116, 186.
60. J. Šrogl, A. Gogoll, and P. Kočovský, 1994, *J. Org. Chem.*, 59, 2246.
61. P. Kočovský and J. Šrogl, 1992, *J. Org. Chem.*, 57, 4565.
62. J. Šrogl and P. Kočovský, 1992, *Tetrahedron Lett.*, 33, 5991.
63. S. Ino, H. Oinuma, R. Yamakado, M. Morisaki, Y. Furukawa, and T. Hata, 1991, *Chem. Pharm. Bull. (Japan)*, 39, 3335.
64. S. Yamada, H. Ishizaka, H. Ishida, and K. Yamamoto, 1995, *J. Chem. Soc., Chem Commun.*, 423.
65. M. Dabovic, M. Bjelakovic, V. Andrejevic, L. Lorenc, and M.Lj. Mihailovic, 1994, *Tetrahedron*, 50, 1833.
66. H. Suginome, M. Nagai, M. Moizumi, K. Kohzuki, K. Fukumoto, and T. Kametani, 1988, *Tetrahedron Lett.*, 29, 4959.
67. A. Boto, C. Betancor, R. Hernández, M.S. Rodríguez, and E. Suárez, 1993, *Tetrahedron Lett.*, 34, 4865.
68. H. Suginome, M. Takemura, N. Shimoyama, and K. Orito, 1991, *J. Chem. Soc., Perkin Trans. 1*, 2721.
69. I. Abe, M. Rohmer, and G.D. Prestwich, 1993, *Chem. Rev.*, 93, 2189.
70. E.J. Corey, S.C. Virgil, H. Cheng, C.H. Baker, S.P.T. Matsuda, V. Singh, and S. Sarshar, 1995, *J. Am. Chem. Soc.*, 117, 11819.
71. J.C. Medina and K.S. Kyler, 1988, *J. Am. Chem. Soc.*, 110, 4818.
72. C. Pale-Grosdemange, C. Feil, M. Rohmer, and K. Poralla, 1998, *Angew. Chem., Int. Ed. Engl.*, 37, 2237.
73. W.S. Johnson, S.D. Lindell, and J. Steele, 1987, *J. Am. Chem. Soc.*, 109, 5852.
74. W.S. Johnson, B. Chenera, F.S. Tham, and R.K. Kullnig, 1993, *J. Am. Chem. Soc.*, 115, 493.
75. W.S. Johnson, M.S. Plummer, S.P. Reddy, and W.R. Bartlett, 1993, *J. Am. Chem. Soc.*, 115, 515.
76. W.S. Johnson, V.R. Fletcher, B. Chenera, W.R. Bartlett, F.S. Tham, and R.K. Kullnig, 1993, *J. Am. Chem. Soc.*, 115, 497.

77. J.A. Clase and T. Money, 1990, *J. Chem. Soc., Chem. Commun.*, 1503.

78. D. Kim, S. Kim, J.J. Lee, and H.S. Kim, 1990, *Tetrahedron Lett.*, 31, 4027.

79. S. Möller, D. Weiss, and R. Beckert, 1995, *Liebigs Ann.*, 1397.

80. M.A. Dombroski, S.A. Kates, and B.B. Snider, 1990, *J. Am. Chem. Soc.*, 112, 2759.

81. P.A. Zoretic, M. Ramchandani, and M.L. Caspar, 1991, *Synth. Commun.*, 21, 923.

82. P.A. Zoretic, X. Weng, M.L. Caspar, and D.G. Davis, 1991, *Tetrahedron Lett.*, 32, 4819.

83. S. Handa and G. Pattenden, 1998, *J. Chem. Soc., Chem. Commun.*, 311.

84. Y. Takahashi, S. Tomida, Y. Sakamoto, and H. Yamada, 1997, *J. Org. Chem.*, 62, 1912.

85. J. Boivin, A.-M. Schiano, and S.Z. Zard, 1992, *Tetrahedron Lett.*, 33, 7849.

86. S.K. Pradhan and S. Sitabkhan, 1993, *Tetrahedron Lett.*, 34, 5615.

87. T. Arencibia, T. Prangé, J.A. Salazar, and E. Suárez, 1995, *Tetrahedron Lett.*, 36, 6337.

88. N. Almirante and A. Cerri, 1997, *J. Org. Chem.*, 62, 3402.

89. D. Batty and D. Crich, 1991, *J. Chem. Soc., Perkin Trans. 1*, 2894.

90. M. Kurosu, L.R. Marcin, T.J. Grinsteiner, and Y. Kishi, 1998, *J. Am. Chem. Soc.*, 120, 6627.

91. T. Zoller, D. Uguen, A. De Cian, J. Fischer, and S. Sablé, 1997, *Tetrahedron Lett.*, 38, 3409.

92. T. Takahashi, K. Shimizu, T. Doi, J. Tsuji, and Y. Fukuzawa, 1988, *J. Am. Chem. Soc.*, 110, 2674.

93. H. Nemoto and K. Fukumoto, 1998, *Tetrahedron*, 54, 4151.

94. H. Nemoto, M. Ando, and K. Fukumoto, 1990, *Tetrahedron Lett.*, 31, 6205.

95. H. Nemoto, M. Nagai, M. Moizumi, K. Kohzuki, K. Fukumoto, and T. Kametani, 1988, *Tetrahedron Lett.*, 29, 4959.

96. H. Nemoto, S. Fujita, M. Nagai, K. Fukumoto, and T. Kametani, 1988, *J. Am. Chem. Soc.*, 110, 2931.

97. H. Nemoto, M. Moizumi, M. Nagai, K. Fukumoto, and T. Kametani, 1988, *J. Chem. Soc., Perkin Trans. 1*, 885.

98. H. Nemoto, N. Matsuhashi, M. Imaizumi, M. Nagai, and K. Fukumoto, 1990, *J. Org. Chem.*, 55, 5625.

99. G. Quinkert, M. del Grosso, A. Bucher, M. Bauch, W. Döring, J.W. Bats, and G. Dürner, 1992, *Tetrahedron Lett.*, 33, 3617.

100. R. Skoda-Földes, G. Jeges, L. Kollár, J. Horváth, and Z. Tuba, 1996, *Tetrahedron Lett.*, 37, 2085.

101. T.L. Gilchrist and R.J. Summersell, 1988, *J. Chem. Soc., Perkin Trans. 1*, 2595.

102. A. Perez-Medrano and P.A. Grieco, 1991, *J. Am. Chem. Soc.*, 113, 1057.

103. K. Takahashi, K. Mikami, and T. Nakai, 1988, *Tetrahedron Lett.*, 29, 5277.

104. K. Mikamai, K. Takahashi, and T. Nakai, 1990, *J. Am. Chem. Soc.*, 112, 4035.

105. T. Ibuka, T. Taga, S. Nishii, and Y. Yamamoto, 1988, *J. Chem. Soc., Chem. Commun.*, 342.

106. T. Honda and M. Katoh, 1998, *Heterocycles*, 47, 481.

107. B.G. Hazra, P.L. Joshi, B.B. Bahule, N.P. Argade, V.S. Pore, and M.D. Chordia, 1994, *Tetrahedron*, 50, 2523.

108. S. Takatsuto, T. Watanabe, S. Fujioka, and A. Sakurai, 1997, *J. Chem. Res. (S)*, 134.

109. T. Kametani, M. Tsubuki, and T. Honda, 1989, *Heterocycles*, 28, 59.

110. D.A. Livingston, J.E. Petre, and C.L. Bergh, 1990, *J. Am. Chem. Soc.*, 112, 6449.

111. P. Bovicelli, P. Lupattelli, D. Fracassi, and E. Mincione, 1994, *Tetrahedron Lett.*, 35, 935.

112. G.-D. Zhu and W.H. Okamura, 1995, *Chem. Rev.*, 95, 1877.

113. A. Kurek-Tyrlik, K. Minksztym, and J. Wicha, 1995, *J. Am. Chem. Soc.*, 117, 1849.

114. S. Marczak, K. Michalak, and J. Wicha, 1995, *Tetrahedron Lett.*, 36, 5425.

115. G. Stork, D. Hutchinson, M. Okabe, D. Parker, C.S. Ra, R. Ribéreau, T. Suzuki, and T. Zebovitz, 1992, *Pure Appl. Chem.*, 64, 1809.

116. P. Grzywacz, S. Marczak, and J. Wicha, 1997, *J. Org. Chem.*, 62, 5293.

117. S. Marczak, K. Michalak, Z. Urbanczyk-Lipkowska, and J. Wicha, 1998, *J. Org. Chem.*, 63, 2218.

118. (a) A. Ganesan, 1996, *Angew. Chem., Int. Ed. Engl.*, 35, 611; (b) K.J. Hale, 1998, 2nd Supp. to 2nd edn. of *Rodd's Chemistry of Carbon Compounds*, vol. IVG/H, (M. Sainsbury, Ed.), Elsevier, Amsterdam, pp. 65-112.

119. G.R. Pettit, M. Inoue, Y. Kamano, D.L. Herald, C. Arm. C. Dufresne, N.D. Christie, J.M. Schmidt, D.L. Doubek, T.S. Krupa, 1988, *J. Am. Chem. Soc.*, 110, 2006.

120. J.U. Jeong, S.C. Sutton, S. Kim, and P.L. Fuchs, 1995, *J. Am. Chem. Soc.*, 117, 10157.

121. S. Kim and P.L. Fuchs, 1994, *Tetrahedron Lett.*, 35, 7163.

122. S. Kim, S.C. Suton, and P.L. Fuchs, 1995, *Tetrahedron Lett.*, 36, 2427.

123. J.U. Jeong and P.L. Fuchs, 1995, *Tetrahedron Lett.*, 36, 2431.

124. S. Bhandaru and P.L. Fuchs, 1995, *Tetrahedron Lett.*, 36, 8351.

125. C. Guo, S. Bhandaru, P.L. Fuchs, and M.R. Boyd, 1996, *J. Am. Chem. Soc.*, 118, 10672.

126. S. Fukuzawa, S. Matsunaga, and N. Fusetani, 1997, *J. Org. Chem.*, 62, 4484.

127. M. Drogemuller, R. Jautelat, and E. Winterfeldt, 1996, *Angew. Chem., Int. Ed. Engl.*, 35, 1572.

128. T.G. LaCour, C. Guo, S. Bhandaru, M.R. Boyd, and P.L. Fuchs, 1998, *J. Am. Chem. Soc.*, 120, 692.

129. C.H. Heathcock and S.C. Smith, 1994, *J. Org. Chem.*, 59, 6828.

130. C. Guo and P.L. Fuchs, 1998, *Tetrahedron Lett.*, 39, 1099.

131. M. Lu, B. Yu, and Y. Hui, 1998, *Tetrahedron Lett.*, 39, 415.

132. M.M. Kabat, 1995, *J. Org. Chem.*, 60, 1823.

133. J.P. Kutney, K. Piotrowska, J. Somerville, S.-P. Huang, and S.J. Rettig, 1989, *Can. J. Chem.*, 67, 580.

134. J. Hynes, L.E. Overman, T. Nasser, and P.V. Rucker, 1998, *Tetrahedron Lett.*, 39, 4647.

135. D.H.R. Barton, R.M. Hesse, S.D. Lindell, and M.M. Pechet, 1991, *J. Chem. Soc., Perkin Trans. 1*, 2855.

136. S. Sato, M. Nakada, and M. Shibasaki, 1996, *Tetrahedron Lett.*, 37, 6141.

137. D.W. Harney and T.A. Macrides, 1997, *J. Chem. Soc., Perkin Trans. 1*, 1353.

138. A.R. Daniewski and E. Piotrowska, 1989, *Liebigs Ann. Chem.*, 571.

139. M. Ihara, Y. Tokunaga, and K. Fukumoto, 1990, *J. Org. Chem.*, 55, 4497.

140. W. Zhou and G. Wei, 1990, *Synthesis*, 822.

141. T.T. Curran, G.A. Flynn, D.E. Rudisill, and P.M. Weintraub, 1995, *Tetrahedron Lett.*, 36, 4761.

142. H. Künzer and M. Thiel, 1994, *Tetrahedron Lett.*, 35, 2329.

143. W.T. Pennington, G. Resnati, and D.D. DesMarteau, 1992, *J. Org. Chem.*, 57, 1536.

144. M. Haase-Held, M. Hatzis, and J. Mann, 1993, *J. Chem. Soc., Perkin Trans. 1*, 2907.

145. M. Tsubuki, K. Kanai, K. Keino, N. Kakinuma, and T. Honda, 1992, *J. Org. Chem.*, 57, 2930.

146. P. Kočovský and R.S. Baines, 1993, *Tetrahedron Lett.*, 34, 6139.

147. L.F. Tietze, T. Nöbel, and M. Spescha, 1998, *J. Am. Chem. Soc.*, 120, 8971.

148. R. Tavares, T. Randoux, J.-C. Braekman, and D. Daloze, 1993, *Tetrahedron*, 49, 5079.

149. D. Garguilo, T.A. Blizzard, and K. Nakanishi, 1989, *Tetrahedron*, 45, 5423.

150. W. Deng, M.S. Jensen, L.E. Overman, P.V. Rucker, and J.-P. Vionnet, 1996, *J. Org. Chem.*, 61, 6760.

151. K. Mori and K. Fukumatsu, 1993, *Proc. Jpn. Acad., Ser. B*, 69, 61 (1994, Chem. Abstr., 119, 139596).

152. Y. Endo, H. Fukasawa, Y. Hashimoto, and K. Shudo, 1994, *Chem. Pharm. Bull. (Japan)*, 42, 462.

153. V.A. Khripach, 1990, *Pure Appl. Chem.*, 62, 1319.

154. T.G. Back, D.L. Baron, W.D. Luo, and S.K. Nakajima, 1997, *J. Org. Chem.*, 62, 1179.

155. M. Yoshikawa, T. Murakami, A. Ikebata, S. Wakao, N. Murakami, H. Matsuda, and J. Yamahara, 1997, *Chem. Pharm. Bull. (Japan)*, 45, 1186.
156. X. Zhang, M.N. Rao, S.R. Jones, B. Shao, P. Feibush, M. McGuigan, N. Tzodikov, B. Feibush, J. Sharkansky, B. Snyder, L.M. Mallis, A. Sarkahian, S. Wilder, J.E. Turse, and W.A. Kinney, 1998, *J. Org. Chem.*, 63, 8599.

9 Synthesis of enediynes and dienediynes

S. Caddick, S. Shanmugathasan and N.J. Smith

9.1 Introduction

In recent years the enediyne and dienediyne families of natural products have occupied a central position in natural product synthesis.[1–6]

The considerable challenge which their unusual structures offer has stimulated many synthetic chemists to embark upon their synthesis. Such research endeavours have led to numerous developments in synthetic chemistry in particular, in the development of new synthetic methodologies.

It is not, however, simply the compelling molecular structures of these molecules which has led to such widespread research activity. These molecules also exhibit a potent biological activity and they have an ability to interact and damage DNA with varying degrees of selectivity. Thus, these compounds can be considered as important leads for the development of novel pharmaceutical agents with applications in gene-targeting, protein engineering, and cancer chemotherapy. Whilst it is outside the scope of this treatment to discuss the detailed biological effects of these compounds it is worth noting that these molecules promote DNA damage by H-atom abstraction by an arene biradical on the target DNA. Illustrative examples of the process by which the enediynes (calicheamicin) and dienediynes (neocarzinostatin) generate their respective biradicals are discussed below.

9.1.1 Calicheamicin

The generally accepted mechanism for activation of the calicheamicins is shown in Scheme 9.1.

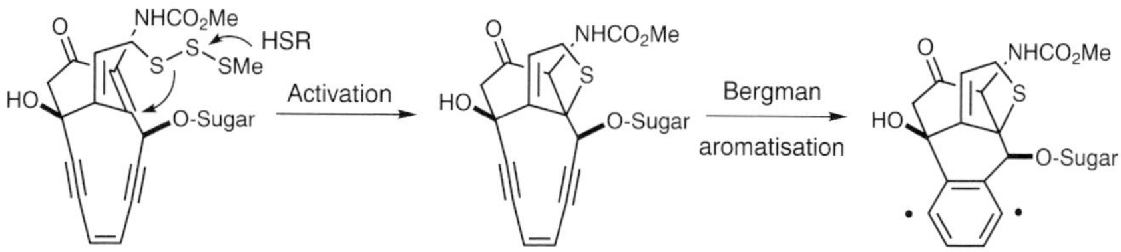

Scheme 9.1.

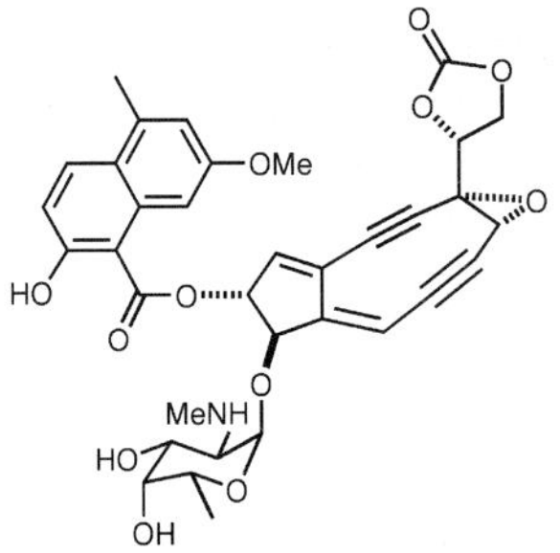

Neocarzinostatin Chromophore A

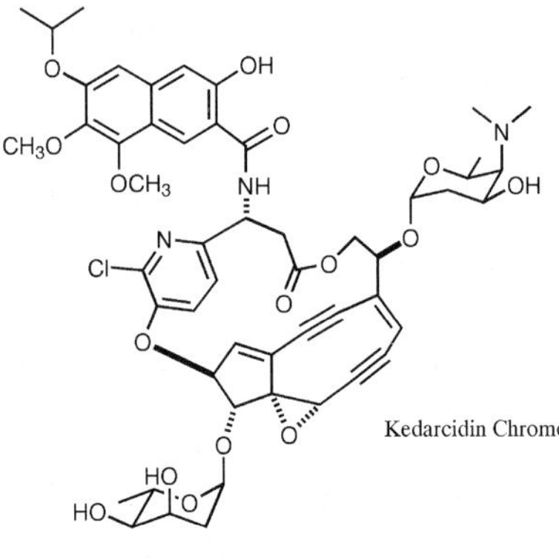

Kedarcidin Chromophore

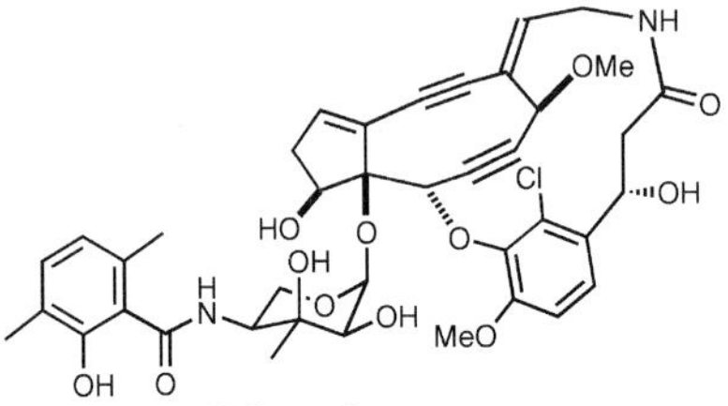

Maduropeptin

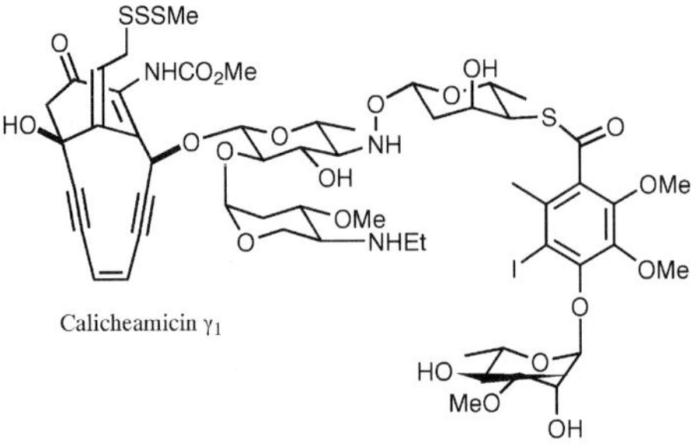

Calicheamicin γ₁

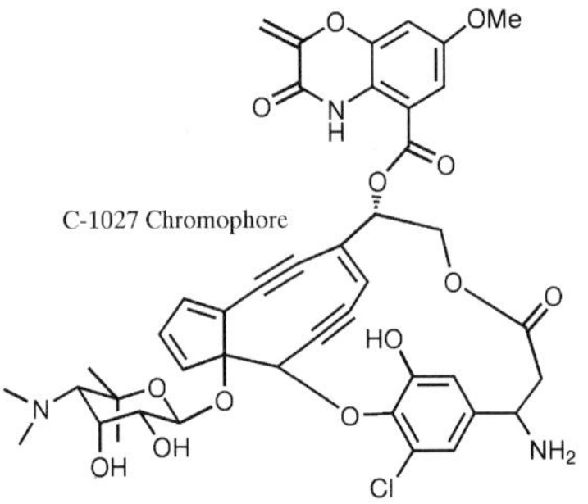

C-1027 Chromophore

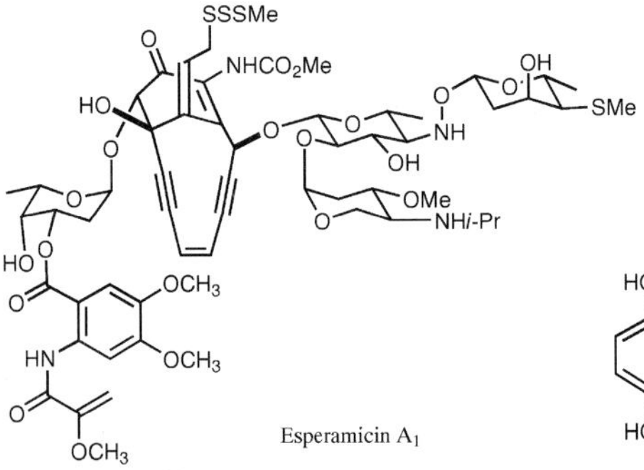

Esperamicin A₁

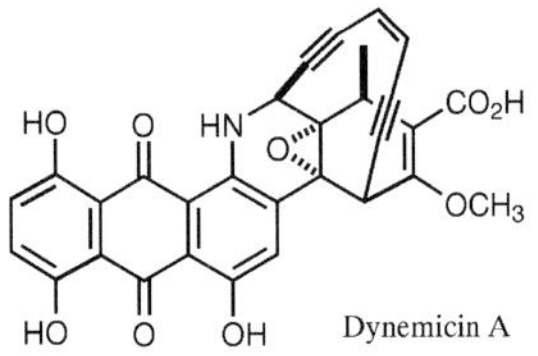

Dynemicin A

Structures for some enediynes and dienediynes.

The sugar portion delivers the natural product to its target DNA sequence; thiol addition to the trisulfide linkage facilitates intramolecular conjugate addition. The strained enediyne ring then undergoes spontaneous Bergman aromatisation to generate a biradical which promotes DNA damage by H-atom abstraction.

9.1.2   Neocarzinostatin chromophore

This operates by a related process but the interaction with DNA is less selective and the molecule shows increased toxicity. The naphthoate (Ar) is generally perceived to confer moderate DNA base-selectivity in binding. The triggering process involves direct addition of thiol to the chromophore which then remains attached to the dienediyne unit; aromatisation generates a biradical which promotes DNA damage by H-atom abstraction (Scheme 9.2).

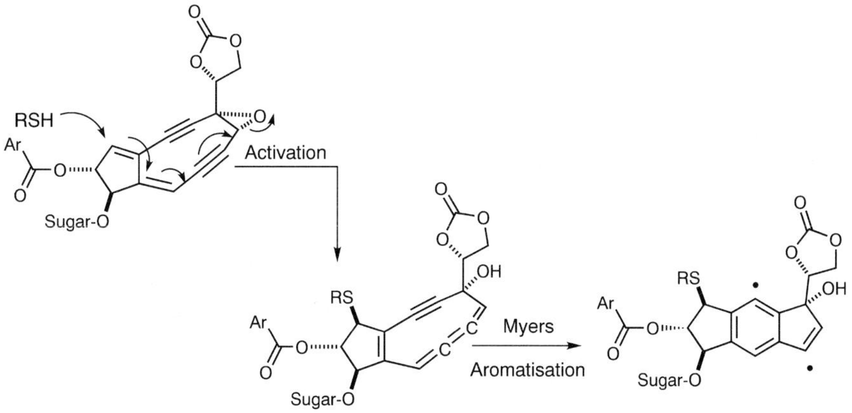

Scheme 9.2.

The various members of the enediyne and dienediyne classes operate in a similar general manner but the details associated with DNA interaction, triggering and biological effect differ and have been well documented.[7]

9.2   Synthesis of enediynes

This chapter will concentrate on selected highlights of chemical synthesis of these natural products; the reader is directed to more detailed review articles[2,3,4,5] for further, more comprehensive, treatment of the area.

9.2.1   Calicheamicins[8, 9]

The calicheamicins and esperamicins are closely related enediynes which have the general structure **1**.

X = H, Calicheamicins,

X = O-Sugar, Esperamicins

1

The main structural difference between the two classes is found in the carbohydrate domain(s). Esperamicins have an additional hydroxyl group (X) attached to the central core from which an additional glycosidic linkage is made. The calicheamicins have been the subject of most synthetic effort, and the majority of this section will concentrate on selected achievements in this area.

9.2.1.1   Model studies

Perhaps unsurprisingly, the first synthetic objective to be addressed was formation of the core enediyne macrocycle. The first published approach to a core structure was that of Schreiber and co-workers, who developed what is still one of the most imaginative strategies ever used for these targets.[10] Assembly of the key precursor was achieved from **2**, **3** and **4** by a sequence which exploited a very efficient palladium-mediated cross-coupling reaction. Intramolecular Diels–Alder (IMDA) reaction of **5** was notable, in that it furnished the tricyclic adduct **6**. Whilst the IMDA reaction did not give the desired ring system directly, carbonate opening, alcohol activation and oxidative removal of the *p*-methoxybenzyl (PMB) group using DDQ yielded **7**. A stereoelectronically controlled pinacol reaction completed the core structure **8**. This general approach is very appealing and differs significantly from other approaches more commonly utilised in the area of enediyne synthesis.

Another early approach, described by Magnus and collaborators, pioneered the use of cobalt-complexed aldehydes to assist intramolecular cyclisation reactions.[11] Conjugate addition of thiophenol anion to enone **9** in the presence of a Lewis acid gave **10** as a single isomer. The stereoselectivity observed may originate from coordination of the Lewis acid to the ketone and aldehyde carbonyl groups in the ring-closure step. Cobalt complexation masks the cyclic enediyne ring system and helps relieve ring strain in the cyclised product **10**. The conversion of **10** into **11**

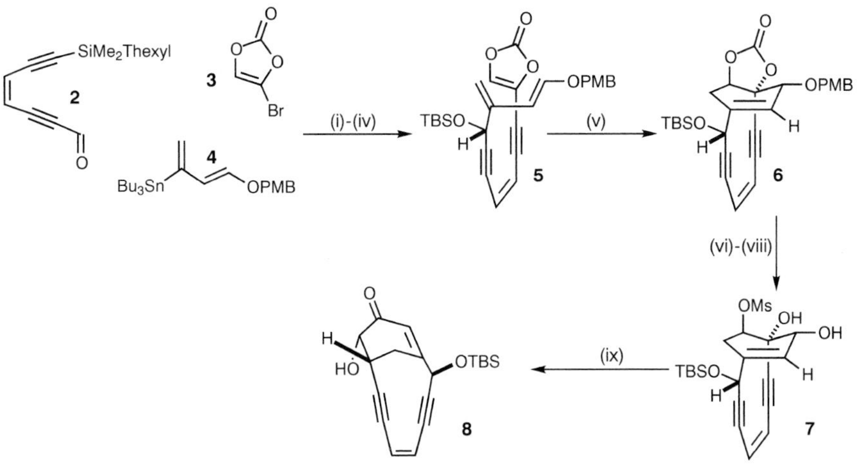

Scheme 9.3 Reagents: (i) **4**, *n*-BuLi, MgBr$_2$, $-78°C$, then **2**, 79%; (ii) *n*-Bu$_4$NF; (iii) TBSOTf, TEA, 80%; (iv) **3**, Pd(CH$_3$CN)$_2$Cl$_2$, CuI, TEA, Bz, 49%; (v) 120°C, 40%; (vi) K$_2$CO$_3$, H$_2$O, MeOH, 83%; (vii) MsCl, 73%; (viii) DDQ, 91%; (ix) Et$_2$AlCl, 65%.

is rich in interesting chemistry. Sulfide oxidation followed by elimination via 2,3-sigmatropic rearrangement installs the double bond. Oxidative decomplexation of the alkyne using a cerium(IV) reagent liberated the enediyne. Secondary alcohol protection allowed allylic oxidation to be achieved, and the sequence was completed with an amination reaction and gave **11** (Scheme 9.4).

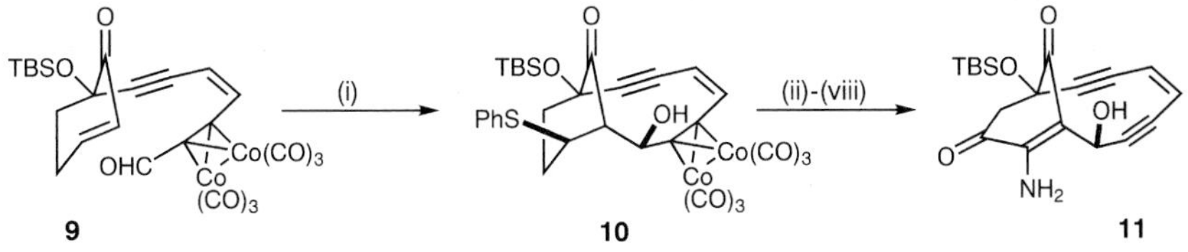

Scheme 9.4 Reagents: (i) Me$_2$AlSPh, CH$_2$Cl$_2$, $-70°C$, Ti(O*i*-Pr)$_4$, 71%; (ii) DMDO, 64%; (iii) CAN, 76%; (iv) TBSOTf, *i*Pr$_2$NEt; (v) PhthalSePh, (vi) DMDO; (vii) TFA, H$_2$O; (viii) Ph$_2$S=NH.

This general approach has been used to prepare numerous members of the enediyne/dienediyne class, but has not to date featured in a completed synthesis of any of these natural products.

The first examples of strategies which eventually became the most widely used in total synthesis were described simultaneously by the groups of Kende[12] and Danishefsky.[13] They involved addition of an acetylide anion to an aldehyde. In the simpler of these two studies, an acetylide anion generated by syringe pump addition of **12** to $LiN(SiMe_3)_2$ added intramolecularly and gave **13** as the major product of a mixture of epimers (3:1) (Scheme 9.5).[12]

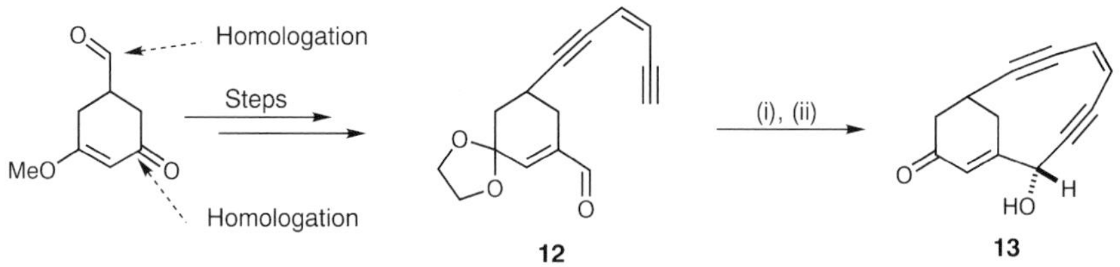

Scheme 9.5   Reagents: (i) $LiN(SiMe_3)_2$, 42%; (ii) 0.2 M HCl, acetone, H_2O, 90%.

The report from Danishefsky's group utilised a similar strategy.[13] Assembly of the key cyclisation precursor was particularly noteworthy as it exploited an intermolecular addition of an acetylide anion to ketone **14** in the presence of an epoxide and an 'in situ' protected aldehyde unit. Intramolecular addition of the anion derived from **15** followed by enol ether hydrolysis finally afforded enone **16** (Scheme 9.6).

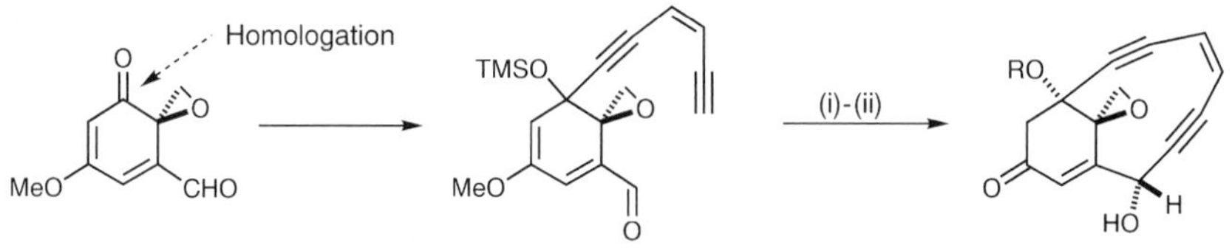

Scheme 9.6   Reagents: (i) $(Me_3Si)_2NK$, PhMe, $-78°C$, 52%; (ii) $(CO_2H)_2$, THF-H_2O, 90%.

Calicheamicinone: there have been four syntheses of Calicheamicinone.

In an extension of the aforementioned alkynyl anion approach, the Danishefsky group completed a landmark first synthesis of racemic calicheamicinone **22**.[14] Using the previously described 'in situ' aldehyde protection method, tricyclic adduct **18** was prepared. The use of potassium-3-ethyl-3-pentoxide to mediate the intramolecular addition process is particularly noteworthy. Elaboration of **18** involved protecting-group exchange, epoxide opening and diol oxidation to deliver the enediyne **19**. This set the stage for introduction of the azido (and eventually amino) moiety. Conjugate addition of azide to **19** using sodium azide and intramolecular Emmons olefination are highlights of the conversion of **19** to **21**. It is notable that the urethane moiety was unmasked subsequent to this latter transformation. A final key synthetic achievement was the installation of the trisulfide moiety by a linear sequence involving introduction of a single thiol unit which was elaborated using phthalimidomethyl disulfide. Simple CSA-mediated ketal cleavage gave **22** (Scheme 9.7).

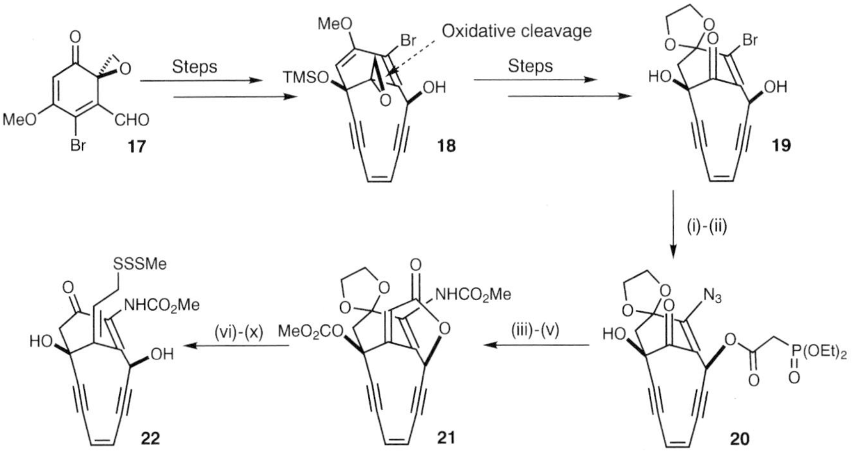

Scheme 9.7   Reagents: (i) NaN₃, MeOH, 82%; (ii) (EtO)₂P(O)CH₂COCl, Pyr; (iii) LiBr, Et₃N, THF, r.t.; (iv) H₂S, MeOH, 95%; (v) COCl₂, MeOH, Pyr, 80%; (vi) DIBAH, then NaBH₄, 43%; (vii) CH₃COSH, PPh₃, DIAD, 45%; (viii) DIBAL-H; (ix) PhthalSSMe, 65%; (x) CSA, THF, r.t., 65%.

The first asymmetric total synthesis of (−)-calicheamicinone **22** was described by Nicolaou and associates.[15] The judicious use of an asymmetric allylation enabled the preparation of oxime **23** with two asymmetric carbon atoms established in an early precursor (Scheme 9.8). These two chiral centres played a key role in the control of relative (and consequently absolute) stereochemistry in many of the synthetic transformations. In contrast to the Danishefsky approach, the amino group was introduced at a very early stage in this synthesis by use of an intramolecular nitrile oxide cycloaddition (selectivity 4:1) which, after a deprotection–oxidation sequence, gave **24**. A fairly lengthy sequence was then used to elaborate **24** into **25** with several noteworthy transformations. A highly stereocontrolled intermolecular acetylide addition was used to introduce the first acetylene fragment which was eventually elaborated to the appropriate enediyne by a palladium-mediated cross-coupling. The intermolecular olefination used to obtain **25** was highly stereoselective. Elaboration of **25** to **26** required unmasking and

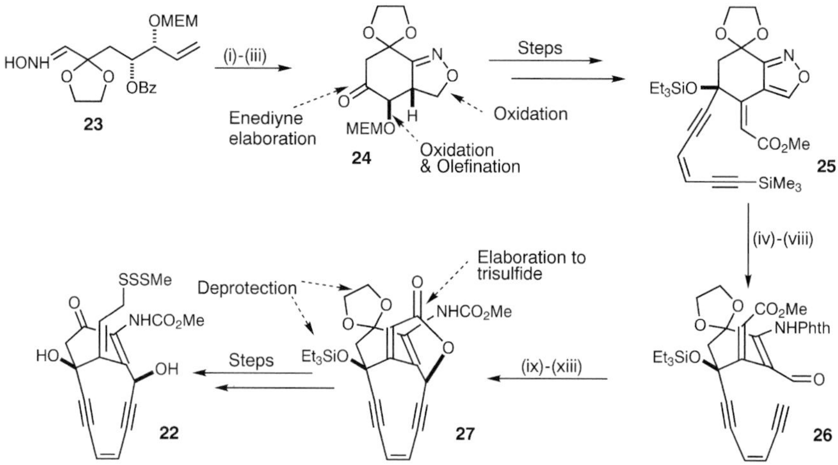

Scheme 9.8   Reagents: (i) NaOCl, CH$_2$Cl$_2$, 0°C, 2 h, 65% (4:1); (ii) NaOMe, MeOH, 0°C, 12 h, 100%; (iii) Jones, acetone, 95%; (iv) Mo(CO)$_6$, MeCN-H$_2$O, 80°C, 76%; (v) NaOMe, 0°C, 92%; (vi) phthaloyl chloride, Pyr, 0°C; (vii) silica gel, 2 h; (viii) Ac$_2$O, MeNO$_2$, 78%; (ix) KHMDS, −90°C, 44%; (x) MsCl, DMAP, 0°C; (xi) silica gel, 90%; (xii) MeNHNH$_2$, 99%; (xiii) triphosgene, then Pyr, MeOH, 82%.

protection of the enamine and was accomplished by reductive cleavage of the isoxazole ring using molybdenum hexacarbonyl and phthalation. Intramolecular acetylide addition of **26** gave the undesired stereoisomer. Fortunately, however, the ester moiety could be used to correct the stereocentre via lactone **27**. The final stages of the synthesis were introduction of the trisulfide moiety and deprotection, which used standard transformations.

The Clive synthesis[16] also introduced the nitrogen functionality early, but initially as a nitro group. The starting intermediate for this work was **28**, which was available via Diels–Alder technology. The proximity of the carboxylic ester and allyl moieties in **28** allowed for their eventual union via oxidation and lactol formation. Other key features of the conversion of **28** to **29** included stereoselective acetylide addition and nitro group reduction using NiCl$_2$ and NaBH$_4$. Intermediate **29** was cleverly manipulated into **30** by a sequence involving oxidation, which gave a diene (presumably via an intermediate imine) and regioselective radical bromination (Scheme 9.9). Silver-assisted hydrolysis of **30** and esterification gave **31** which was converted into **32a** (X = H) by acetylide addition and lactonisation. Preparation of bis-iodide **32b** by alkyne iodination set the stage for the key cyclisation reaction which involved an impressive one-pot palladium-mediated coupling sequence using (Z)-1,2-bis(trimethylstannyl)ethene as a coupling partner. Completion of the synthesis

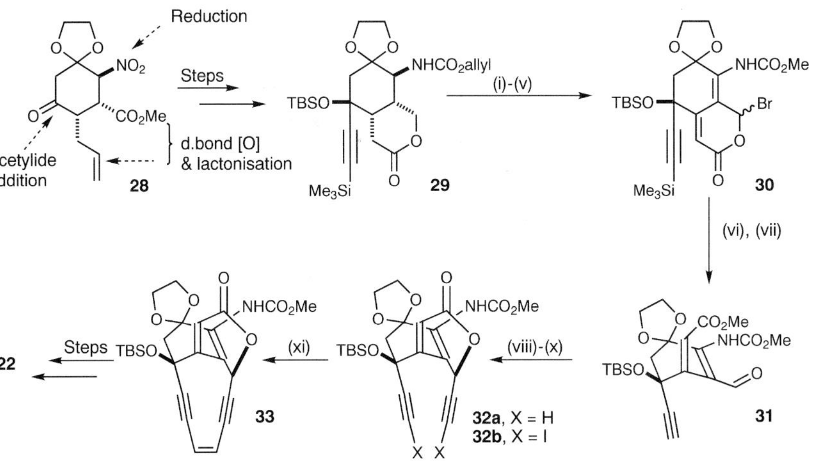

Scheme 9.9 Reagents: (i) LDA, PhSeBr, DMDO, 85%; (ii) Pd(PPh₃)₄, dimedone, 93%; (iii) *t*-BuOCl, DBU, 81%; (iv) triphosgene, pyr, MeOH, 91%; (v) NBS, (PhCO)₂O₂; (vi) H₂O, AgNO₃, (vii) CH₂N₂, 77%; (viii) TMSCCLi, CeCl₃, THF, −78°C, 91%; (ix) TBAF, 46%; (x) NIS, AgNO₃, 89%; (xi) (*Z*)-1,2-bis(trimethylstannyl)ethene, Pd(PPh₃)₄, 60°C, 72%.

was reliant on standard methods analogous to previously described work. It is notable that these workers have shown that if **28** were available in the (*S*) configuration, it could be transformed into either enantiomer of calicheamicinone via a slightly modified synthetic route.

The most recent approach is from the Magnus laboratory[17] and utilised intermediate **35** which had been prepared previously using a cobalt-assisted aldol strategy. However, the use of a more conventional acetylide approach from **34** was found to be more practicable for the preparation of **35** which could be elaborated to **36** using established procedures (Scheme 9.10). The use of the Wadsworth-Emmons procedure delivered the key olefin (and thus lactone moiety) with the correct stereochemistry and conversion to the trisulfide completed the most efficient synthesis of the target (racemic) calicheamicinone (28 steps, overall yield 1.4%).

Oligosaccharide synthesis: the first synthesis of the methyl glycoside of the oligosaccharide fragment was completed by Nicolaou and collaborators.[18] The glycosidic linkages of the key fragments **39** and **42** were established using Schmidt and glycosyl fluoride technologies, respectively (Schemes 9.11 and 9.12). Notable features include a stannoxane-mediated protocol for selective C(4) oxidation in the preparation of **39**, and the excellent yield and stereoselectivity in the glycosylation reaction that led to **42** (95% yield, α-anomer only).

However, the real highlights of this work are to be found in the latter stages of the synthesis. In order to install the NO bond an oxime was

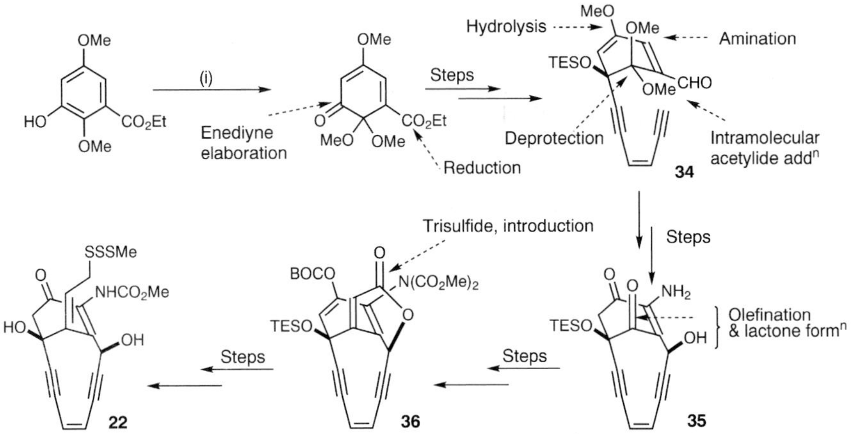

Scheme 9.10   Reagents: (i) PhI(OAc)$_2$, 87%.

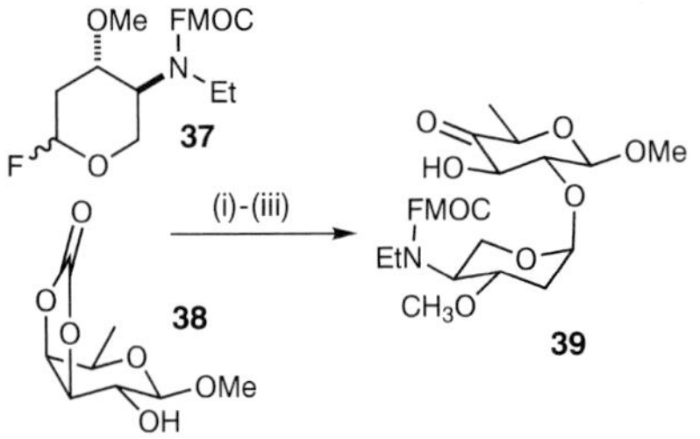

Scheme 9.11   Reagents: (i) AgClO$_4$, SnCl$_4$, THF, $-78°C–20°C$, 3 h, 86% (α:β, 4.5:1); (ii) NaH (0.01 equiv.), HOCH$_2$CH$_2$OH, 93%; (iii) Bu$_2$SnO, MeOH, 65°C, then Br$_2$ (1 equiv.), Bu$_3$SnOMe, (1 equiv.), CH$_2$Cl$_2$, 70%.

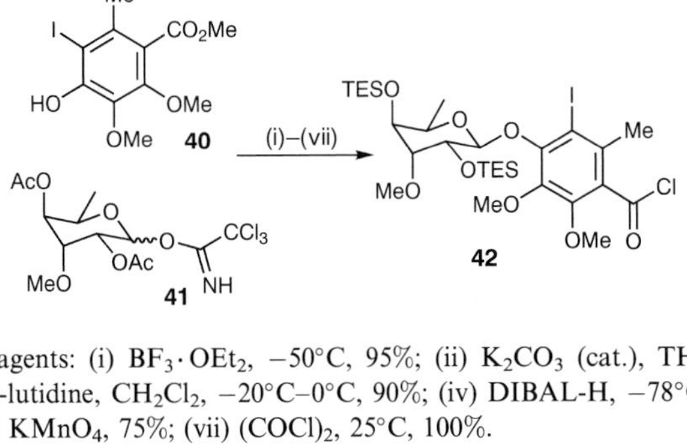

Scheme 9.12   Reagents: (i) BF$_3$·OEt$_2$, $-50°C$, 95%; (ii) K$_2$CO$_3$ (cat.), THF, MeOH, 93%; (iii) Et$_3$SiOTf, 2,6-lutidine, CH$_2$Cl$_2$, $-20°C–0°C$, 90%; (iv) DIBAL-H, $-78°C$, 83%; (v) PDC, CH$_2$Cl$_2$, 80%; (vi) KMnO$_4$, 75%; (vii) (COCl)$_2$, 25°C, 100%.

generated by coupling of **39** with fucose derivative **43** under acidic conditions (Scheme 9.13). Further elaboration gave **44** which underwent a 3,3-sigmatropic rearrangement of the allyl thiocarbonyl imidazole upon heating. This was a particularly effective strategy for positioning the thiol moiety within the central 2-deoxy sugar moiety of **45**. The coupling of thiol **45** with **42** proceeded under basic conditions to give **46**. The final stages of the synthesis involved deprotection and oxime reduction to obtain **47** (NaCNBH$_3$, MeOH, pH 3, 90% yield, 2:1 ratio).

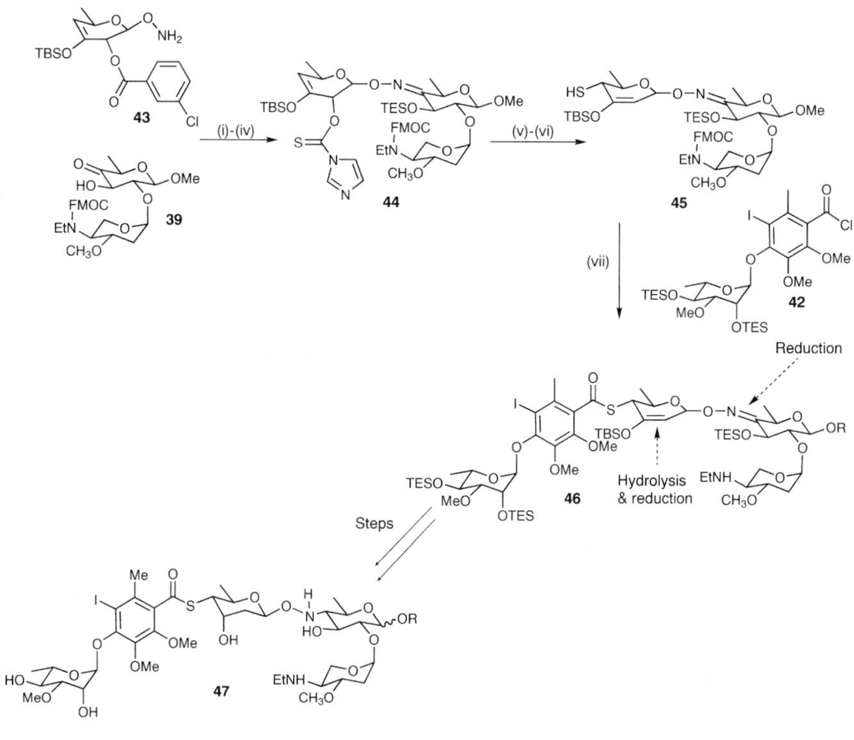

Scheme 9.13   Reagents: (i) PPTS, 25°C, 83%; (ii) Et$_3$SiOTf, lutidine, 0°C–25°C, 100%; (iii) DIBAL, CH$_2$Cl$_2$, −78°C, 91%; (iv) thiocarbonylimidazole, 87%; (v) toluene, 110°C, 98%; (vi) NaSMe, EtSH, 0°C, 95%; (vii) Et$_3$N, DMAP, 80%.

The Danishefsky group were the first to describe a synthesis of an appropriately functionalised oligosaccharide donor, **54**.[19] As with the Nicolaou contribution this group made good use of the Schmidt technology to prepare **50** (Scheme 9.14). Direct coupling of the acyl halide **50** with thiol **51** gave **52** ready for NO bond formation. The key NO glycoside linkage was made using a displacement of triflate **53** by

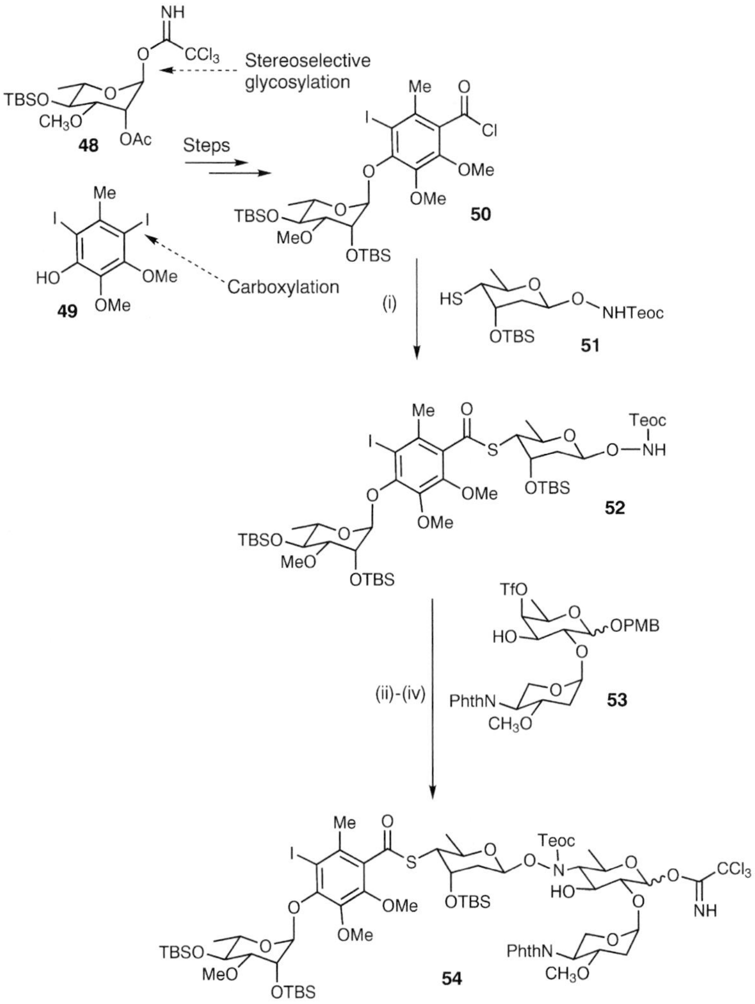

Scheme 9.14 Reagents: (i) Et₃N, DMAP, 85%; (ii) NaH, **52**, DMF, 0°C to r.t., then **53**, DMF, 0°C, 80%; (iii) DDQ, CH₂Cl₂, H₂O, 79%; (iv) Cl₃CCN, DBU, CH₂Cl₂, r.t., 73%.

hydroxylamine **52**. Deprotection of the PMB groups using DDQ and conversion into the trichloroacetimidate portion **54** proceeded to give the first glycosyl donor ready for coupling. In this report, model coupling reactions established the feasibility of Lewis-acid-mediated glycosylation to aglycon-like molecules. These studies were important as this glycosylation methodology was eventually employed in the successful total synthesis.

9.2.1.2   Total synthesis

There have been two completed total syntheses of calicheamicin.[20] The first synthesis was achieved by the Nicolaou group.[21] Their approach was a bold one, uniting two key fragments with a significant amount of functional group interconversion still needing to be carried out. Thus, the coupling of trichloroacetimidate **56** with **55** gave the desired product **57** as a single isomer in 76% yield (Scheme 9.15). Reductive cleavage of the benzoate group with DIBAL (*i*-Bu$_2$AlH) in the presence of the thiobenzoate group was particularly notable, the selectivity attained being attributed to steric factors. Mitsunobu chemistry was then used to install the first sulfur atom of the trisulfide moiety, with reduction of the oxime by NaCNBH$_3$ under Lewis-acid-mediated conditions liberating the hydroxylamine unit in **58** (ratio of isomers 2:1). The use of Harpps reagent completed the trisulfide synthesis, and global deprotection finalised this magnificent synthetic achievement.

An extraordinarily beautiful endgame for calicheamicin was realised by the Danishefsky team. It featured the remarkable glycosylation of **60** with **61** (Scheme 9.16).[22] The compatibility of the trisulfide with these reaction conditions was particularly surprising. Removal of the protecting groups from the glycosylation product completed their outstanding synthesis.

9.2.2   Esperamicins

The main structural difference between the calicheamicins and esperamicins[23] lies in the nature of the carbohydrate domains and in the presence of an additional oxygen functionality in the central core. To date there has been no completed synthesis of the esperamicin class. In fact the majority of work on model systems has been directed primarily towards calicheamicin synthesis and so there is still much to be achieved in this arena.[24–26]

Nicolaou and co-workers have described studies toward a functionalised cyclohexanone fragment of esperamicin based on the retrosynthetic analysis shown in Scheme 9.17.[27] Their synthesis started with a preparation of the monprotected quinone **66** from **64** and involved the key oxidative cyclisation of **65** using CAN (Scheme 9.18). An unusual asymmetric epoxidation of the electron-deficient allyl alcohol **67** followed by Dess–Martin periodinane oxidation and alkynylation using (MeO)$_2$P(O)CHN$_2$ gave epoxide **68**. Stereocontrolled acid-assisted epoxide opening gave **69** in which the key *trans* diol portion had been established. This highly functionalised intermediate may be of value in a final esperamicin synthesis.

Danishefsky and associates have demonstrated that their intramolecular alkynyl addition chemistry could also be used for the

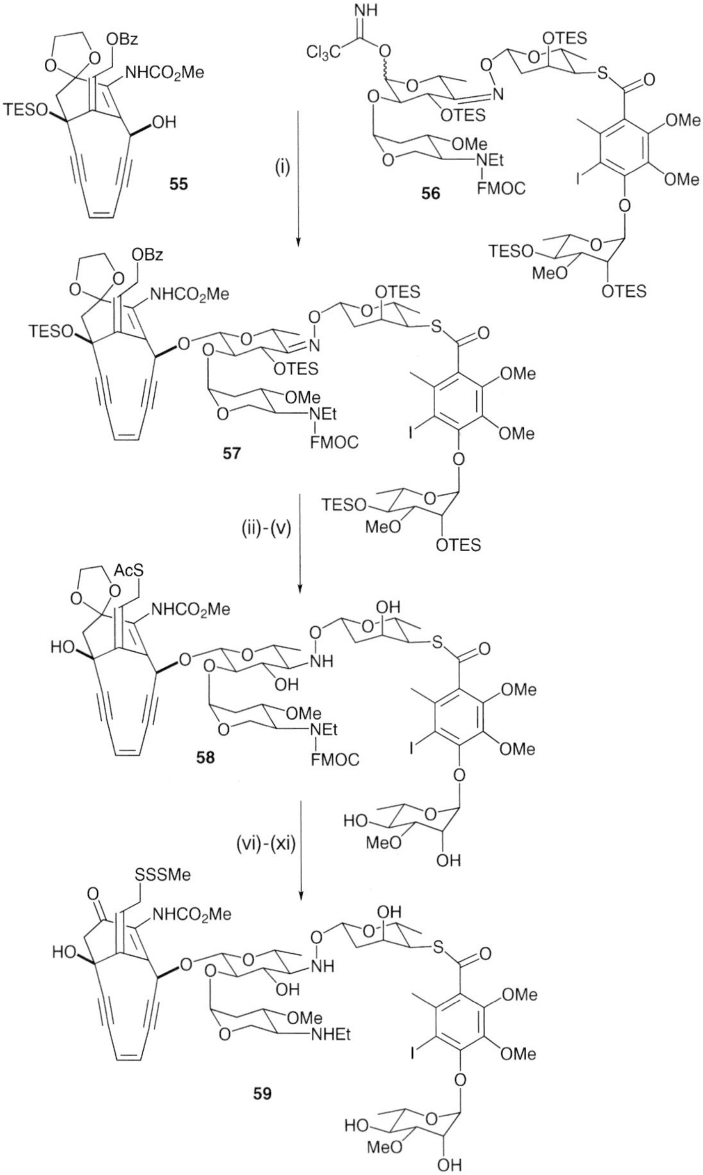

Scheme 9.15 Reagents: (i) $BF_3 \cdot OEt_2$, CH_2Cl_2, $-40°C$, 76%; (ii) DIBAL-H, CH_2Cl_2, $-78°C$, 91%; (iii) Ph_3P, DEAD, AcSH, THF, 0°C, 96%; (iv) $HF \cdot Pyr$, THF-CH_2Cl_2, 0°C to 25°C, 94%; (v) $NaCNBH_3$, $BF_3 \cdot OEt_2$, THF, $-40°C$, 96%; (vi) Et_3SiOTf, i-Pr_2NEt, CH_2Cl_2, 0°C, then AcOH, EtOAc-H_2O, 25°C, 75%; (vii) i-Bu_2AlH, CH_2Cl_2, $-90°C$; (viii) Phthal-SSMe, CH_2Cl_2, 0°C to 25°C, 57%; (ix) $HF \cdot Pyr$, THF-CH_2Cl_2, 0°C to 25°C, 90%; (x) $TsOH \cdot H_2O$, THF-H_2O, 25°C, 69%; (xi) Et_2NH-THF-H_2O, 25°C, 90%.

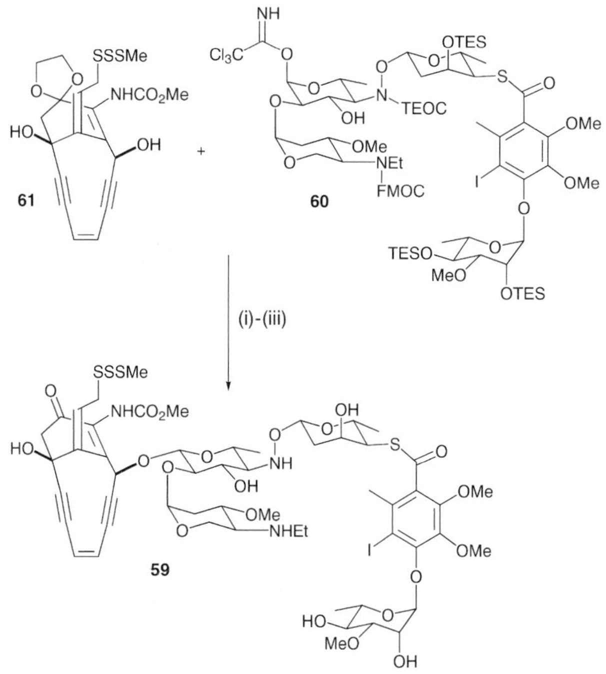

Scheme 9.16 Reagents: (i) AgOTf, molecular sieves, CH$_2$Cl$_2$, 25°C, 12 h, 34%; (ii) CSA, THF, H$_2$O, 25°C, 96 h; (iii) TBAF, THF, 15 min, 32%.

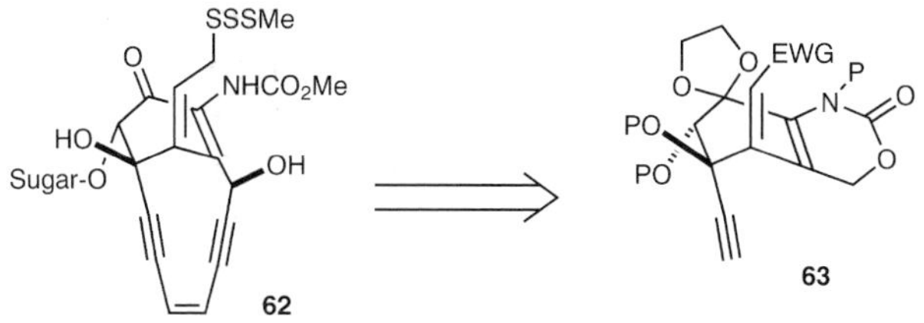

Scheme 9.17.

preparation of functionalised intermediates toward esperamicin. In work which is analogous to their studies in the calicheamicin area they showed that intermediate **70** underwent an *m*-CPBA oxidation to give **71** in 50% yield (Scheme 9.19). This provides an advanced intermediate with the desired diol functionality although this stereochemical arrangement would require modification for an esperamicin synthesis.

The most advanced synthetic studies directed specifically toward esperamicin have been carried out by Grierson and co-workers.[28, 29] A key intermediate in their work was the deconjugated enone **72** which

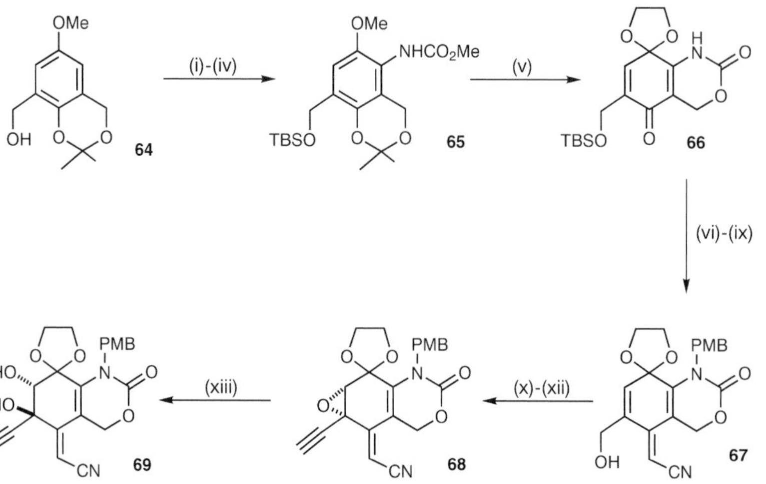

Scheme 9.18   Reagents: (i) HNO₃, AcOH, 91%; (ii) H₂, PtO₂, K₂CO₃, MeOH, 80%; (iii) TBSCl, Im, DMAP, 95%; (iv) COCl₂, Et₃N, DMAP, 0°C, then MeOH, 72%; (v) Ce(NH₄)₂(NO₃)₆, 1.5 h, then silica gel, 16 h, 30%; (vi) PMBBr, K₂CO₃, 3.5 h, 90%; (vii) *t*-BuLi, CH₃CN, THF, −78°C, 82%; (viii) (CF₃CO)₂O, Et₃N, DMAP, CH₂Cl₂, 83%; (ix) HF · Pyr, THF, 0°C, 1 h, 96%; (x) TiO(*i*-Pr)₄, DIPT, *t*-BuOOH, sieves, 86%; (xi) Dess–Martin periodinane, 36 h, 99%; (xii) (MeO)₂P(O)CHN₂, *n*-BuLi, THF, −78°C, 50%; (xiii) H₂SO₄, *t*-BuOH-H₂O, 90°C, 60%.

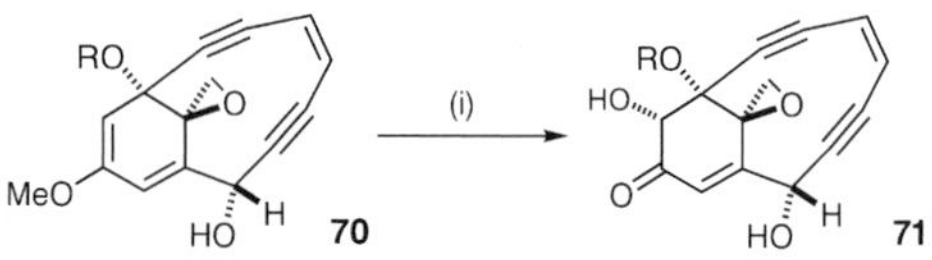

Scheme 9.19   Reagents: (i) *m*-CPBA, THF-H₂O, 50%.

was prepared in both enantiomeric forms from (−)-quinic acid **74** (Scheme 9.20). The enediyne portion was installed by a sequence involving stereoselective addition of an organocerium reagent derived from trimethylsilyl acetylene and a palladium coupling using *cis*-(1-chloro-4-(trimethylsilyl)-1-buten-3-yne. Ring closure was effected on **75** using standard base-mediated protocols, as the authors found that fluoride-mediated desilylative cyclisation was unsuccessful. Removal of the ketal protecting group was problematic under aqueous acidic conditions, but success was eventually forthcoming when EtSH was used in a transketalisation protocol. Installation of the urethane portion was achieved after aziridination of enone **76** with Ph₂S=NH (Scheme 9.21). This intermediate is the most advanced so far reported, and this group is in a good position to complete an esperamicinone synthesis.

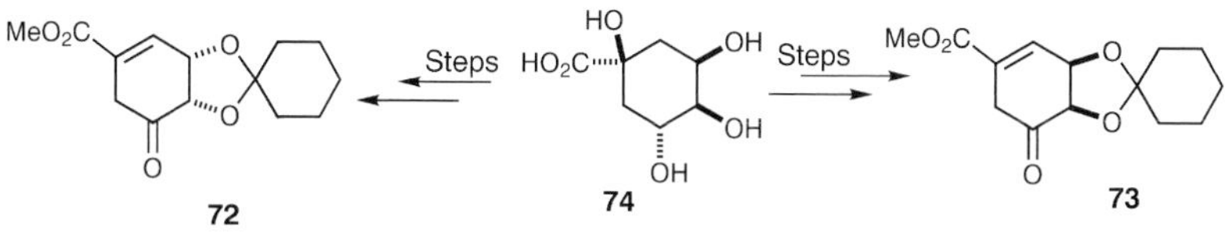

Scheme 9.20.

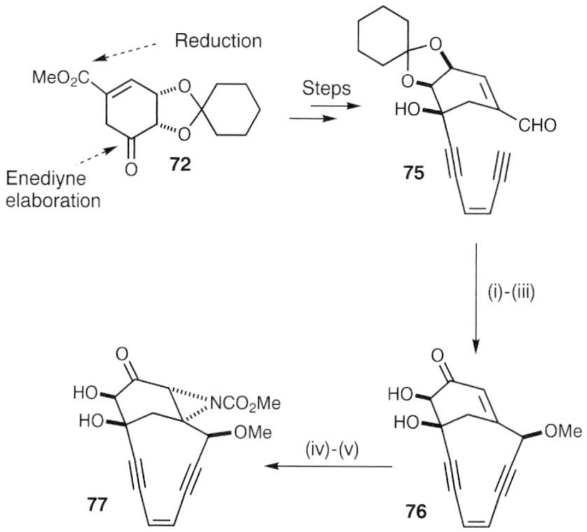

Scheme 9.21   Reagents: (i) KHMDS, THF, MeI, $-78°C$, 60%; (ii) EtSH, TFA, 83%; (iii) MnO$_2$, 79%; (iv) Ph$_2$S=NH, H$_2$O, CH$_2$Cl$_2$, 20°C, 60%; (v) ClCO$_2$Me, 65%.

9.2.3 Dynemicin[30]

9.2.3.1 Model studies

Cobalt-stabilised propargylic cations have featured in a remarkably concise synthetic approach to the core structure **81** of dynemicin A.[31] The enediyne portion was introduced into the quinoline unit using the Yamaguchi protocol which involves *N*-acylation with methyl chloroformate and nucleophilic addition with the Grignard reagent derived from enediyne **78a** (Scheme 9.22). Selective removal of the THP group and regiospecific complexation gave cyclisation precursor **80** which underwent smooth cyclisation in the presence of triflic anhydride. Oxidative decomplexation with iodine gave the azabicyclo[7.3.1]diynene core structure **81**.

Approaches involving more conventional cyclisation protocols have also been developed by others in this area. Nicolaou and colleagues have reported a route to a functionalised core of dynemicin starting from

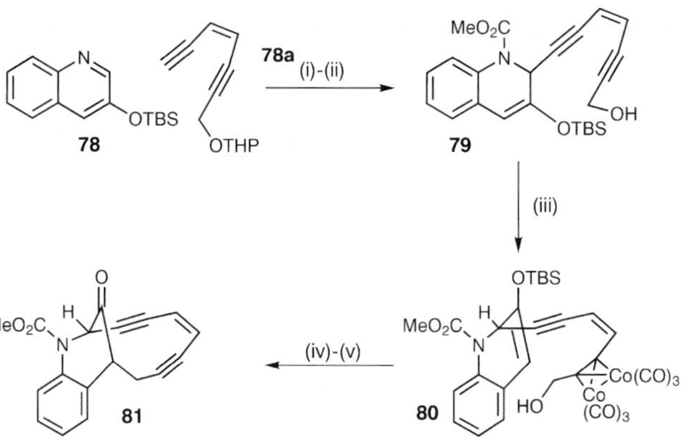

Scheme 9.22 Reagents: (i) **78**, ClCO$_2$Me, then EtMgBr, **78a**, 64%; (ii) PyH$^+$TsO$^-$, 88%; (iii) Co$_2$(CO)$_8$, 54%; (iv) Tf$_2$O, $-10°$C, CH$_2$Cl$_2$, MeNO$_2$, DBMP, 43%; (v) I$_2$, 52%.

quinoline derivative **82** (Scheme 9.23).[32] This species was converted into **83** by conventional procedures. The key ring-forming step was an LDA-mediated ring-closure. Deoxygenation of the resulting tertiary alcohol was accomplished via the thiocarbonylimidazole derivative **84**. This gave **85** in a reaction which is notable for the stability of the epoxide to the reaction conditions.

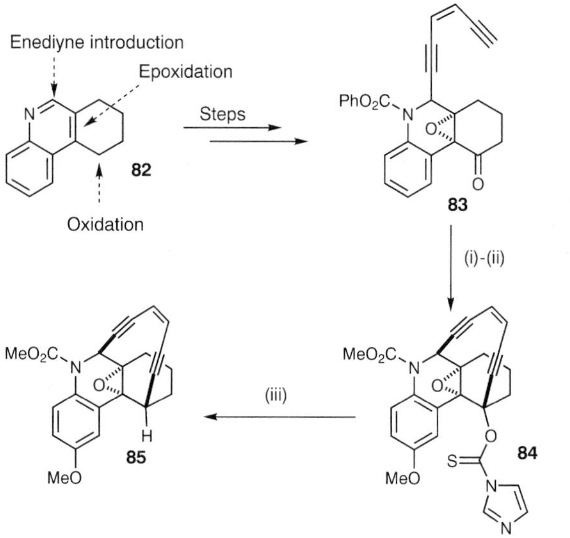

Scheme 9.23 Reagents: (i) LDA, toluene, $-78°$C, 80%; (ii) thiocarbonylimidazole, DMAP, CH$_2$Cl$_2$, 25°C, 48 h, 91%; (iii) Bu$_3$SnH, AIBN, toluene, 75°C, 2 h, 75%.

Wender and associates have described a related model study, but rather than relying upon direct deprotonation to effect closure of **86** to **88**, they used a desilylative method to generate the requisite alkynyl anion. This delivered compound **88** as the major component of a 2:1 mixture of diastereoisomers (Scheme 9.24).[33, 34]

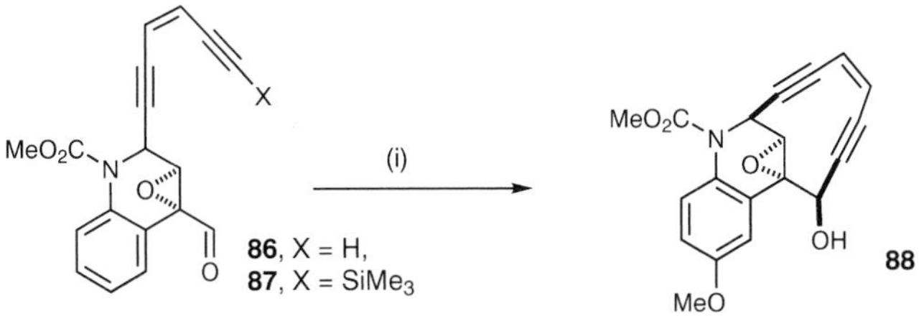

86, X = H,
87, X = SiMe$_3$

Scheme 9.24 Reagents: (i) CsF, CH$_3$CN, 0°C, 3 h, 21%.

Schreiber and co-workers reported one of the earliest approaches to the core structure of dynemicin A using a novel transannular Diels–Alder reaction. Assembly of the key precursor **90** from **89** was achieved by standard methods (Scheme 9.25). Macrocyclisation with PyBroP instigated an immediate transannular cycloaddition to give product **91** after epimerisation (DBU) and oxidation (CAN). Ionisation of the alcohol and trapping of the intermediary carbonium ion with mesitylsulfonylhydrazide gave the reduced product **93**, presumably via diazene intermediate **92**. Further manipulation of lactone **93** gave **94**.[35]

9.2.3.2 Total synthesis of dynemicins
Schreiber and co-workers used an analogue of the aforementioned species **94** to carry out the synthesis of synthetic dynemicin A methyl ester.[36] Thus, regioselective Friedel–Crafts coupling of **95** with 3-bromo-4,7-dimethoxyphthalide in the presence of AgOTf, followed by methylation, gave **97** (Scheme 9.26). Ionic reduction of benzylic ester **97** with Et$_3$SiH provided carboxylic acid **98**. An intramolecular Friedel–Crafts reaction with the acid chloride derived from **98** followed by oxidation furnished **99** as a single (unknown) diastereomer. The final stages of the synthesis were epoxidation with *m*-CPBA, urethane deprotection, oxidation and methylation of the intermediary imine **100** which gave the tri-O-methyl ether of dynemicin methyl ester **101**.

The first completed synthesis of dynemicin A was described by Myers and co-workers. It commenced from cyclohexanedione **102**, itself obtainable by optical resolution on large scale. Transformation of **102** into **103** was achieved by palladium-mediated coupling of an enol triflate with *tert*-butyl-2-borono-4-methoxycarbanilate (Scheme 9.27). Heating of

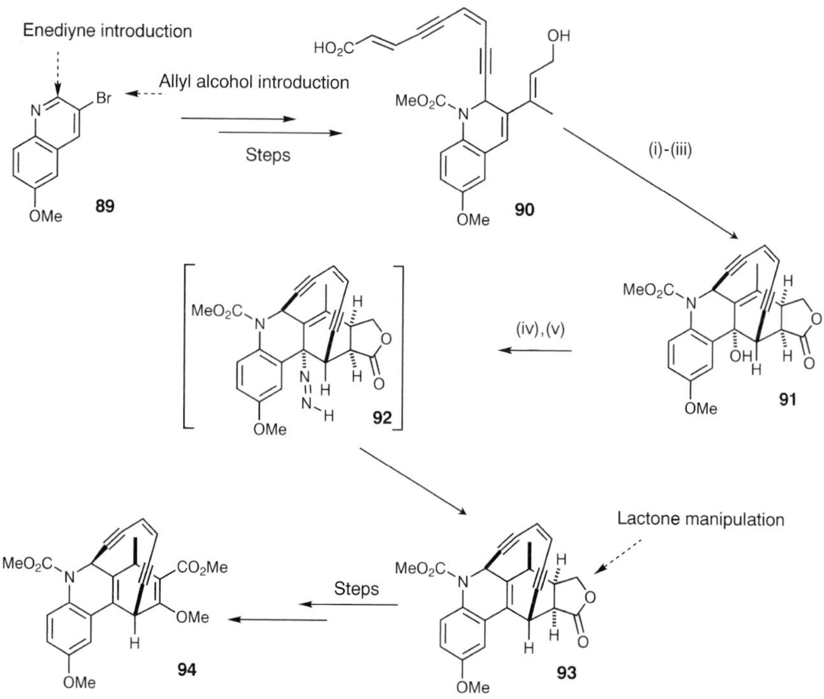

Scheme 9.25   Reagents: (i) PyBroP, Et$_3$N, CH$_2$Cl$_2$, 51%; (ii) DBU, THF, 92%; (iii) CAN, CH$_3$CN, 97%; (iv) MeAlCl$_2$, CH$_2$Cl$_2$, $-78°$C to 35°C; (v) ArSO$_2$NHNH$_2$, 0°C, 92%.

103 in 2-chlorophenol led to ring closure via N-deprotection and displacement of the menthyl group. Compound **104** was obtained in 84% yield. This seemingly simple addition–elimination reaction was found to be solvent-dependent. The authors suggest that the solvent might be acting as a weak Brønsted acid since the reaction was unsuccessful in o-dichlorobenzene at the same temperature. Conversion of **104** into **105** required a palladium-assisted quinolone reduction via an amido-triflate intemediate, and m-CPBA oxidation in methanol to establish the appropriate level of oxygenation. After protecting-group exchange, the enediyne unit of **106** was introduced by subjecting **105** to the ubiquitous Yamaguchi protocol which, in this instance, proceeded with high stereocontrol (greater than 20:1). Ring closure of **106** was effected with KN(SiMe$_3$)$_2$, and the cyclic thionocarbonate **107** prepared using excess 1,1′-thiocarbonyldiimidazole and DMAP. Treatment of the resulting adduct with TBTH under free-radical chain conditions caused reductive cleavage and gave ketone **108**. The authors reported that the appropriate conditions for α-functionalisation were difficult to identify. Success was eventually attained by stirring **108** under an atmosphere of

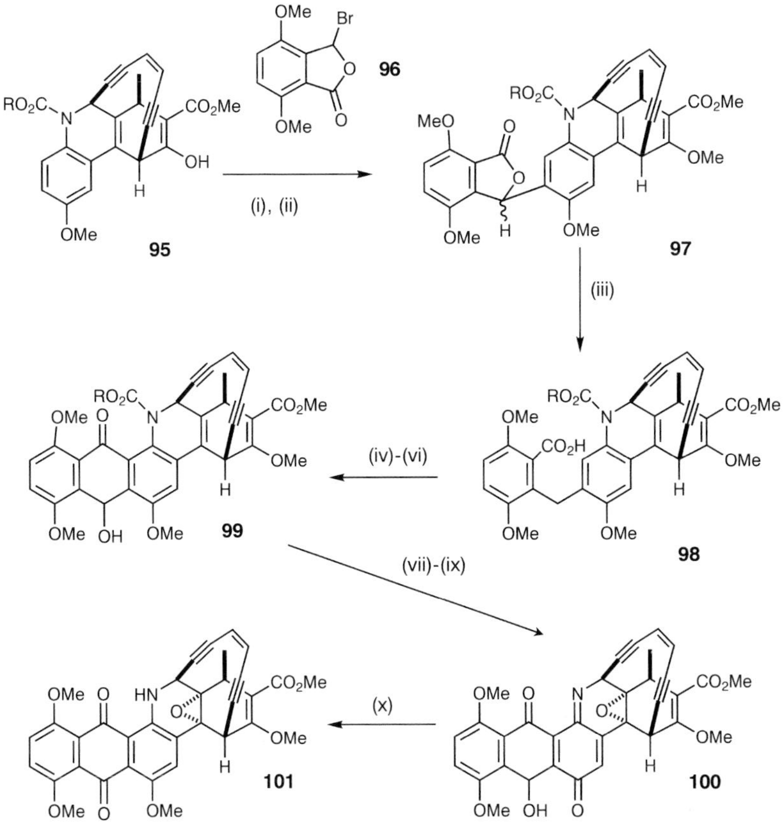

Scheme 9.26 Reagents: (i) AgOTf, sieves, CH_2Cl_2; (ii) K_2CO_3, Me_2SO_4, acetone, 57%; (iii) MeAlCl$_2$, Et$_3$SiH, CH_2Cl_2, $-78°C$ to $-50°C$, 82%; (iv) SOCl$_2$, (v) TMSOTf, $-78°C$ to $0°C$; (vi) DDQ, $0°C$, 51%; (vii) m-CPBA, pH 7, 73%; (viii) DBU, MeOH; (ix) CAN, MeCN, H_2O, $0°C$; (x) Cs_2CO_3, MeI, acetone, 50%.

CO_2 in the presence of MgBr$_2$ and Et$_3$N. Protecting-group manipulations and oxidation finally afforded **109**.[37]

The final stages of the synthesis focused on introducing the anthraquinone unit into the target. This was readily achieved by the Diels–Alder reaction of **109** with isobenzofuran **110** (Scheme 9.28). After 5 min heating at 55°C adduct **111** was isolated in 75% yield. The synthesis was completed by treating **111** with MnO$_2$ in the presence of Et$_3$N · HF in THF for 9 min; this gave **112** in reasonable yield. It is clear that careful control of the reaction conditions is critical for success in this step, which highlights the sensitivity of these molecules and the skill required to bring their total syntheses to fruition.[38]

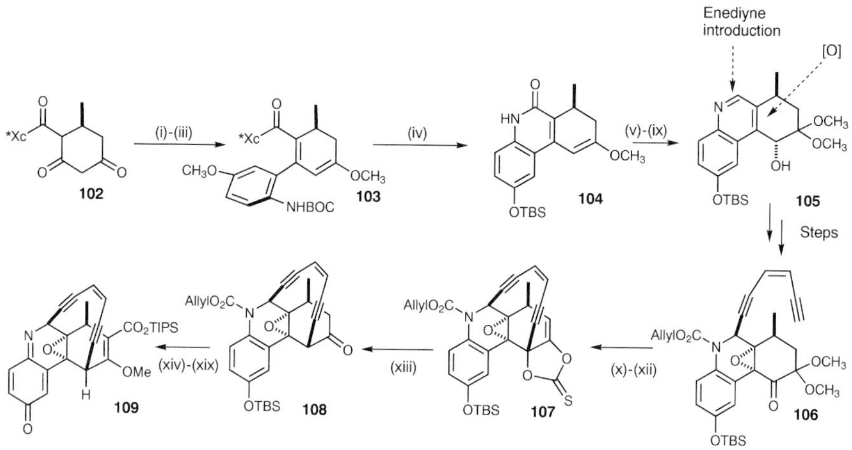

Scheme 9.27 Reagents: (i) CSA, MeOH, 71%; (ii) NaH, Tf$_2$O, $-78°$C, 95%; (iii) Pd(PPh$_3$)$_4$, Na$_2$CO$_3$, dioxane, *tert*-butyl-2-borono-4-methoxycarbanilate, 90%; (iv) 180°C, 4-chloro-phenol, 84%; (v) Tf$_2$O, 2,6-di-*tert*-butylpyridine, CH$_2$Cl$_2$, $-78°$C to 0°C, 86%; (vi) *m*-CPBA, MeOH, 67%; (vii) HCO$_2$H, Et$_3$N, Pd(PPh$_3$)$_4$, 97%; (viii) EtMgBr, THF, NaSEt, DMF, 71%; (ix) TBSCl, Im, DMF, 91%; (x) KN(SiMe$_3$)$_2$, CeCl$_3$, $-78°$C, 94%; (xi) *p*-TSA, acetone, 83%; (xii) thiocarbonyldiimidazole, DMAP, CH$_2$Cl$_2$, 85%; (xiii) TBTH, AIBN, toluene, 70°C, 97%; (xiv) MgBr$_2$, Et$_3$N, CO$_2$; (xv) MeOTf, KOBu-*t*, toluene, $-78°$C, 49%; (xvi) Et$_3$N · HF, MeCN; (xvii) ArI(OAc)$_2$, MeOH, 89%; (xviii) TIPSOTf, Et$_3$N, THF, $-78°$C to 0°C, 85%; (xix) TBTH, Pd(PPh$_3$)$_2$Cl$_2$, benzene, 60%.

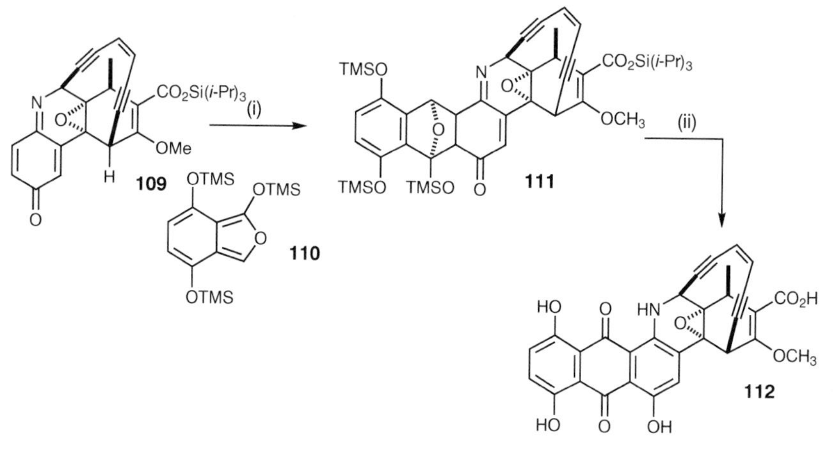

Scheme 9.28 Reagents: (i) THF, $-20°$C to 55°C, 5 min, 75%; (ii) MnO$_2$, 3HF · Et$_3$N, THF, 23°C, 9 min, 53%.

Danishefsky and associates have also completed a synthesis of racemic dynemicin.[39,40] Formation of the key quinoline **115** was achieved by a series of reactions that included a number of key steps. Intramolecular Diels–Alder of **113**, oxidative cleavage of **114** with CAN and cyclo-condensation of the intermediary quinone with ammonium acetate and protection, gave **115** (Scheme 9.29). Standard manipulations on **115**

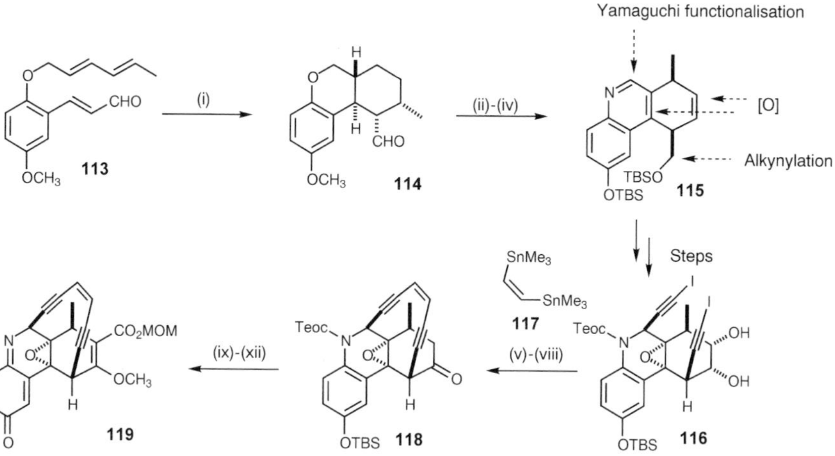

Scheme 9.29 Reagents: (i) ZnCl₂, CH₂Cl₂, 25°C, 60%; (ii) CAN, MeCN, H₂O, 90%; (iii) NH₄OAc, HOAc, 100°C, 89%; (iv) TBSCl, Im, CH₂Cl₂, 98%; (v) **117**, Pd(PPh₃)₄, DMF, 75°C, 81%; (vi) Tf₂O, Pyr, CH₂Cl₂, 95%; (vii) Dess–Martin periodinane, CH₂Cl₂, 95%; (viii) CrCl₂, THF, 75%; (ix) MgBr₂, CO₂, Et₃N, MeCN, then MOMCl, i-Pr₂NEt, THF, 61%; (x) CH₂N₂, MeOH, 0°C, 70%; (xi) TBAF, THF, 0°C, (xii) PhI(OAc)₂, 0°C, 60%.

(osmylation; stereocontrolled Yamaguchi alkynylation; conversion of the aldehyde into the alkyne; and iodination) gave **116** after epoxidation. In a very bold olefination procedure, a double cross-coupling protocol established the cyclic enediyne unit. The two hydroxyl groups were differentiated by selective triflation. This was followed by Dess–Martin periodinane oxidation and triflate reduction using CrCl₂ which gave ketone **118**.

With a sequence analogous to that described by Myers, **118** was converted into **119**. The final stages of this synthesis required the installation of the anthraquinone ring system. In this instance an alternative Diels–Alder approach was investigated using carboxylate anion **121** which on treatment with **120** gave **122** (Scheme 9.30). Oxidation and deprotection of the resulting adduct gave dynemicin A (**112**).

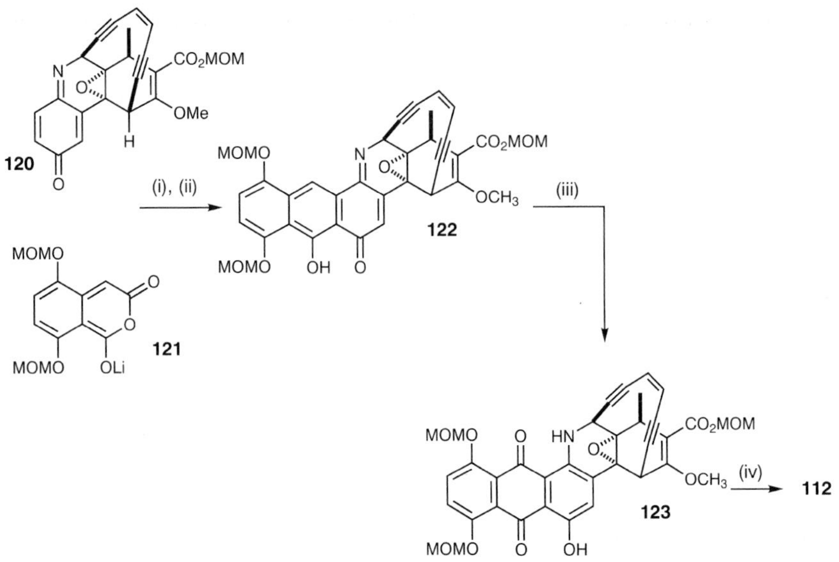

Scheme 9.30 Reagents: (i) 0°C, 1 h; (ii) PhI(OCOCF$_3$)$_2$, THF, 0°C, 5 min; (iii) air, 25°C, daylight, THF, 24 h; (iv) MgBr$_2$, Et$_2$O, 0°C–25°C, 36 h (overall yield, 15%).

9.3   Synthesis of dienediynes

9.3.1   Neocarzinostatin chromophore[41]

9.3.1.1   Model studies

Neocarzinostatin chromophore is the oldest member of the class and the molecule that has attracted most attention to date.[42–47] There have been many general approaches to the strained nine-membered ring system, which offers an outstanding challenge to organic chemists interested in chemical synthesis. The first study reported by Wender and associates involved a ring-contraction approach. 2-Cyclopentenone was converted into sulfone **125** by a sequence which made extensive use of palladium-catalysed transformations (Scheme 9.31). Photolysis of sulfone **125** followed by dehydration gave the parent ring system **126**.[48]

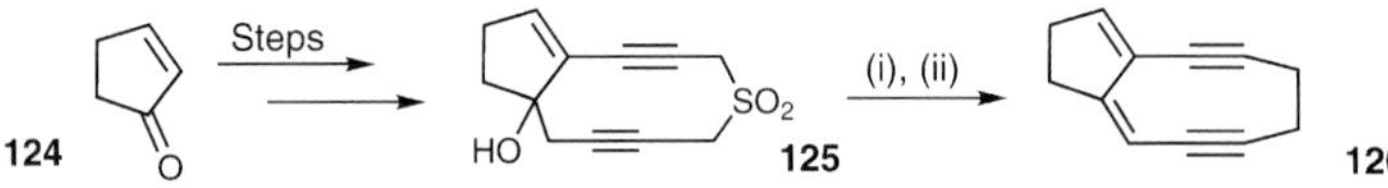

Scheme 9.31 Reagents: (i) PhCOPh, MeCN/C$_6$H$_6$, $h\nu$, r.t., 9%–15%; (ii) MsCl, DMAP, CH$_2$Cl$_2$.

Further model studies from the Wender group have established two alternative procedures.[49] The first involved ring closure of aldehyde 127 using a chromium-mediated coupling which proceeded in good yield and delivered the new carbon–carbon bond with a high level of *cis*-stereoselectivity (about 6:1; Scheme 9.32). The second involved a

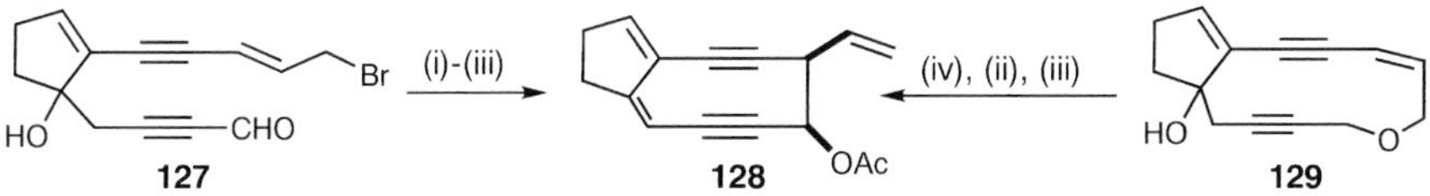

Scheme 9.32   Reagents: (i) $CrCl_2$, THF, 77%–88% (1:1); (ii) Ac_2O, Et_3N, DMAP, r.t., 52%; (iii) DMAP, MsCl, Et_3N, 0°C, 25%; (iv) *n*-BuLi, HMPA, −78°C, 29%.

sigmatropic rearrangement of 129 which was much less satisfactory and proceeded in modest yields of less than 29% (Scheme 9.32). Both approaches gave a common intermediate which was readily transformed into 128 by protection and base-mediated elimination.

Another sigmatropic rearrangement approach has been exploited by Takahashi and co-workers who prepared intermediate 132 from 130 and 131 (Scheme 9.33).[50] Palladium-catalysed ring opening of 132 with benzoic acid gave a mixture of C(9) and C(12) benzoates presumably via a π-allyl-palladium intermediate. Production of a mixture at this juncture was of little consequence because all isomers were cleverly converted to 133 by protection–deprotection and oxidation. Elaboration to Wittig

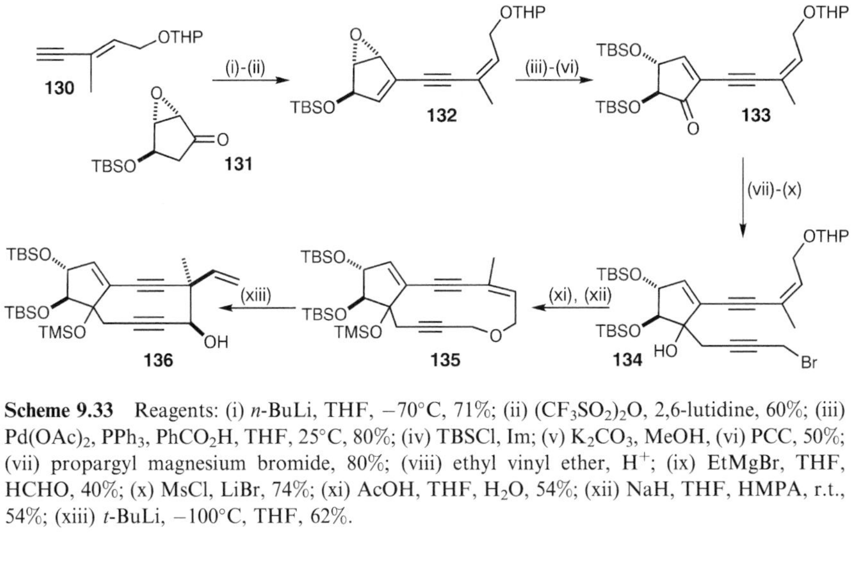

Scheme 9.33   Reagents: (i) *n*-BuLi, THF, −70°C, 71%; (ii) $(CF_3SO_2)_2O$, 2,6-lutidine, 60%; (iii) $Pd(OAc)_2$, PPh_3, $PhCO_2H$, THF, 25°C, 80%; (iv) TBSCl, Im; (v) K_2CO_3, MeOH, (vi) PCC, 50%; (vii) propargyl magnesium bromide, 80%; (viii) ethyl vinyl ether, H^+; (ix) EtMgBr, THF, HCHO, 40%; (x) MsCl, LiBr, 74%; (xi) AcOH, THF, H_2O, 54%; (xii) NaH, THF, HMPA, r.t., 54%; (xiii) *t*-BuLi, −100°C, THF, 62%.

precursor **135** required introduction of a four-carbon fragment via propargyl magnesium bromide and alkyne homologation using standard
protocols. Treatment of **135** with *tert*-butyllithium at −100°C provided
136 as a single isomer in good yield via Wittig rearrangement.

A very different strategy was described by Magnus and colleagues
who made use of their cobalt-assisted aldol methodology in the construction of functionalised intermediates.[51, 52] Assembly of precursor
138 was achieved from **137** by enediyne introduction and Sharpless
epoxidation (Scheme 9.34). Regioselective alkyne complexation and acetal
cleavage led to **139**, which on treatment with *n*-Bu$_2$BOTf under basic

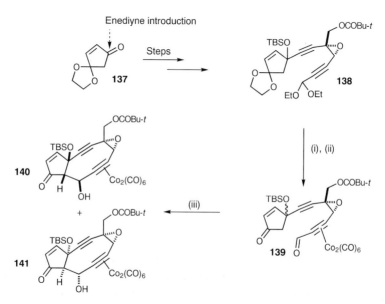

Scheme 9.34 Reagents: (i) Co$_2$(CO)$_8$, heptane, 25°C, 80%; (ii) CF$_3$CO$_2$H, CHCl$_3$, H$_2$O, 56%;
(iii) *n*-Bu$_2$BOTf, Et$_3$N, CH$_2$Cl$_2$, −78°C to 0°C, 57% (1:1).

conditions gave **140** and **141** as a 1:1 mixture originating from the diastereomeric mixture of starting materials. Thus this mode of ring closure
established the new carbon–carbon bond with complete stereocontrol.

In a related approach, ring closure of **145** was reported to give **146**
as the major isomer in good yield (Scheme 9.35).[53] Intermediate **145**
was readily assembled by a convergent strategy that relied upon the conjugate addition of enediyne **142** to enone **143**. It is likely that the free
hydroxyl in **143** chelated with an organoaluminium reagent to deliver
144. It is notable that this synthesis established the appropriate oxygen

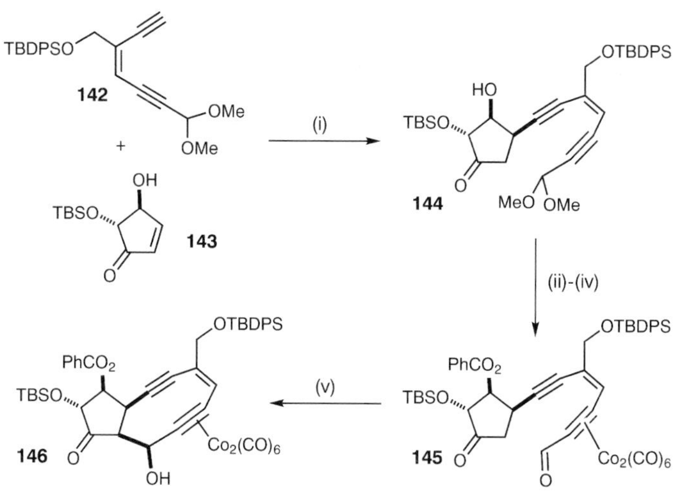

Scheme 9.35 Reagents: (i) **142**, *n*-BuLi, THF, −78°C, 30 min, then Et$_2$AlCl, r.t., 1 h, then **143**, 47%–57%; (ii) PhCOCl, Pyr, DMAP, CH$_2$Cl$_2$, 79%; (iii) Co$_2$(CO)$_8$, CH$_2$Cl$_2$, 0°C, 92%; (iv) TFA, 30%, CH$_2$Cl$_2$, 98%; (v) *n*-Bu$_2$BOTf, Et$_3$N, CH$_2$Cl$_2$, −78°C to 0°C, 68%–75%.

functionality on the five-membered ring prior to nine-membered ring formation.

Intramolecular alkynyl anion addition to an aldehyde has also been used as a general strategy for the construction of the NCS ring system. Myers and colleagues described such an approach to the epoxy diyne core of NCS.[54] Their route exploited an intermolecular addition of the anion derived from **148** (derived via Sharpless asymmetric epoxidation (AE) of allyl alcohol) to **147** to obtain **149** in 40% yield (Scheme 9.36). Sulfoxide elimination, desilylation and acetal cleavage were all achieved in the presence of the epoxide functionality which was remarkably tolerant of a range of reaction conditions. Intramolecular cyclisation onto the *s*-trans isomer of **150** proceeded with LiN(TMS)$_2$/CeCl$_3$. It gave **151** in 87% yield. The authors report that without CeCl$_3$ these reactions were far less satisfactory.

In a later report[55] the same workers describe the conversion of a related intermediate **152** into **154** by allylic transposition involving S$_N$2′ displacement of **152** with trifluoroacetic acid. Selective deprotection followed by elimination gave **154** (Scheme 9.37). The authors note the difficulty in identifying conditions for this sequence of steps.

9.3.1.2 Total synthesis of neocarzinostatin aglycon

Myers and co-workers have completed the only aglycon synthesis to date.[56] It transpired that they were unable to employ their previously

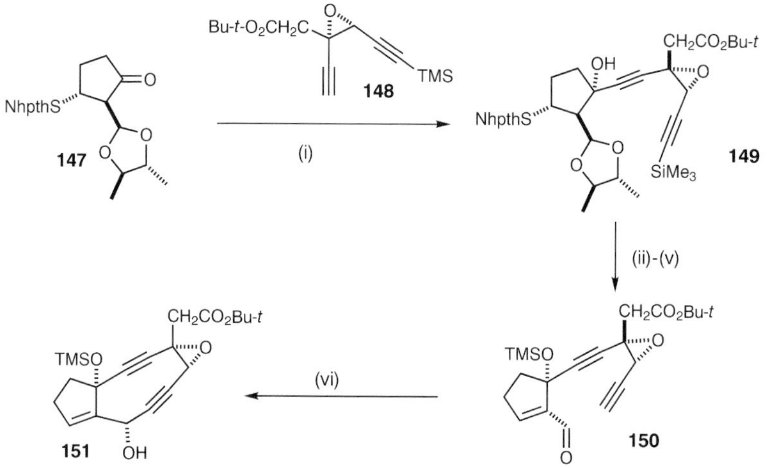

Scheme 9.36 Reagents (i) **148**, NaN(TMS)$_2$, toluene, $-78°C$, 40%; (ii) *m*-CPBA, CH$_2$Cl$_2$, $-78°C$, then *i*-Pr$_2$NEt, toluene, reflux, 4 h, 84%; (iii) KF·2H$_2$O, MeOH, 3 h, 100%; (iv) CSA, MeCN, H$_2$O, 0°C, 20 h; (v) (CH$_3$)$_3$SiOSO$_2$CF$_3$ (8 equiv.), CH$_2$Cl$_2$, $-78°C$, 80%; (vi) LiN(TMS)$_2$ (25 equiv.), CeCl$_3$ (3 equiv.), THF, $-78°C$ to 0°C, 87%.

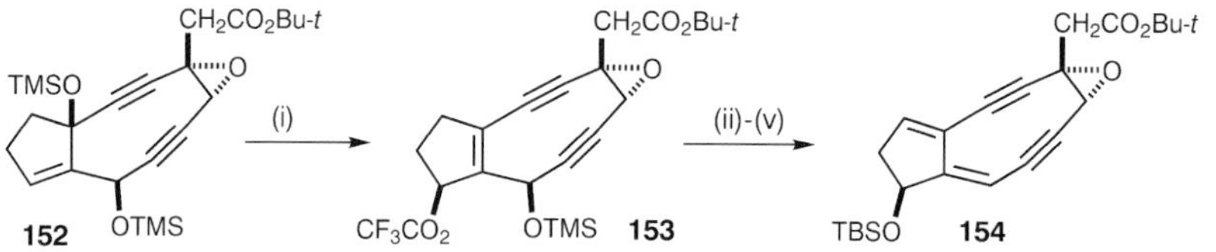

Scheme 9.37 Reagents: (i) TFA, CH$_2$Cl$_2$, 0°C, 49%; (ii) MeOH, Et$_3$N, toluene, 0°C; (iii) TBSOTf, lutidine, $-78°C$; (iv) HF·Et$_3$N, MeCN; (v) methanesulfonic anhydride, pyridine, 22%.

described allylic transposition sequence. However, precursor **155** was assembled by a sequence similar to their previously published work. Of particular note was the addition of the anion derived from an epoxy-diyne to establish the C(1)–C(2) bond. Deprotection of the acetal and reduction allowed introduction of a second epoxide moiety via Sharpless AE. Reoxidation using the Dess–Martin reagent gave **156** which underwent smooth cyclisation to **157** under carefully controlled base-mediated conditions (Scheme 9.38); the presence of LiCl was crucial for success. This material was, rather unexpectedly, very stable to general manipulations (chromatography, etc.). Introduction of the naphthoate fragment was achieved via a DCC coupling. Cleavage of the acetal and installation of the cyclic carbonate proceeded without event and, after appropriate protecting group manipulations, the tertiary alcohol was

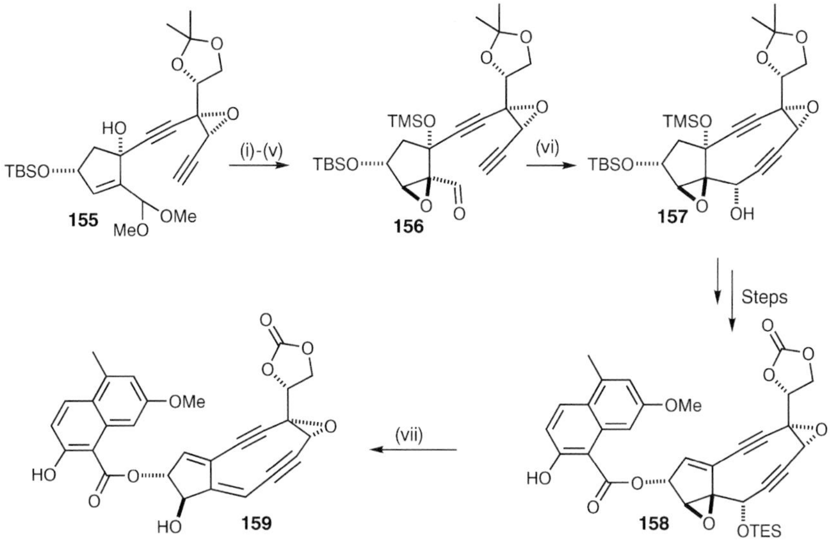

Scheme 9.38 Reagents: (i) TsOH, acetone, 90%; (ii) TMSOTf, 2,6-lutidine, CH$_2$Cl$_2$, −78°C, 88%; (iii) DIBAL, PhCH$_3$, −78°C, 92%; (iv) (+)-DET, Ti(O-i-Pr)$_4$, TBHP, 4Å sieves, CH$_2$Cl$_2$, −20°C, 91%; (v) Dess–Martin periodinane, Pyr, CH$_2$Cl$_2$, 97%; (vi) LiN(SiPh(CH$_3$)$_2$)$_2$, LiCl, THF, −78°C, 79%; (vii) PPh$_3$, I$_2$, imidazole, CH$_2$Cl$_2$, −17°C, 15%–30%.

dehydrated using the Martin sulfurane reagent to obtain **158**. The conversion of **158** to NCS aglycon was accomplished with PPh$_3$, I$_2$ and imidazole. Compound **159** was found to bind to the apoprotein.

9.3.1.3   Total synthesis of neocarzinostatin chromophore

In an outstanding achievement the Myers group completed the total synthesis of NCS chromophore by coupling aglycon **159** to sugar **160** (Scheme 9.39).[57] Employment of the ubiquitous Schmidt trichloroacetimidate methodology allowed the successful realisation of this endgame. What is notable about this particular glycosylation is the presence of the free secondary amine in the fucosamine portion **160**, and the compatibility of the sensitive aglycon with the Lewis-acid conditions employed. The authors postulate that the α-selectivity of this coupling is due to hydrogen-bond mediated intramolecular delivery of the aglycon to oxonium **162** to give **161**. Deprotection under acidic conditions gave the natural product **163**. This is the first and, to date, only synthesis of NCS chromophore. In addition to being a magnificent synthetic achievement this work is particularly noteworthy in that the final glycosylation step was also carried out with radiolabelled fucosamine to

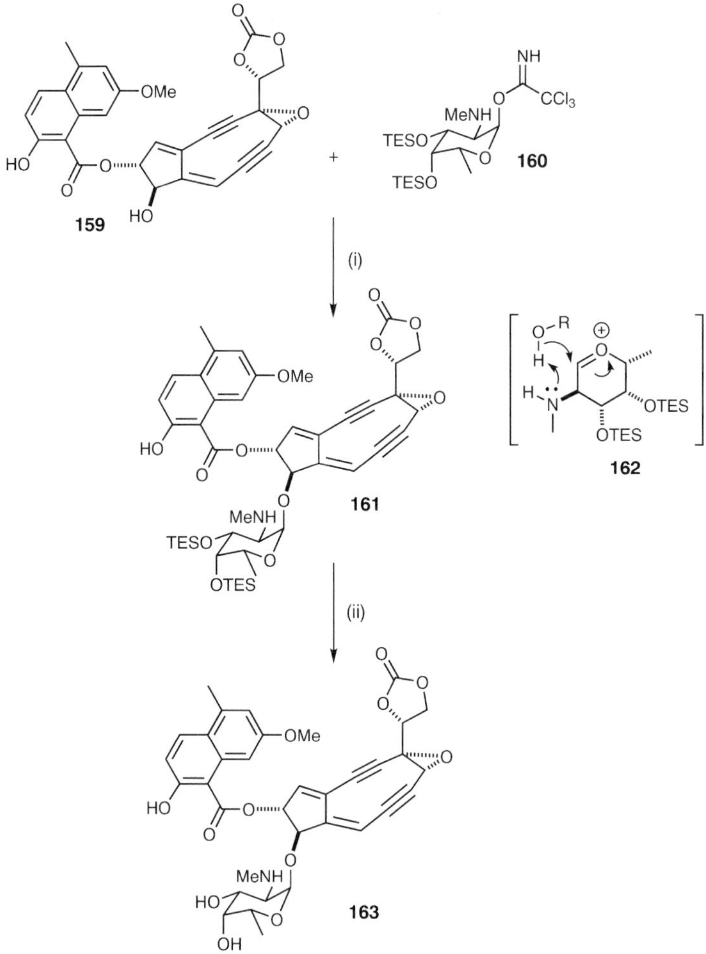

Scheme 9.39   Reagents: (i) BF$_3 \cdot$OEt$_2$, sieves, PhCH$_3$, -30°C, 51%; (ii) HF, Pyridine, THF, 49%.

provide the natural product **163** in radiolabelled form. This should open up the way for biological investigations to be carried out *in vivo*.

9.3.2   *Kedarcidin and C-1027*

There has been far less work on the synthesis of these natural products [58–60] although aromatic and carbohydrate fragments have been described by a number of groups.[61, 62] As can be seen from their structures these molecules are similar to neocarzinostatin chromophore and there have been some attempts to devise synthetic routes.[63] One appealing effort described by Grierson and co-workers involved the elaboration of a key precursor using a cobalt-assisted carbonyl ene reaction.[64] Treatment of **164** with **165** under Lewis-acid conditions gave **166** in 91% yield with good *erythro* selectivity (Scheme 9.40). Whilst standard manipulations afforded cyclisation precursor **167**, base-mediated cyclisation only proceeded in poor yield and delivered **170** presumably via aromatisation of nine-membered intermediates **168** and **169**.

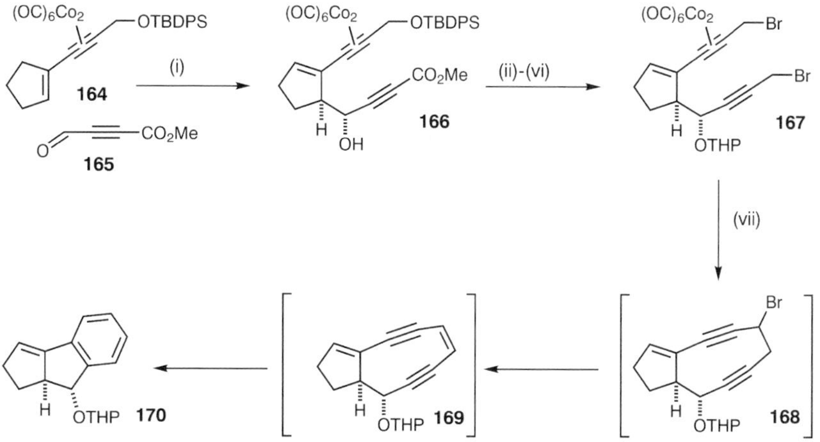

Scheme 9.40 Reagents: (i) (MeO)MeAlCl, −40°C to 0°C, 91%; (ii) CAN, MeOH, 0°C, 84%; (iii) DHP, (TMSO)$_2$SO$_2$, CH$_2$Cl$_2$, 0°C, 87%; (iv) DIBAL, 0°C; (v) TBAF, THF, 0°C; (vi) CBr$_4$, PPh$_3$, 1,2-lutidine, MeCN, 0°C, 85%; (vii) LiHMDS, HMPA, THF, −40°C, 10%.

A more advanced and extremely promising strategy has been described by Hirama and co-workers.[65] Conversion of enantiomerically enriched enone **171** into **172** was realised by manipulations that included a very effective palladium-mediated carbonylation reaction and an osmium-mediated oxidation to install the diol unit (Scheme 9.41). Selective dibutylstannylene oxidation of the secondary alcohol gave **173** which coupled to diyne **174** after the latter was deprotonated. Tertiary hydroxyl group elimination followed by protecting-group manipulations gave **175**, which underwent LiHMDS/CeCl$_3$-mediated cyclisation. The product was immediately converted into epoxy alcohol **176** by desilylation and intramolecular substitution (epoxidation) of an intermediate mesylate. Elimination gave **177** which the authors calculate has a half-life of around 30 min.

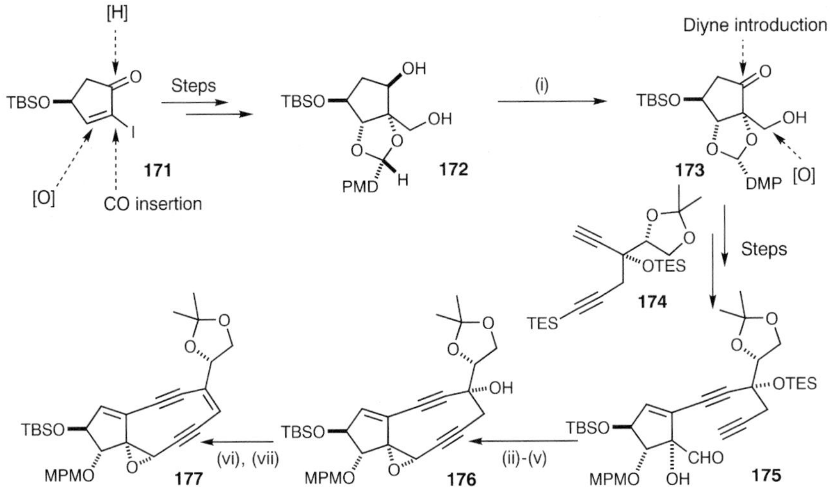

Scheme 9.41 Reagents: (i) Bu₂SnO, toluene, then NBS, CHCl₃, 83%; (ii) LiHMDS, CeCl₃, THF, −20°C, to r.t., 65%; (iii) MsCl, Et₃N, DMAP, CH₂Cl₂, 67%; (iv) TBAF, THF, 0°C, 1 h, 69%; (v) TBSOTf, 2,6-lutidine, THF, −78°C, 76%; (vi) MsCl, Et₃N, CH₂Cl₂; (vii) DBU, CD₂Cl₂, CH₂Cl₂, 32%.

The same group have used a similar strategy to obtain C-1027 models via the related cyclisations of **178** and **180** to **179** and **181**, respectively, using the CeCl₃-assisted LHMDS methodology (Schemes 9.42 and 9.43).[66]

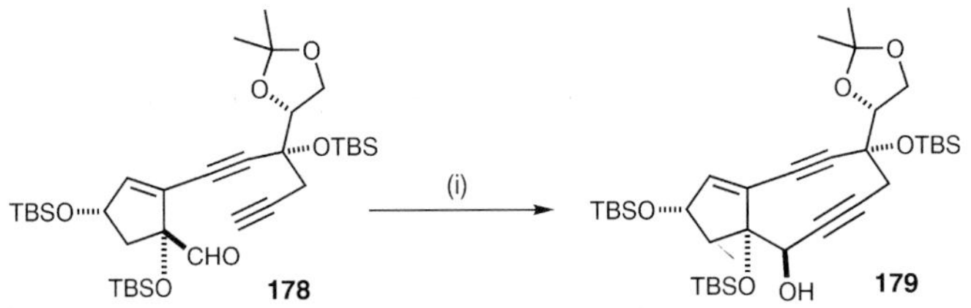

Scheme 9.42 Reagents: (i) LiN(TMS)₂, CeCl₃, THF, −40°C to r.t., 78%.

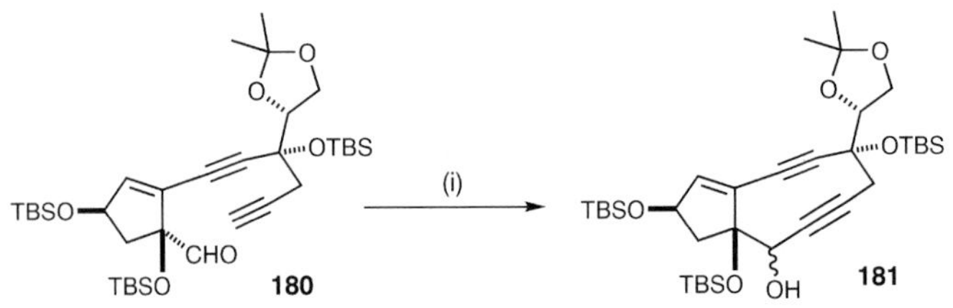

Scheme 9.43 Reagents: (i) LiN(TMS)₂, CeCl₃, THF, −40°C to r.t., 1 h, 39%, 1:1 mixture.

9.4 Summary and outlook

The enediyne and dienediynes have proven to be an enormously stimulating class of natural products to a wide range of scientists. There have been a remarkable number of new methodologies developed for synthetic chemistry and whilst many of the innovative model studies have yet to deliver completed total syntheses it is likely that many will do so. Moreover, the ever increasing number of new related natural products[67] will surely provide further challenges for the synthetic chemist in the twenty-first century. It is interesting to note that much of the synthetic chemistry has delivered new classes of enediyne and dienediyne analogue as potential DNA modifiers[68-74] and further developments in this area of activity are likely to continue. It is then that we will begin to realise the full biological potential of this remarkable class of molecules.

Acknowledgements

We wish to acknowledge financial support for our research programme from the following sources: BBSRC, EPSRC, Royal Society, Astra Charnwood, Glaxo-Wellcome, Parke-Davis, Rhône-Poulenc Rorer, SmithKline Beecham, and Zeneca. We also acknowledge P. Caddick for assistance in preparation of the manuscript and L. Frost for proof-reading the manuscript.

References

1. K.C. Nicolaou, A.L. Smith, and E.W. Yue, 1993, *Proc. Natl. Acad. Sci., USA*, 90, 5881-5888.
2. S.J. Danishefsky and M.D. Shair, 1996, *J. Org. Chem.*, 61, 16-44.
3. H. Lhermitte and D.S. Grierson, 1996, *Contemporary Organic Synthesis*, 41-63.
4. H. Lhermitte and D.S. Grierson, 1996, *Contemporary Organic Synthesis*, 93-106.
5. K.C. Nicolaou and W.-M. Dai, 1991, *Angew. Chem., Int. Ed. Engl.*, 30, 1387-1416.
6. A.L. Smith and K.C. Nicolaou, 1996, *J. Med. Chem.*, 39, 2103.
7. I.H. Goldberg, 1993, *Frontiers in Pharmacology and Therapeutics: Cancer Chemotherapy*, Blackwell, Oxford.
8. J. Golik, G. Dubay, G. Groenewold, H. Kawaguchi, M. Konishi, B. Krishnan, H. Ohkuma, K. Saitoh, and T. Doyle, 1987, *J. Amer. Chem. Soc.*, 109, 3462.
9. M. Lee, T. Dunne, M. Siegel, C. Chang, G. Morton, and D. Borders, 1987, *J. Amer. Chem. Soc.*, 109, 3464.
10. F.J. Schoenen, J.A. Porco Jr., S.L. Schreiber, G.D.VanDuyne, and J. Clardy, 1989, *Tetrahedron Lett.*, 30, 3765-3768.
11. P. Magnus, 1994, *Tetrahedron*, **50**, 5, 1397-1418.
12. A.S. Kende and C.A. Smith, 1988, *Tetrahedron Lett.*, 29, 4217-4220.
13. S.J. Danishefsky, N.B. Mantlo, and D.S. Yamashita, 1988, *J. Am. Chem. Soc.*, 110, 6890-6892.

14. M.P. Cabal, R.S. Coleman, and S.J. Danishefsky, 1990, *J. Am. Chem. Soc.*, 112, 3253-3255.
15. A.S. Smith, C.K. Hwang, E. Pitsinos, G.R. Scarlato, and K.C. Nicolaou, 1992, *J. Am. Chem. Soc.*, 114, 3134-3136.
16. D.L.J. Clive, Y. Bo, Y. Tao, S. Daigneault, Y-J. Wu, and G. Meigan, 1996, *J. Am. Chem. Soc.*, 118, 4904-4905.
17. I. Churcher, D. Hallett, and P. Magnus, 1998, *J. Am. Chem. Soc.*, 120, 3518-3519.
18. K.C. Nicolaou, R.D. Groneberg, T. Miyyazaki, N.A. Stylianides, T.J. Sculze, and W. Stahl, 1990, *J. Am. Chem. Soc.*, 112, 8192-8193.
19. R.L. Halcomb, S.H. Boyer, and S.J. Danishefsky, 1992, *Angew. Chem., Int. Ed. Engl.*, 31, 338-340.
20. K.C. Nicolaou, 1993, *Angew. Chem., Int. Ed. Engl.*, 32, 1377-1385.
21. K.C. Nicolaou, C.W. Hummel, M. Nakada, K. Shibayame, E.N. Pitsinos, H. Saimoto, Y. Mizuno, K.-U. Baldeius, and A.L. Smith, 1993, *J. Am. Chem. Soc.*, 115, 7625-7635.
22. S.A. Hitchcock, S.H. Boyer, M.Y. Chu-Moyer, S.H. Olsen, and S.J. Danishefsky, 1994, *Angew. Chem., Int. Ed. Engl.* 33, 858-861.
23. M. Konishi, H. Ohkuma, K.-I. Saitoh, and H. Kawaguchi, 1985, *J. Antibiotics*, XXXVIII, 11, 1605-1609.
24. H. Mastalerz, T. Doyle, J. Kadow, K. Leung, and D. Vyas, 1995, *Tetrahedron Lett.*, 36, 4927-4930.
25. H. Masterlerz, T. Doyle, J.F. Kadow, and D.M. Vyas, 1996, *Tetrahedron Lett.*, 37, 8683-8686.
26. E. de Silva, J. Prandi, and J.-M. Beau, 1994, *J. Chem. Soc., Chem. Commun.*, 2127-2128.
27. D.A. Clark, F. de Richards, and K.C. Nicolaou, 1983, *Tetrahedron*, 39, 11391-11426.
28. G. Ulibarri, W. Nadler, T. Skrydstrup, H. Audrain, A. Chiaroni, C. Riche, and D.S. Grierson, 1995, *J. Org. Chem.*, 60, 2753-2761.
29. G. Ulibarri, H. Audrain, W. Nadler, H. Lhermitte, and D. Grierson, 1996, *Pure Appl. Chem.*, 68, 3, 601-604.
30. M. Konishi, H. Ohkuma, T. Tsuno, T. Oki, G.D. VanDuyne, and J. Clardy, 1990, *J. Am. Chem. Soc.*, 112, 9, 3715-3716.
31. P. Magnus, and S.M. Fortt., 1991, *J. Chem. Soc., Chem. Commun.*, 544-546.
32. K.C. Nicolau, C.-K. Hwang, A.L. Smith, and S.V. Wendborn, 1990, *J. Am. Chem. Soc.*, 112, 7416-7418.
33. P.A. Wender and C.K. Zercher, 1991, *J. Am. Chem. Soc.*, 113, 2311-2313.
34. For a closely related approach see: (a) T. Nishikawa, M. Yoshikai, K. Obi, and M. Isobe, 1994, *Tetrahedron Lett.*, 35, 7997-8000; (b) T. Nishikawa, A. Ino, M. Isobe, and T. Goto., 1991, *Chemistry Letters*, 1271-1274.
35. J.L. Wood, J.A. Porco Jr., J. Taunton, A.Y. Lee, J. Clardy, and S.L. Schreiber, 1992, *J. Am. Chem. Soc.* 114, 14, 5898-5900.
36. J. Taunton, J.L. Wood, and S.L. Schreiber, 1993, *J. Am. Chem. Soc.*, 115, 10378-10379.
37. A.G. Myers, M.E. Fraley, and N.J. Tom, 1994, *J. Am. Chem. Soc.*, 116, 11556-11557.
38. A.G. Myers, N.J. Tom, M.E. Fraley, S.B. Cohen, and D.J. Madar, 1997, *J. Am. Chem. Soc.*, 119, 6072-6094.
39. M.D. Shair, T.Y. Yoon, K.K. Mosny, T.C. Chou, and S.J. Danishefsky, 1996, *J. Am. Chem. Soc.*, 118, 9509-9525.
40. M.D. Shair, T.Y. Yoon, and S.J. Danishefsky, 1995, *Angew. Chem., Int. Ed. Engl.*, 34, 16, 1721-1723.
41. N. Ishida, K. Miyazaki, K. Kumagi, and M. Rikimaru, 1965, *J. Antibiot.*, 18, 68.
42. S. Torii, H. Okumoto, T. Tadokoro, A. Nishimura, and M.A. Rashid, 1993, *Tetrahedron Lett.*, 34, 2139–2142.
43. J.M. Nuss, R.A. Rennels, and B.M. Levine, 1993, *J. Am. Chem. Soc.*, 115, 6991-6992.

44. P.A. Wender, J. Wisniewski Grissom, U. Hoffman, and R. Mah, 1990, *Tetrahedron Lett.*, 31, 6605-6608.

45. A.G. Myers, V. Subramanian, and M. Hammond, 1996, *Tetrahedron Lett.*, 37, 587-590.

46. K. Takahashi, T. Suzuki, and M. Hirama, 1992, *Tetrahedron Lett.*, 33, 4603-4604.

47. N. Petasis and K.A. Teets, 1993, *Tetrahedron Lett.*, 34, 805-808.

48. P.A. Wender, M. Harmata, D. Jeffrey, C. Mukai, and J. Suffert, 1988, *Tetrahedron Lett.*, 29, 909-912.

49. P.A. Wender, J.A. McKinney, and C. Mukai, 1990, *J. Am. Chem. Soc.*, 112, 5369-5370.

50. T. Takahashi, H. Tanaka, Y. Hirai, T. Doi, H. Yamada, T. Shiraki, and Y. Sugiura, 1993, *Angew. Chem., Int. Ed. Engl.*, 32, 1657-1659.

51. P. Magnus, R. Carter, M. Davies, J. Elliott, and T. Patterna, 1996, *Tetrahedron*, 52, 18, 6283-6306.

52. P. Magnus and T. Pitterna, 1991, *J. Chem. Soc., Chem. Commun.*, 541-543.

53. S. Caddick and V.M. Delisser, 1997, *Tetrahedron Lett.*, 38, 2355-2358.

54. A.G. Myers, P.M. Harrington, and E.Y. Kuo, 1991, *J. Am. Chem. Soc.*, 113, 694-695.

55. A.G. Myers, P.M. Harrington, and B.-M. Kwon, 1992, *J. Am. Chem. Soc.*, 114, 1086-1087.

56. A.G. Myers, M. Hammond, Y. Wu, J.-N. Xiang, P.M. Harrington, and E.Y. Kuo, 1996, *J. Am. Chem. Soc.*, 118, 10006-10007.

57. A.G. Myers, J. Liang, M. Hammond, P.M. Harrington, Y. Wu, and E.Y. Kuo, 1998, *J. Am. Chem. Soc.*, 120, 5319-5320.

58. S. Kawata, S. Ashizawa, and M. Hirama, 1997, *J. Am. Chem. Soc.*, 119, 12012-12013.

59. J.E. Leet, J. Golik, S.J. Hofstead, and J.A. Matson, 1992, *Tetrahedron Lett.*, 33, 6107-6110.

60. K.-I. Iida, T. Ishii, M. Hirama, T. Otani, Y. Minami, and K.-I.Y. Ohsida, 1993, *Tetrahedron Lett.*, 34, 4079-4082.

61. M. Hornyak, I.F. Pelyvas, and F.J. Sztaricskai, 1993, *Tetrahedron Lett.*, 34, 4087-4090.

62. T. Vuljanic, J. Kihlberg, and P. Somfai, 1994, *Tetrahedron Lett.*, 35, 6937-6940.

63. S. Torii, H. Okumoto, T. Tadokoro, A. Nishimura, and M.A. Rashid, 1993, *Tetrahedron Lett.*, 34, 2139-2142.

64. K. Mikami, F. Feng, H. Matsueda, A. Yoshida, and D.S. Grierson, 1996, *Synlett*, 833-836.

65. S. Kawata, F. Yoshimura, J. Irie, H. Ehara, and M. Hirama, 1997, *Synlett*, 250-252.

66. K.-I. Iida and M. Hirama, 1994, *J. Am. Chem. Soc.*, 116, 10310-10311.

67. T. Ando, M. Ishii, T. Kajura, T. Kameyama, K. Miwa, and Y. Sugiura, 1998, *Tetrahedron Lett.*, 39, 6495-6498.

68. P.A. Wender, C.K. Zercher, S. Beckham, and E.-M. Haubold, 1993, *J. Org. Chem.*, 58, 22, 5867-5869.

69. M.D. Shair, T. Yoon, T.-C. Chou, and S.J. Danishefsky, 1994, *Angew. Chem., Int. Ed. Engl.*, 33, 23/24, 2477-2479.

70. M. Hirama, T. Gomibuchi, K. Fujiwara, Y. Sugiura, and M. Uesugi, 1991, *J. Am. Chem. Soc.*, 113, 9851-9853.

71. K. Nakatani, K. Arai, N. Hirayama, F. Matsuda, and S. Terashima, 1990, *Tetrahedron Lett.*, 31, 2323-2326.

72. P.A. Wender and M.J. Tebbe, 1994, *Tetrahedron*, 50, 5, 1419-1434.

73. A.G. Myers, M.E. Kort, and M. Hammond, 1997, *J. Am. Chem. Soc.*, 119, 13, 2965-2972.

74. M. Hirama, 1993, Synthesis and Chemistry of Neocarzinostatin Analogs, in *Recent Progress in the Chemical Synthesis of Antibiotics and Related Microbial Products*, 2 (G. Lukacs, Ed.), Springer Verlag, Berlin, pp. 293-319.

10 The chemical synthesis of linear peptides and amino acids

C.M. Bladon and P.B. Wyatt

10.1 Introduction

The first half of this chapter reviews progress in peptide synthesis and starts with a section in which significant developments in stepwise solid-phase peptide synthesis (SPPS) are highlighted. This is followed by a discussion of the chemoselective ligation methodology for the preparation of small proteins, and then the combinatorial approach to peptide synthesis is briefly surveyed. The second half of the chapter provides an account of the main chemical methods for preparing enantiomerically pure α-amino acids, illustrated with examples of natural products which have been synthesised. Finally, some leading references to the stereoselective synthesis of β- and γ-amino acids are included.

10.2 Improvements to solid-phase peptide synthesis chemistries

By the late 1980s highly optimised SPPS protocols allowed polypeptides of up to 30–40 amino acids in length to be routinely constructed. Two strategies predominated: Merrifield's pioneering *tert*-butoxycarbonyl (Boc)/benzyl chemistry and the 9-fluorenylmethoxycarbonyl (Fmoc)/ *tert*-butyl (Bu^t) technique developed by Sheppard. The latter method is chemically less complex than the Boc procedure, uses milder reagents, and has become the method of choice in many laboratories for the solid-phase synthesis of peptides. A general overview of the SPPS strategy is outlined in Scheme 10.1.

This chapter cannot do justice to the many new reagents and methods developed over the past 10 years and the interested reader is directed to a number of books[1] and review articles[2] of the field for detailed discussions of various aspects of the solid-phase technique. The book by Atherton and Sheppard[1a] and the multiauthored volume in the '*Methods in Enzymology*' series[1f] contain extensive experimental procedures which are helpful to a newcomer entering the field.

10.2.1 Resins and linkers

New and improved resins based on composite polymers have rapidly been adopted into routine SPPS. By grafting polyethylene glycol (PEG) onto

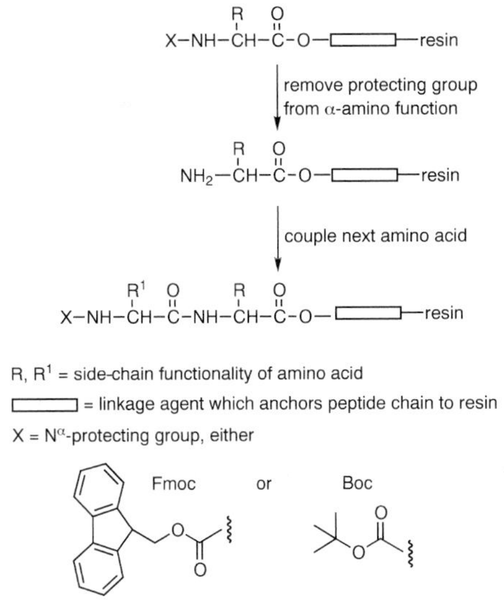

R, R^1 = side-chain functionality of amino acid

▭▭ = linkage agent which anchors peptide chain to resin

X = N$^\alpha$-protecting group, either

Scheme 10.1 Assembly of peptide chain: the two-step cycle of deprotection and coupling is repeated until the chain assembly is complete. The peptide chain is then cleaved from the resin and purified.

polystyrene, Rapp[3] developed TentaGel, a hydrophilic copolymer better suited to the assembly of polar peptide chains. A mechanically stable resin, TentaGel is compatible with both batch and continuous-flow protocols in Fmoc synthesis and it is commercially available either in the amino functionalised form, derivatised with a linker, or with the first amino acid already attached. Favourable physicochemical properties are also reported for Meldal's PEG-polyacrylamide (PEGA) resins[4] and the cross-linked ethoxylate acrylate (CLEAR) supports of Kempe and Barany.[5]

An *o*-chlorotrityl chloride resin[6] was designed for the preparation of protected fragments by the Fmoc method and complements the use of the hyper-acid-labile handles such as **1a**[1a] and **1b**.[7] Today, Sheppard's

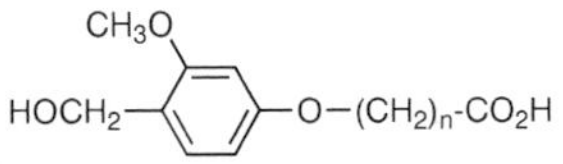

(**1a**) n = 1, 4-(4-hydroxymethyl-3-methoxyphenoxy)acetic acid

(**b**) n = 3, 4-(4-hydroxymethyl-3-methoxyphenoxy)butyric acid

acetic acid derivative is often replaced by the more acid-labile butyric acid analogue of Riniker. Photo-labile *ortho*-nitrobenzyl, fluoride-labile silicon-containing, and Pd(0)-sensitive allyl-based linkers[8] are also compatible with the preparation of protected peptides using the Fmoc/Bu^t strategy. The allylic linkers are, moreover, particularly suited to the Boc/benzyl methodology. The standard linker for the preparation of fully deprotected peptide acids using Fmoc chemistry continues to be 4-(hydroxymethyl)phenoxyacetic acid.[1a] Fine tuning of the anchoring structure and the corresponding removal conditions has also resulted in the development of Rink's acid-labile benzhydrylamine 2[9] which has

(2) Fmoc-Rink linker

largely superseded 4-(hydroxymethyl)benzoic acid[1a] as the preferred linker for the preparation of peptide amides by the Fmoc approach. However, the benzoic acid derivative remains one of the most versatile peptide linkers as changing the nature of the nucleophile used for cleavage can yield a variety of C-terminus modified products, for example amides with NH_3 and hydrazides with NH_2NH_2.

Demand for non-standard C-termini has led to the development of a variety of linkers[8] which can yield, for example, alcohols, aldehydes, esters, N-alkylamides, sulfonamides and thioacids directly upon cleavage, and this advance is, in part, a result of a renaissance of interest in the solid-phase synthesis of small organic molecules.

10.2.2   Protecting groups

The basis of the Fmoc/Bu^t strategy is the orthogonal protection scheme; base-labile Fmoc group (removed with piperidine) for N^α-protection and tert-butyl-based moieties cleaved by mild acid [trifluoroacetic acid (TFA)] for side-chain functionality. In routine syntheses the hydroxyls of Ser, Thr and Tyr are protected as tert-butyl ethers, the carboxyls of Asp and Glu as tert-butyl esters while Boc is the standard choice for masking the His imidazole and Lys ε-amino moieties. Incorporation of Trp does not normally require the indole NH to be protected, while acetamidomethyl (Acm) is one of several groups available to mask the cysteine thiol. The side-chain guanidino group of Arg has proved problematic. The 4-methoxy-2,3,6-trimethylbenzenesulfonyl (Mtr) is a suitable but not wholly satisfactory protecting group, as cleavage requires several hours with TFA, and prolonged deprotection times can lead to side reactions, particularly in Trp-containing peptides. By increasing the substitution on the aromatic ring Ramage and co-workers devised the 2,2,5,7,8-pentamethylchroman-6-sulfonyl (Pmc, **3a**) group[10] with acid sensitivity comparable to *tert*-butyl. Pmc and the dihydrobenzofuran analogue (Pbf, **3b**),[11] which has slightly faster reaction kinetics and consequently

(**3a**) n = 2, Fmoc-Arg (Pmc)-OH

(**b**) n = 1, Fmoc-Arg(Pbf)-OH

fewer problems associated with side reactions, have effectively solved the arginine problem in solid-phase synthesis.

Other groups compatible with Fmoc/Bu^t chemistry that require different chemical methods for removal have gained in prominence. Allyl esters[12] and the urethane derivative *N*-allyloxycarbonyl[13] are removed under mild conditions by Pd(PPh$_3$)$_4$-catalysed allyl transfer, whereas the amino protecting group 1-(4,4-dimethyl-2,6-dioxocyclohexylidene)ethyl (Dde, **4a**) developed by the Bycroft group,[14] and the more hindered *iso*-valeryl variant ivDde (**4b**),[15] are removed by 2% (vol/vol) hydrazine in

(**4a**) R = CH$_3$, Fmoc-Lys(Dde)-OH

(**b**) R = CH$_2$-CH(CH$_3$)$_2$, Fmoc-Lys(ivDde)-OH

DMF. The added dimension of orthogonality for the amino groups of Lys provided by Dde/ivDde is complemented by the analogous 4-{*N*-[1-(4,4-dimethyl-2,6-dioxocyclohexylidene)-3-methylbutyl]amino}benzyl ester (ODmab, **5**)[16] protection for the carboxylic acid function of Asp and

(5a) n = 1, Fmoc-Asp(ODmab)-OH
(b) n = 2, Fmoc-Glu(ODmab)-OH

Glu residues. The combination of Fmoc/Bu-t/Dde-ODmab has proved a valuable tool in the synthesis of cyclic, branched and side-chain modified peptides.

Clustering of peptides within the polymeric network during chain assembly often causes coupling difficulties and frequently leads to incomplete acylations. Several factors are believed to contribute to this aggregation, including the peptide sequence itself, sterically hindered amino acids and intrachain or interchain hydrogen bonding interactions. Formation of β-sheet and other secondary structures can be prevented by the introduction of tertiary amide bonds, and the Medical Research Council (MRC) group in Cambridge has developed the Fmoc/Bu-t compatible 2-hydroxy-4-methoxybenzyl (Hmb, **6a**) protecting group for

(6a) R^1 = H, Hmb
(b) R^1 = CH$_3$CO, Ac-Hmb

amides.[17] Introducing an Hmb group onto the nitrogen of a peptide bond prevents this atom from acting as a hydrogen bond donor, thus precluding the adoption of a secondary structure. A Hmb moiety every five or six residues of the sequence is sufficient to inhibit chain interactions and the group is easily removed under the normal conditions (95% TFA) required for final cleavage and deprotection of the peptide. Another aspect of the Hmb group is that acylation of the hydroxyl function with acetic anhydride and N,N-diisopropylethylamine (DIPEA)[18] results in increased acid stability with the effect that peptides are able to retain the Ac-Hmb (**6b**) protection upon cleavage from the resin. These backbone protected peptides exhibit improved solubility properties and purification by reversed-phase high-pressure liquid chromatography (RP-HPLC) of otherwise intractable sequences is facilitated.[19] The free peptides can be regenerated by deacylation with 5% hydrazine in DMF followed by cleavage of the Hmb group with TFA.

An alternative strategy in preventing peptide aggregation has been reported by the Mutter group.[20] Their approach is to introduce a kink into the peptide backbone by incorporating oxazolidine or thiazolidine (pseudo-proline, Ψ Pro) units (**7**) into the growing peptide chain thus

(**7a**) R = H, X = O (serine)
(**b**) R = CH_3, X = O (threonine)
(**c**) R = H, X = S (cysteine)

disrupting β-sheet structures. The pseudo-prolines are, moreover, reversible protecting groups for Ser, Thr and Cys and this feature combined with the solubilising effect of the heterocycles has proven a powerful tool for preparing difficult hydrophobic transmembrane sequences. Pseudo-prolines are incorporated as preformed dipeptide building blocks Fmoc-Xxx-Ser/Thr/Cys(Ψ^{R^1,R^2}Pro)-OH into the chain and the coupling proceeds without racemisation. The nature of the C(2) substituent on the heterocycle directly affects the acid lability of the ring system. Whereas derivatives **7a, 7b** (R^1, $R^2 = CH_3$) and **7c** ($R^1 = H$, $R^2 = $ 2,4-dimethoxyphenyl) are cleavable with 90% TFA and are useful in standard Fmoc syntheses, oxazolidines **7a** and **7b** (where $R^1, R^2 = H$) and thiazolidine (**7c**, R^1, $R^2 = CH_3$) are more acid stable and reaction with trifluoromethansulfonic acid is required to regenerate the amino acids. These latter derivatives confer substantially improved solubility properties on protected peptide segments and have consequently found application in convergent syntheses and in chemoselective ligation procedures.

Recent significant developments in Boc/benzyl chemistry are predominantly associated with the introduction of chemoselective ligation procedures (see section 10.3.2).

10.2.3 Activation and coupling reagents

Standard synthetic protocols for the formation of the amide bond in simple peptides frequently involve the use of *N*-hydroxybenzotriazole

(HOBt) either in combination with a carbodiimide or incorporated into a stand-alone reagent. Most of the HOBt-based reagents are phosphonium or aminium salts, for example PyBOP (**8**)[21] and HBTU (**9**),[22, 23] and in the

(**8**) Benzotriazol-1-yloxytris(pyrrolidino)-
phosphonium hexafluorophosphate (PyBOP)

(**9**) N-[(1H-benzotriazol-1-yl)-(dimethylamino)methylene]-N-
methylmethanaminium hexafluorophosphate
N-oxide (HBTU)

presence of a tertiary base DIPEA or N-methylmorpholine (NMM) the carbonyl group of the amino acid is smoothly converted *in situ* into the oxybenzotriazole ester. Ease of use, inhibition of side reactions and a reduction in racemisation have led to the wide adoption of these reagents in routine SPPS. Recently, the aza analogues 1-hydroxy-7-azabenzotriazole (HOAt, **10**) and HATU (**11**)[23, 24] have been reported to be superior

(**10**) 1-Hydroxy-7-azabenzotriazole (HOAt)

(**11**) N-{(dimethylamino)-1H-1,2,3-triazolo[4,5-b]pyridin-1-yl-
methylene}-N-methylmethanaminium hexafluorophosphate
N-oxide (HATU)

coupling reagents with enhanced coupling rates and reduced risk of racemisation relative to HOBt and HBTU. The aza analogues incorporate a nitrogen atom at position 7 of the aromatic ring, and it has been proposed that the increased reactivity and reduced risk of race-misation associated with these reagents could be the result (in part) of the neighbouring-group effect shown in Scheme 10.2. In the synthesis of the test peptide H-Tyr-Aib-Aib-Phe-Leu-NH$_2$ with the difficult Aib-Aib coupling, the pentapeptide was obtained with a purity of 94% for the HATU synthesis compared with only 43% for HBTU.[25]

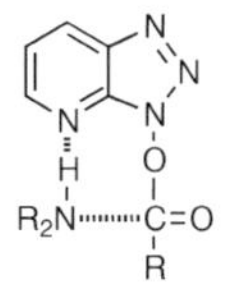

Scheme 10.2.

Fmoc amino-acid fluorides[26] have also emerged as rapid-acting acylating agents in peptide synthesis and are especially suitable for the coupling of hindered α,α-dialkyl amino acids such as α-aminoisobutyric acid (Aib). Their high reactivity is partly a consequence of the small size of the fluoride leaving group, and coupling reactions proceed well either in the presence of base (DIPEA)[27] or under neutral conditions.[28] Alamethicin (**12**), one of the naturally occurring peptaibol group of

Ac-Aib-Pro-Aib-Ala-Aib-Ala-Gln-Aib-Val-Aib-Gly-Leu-Aib-Pro-Val-Aib-Aib-Glu-Gln-Phe-ol

(**12**) Alamethicin F30

antibiotics, is a 20-residue peptide containing eight Aib residues. The first solid-phase syntheses of this and a number of related compounds have been achieved by using the fluoride method.[29] The Fmoc or Boc protected amino acid fluorides are stable crystalline solids that can be prepared by reaction of the acid with cyanuric fluoride in the presence of pyridine,[30] or with diethylamino sulfur trifluoride[31] in CH$_2$Cl$_2$. Alternatively, fluoride derivatives can be prepared *in situ* by reaction of Fmoc amino acids with tetramethylfluoroformamidinium hexafluorophosphate.[32]

Alamethicin and the related peptaibols terminate in an amino alcohol, and in their synthesis of these peptides Carpino and co-workers adopted the *o*-chlorotrityl chloride resin to allow direct anchoring of Fmoc-Phe-ol.

10.2.4 *Phosphorylated and glycosylated peptides*

Glycosylation and phosphorylation of proteins are important post-translational modifications, and the oligosaccharides, linked predomi-nantly to Ser, Thr and Asn residues, and phosphate groups, on the side-chains of Ser, Thr and Tyr, play critical roles in a diverse range of biological processes.

Phosphopeptides are prepared either by incorporating protected phosphoamino acid building blocks, many of which are commercially available, into chain assembly or phosphorylating the OH side-chain groups after peptide assembly is complete.[33] Fmoc synthesis of many phosphotyrosine [Tyr(P)] containing peptides is relatively straightforward using preassembly phophorylation methods and Fmoc-Tyr(PO_3R_2)-OH (R = Bzl, Bu')[34, 35] with PyBOP or HBTU activation. In some syntheses, however, protected tyrosines have not proven to be the most suitable derivatives for coupling. For example, Barany and co-workers[36] found that the unprotected phosphoamino acid Fmoc-Tyr[PO_3H_2]-OH was more appropriate for the preparation of 4-to-10 residue Tyr(P) peptides based on the phosphoALBP sequence. Furthermore, Meutermans and Alewood[37] prepared Stat 91[695–708] using an interassembly strategy in which phosphorylation was carried out immediately after introduction of an unprotected tyrosine residue. The monoprotected derivative Fmoc-Tyr[PO(OBzl)OH]-OH[38] has recently been reported as an excellent building block for the synthesis of Tyr(P) peptides and may attain prominence in the future.

Boc-Tyr(PO_3Me_2)-OH[39] and Boc-Ser/Thr(PO_3Me_2)OH[40] are compatible with the harsher reaction conditions used in the Boc mode of synthesis, and both have been successfully used in the preparation of a doubly phosphorylated 19-residue MAP kinase peptide.[41]

The repetitive base deprotection steps needed in Fmoc synthesis have precluded the use of the bis-protected Ser(P) and Thr(P) building blocks, as β-elimination occurs under such conditions, resulting in loss of the phosphate and formation of the dehydroalaninyl peptide. However, once incorporated into peptides the mono-benzyl compounds Fmoc-Ser/Thr[PO(OBzl)OH]-OH developed by Wakamiya and co-workers[42] are stable to base and have been used to prepare a number of Ser(P) and Thr(P) peptides.[38] Post-assembly phosphorylation has been the preferred strategy for the synthesis of Ser(P) and Thr(P) peptides using the Fmoc approach. In the synthesis of the difficult bis phosphopeptide **13**[43] Thr

$$\overset{PO_3H}{\underset{|}{}} \quad \overset{PO_3H}{\underset{|}{}}$$

H-His[178]-Thr-Gly-Phe-Leu-Thr-Glu-Tyr-Val-Ala-Thr[188]-OH

(**13**) MAP kinase ERK 2 [178-188; Thr(PO_3H)[183]; Tyr(PO_3H)[185]]

was introduced without side-chain protection using Fmoc-Thr(H)-OH and *in situ* activation. A phosphitylation reaction with a dialkylphosphoramidite, followed by oxidation with I_2 or Bu'OOH, yielded the Thr(OPO_3H_2) moiety. The recently reported bis(pentafluorophenyl) chlorophosphate may also provide a useful alternative, and more direct method, for phosphorylation.[44]

In a series of elegant studies published by several groups, many of the difficulties previously associated with the synthesis of glycopeptides now appear to have been effectively overcome. Two main strategies now appear to predominate in this field: the building-block approach, involving incorporation of preformed N^α-Fmoc-glycosyl amino acids; and the convergent approach, in which one or more saccharide moieties are linked at defined positions to a protected peptide.[45] In the preparation of N-linked glycopeptides by the latter approach, a side-chain-activated aspartyl residue is coupled with an unprotected glycosylamine, but this reaction is in competition with intramolecular cyclisation to the aspartimide (Scheme 10.3). By careful control of experimental conditions

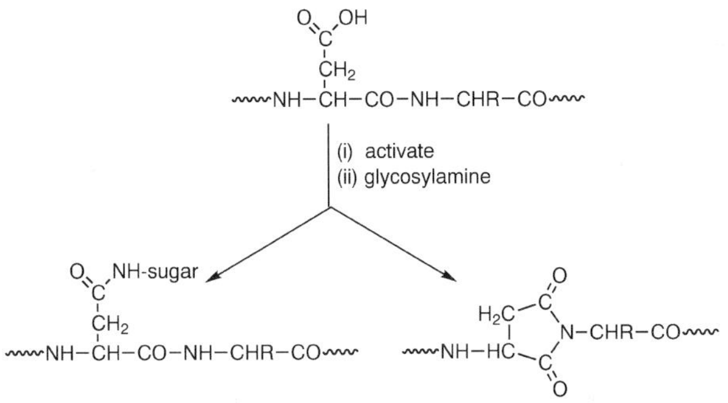

Scheme 10.3   Aspartimide formation during convergent N-glycopeptide synthesis.

Anisfeld and Lansbury[46] were able to minimise the aspartimide reaction to obtain small glycopeptides in good yield. However, when Johnson and co-workers substituted the aspartyl amide proton with their Hmb protecting group the aspartimide reaction was completely suppressed and condensation of the partially protected resin-bound hexapeptide **14** with N,N'-diacetylchitobiosylamine yielded only the required glycosylated product (**15**; Scheme 10.4).[47] Moreover, the presence of the Hmb group was thought to improve the solubility of the peptide resin through disruption of interchain hydrogen bonding.

The building-block approach has been applied to the synthesis of both O- and N-linked glycopeptides. Fmoc/Bu^t is the preferred methodology as the harsh acidic conditions used in the Boc/benzyl protocols are frequently incompatible with the acid-sensitive glycosidic bonds. The general picture emerging is that the sugar hydroxyls should normally be protected with acetyl groups. The coupling of the glycosyl amino acids

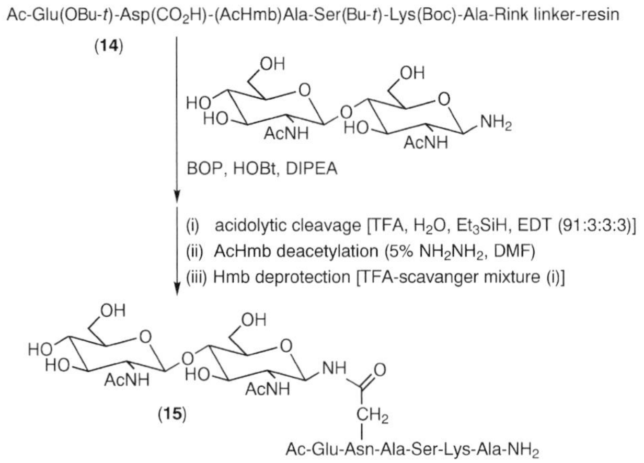

Ac-Glu(OBu-*t*)-Asp(CO$_2$H)-(AcHmb)Ala-Ser(Bu-*t*)-Lys(Boc)-Ala-Rink linker-resin

(14)

BOP, HOBt, DIPEA

(i)   acidolytic cleavage [TFA, H$_2$O, Et$_3$SiH, EDT (91:3:3:3)]
(ii) AcHmb deacetylation (5% NH$_2$NH$_2$, DMF)
(iii) Hmb deprotection [TFA-scavanger mixture (i)]

(15)

CH$_2$

Ac-Glu-Asn-Ala-Ser-Lys-Ala-NH$_2$

Scheme 10.4.

does not appear to be affected by the size and complexity of the sugar, and fears that the base-mediated deprotection of the N^α-Fmoc groups would promote β-elimination of the carbohydrate *O*-linked to serine/ threonine residues have now proven unfounded. The compatibility of the glycosyl amino acids with standard solid-phase protocols has also resulted in the use of multiple synthesis techniques. With N^α-Fmoc-Ser/ Thr(Ac$_3$-α-D-GalNAc)-OPfp as the glycosyl building block, Meldal and co-workers assembled 40 mucin-type glyco-octapeptides in parallel[48] in which different threonine residues of Ac-Pro-Thr-Thr-Thr-Pro-Ile-Ser-Thr-NH$_2$ and serine residues of Ac-Gly-Ser-Ser-Ser-Gly-Ser-Pro-Gly-NH$_2$ were systematically glycosylated with *N*-acetyl galactose. One equivalent of 3,4-dihydro-3-hydroxy-4-oxo-1,2,3-benzotriazine (Dhbt-OH) was added to the acylating mixture, in order to monitor the progress of the coupling reaction, by visually noting the disappearance of the bright yellow colour caused by ion-pairing between the free α-amino groups and the benzotriazine. A further series of 60 mucin *O*-glycodecapeptides containing di- and tri-saccharides such as β-D-Gal(1 → 3)-α-D-GalNAc and β-D-GlcNAc-(1 → 3)-[β-D-GlcNAc-(1 → 6)]-α-D-GalNAc was synthesised in an analogous manner, with threonine building blocks bearing the more complex oligosaccharide structures.[49]

In an alternative strategy to the preparation of the highly glycosylated mucin-type glycopeptides the Meldal group has incorporated a partially protected galactose threonine building block into the peptide chain. The saccharide chain was then extended by reaction of the unsubstituted 3-OH with perbenzoylated glycosyl trichloroacetimidates in the presence of a catalytic amount of trimethylsilyl trifluoromethanesulfonate (TMSOTf) (Scheme 10.5).[50]

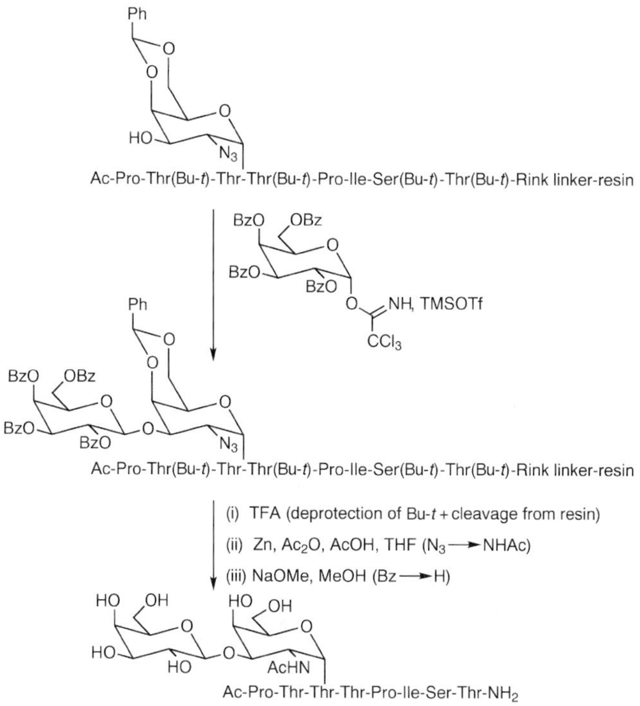

Ac-Pro-Thr(Bu-*t*)-Thr-Thr(Bu-*t*)-Pro-Ile-Ser(Bu-*t*)-Thr(Bu-*t*)-Rink linker-resin

BzO, OBz / NH, TMSOTf / CCl₃

Ac-Pro-Thr(Bu-*t*)-Thr-Thr(Bu-*t*)-Pro-Ile-Ser(Bu-*t*)-Thr(Bu-*t*)-Rink linker-resin

(i)   TFA (deprotection of Bu-*t* + cleavage from resin)
(ii) Zn, Ac₂O, AcOH, THF (N₃ ⟶ NHAc)
(iii) NaOMe, MeOH (Bz ⟶ H)

Ac-Pro-Thr-Thr-Thr-Pro-Ile-Ser-Thr-NH₂

Scheme 10.5.

Meldal and collaborators have also applied the building-block approach to the synthesis of highly glycosylated *N*-linked glycopeptides.[51] A variety of sugars, including complex oligosaccharides isolated from natural glycoproteins, were converted to the glycosylamines and then coupled to the side-chains of activated Fmoc-Asp(ODhbt)-OBu-*t*. *O*-Acetylation of the carbohydrate moiety and cleavage of the *tert*-butyl group afforded building blocks with for example di- and tri-antennary oligosaccharide structures (**16**). These compounds were reported to be

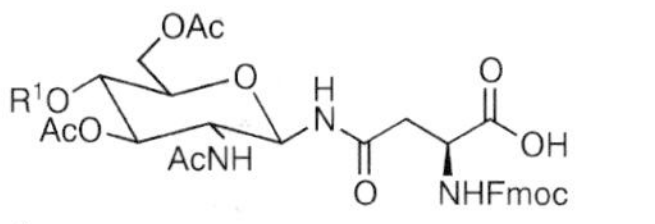

R^1 = (Ac₄Gal)₂(Ac₂GlcNAc)₃(Ac₃Man)₂(Ac₂Man)
        (Ac₄Gal)₃(Ac₂GlcNAc)₄(Ac₃Man)₂(Ac₂Man)

(**16**)

fully compatible with solid-phase protocols, and the large carbohydrate groups did not appear to hinder the coupling reactions between the asparagine derivative and the N-terminal amino group of the resin-bound peptide.

10.2.5 Purification methods

High-pressure liquid chromatography (HPLC) is a powerful tool for the purification of biomolecules[52] but separation of the target peptide from crude SPPS mixtures can be tedious and time-consuming. A strategy to facilitate the purification process has been explored by several groups and involves attaching a removable tag to the N-terminus of the resin-bound peptide as the final step in solid-phase chain assembly. The tag differentiates the full-length peptide from capped truncated or terminated sequences, and after cleavage from the resin the tagged peptide is separated from the crude mixture by covalent or affinity chromatography. The tag is removed following purification or as part of the purification protocol. Examples of tags include the tetrabenzo[a,c,g,i]fluorenyl-17-methoxycarbonyl (Tbfmoc) group with affinity for porous graphitised carbon,[53] histidine tails which bind to Ni-agarose columns[54] and the $R—SO_2—CH_2CH_2—O—CO—$ moiety which can be hydrophobic (R=long-chain n-alkyl)[55] or functionalised for covalent chemoselective reactions.[56]

The solubilising effect of Ac-Hmb and pseudo-proline moieties on sparingly soluble peptides was discussed in section 10.2.2. An alternative method of conferring solubility on hydrophobic peptides has been to incorporate a 'solubilising tail' sequence and assemble constructs of the form [hydrophobic peptide]–[orthogonal linker]–[solubilising peptide]. Solubilising tails suitable for both the Fmoc/Bu^t[57] or Boc/benzyl[58] strategies have been developed. The water-soluble constructs were readily purified by using standard HPLC procedures and then the solubilising tail sequences removed to give the desired hydrophobic peptides.

10.3   Convergent strategies

Convergent synthesis involves the coupling of peptide fragments to prepare longer sequences. In the more traditional method, known as fragment condensation, protected peptide fragments are assembled on the solid phase, cleaved, purified to homogeneity and then coupled either in solution or on the solid phase. A new technology of chemical ligation has emerged in recent years in which unprotected segments are coupled in aqueous solution by means of the chemoselective reaction of unique, mutually reactive functional groups.

10.3.1 Fragment condensation

Protected fragments are routinely obtained using the Fmoc/Bu-*t* strategy plus acid-sensitive *o*-chlorotrityl or super acid-sensitive (SASRIN) resins and/or hyper-acid-labile linkers such as (**1a** and **1b**) or through a combination of an oxime-based resin and Boc/benzyl protection.[59] Photo-labile and allyl linkers provide alternative modes of peptide–resin anchoring which are compatible with both protection strategies. Two frequently encountered problems in fragment condensation are difficulties during the purification of protected peptides and slow segment coupling. Both stem from the poor solubility of protected peptides in a variety of solvents but the advent of Ac-Hmb backbone protection and use of pseudoprolines (see section 10.2.2) may provide a general solution to these issues. Coupling of a protected fragment is more demanding than a protected amino acid as the *C*-terminal residue of a peptide segment does not have a urethane group and is consequently more prone to epimerisation. The Cambridge team have recently reported that alkylation of the *C*-terminal amide bond of a protected segment with 6-hydroxy-5-methyl-1,3-benzoxathiolyl (**17**) dramatically suppressed epi-

(**17**)

merisation during activation and coupling.[60] The group is removed by treating with $NH_4I/(CH_3)_2S$ to reduce the sulfoxide to sulfide which is acid-labile and cleaved with TFA. Although the study focused on coupling the dipeptide Ac-Asp(Bu-*t*)-Phe-OH to H-Lys(Boc)-resin the method has the potential to become a general solution to the epimerisation problem.

10.3.2 Chemoselective ligation

Chemoselective ligation of unprotected peptide segments is a simple and elegant procedure which has opened up new routes for the synthesis of proteins.[61] The underlying concept is to incorporate into unprotected peptides unique and mutually reactive functional groups (one type on each segment) and then selectively to couple the fragments in an aqueous environment. The advantage of using unprotected peptides is that sequences up to about 60 amino acids can be readily prepared, purified to homogeneity and characterised by standard techniques. Furthermore, the unprotected peptides are frequently soluble in aqueous solutions.

Schnölzer and Kent[62] were the first to apply the concept of chemoselective ligation to large peptide fragments in a report describing the synthesis of a HIV-1 protease analogue. Two approximately 50-residue unprotected peptides, one bearing an $^{\alpha}$COSH at the carboxyl terminus and the other a bromoacetyl function at the amino terminus, reacted to form a thioester bond at the ligation site. Although the product $[(NHCH_2COSCH_2CO)^{51-52}\text{-}Aba^{67,\,95}]$HIV-1 protease featured a nonamide bond at the ligation site the folded dimer had enzymatic properties identical to the natural protein. Subsequent examples of the technique included the preparation of a second analogue of HIV-1 protease,[63] a 94-residue fibronectin ^{10}F3 module[64] and a four-helix bundle template-assembled synthetic protein (TASP).[65] The analogous S_N2 reaction using the thiol group, of for example the side-chain of cysteine, to displace bromine from the alkyl bromide component and give the thio–ether linked products has been exploited in the synthesis of a HIV-1 protease analogue,[66] a transmembrane peptide bearing solubilising tail sequences,[67] an integrin $\alpha_{IIb}\beta_3$ neoprotein[68] and four-helix bundle TASPs.[69]

Unnatural functionality can readily be accommodated in model proteins, and in the synthesis of neoglycopeptide conjugates[70] and peptide dendrimers,[71] unprotected segments have been joined together through formation of an oxime or a hydrazone bond or a thiazolidine ring. The latter strategy, introduced by Liu and Tam,[72] is particularly attractive as the thiazolidine ring is a pseudoproline residue and the ligated product has an all-amide-bond backbone. The chemoselective elements are the 1,2-aminothiol moiety of an N-terminal cysteine and a C-terminal glycolaldehyde which condense at pH 4 to form a nonamide thiazolidine ring. This capture step brings the α-amino and α-carboxyl moieties into proximity and, at a slightly more alkaline pH, the intermediate thiazolidine ring undergoes an intramolecular O → N acyl migration to form the pseudoprolyl imide bond (Scheme 10.6). Active HIV-1 protease analogues[73] and cell-permeable peptides[74] have also been prepared by means of this ligation route. By incorporating the chemoselective elements into both termini of a single unprotected sequence the capture step results in cyclisation and formation of cyclic peptides.[75]

In the synthesis of large protein analogues with masses in excess of 20 kDa a useful approach is to employ two or more mutually compatible ligation chemistries and join several unprotected peptide segments together in either a sequential or convergent fashion.[76]

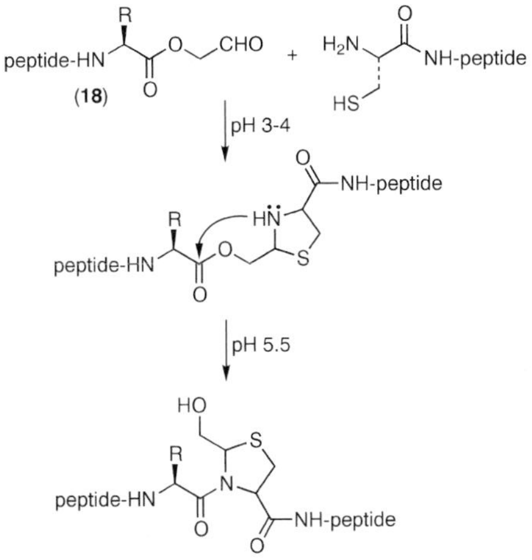

Scheme 10.6 Thiazolidine ring ligation strategy. The aldehyde functionality in **18** is introduced as the acetal after assembly of the peptide chain by esterifying the caesium salt of the peptide acid with bromoacetaldehyde dimethyl acetal; the acetal is then hydrolysed with TFA.

The ultimate goal of native chemical ligation was achieved in 1994 by the Kent group,[77] who used an ingeniously simple and elegant reaction between an *N*-terminal cysteine residue and a *C*-terminal thioester moiety. The ligation is initiated by a chemoselective transthioesterification reaction to generate a thioester-linked intermediate which then undergoes spontaneous intramolecular S → N acyl rearrangement to form an amide bond between the two polypeptides and regeneration of the Cys side-chain (Scheme 10.7). The peptides were prepared by Boc/benzyl

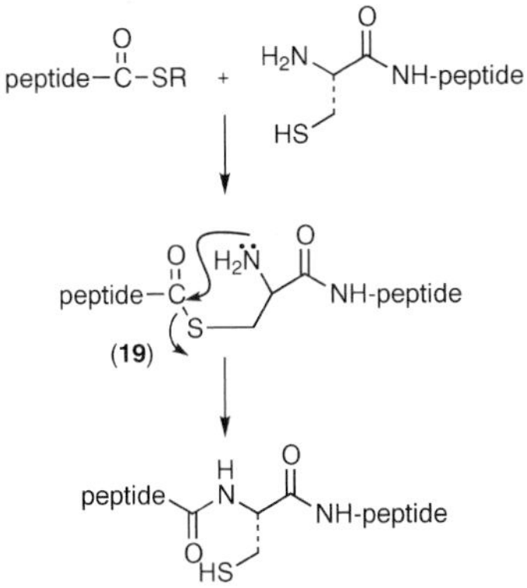

Scheme 10.7 Native chemical ligation to yield polypeptides with an amide bond at the Xxx-Cys site.

chemistry and the α-thiocarboxylic acid, generated directly upon HF cleavage from a thioester linker,[78] was converted to the activated $^\alpha$COSCH$_2$Ph moiety prior to the ligation reaction. The process was initially illustrated with the synthesis of the cytokine [Ala^{33}]-interleukin 8 (IL8), a 72-residue polypeptide containing two disulfide bonds. Ligation was carried out for three days in phosphate buffer (pH 7.6, containing 6 M guanidine hydrochloride) in the presence of benzyl mercaptan to prevent disulfide formation. The purified ligation product in the fully reduced form was oxidised to form the folded disulfide-bridged molecule which was identical to an authentic sample of the IL8 polypeptide. Subsequently, the methodology was optimised and used to prepare a number of proteins with full biological activity. Examples include the enzymes [Cys^{49}]barnase[79] and human secretory phospholipase A$_2$,[80] the serine protease inhibitor turkey ovomucoid third domain,[81] bovine pancreatic trypsin inhibitor,[82] several cytokines[83] and peptides with putative α-helical coiled-coil structures.[84] Cyclic peptides can be prepared by an intramolecular version of the ligation reaction.[85]

Native chemical ligation can also be effected by using the α-thiocarboxylic acid of the acyl segment as the nucleophile and reversing the direction of the initial attack. The thiocarboxylic acid can either S-alkylate an N-terminal β-bromoalanyl moiety on the amino segment[86] to form the intermediate thioester (19) which spontaneously rearranges as depicted in Scheme 10.7. Alternatively it can form a mixed acyl disulfide by reaction with an activated thiol side chain of an N-terminal Cys residue.[87] The S,N-acyl transfer proceeds in an analogous manner but via a six-membered cyclic intermediate containing a disulfide to yield a product bearing a hydrodisulfide (S—SH) on the cysteine. Reduction of the S—SH group with H$_2$S gives the native Cys residue at the ligation site.

An initial limitation of the native ligation technique was the requirement for the new amide bond to be formed at a Xxx-Cys site in the target polypeptide. Ligation at Xxx-Gly sequences is now also possible through the use of a removable auxiliary group to mimic the side-chain of cysteine[88] and at Xxx-His sites by coupling a thiocarboxylic acid with an N-terminal histidyl residue in the presence of Ellman's reagent.[89] In the Xxxx-Gly ligation method, an R-NH-OCH$_2$CH$_2$SH unit is prepared from a terminal N-bromoacetamide residue in a peptide fragment. A transthioesterification is then performed with the terminal thioester of another peptide fragment, R'-COSR''. The new coupled thioester then undergoes spontaneous $5 \rightarrow$ N migration to establish the desired amide bond [R'-CON(OCH$_2$CH$_2$SH)-R] between the two fragments. The N-alkoxythiol group is then cleaved by reduction with Zn/H$^+$ to give the required coupled peptide sequence linked via a glycine residue. This extended ligation procedure has also been applied to the preparation of cyclic peptides.[90]

Synthesis of peptide-α-thioacids is straightforward in Boc/benzyl SPPS using thioester linkers.[78] Direct preparation of the thioester segment using Fmoc/Bu^t chemistry is difficult as a CO—S peptide–resin link is labile towards the piperidine used to remove Fmoc protection. However a practical process has been devised of preparing C-terminal thioesters compatible with Fmoc synthesis in which protected peptide acids, constructed by using an o-chlorotrityl resin, are coupled with HS—$(CH_2)_2$—CO_2Et and then deprotected with 95% TFA.[91]

The development of a method of producing recombinant protein thioesters has paved the way for an expansion of native chemical ligation into the field of protein engineering.[92] A C-terminal thioester is an intermediate generated during protein splicing and this species can be intercepted and ligated to a synthetic peptide to give a semisynthetic protein. The technique, termed 'expressed protein ligation', has already been used to prepare semisynthetic constructs of the approximately 600-amino-acid σ^{70}subunit of $E.$ $coli$ RNA polymerase and of the 50 kDa Src kinase protein and has the potential for widespread application.

10.4   Combinatorial chemistry

Since the introduction of combinatorial synthesis,[93] the technique has attracted enormous attention and this is reflected in an ever-expanding literature on the subject. Entire journals are now devoted to progress in the field, and the present discussion is limited to a brief overview of peptide-library synthesis. The interested reader is directed to the many books[94] and reviews[95] on combinatorial chemistry to gain further insight into the field. Combinatorial chemistry is of immense interest to pharmaceutical companies as it has the potential to be a rapid method for the identification and optimisation of novel lead compounds, thus streamlining the drug discovery process.

Combinatorial methods were initially applied to the preparation of peptides and oligonucleotides but the technique has now evolved to cover the synthesis of nonoligomeric small molecule libraries. Combinatorial chemistry involves three main phases: synthesis, screening, and analysis. The libraries (collections of molecular diversity) are first generated by the systematic connection of different building blocks, and in the case of peptides, this is effectively SPPS. The peptides are usually fairly short and consequently their assembly is relatively straightforward. The Fmoc strategy predominates, although in the construction of some libraries alternative linkers and protecting groups are used in preference to those normally associated with the standard protocols for single-compound synthesis. The libraries are then screened with biological target molecules to detect novel compounds that interact with them and finally these 'hits' are isolated and identified.

The generation of libraries is conceptually simple and falls into two general categories: random mixtures or arrays. In array or spatially addressable synthesis, individual compounds are assembled at defined locations on the solid-support, for example a glass plate or a matrix of plastic pins, and thus the composition of each member of the library is identified by its position in the array. The split-synthesis procedure allows for the rapid generation of huge numbers of peptide sequences in a single mixture through the repetition of a simple divide, couple and recombine process. For example the complete library of all tetrapeptides contains 20^4 (160 000) different sequences and is prepared as follows. A batch of resin is divided into 20 equal portions and each of the 20 proteinogenic amino acids is coupled to a different aliquot. When coupling is complete excess reagent is removed by a series of washing steps and then the resins are recombined and thoroughly mixed. The N^α protecting group is then removed and the pool of resin, now containing 20^1 different species of H-Xxx-resin, is again washed. The cycle of dividing the resin into equal portions, quantitatively coupling the amino acids separately onto each aliquot, washing the resin, recombining and mixing the resin aliquots and then cleaving the N^α protecting group is repeated three more times to give a combinatorial library containing 20^4 members. The final step of the process is removal of side-chain protecting groups and, if an appropriate linker is present, cleavage from the resin. The key outcome of the 'split-synthesis' process is that each component bead of the resin contains only a single member of the library, the 'one bead–one peptide' concept, although there may be multiple copies of the same compound on an individual bead. The process can be performed manually or it may be automated by using sophisticated instrumentation. The isolation and identification of active compounds from the mixture is more complex than from the array method and requires one of a variety of intricate deconvolution or encoding strategies, sorting or microsequencing procedures.

The intentional generation of mixtures of organic compounds is an anathema to traditional chemists and the approach of parallel synthesis in which compounds are produced individually by using advanced robotic techniques and then characterised by standard methods, may ultimately prove to be the preferred route for lead compound identification and optimisation.

10.5 The stereoselective synthesis of α-amino acids

10.5.1 Overview

The synthesis of α-amino acids in enantiomerically pure form has been the subject of prodigious effort by chemists. The state of the art in 1989 was described in the monograph by Williams.[96] Another excellent review, compiled by Duthaler in 1994,[97] focuses on subsequent developments in the field, and progress in amino-acid synthesis is comprehensively reviewed by Barrett and colleagues in the series of annual Specialist Periodical Reports.[1e] Reviews on particular aspects of amino-acid synthesis have covered topics such as the asymmetric synthesis of arylglycines,[98] the incorporation of stable isotopes[99] and uses of free-radical reactions[100] and of chromium carbene complexes.[101] The present discussion will emphasise those methods which are of proven utility and will review some recent highlights; coverage is necessarily selective. Some of the more widely used general approaches for defining the chiral centre at the α-carbon atom of the α-amino acids are indicated in Scheme 10.8.

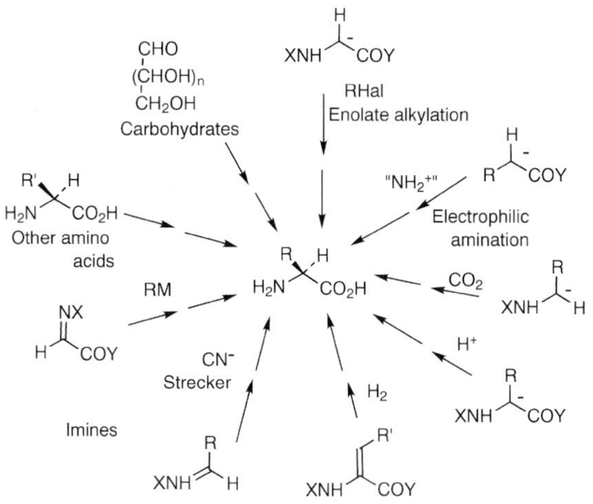

Scheme 10.8   General approaches for enantioselective synthesis of α-amino acids.

10.5.2 Asymmetric alkylations of glycine enolate equivalents

The chiral auxiliary-based alkylation of glycine enolates devised by Schöllkopf is a very long-established and widely-used method for the

asymmetric synthesis of α-amino acids (Scheme 10.9).[102] An enantiomerically pure amino acid (commonly valine) is coupled to glycine to

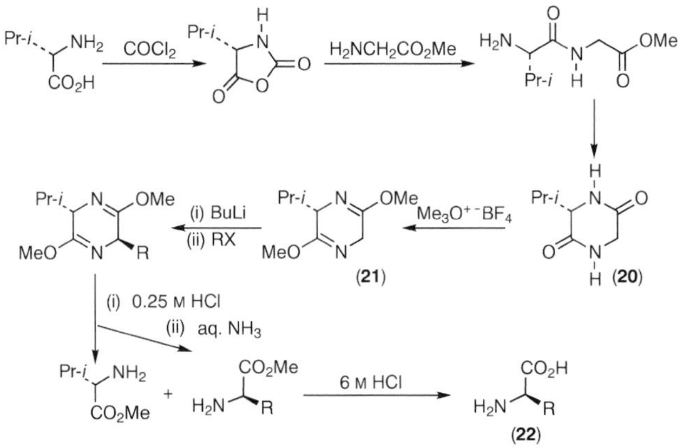

Scheme 10.9   Glycine enolate alkylation using Schöllkopf's bis-lactim ether.

give the corresponding diketopiperazine (20). Methylation of both oxygen
atoms gives the bis-lactim ether (21), which is commercially available in
both enantiomeric forms and can be prepared reliably on the 45 g scale.[103]
Treatment with a strong base leads to regioselective deprotonation of
the glycine fragment to give an enolate equivalent; this is then able to
undergo diastereoselective reaction with various electrophiles, showing
a strong preference for reaction on that face of the anion which is *anti*
to the sterically demanding *iso*-propyl group. Thus the use of L-valine as
the chiral auxiliary leads to the selective synthesis of D-amino acids (22);
these are produced with enantiomeric excess (ee) values ranging from
91%–95% for phenylalanine to 60%–65% for propargylglycine. Greater
selectivity may be achieved by replacing the valine residue by the bulkier,
but more costly, *tert*-leucine. The alkylation product is readily hydrolysed
by using mild acid to give a mixture of the methyl esters of valine and
the newly synthesised amino acid. The separation of these two products
can sometimes be accomplished by distillation, but in other cases a tedious
chromatographic separation may be necessary. Fairly vigorous acid hydrolysis is normally used to cleave the methyl ester and so Schöllkopf's
approach is not suitable for the synthesis of amino acids with acid-
sensitive side-chains or which are liable to acid-induced racemisation. The
use of alanine in place of glycine in the Schöllkopf approach is a very
good way of making α-alkylalanines; these amino acids do not have an
enolisable hydrogen atom at the chiral centre and so cannot racemise
during the final hydrolysis.

Some other synthetic equivalents of the glycine α-anion are **23–30** and are compared in Table 10.1. Deprotonation of the parent compounds

Precursors to chiral glycine enolate equivalents.

(**23–30**, R = H), followed by reaction with alkyl halides or other electrophiles, leads to the diastereoselective formation of the alkylation products (**23–30**, R = alkyl) which arise by attack on the more accessible faces of the enolate-type intermediates. In many cases these products may be subjected to further deprotonation and alkylation steps, leading to α,α-dialkylglycine derivatives in which the *second* alkyl group enters from the more accessible face of the new enolate.

Seebach's imidazolidinones (**23**)[104] have been used widely and successfully but suffer from the limitation that hydrolytic conversion of the alkylation product into the amino acid typically requires the use of hot mineral acid. The analogous oxazolidinones (**31**) are useful intermediates for the α-alkylation of amino acids using the principle of 'self-regeneration' of stereocentres (Scheme 10.10).[105] Condensation of the sodium salt of an α-amino acid with trimethylacetaldehyde, followed by treatment with benzoyl chloride, gives a high yield of the *cis*-1,3-oxazolidin-5-one. Formation of the corresponding enolate destroys the original chiral centre at C(4) but stereoselective alkylation can then be

Table 10.1 Comparison of diastereoselective benzylation reactions of some glycine enolate equivalents. Reactions were performed by using BnBr unless indicated otherwise

Starting material	Base/ solvent	Diastereoisomeric excess of crude benzylation product	Isolated yield of major benzylation product (%)	Cleavage conditions[a] to give phenylalanine	Ref.
21, Scheme 10.9	BuLi THF	91–95[b]	81[b]	0.25 M HCl, 10h (to give H-Phe-OMe)	102
25	NaN(SiMe$_3$)$_2$ THF	Not reported	70	CF$_3$CO$_2$H/CH$_2$Cl$_2$ then H$_2$/Pd	108
27	BuLi THF- HMPA	95[c]	93	0.5 M HCl then LiOH/THF/H$_2$O	109
28	LiN(SiMe$_3$)$_2$ THF	98	88	Ce^{4+}/MeCN/H$_2$O then 6 M HCl, 24 h, reflux	111
29	LDA THF/LiCl	91	79[d]	0.5 M NaOH, 2 h, reflux	112
30	LiN(SiMe$_3$)$_2$ THF/LiCl	99	85	6 M HCl, reflux	113

[a] Room temperature unless stated otherwise.
[b] Diastereoselectivity in reference 102 was estimated from the enantiomeric excess of the phenylalanine obtained by hydrolysis.
[c] Using BnI.
[d] 65% on recrystallisation.

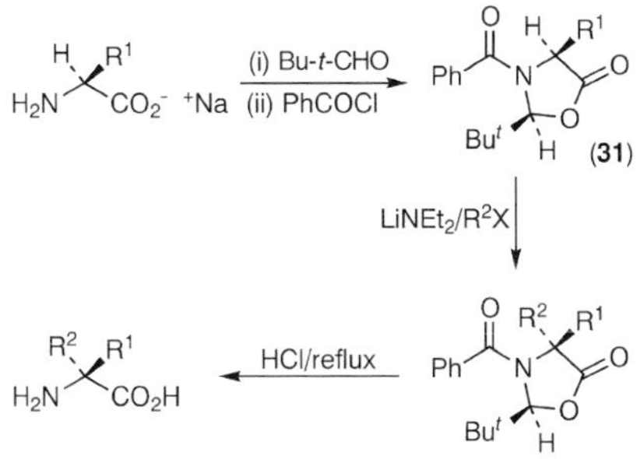

Scheme 10.10 α-Alkylation of α-amino acids with 'self-regeneration' of the stereocentre.

performed with about 80%–96% diastereoisomeric excess (de) under control by the C(2) chiral centre. The overall effect is of substitution of the α-hydrogen of the amino acid by an alkyl group with retention of configuration. Modifications to this procedure, including the use of ferrocenecarboxaldehyde in place of trimethylacetaldehyde, have allowed alkylations to be performed in greater than 98% de.[106] Seebach has also introduced the dihydroimidazoles (**24**), which can be hydrolysed to amino-acid methyl esters by using cold, dilute, aqueous trifluoroacetic acid.[107]

Williams's oxazinones (**25**)[108] may be reductively cleaved at both benzylic positions, for example by lithium in liquid ammonia, to give N-Boc-protected amino acids. If the corresponding N-benzyloxycarbonyl-oxazinones (**26**) are employed then reductive cleavage and N-deprotection may be achieved in a single step by catalytic hydrogenolysis.

Oppolzer's camphorsultams[109] (**27**) are highly amenable to purification by crystallisation and undergo cleavage under mild conditions. A direct comparison showed that Oppolzer's method yielded ^{13}C-labelled Boc-Leu-OH of 99.7% ee, compared with 97.2%–97.4% for the Schöllkopf approach; the overall yields were similar.[110] A disadvantage of the Oppolzer method is that alkylations of lithiated **27** usually require a co-solvent such as the toxic hexamethylphosphoramide (HMPA). However, examples using the less hazardous N,N'-dimethylpropyleneurea (DMPU) have been reported and reactive alkylating agents (e.g. PhCH₂I) can give excellent results in a dichloromethane–water solvent system with phase-transfer catalysis.

The diketopiperazines (**28**) were designed by Davies to show greater crystallinity than the bis-lactim ethers (**21**).[111] The 4-methoxybenzyl groups are considered to have an important role in transmitting stereochemical information from the valine chiral centre to ensure excellent facial selectivity in the attack of the electrophile on the glycine enolate equivalent. Conversion of the alkylation products of **28** into amino acids involves oxidative cleavage of the 4-methoxybenzyl groups by Ce^{4+}, followed by vigorous acid hydrolysis of the amide bonds. This produces the new amino acid as a mixture with valine, the chiral auxiliary, so that as in the Schöllkopf approach a careful separation is needed at a late stage in the synthesis.

The pseudoephedrine glycinamides (**29**) of Myers are a recent and attractive development.[112] Hydrolysis in aqueous alkali occurs more readily than for simple amides and this is attributed to neighbouring-group participation by the hydroxyl group of the pseudo-ephedrine auxiliary. A process research group from Merck has reported that the *cis*-aminoindanol derivatives (**30**) are readily available and undergo highly diastereoselective alkylation. Addition of lithium chloride significantly increases the yield of alkylated product; this was thought to be because the enolates otherwise exist as relatively unreactive aggregates.[113]

Most of the recorded alkylations of glycine enolate equivalents involve the use of primary, benzylic or allylic halides. Secondary alkyl halides have been used, but reactions of tertiary alkyl or aryl halides are unlikely to be useful because the former are prone to elimination and the latter are poor electrophiles.

The electrophilic attack of an aldehyde upon a chiral glycine enolate equivalent allows the possibility of forming two new chiral centres in a selective way. An early example of such a reaction was Schöllkopf's synthesis of D-threonine (**33**; Scheme 10.11) in which the use of a titanium

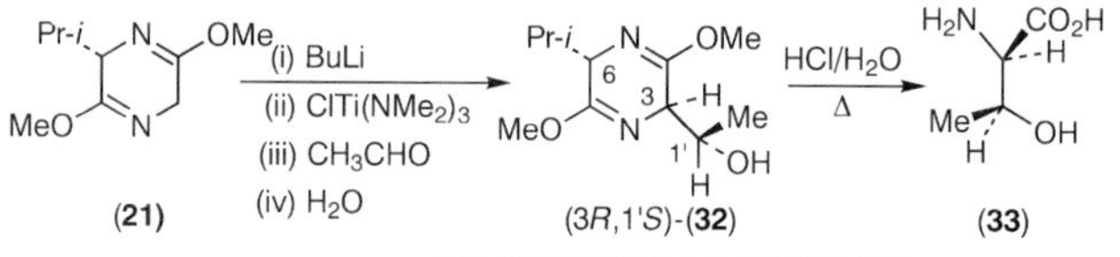

Scheme 10.11   Synthesis of D-threonine by means of Schöllkopf's bis-lactim ether.

enolate ensured good control of the configuration at the β-carbon of threonine.[114] The stereoselectivity was accounted for by the preference for a chair-like, six-membered transition state in which the methyl group of acetaldehyde occupies a pseudo-equatorial position.

The amino acid MeBmt (**35**) is an unusual derivative of threonine which has a crucial role in the biological activity of the immunosuppressive undecapeptide cyclosporine. It has been synthesised by several research groups, including those of Evans[115] and Seebach,[116] both of whom chose to use asymmetric aldol reactions of the enal (**34**) with chiral glycine enolate equivalents (Scheme 10.12). A review of MeBmt syntheses has been published.[117]

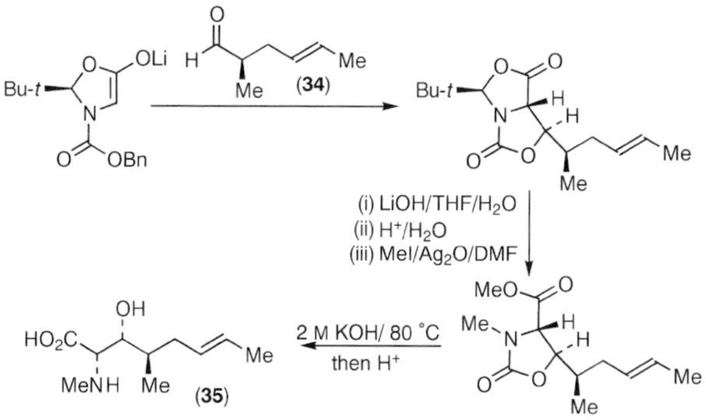

Scheme 10.12 Seebach's synthesis of MeBmt by means of an asymmetric aldol reaction.

Although chiral auxiliaries have proved to be very useful for controlling asymmetric carbon–carbon bond-forming reactions, it is desirable to develop highly selective *catalytic* asymmetric processes. These can effect the transformation of an achiral starting material into an enantiomerically enriched product whilst avoiding the need to attach a stoichiometric amount of enantiomerically pure auxiliary, so reducing both the number of synthetic steps and the required amount of chiral reagent. Such processes are difficult to devise, since it is necessary that there be a high number of turnovers. The desired catalysed process should also be much more rapid than the alternative catalysed process which leads to the unwanted enantiomer, and the uncatalysed reaction to yield the racemate. Several examples of systems which can provide synthetically useful levels of asymmetric induction are now known. The protected glycine imine **36** undergoes highly enantioselective Pd-catalysed asymmetric allylation (Scheme 10.13);[118] however, the scope of this reaction appears to be limited. Chiral phase-transfer catalysed alkylations of the same glycine imine (**36**), using quaternary ammonium derivatives of the cinchona alkaloids, are effective in a variety of cases where the electrophile is either an alkyl halide or a Michael acceptor. For example, Corey *et al.* used methyl acrylate to prepare the glutamic acid derivative **37**.[119] Lygo and Wainwright have benzylated **36** in 91% ee by using benzyl bromide in conjunction with a similar phase-transfer catalyst.[120] The experimental simplicity and the versatility of this approach should ensure that it has a significant future role in amino-acid synthesis.

Trost has explored the use of asymmetric palladium-catalysed allylation to distinguish between the enantiotopic acetate groups of

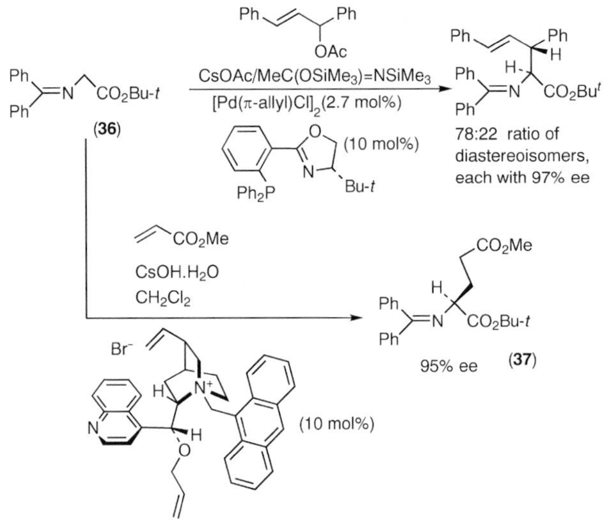

Scheme 10.13 Catalytic asymmetric allylation and alkylation of a glycine imine.

1,1-diacetates [e.g. **38** (Scheme 10.14)]. Reaction with the alanine α-anion equivalent **39** established the configuration of the difficult quaternary chiral centre in a total synthesis of sphingofungin F (**40**), a naturally occurring serine palmitoyltransferase inhibitor with broad-spectrum antifungal activity.[121]

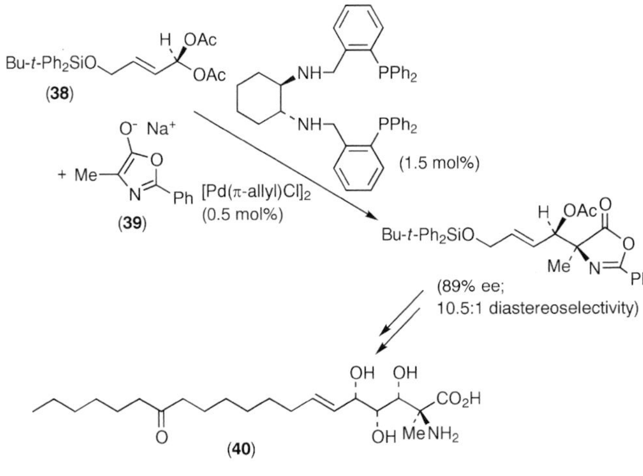

Scheme 10.14 Trost's synthesis of sphingofungin F (**40**).

Methyl isocyanoacetate has been used successfully as a glycine enolate equivalent in asymmetric aldol reactions catalysed by chiral gold–phosphine complexes.[122] The product hydroxyisonitriles spontaneously cyclise to 4,5-dihydro-1,3-oxazoles, acidic hydrolysis of which yields α-amino-β-hydroxy acids.

10.5.3 Asymmetric amination of enolates and related reactions

This approach can be used to form amino acids even where no suitable electrophilic derivative of the side-chain is available, as is the case with the α-arylglycines. However, there are only a few nitrogen electrophiles which can be used for enolate amination (41–43) and all of these compounds are somewhat unstable.

Di-*tert*-butyl azodicarboxylate was one of the first reagents to be used for electrophilic amination. It is a commercially available, relatively stable, yellow solid which gives protected hydrazino acids (44) in high diastereo-

isomeric excesses (e.g. 96% for $R = PhCH_2$) upon reaction with chiral enolates derived from Oppolzer's chiral sulfonamides.[123] Conversion of the products (44) into the corresponding amino acids requires several steps, including acid hydrolysis of the N-Boc groups and hydrogenolysis of the N—N bond. Oppolzer later used the blue 1-chloro-1-nitrosocyclohexane in conjunction with enolates containing his camphorsultam auxiliary (Scheme 10.15).[124] Acidic work-up yielded crude N-hydroxyamino acid derivatives of at least 99% de which were easily converted into free amino acids by reductive cleavage of the N—O bond (Zn/H^+), followed by chiral auxiliary removal $(LiOH/H_2O)$. All of the amino acids which were studied, including phenylalanine and

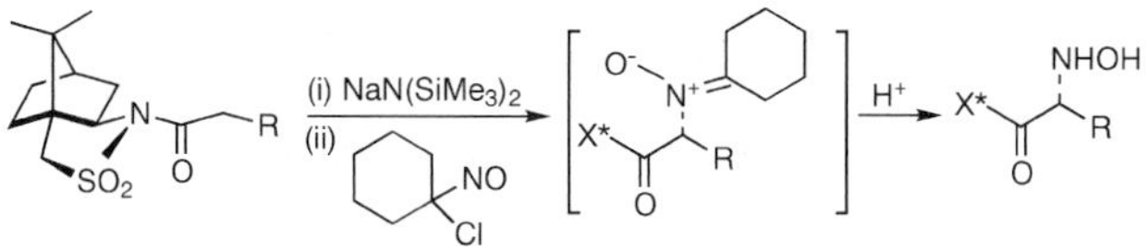

Scheme 10.15.

phenylglycine, were obtained in at least 99% ee with minimal need for purification and no separation of stereoisomers.

The azido (N_3) group can be reduced to the amino group under a variety of mild conditions, for example by catalytic hydrogenation. Extensive studies by Evans showed that treatment with 2,4,6-tri(*iso*-propylbenzene-sulfonyl) azide, followed by work-up with acetic acid, was the preferred procedure for electrophilic azido transfer to chiral imide enolates (Scheme 10.16).[125] This direct azidation seems preferable to the alternative two-step procedure, involving asymmetric electrophilic bromination of the corresponding boron enolates, followed by S_N2 displacement with azide anion.

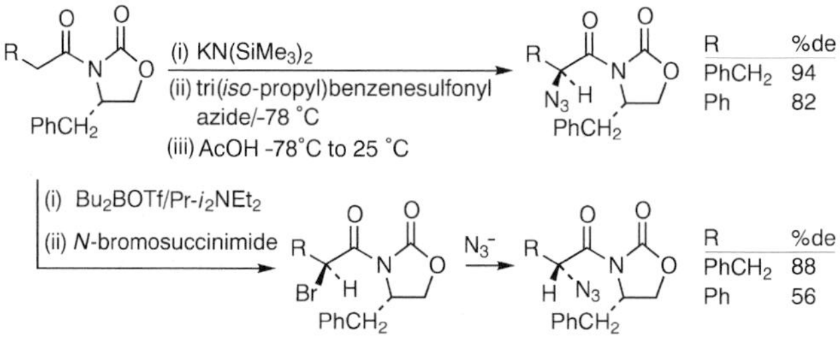

Scheme 10.16.

The Merck group has reported the synthesis of amino acids via reaction of the BocN(Li)OTs reagent with enolates bearing the *cis*-aminoindanol-based chiral auxiliary (Scheme 10.17).[126] The products arising from the asymmetric step were isolated as pure diastereoisomers. The deprotection step was a simple acid hydrolysis, albeit under rather harsh conditions,

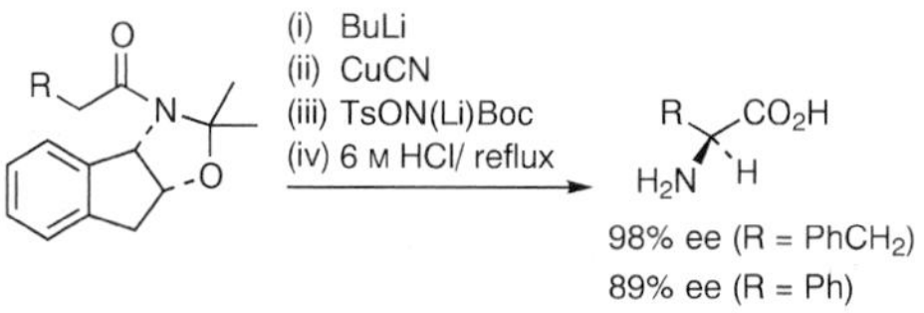

Scheme 10.17.

which may account for the apparent partial racemisation of the readily enolisable phenylglycine.

N-Chlorocarbamates may be used as electrophilic aminating agents for styrenes by using Sharpless's osmium-catalysed asymmetric amino-hydroxylation reaction. The reaction has very good enantioselectivity and shows a regiochemical preference for forming primary alcohols, oxidation of which yields N-protected arylglycines (Scheme 10.18).[127]

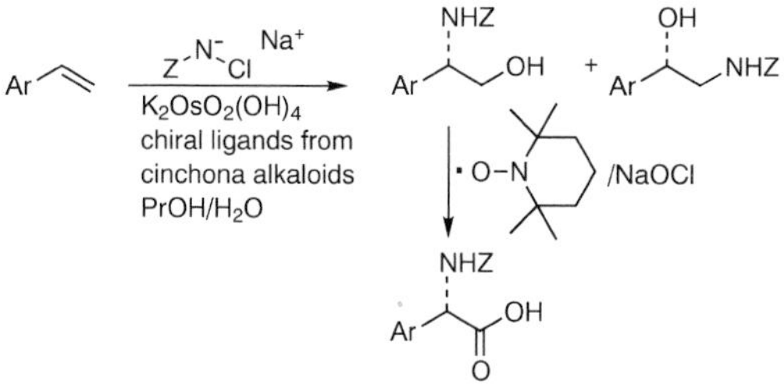

Scheme 10.18.

10.5.4   Asymmetric carboxylation of carbanions

So far this approach has not been much used. A source of difficulty is the need to convert a suitably functionalised precursor into a carbanion which does not possess enolate character. This may be achieved by tin–lithium exchange on an organostannane, but many organotin compounds are toxic and several steps may be required to form the stannane. Fournet *et al.* have devised a route (Scheme 10.19) which converts CO_2 into L-methionine in at least 95% ee within 40 min and so could be used to incorporate ^{11}C (half-life 20 min) which decays by positron emission and is used in medical imaging.[128]

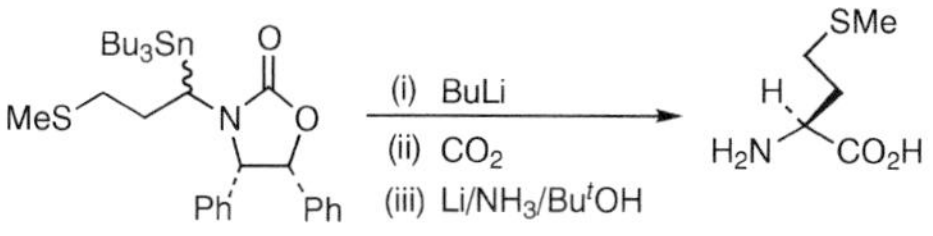

Scheme 10.19   Rapid synthesis of L-methionine to allow use of $^{11}CO_2$.

Park and Beak have generated benzylic carbanions by using the BuLi/(−)-sparteine chiral base (Scheme 10.20).[129] Carboxylation with CO_2 gave a protected derivative of (R)-phenylglycine. Intriguingly, the use of methyl chloroformate as the electrophile yielded predominantly an amino ester of the opposite S-configuration. By using other electrophiles these authors were also able to prepare β- and γ-amino acid derivatives.

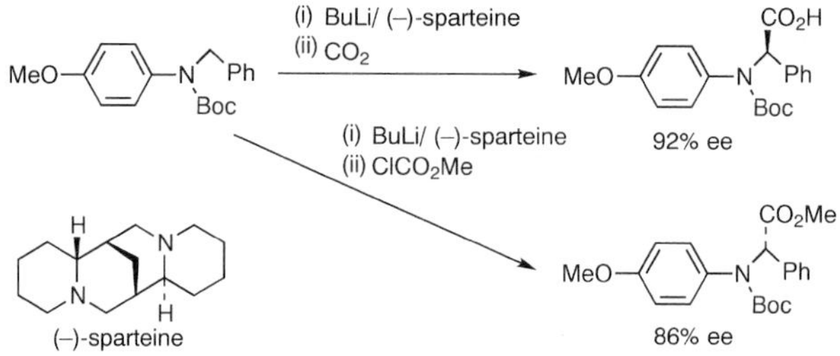

Scheme 10.20 Asymmetric carboxylation of a benzylic anion generated by means of a chiral base.

10.5.5 Asymmetric protonation of enolates

Asymmetric protonation of enolates has the potential, so far largely unrealised, to provide a simple and direct way of generating enantiomerically enriched amino acids from racemic precursors. A difficulty is that proton transfer between heteroatoms can be extremely rapid, so that O-protonation of an enolate by a chiral acid may initially give an enol. Furthermore, if protonation is reversible then thermodynamic control will take over.[130] Duhamel *et al.* conducted pioneering studies involving enolate formation from ($\pm$)-N-(benzylidene)amino esters with lithium amides, followed by reprotonation using O,O-diacyl tartaric acids; unfortunately, ee values did not exceed 70%.[131] Tabcheh *et al.* have reported that imines (**45**; Scheme 10.21), prepared from racemic amino esters and an enantiomerically pure ketone, could be obtained in diastereoisomerically enriched form (79% to 98% or more de, prior to separation of the diastereoisomers) by treatment with base and then acid.[132]

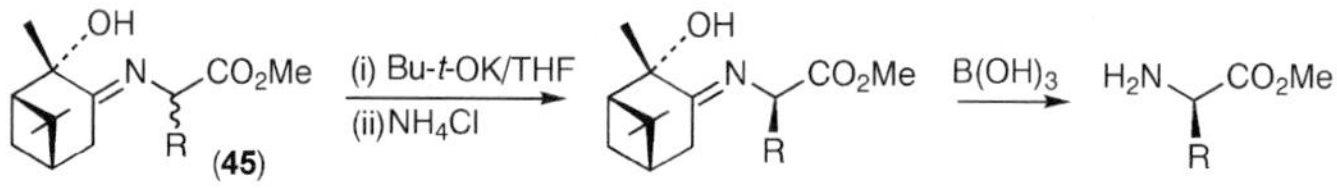

Scheme 10.21 Asymmetric protonation of a chiral enolate.

10.5.6 Asymmetric hydrogenation

The hydrogenation of dehydro amino-acid derivatives in the presence of a modified Wilkinson's catalyst with chiral, bidentate phosphine ligands was one of the early triumphs of catalytic asymmetric synthesis and became the basis of the industrial synthesis of (S)-3,4-dihydroxyphenylalanine (L-DOPA), used in the treatment of Parkinson's disease.[133] The very thorough early studies concerning the scope of such hydrogenations have already been much reviewed.[96, 134] Ghosh and Liu used an asymmetric hydrogenation to establish the configuration of the α-amino-acid unit in the synthesis of the nucleoside antibiotic (+)-sinefungin (48) from D-ribose (Scheme 10.22).[135] The enamide (46) was

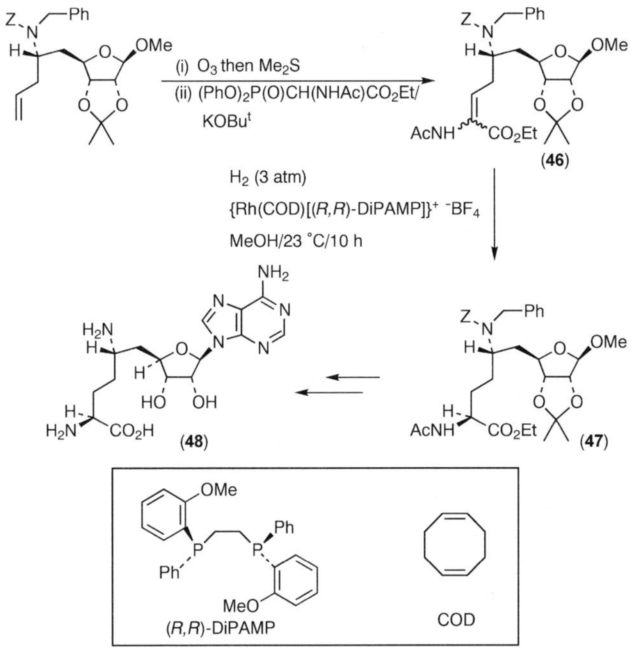

Scheme 10.22 Synthesis of (+)-sinefungin by means of catalytic asymmetric hydrogenation.

obtained as a 1:5 mixture of (E)- and (Z)-geometrical isomers by a Horner–Emmons reaction. Hydrogenation of this mixture by using a 'Rh-DiPAMP'-based catalyst gave (47) as a single stereoisomer, as determined by HPLC and nuclear magnetic resonance (NMR).

The asymmetric hydrogenation method has been extended to the synthesis of 1-azacycloalkane-2-carboxylic acids of ring sizes 5–16; ee values are low for the proline and pipecolic acid derivatives, but are at least 94% for the seven-, eight- and nine-membered homologues.[136]

10.5.7 Asymmetric addition of carbon nucleophiles to imines and related species

Auxiliary controlled nucleophilic addition of cyanide or its equivalent to imines in an asymmetric Strecker-type reaction continues to attract attention. Stereodifferentiating templates include glycosyl-amines,[137] α-phenylglycinol[138] and D-glyceraldehyde,[139] and asymmetric inductions up to 86% de have been reported for the addition of trimethylsilyl cyanide to the enantiopure imines. Davis and co-workers have developed a sulfinimine-mediated asymmetric route[140] which led to a concise preparation of (R)-(4-methoxy-3,5-dihydroxyphenyl)glycine (49; Scheme 10.23),[141] a valuable synthetic derivative of the central amino acid

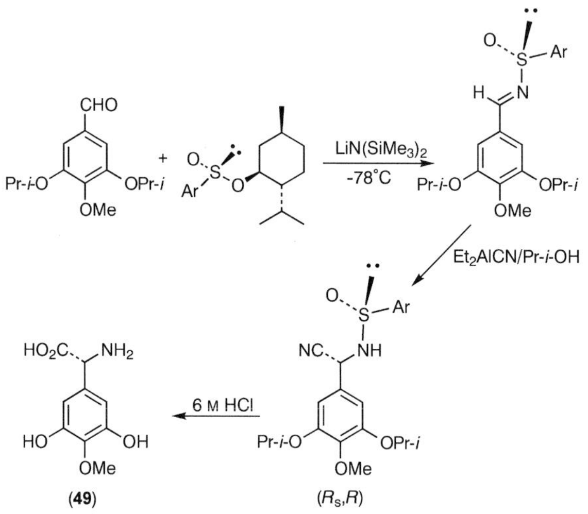

Scheme 10.23.

of vancomycin. The reaction of Et_2AlCN/Pr-i-OH with the sulfinimine is highly diastereoselective (greater than 92% de) and is believed to proceed via a six-membered chair-like transition state in which the alkoxide $EtAl(OPr^i)CN$ coordinates to the sulfinyl oxygen, enabling intramolecular delivery of cyanide to the *Si*-face of the C=N and formation of the (R)-aminonitrile.

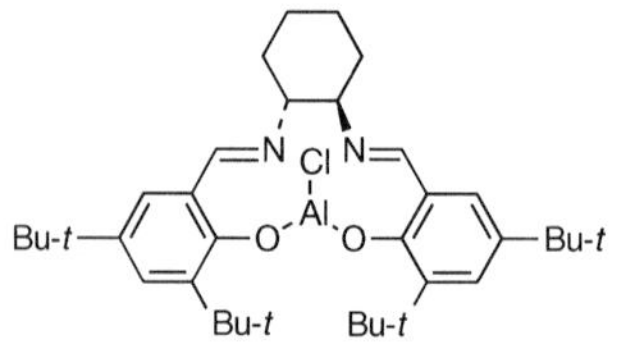

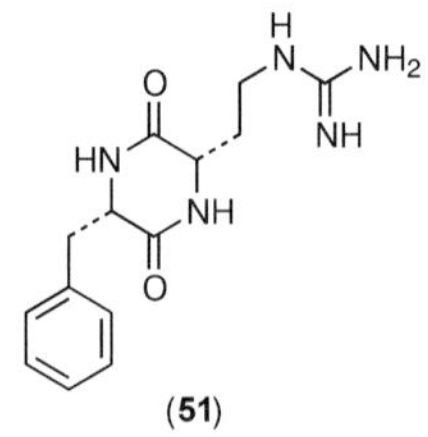

Asymmetric induction in the Strecker reaction can also be achieved using chiral catalysis. A chiral (salen)Al(III) complex (50)[142] and a diketopiperazine (51)[143] have both been reported to catalyse the stereoselective addition of HCN to N-substituted imines, and the (S)-amino nitrile products were obtained in good yield with generally moderate to good enantiomeric purity.

Several optically active α,α-disubstituted amino acids have been prepared via an intramolecular Strecker reaction.[144] The stereochemistry of the disubstituted compounds is induced by the chirality of an optically pure α-amino acid which is used as the amine component in the Strecker reaction. In, for example, the synthesis of α-benzylserine (Scheme 10.24),

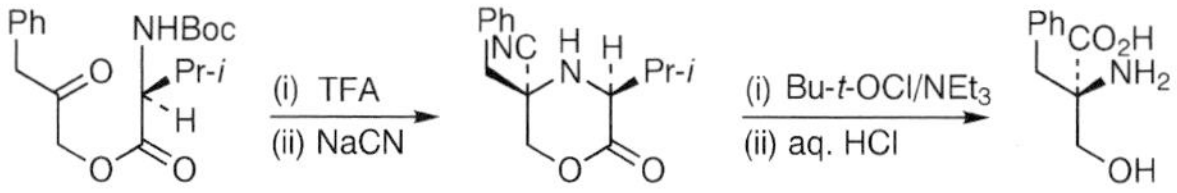

Scheme 10.24.

phenyl acetol (PhCH$_2$COCH$_2$OH) is condensed with Boc-L-valine to form an amino ester which reacts in the presence of NaCN to form the cyclic amino nitrile. The attack of cyanide is to the less-hindered Si-face of the C=N bond to form the (S,S)-diastereoisomer as the major product (88% de).

In a variant of the Mannich reaction, an organoboronic acid is added to the iminium salt formed between a secondary amine and an α-keto acid to yield N-substituted α-amino acids.[145] The reaction can be carried out stereoselectively if a chiral amine is used to form the iminium salt: enantiopure β,γ-unsaturated amino acids have been obtained by this route. Introduction of an alkenyl side-chain with the unsaturation at the γ,δ-position can be achieved by reaction of the iminium ion derived from condensation of glyoxal and N-methyl (R)-phenylglycinol with allylic silanes.[146] The reaction proceeds with complete stereoselectivity to give the products with a *trans* arrangement of the phenyl and allyl substituents. The analogous reaction with enoxysilanes [CH$_2$=C(CH$_3$)—OSiMe$_3$] yields, after cleavage of the chiral template, the N-methyl-γ-keto-α-amino acid. An example of a highly diastereoselective addition of a Grignard reagent to a carbohydrate-derived imine will be discussed later (section 10.5.9, Scheme 10.32).

10.5.8   Synthetic transformations of amino-acid starting materials

Some common amino acids with side-chain functionality, such as L-serine, L-aspartic and L-glutamic acids, are particularly suitable for manipulation to give more unusual amino acids in enantiomerically pure form. Jackson *et al.* have used organometallic derivatives of serine as synthetic equivalents of the alanine β-anion in reactions with electrophiles such as acyl, aryl and allyl halides (Scheme 10.25).[147] Use of an aryl iodide under a balloon of carbon monoxide leads to a carbonyl- ative coupling to form a ketone; this allowed the preparation of the tryptophan metabolite kynurenine [L-β(2-aminobenzoyl)alanine] in two steps from 2-iodoaniline without the need to protect the aromatic amino group.[148]

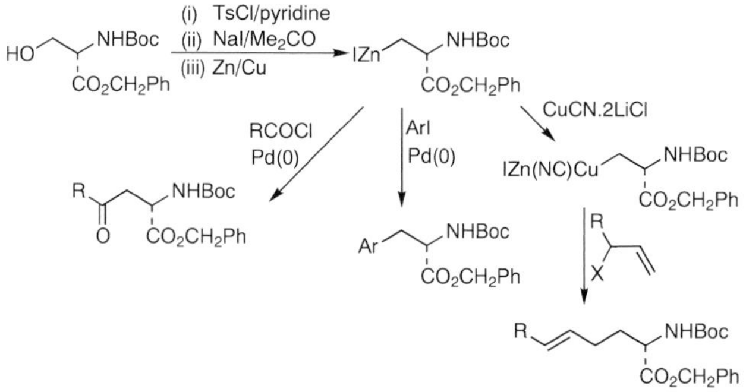

Scheme 10.25   Organometallic derivatives of serine used as alanine β-anion equivalents.

Replacement of the hydroxyl group of a suitably protected serine derivative by a good leaving group can allow nucleophilic substitution reactions to be performed at the β-carbon atom, thus providing a synthetic equivalent of the alanine β-cation. One pitfall in this approach is a possible elimination of both the α-proton and the leaving group to give a dehydroamino acid. Dugave and Ménez have employed the N-trityl protecting group to prevent α-proton abstraction and performed substitution using the anions of dialkyl malonates and similar soft carbon nucleophiles to displace iodide ion.[149] Use of thiolate nucleophiles allowed the synthesis of differentially protected lanthionines (**53**; Scheme 10.26).[150] These monosulfide analogues of cystine are present in a group of antibacterial, cytotoxic and immunomodulating peptides which are known as the lantibiotics. The cyclisation of the iodide (**52**) to the

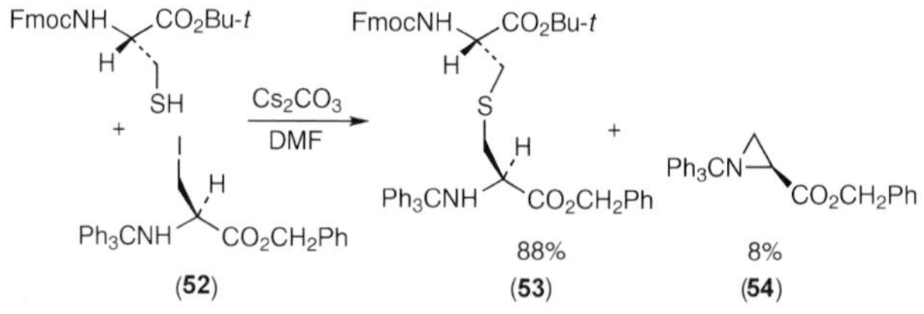

Scheme 10.26.

aziridine (54) competed with intermolecular nucleophilic displacement and became the dominant pathway when the use of harder nucleophiles (e.g. F^-, N_3^-, CN^- or $PhCH_2NH_2$) was attempted.

The serine-derived β-lactones[151] (55; Scheme 10.27) and N-Boc-activated aziridines[152] (56; Scheme 10.28) are elegantly designed alanine

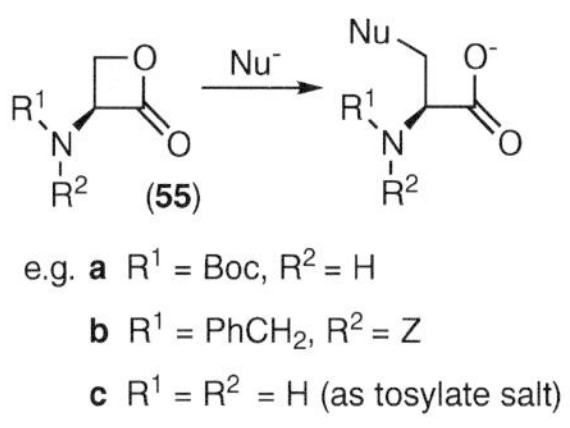

e.g. **a** R^1 = Boc, R^2 = H

   b R^1 = $PhCH_2$, R^2 = Z

   c R^1 = R^2 = H (as tosylate salt)

Scheme 10.27.

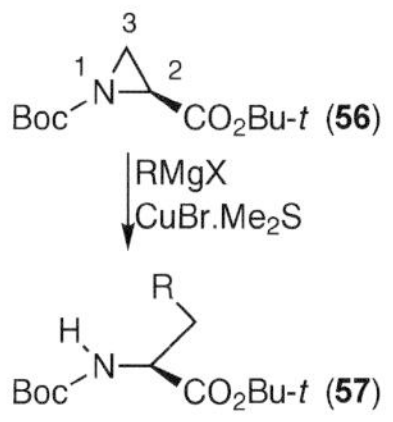

Scheme 10.28.

β-cation equivalents. They are disfavoured from elimination on stereo-electronic grounds and undergo release of ring strain on opening by nucleophiles. β-Lactone formation from N-protected serine derivatives may be achieved by the Mitsunobu reaction in 40%–50% yield;[153] unfortunately, decarboxylative elimination to form enamines competes with the desired cyclisation.

The β-lactones gave high yields in reactions with thiolate, azide, cyanide, amine and halide nucleophiles; with these weakly basic nucleophiles the α-amino group could even be 'protected' as its p-toluenesulfonate salt (55c), so that nucleophilic ring-opening led directly to free amino acids. The N-alkoxycarbonyl-protected β-lactones could also be opened satisfactorily by Grignard reagents in the presence of copper(I) salts.

Baldwin et al. dehydrated Ser-OBu-t by using diethoxytriphenylphosphorane to give the aziridine (56) in somewhat variable yield (25%–69%, depending on the batch of reagent used).[152] The attack on the aziridine ring by Grignard reagents in the presence of copper(I) occurred almost exclusively at C(3), provided that the temperature was kept sufficiently low. Yields of protected alkylglycines were typically in the range 50%–85%.

Aldehydes derived from serine can act as equivalents of the serine β-cation. In manipulating such compounds it is necessary to avoid using any synthetic intermediates which are 1,3-dicarbonyl compounds, as these would readily racemise by enolisation. One solution is to protect the carboxyl function as a bicyclic orthoester, which may later be cleaved by treatment with trifluoroacetic acid followed by hydrolysis using caesium carbonate in aqueous methanol (Scheme 10.29).[154] Addition of

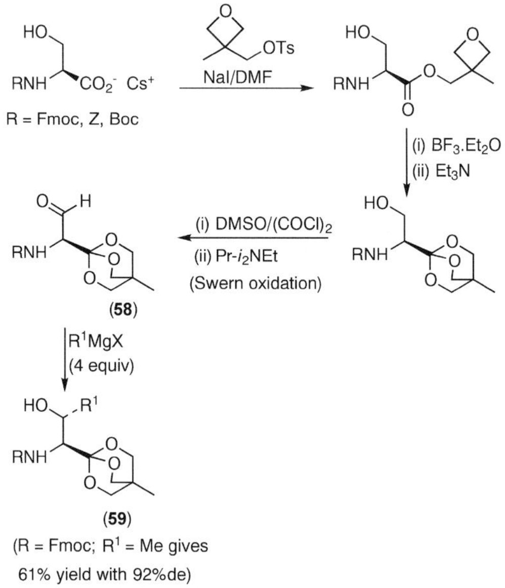

Scheme 10.29.

Grignard reagents to the carbonyl groups of the aldehydes (**58**) occurs with generally moderate yields and moderate to good stereoselectivity in favour of the *threo*-(2*S*,3*R*) diastereoisomer (**59**), consistent with the nonchelated Felkin–Anh model of attack (Figure 10.1). Complete

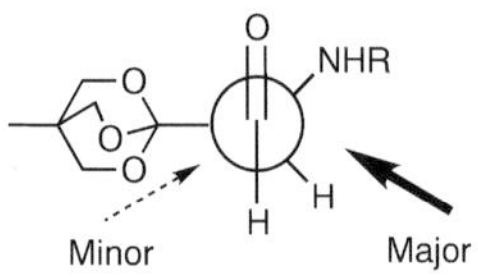

Figure 10.1.

separation of the diastereoisomeric products is not always straightforward. Inversion of the configuration at the centre bearing the hydroxyl
group affords the *erythro*-(2S,3S) diastereoisomers and can be achieved
by oxidation to the corresponding ketone under Swern conditions,
followed by reduction using $LiBH_4$.

Several research groups have employed aldehydes in which the
carbonyl group is derived not from the side-chain of serine but from
the carboxylic acid group. In particular, the 'Garner aldehyde' (**60**),
whose preparation has been described in *Organic Syntheses*, has been

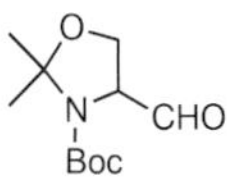

(60)

widely used in natural product synthesis.[155] Zhu *et al.* have commented
that although diastereoselectivities observed in additions to the Garner
aldehyde are generally moderate to good, the preferred face of attack is
somewhat reagent dependent. By contrast aldehyde **61** (Scheme 10.30)
showed strong tendencies to undergo nucleophilic attack in accord with

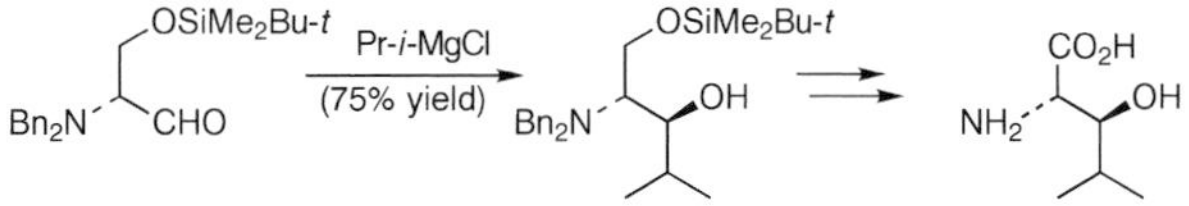

Scheme 10.30.

the Felkin–Anh model.[156] Derivatives of D-serine are used to prepare
L-amino acids [e.g. (2S,3S)-β-hydroxyleucine (**63**)]; this is because the
side-chain hydroxymethyl group [C(3) of serine] has to be oxidised to give
the new carboxylic acid group [C(1) of the product], so that in effect
two of the substituents at the α-carbon are interchanged.

The kainoids are a group of naturally occurring pyrrolidine derivatives
which are conformationally restricted analogues of glutamic acid and
which have powerful neuro-excitatory effects. They are useful tools for
neurological research and their synthesis was reviewed in 1996.[157] One
approach to these compounds (Scheme 10.31) involves the use of the
readily available *trans*-4-hydroxy-L-proline (**64**) as a starting material.
This was the tactic used for the preparation of acromelic acid A (**66**),
found in the poisonous Japanese mushroom *Clitocybe acromelalga*.[158]
The key features of this synthesis are the use of enol triflate (**65**) in a

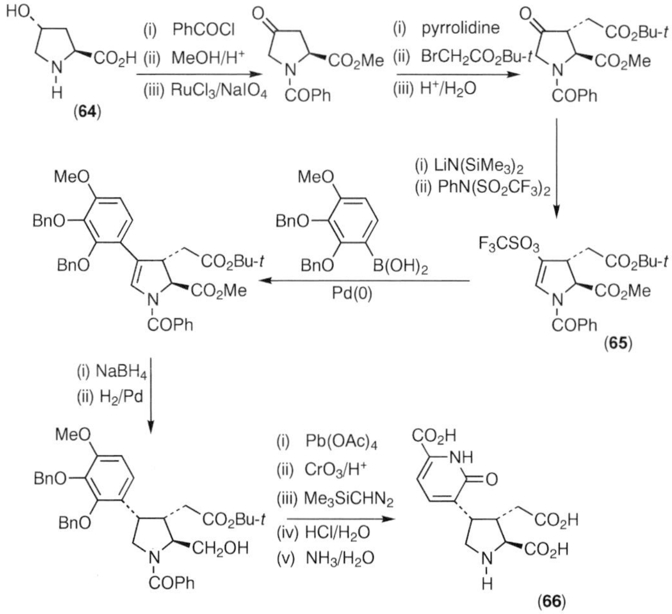

Scheme 10.31 Baldwin's synthesis of acromelic acid A, **66**.

Suzuki coupling, the temporary conversion of the carboxyl group at C(2) into a hydroxymethyl group so as to direct the hydrogenation in favour of the (4*S*)-diastereoisomer and a biomimetic route to the heterocyclic substituent at C(4), which is thought to be biosynthesised from L-DOPA.

10.5.9 Synthetic transformations of carbohydrate starting materials

Carbohydrates provide another useful source of enantiomerically pure starting materials. The chiral imine (**67**), derived from D-mannitol, can act as a glycine α-cation equivalent and has been used in a very efficient synthesis of L-vinylglycine (Scheme 10.32; 73% yield overall).[159] D-Mannitol was also used as a starting material for the synthesis of (−)-*allo*-coronamic acid (**70**; Scheme 10.33), a cyclopropane-containing plant metabolite.[160] The formation of the pyrazoline intermediate (**68**) and the subsequent photolytic extrusion of dinitrogen to form the cyclopropane (**69**) both occurred in quantitative yield to give single stereoisomers. A Corey–Winter reaction was used to remove two of the oxygen atoms from **69**.

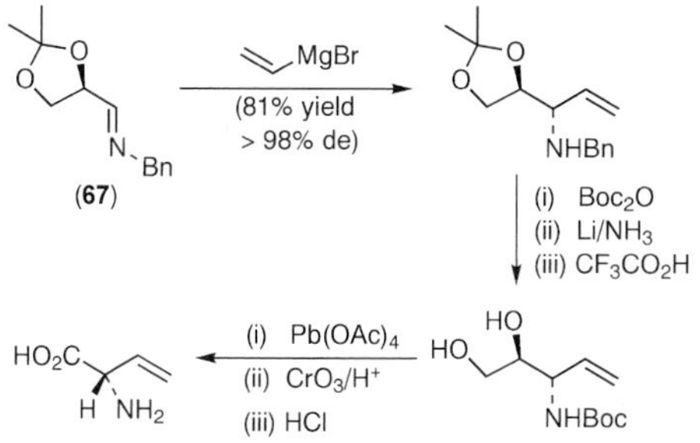

Scheme 10.32   Synthesis of vinylglycine from a D-mannitol derivative.

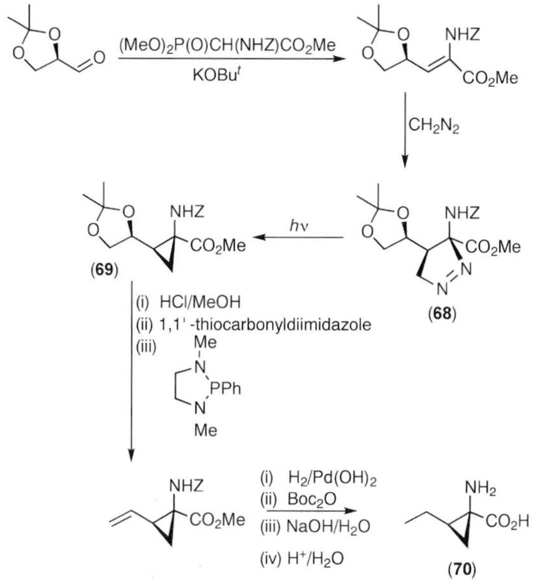

Scheme 10.33   Synthesis of (−)-*allo*-coronamic acid (**70**) from a D-mannitol derivative.

10.6   The stereoselective synthesis of β- and γ-amino acids

Both β- and γ-amino acids occur naturally and substantial reviews of β-amino-acid synthesis have appeared.[161] Work in our own laboratories has led to the preparation of α-substituted derivatives of β-alanine and γ-aminobutyric acid by the reaction of chiral enolates of the Evans type with nitrogen-containing electrophiles.[162] We used the benzotriazole derivative (**71**) as a synthetic equivalent of the Mannich reagent $[CH_2\!\!=\!\!NH_2]^+$

(71)

for β-amino-acid synthesis; bromoacetonitrile supplied the $CH_2CH_2NH_2$ group of the γ-homologues in a masked form.

N-Benzoyl-(2R,3S)-3-phenylisoserine (72) is the amino-acid side-chain of the anticancer drug taxol, present in the Pacific yew (*Taxus brevifolia*).

(72)

At present the most practical approach to the production of taxol is by attachment of synthetic 72 to the baccatin III nucleus, which is readily available from the needles of the European yew (*T. baccata*). Consequently many syntheses of the amino acid 72 have been reported. Some of these construct the bond between C(2) and C(3), for example, by the addition of chiral boron enolates of thioesters to N-trimethylsilylimines.[163] The most successful catalytic asymmetric routes have taken the carbon skeleton of cinnamic acid and added functionality to the C=C bond, for example by Jacobsen's epoxidation[164] or by the Sharpless aminohydroxylation.[165]

Amongst the natural γ-amino acids, statine (73) has attracted particular attention. It is an essential component of pepstatin, a hexapeptide inhibitor of aspartic acid proteases which is able to mimic the tetrahedral intermediate for amide bond hydrolysis. Numerous ways of making statine are known, including elaboration of L-leucine.[166] Greene and Nebois chose to use an asymmetric [2 + 2] cycloaddition of dichloroketene to a vinyl ether, followed by a Beckmann rearrangement as key steps in their preparation of (−)-statine (73; Scheme 10.34).[167]

10.7   Conclusions

More than 20 years of concentrated effort by research groups from all over the world has produced a variety of imaginative approaches to the synthesis of amino acids. For the majority of the nonprotein amino acids these preparative methods will be preferable to isolation from natural sources; however, there is still plenty of room for further innovation.

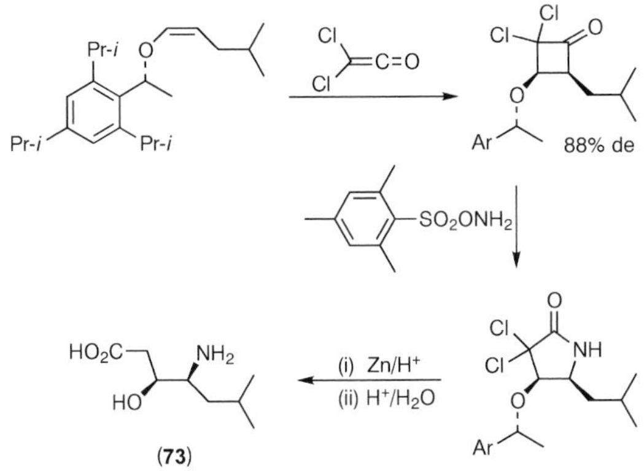

Scheme 10.34 Synthesis of $(-)$-statine, **73**.

Much of the present technology of amino-acid synthesis makes extensive use of protecting groups and chiral auxiliaries, leading to long synthetic sequences, often involving harsh deprotection conditions and a demoralising loss in relative molecular mass at the end of the synthesis (poor 'atom economy'). Many asymmetric reactions have been developed which can generate individual chiral centres as single stereoisomers within the limits of analysis, but difficulties still exist in controlling the formation of contiguous chiral centres. It is likely that future developments, including the application of combinatorial techniques and mass screening methods, will lead to better chiral catalysts: there is a particular need for new catalytic carbon–carbon bond-forming reactions. The availability of highly selective catalysts might enable synthetic chemists to dispense with protecting groups and to operate in water, the most environmentally acceptable of solvents. Living systems have managed to achieve this using only a fraction of the periodic table; chemists are lucky in that when they synthesise natural products *in vitro* they may make much fuller use of the fascinating elements in the heavier transition series.

References

1. (a) E. Atherton and R.C. Sheppard, 1989, *Solid Phase Peptide Synthesis: A Practical Approach*, IRL Press, Oxford; (b) J. Jones, 1991, *The Chemical Synthesis of Peptides*, Clarendon Press, Oxford; (c) M. Bodansky, 1993, *Principles of Peptide Synthesis*, 2nd edn, Springer-Verlag, Berlin; (d) G.A. Grant (ed.), 1992, *Synthetic Peptides, A User's Guide*, Freeman, New York; (e) J.S. Davies (series ed.), *Amino Acids, Peptides and Proteins*, The Royal Society of Chemistry, London; (f) G.B. Fields (ed.), 1997, *Methods in Enzymology*, vol. 289, Academic Press, San Diego.

2. (a) G. Barany, N. Kneib-Cordonier, and D.G. Mullen, 1987, *Int. J. Pept. Protein Res.*, 30, 705; (b) S.B.H. Kent, 1988, *Ann. Rev. Biochem.*, 57, 957; (c) J.M. Humphrey and A.R. Chamberlin, 1997, *Chem. Rev.*, 97, 2243.

3. (a) W. Rapp, L. Zhang, R. Häbich, and E. Bayer, 1988, in *Peptides, Proc. of 20th Eur. Pept. Symp.* (G. Jung and E. Bayer, eds.), Walter de Gruyter, Berlin, pp. 199; (b) E. Bayer, 1991, *Angew. Chem., Int. Ed. Engl.*, 30, 113.

4. M. Meldal, 1992, *Tetrahedron Lett.*, 33, 3077.

5. M. Kempe and G. Barany, 1996, *J. Am. Chem. Soc.*, 118, 7083.

6. (a) K. Barlos, D. Gatos, J. Kallitsis, G. Papaphotiu, P. Sotiriu, Y. Wenqing, and W. Schäfer, 1989, *Tetrahedron Lett.*, 30, 3943; (b) K. Barlos, D. Gatos, S. Kapolos, G. Papaphotiu, W. Schäfer, and Y. Wenqing, 1989, *Tetrahedron Lett.*, 30, 3947.

7. (a) A. Flörsheimer and B. Riniker, 1990, in *Peptides Proc. of 21st Eur. Pept. Symp.* (E. Giralt and D. Andreu, eds), ESCOM, Leiden, p. 131; (b) B. Riniker, A. Flörsheimer, H. Fretz, P. Sieber, and B. Kamber, 1993, *Tetrahedron*, 49, 9307.

8. M.F. Songster and G. Barany, 1997, in *Methods in Enzymology* (G.B. Fields, ed.), vol. 289, Academic Press, San Diego, p. 126.

9. H. Rink, 1987, *Tetrahedron Lett.*, 28, 3787.

10. (a) J. Green, O.M. Ogunjobi, R. Ramage, A.S.J. Stewart, S. McCurdy, and R. Noble, 1988, *Tetrahedron Lett.*, 29, 4341; (b) R. Ramage, J. Green, and A.J. Blake, 1991, *Tetrahedron*, 47, 6353.

11. L.A. Carpino, H. Shroff, S.A. Triolo, E.-S.M.E. Mansour, H. Wenschuh, and F. Albericio, 1993, *Tetrahedron Lett.*, 34, 7829.

12. (a) H. Kunz and H. Waldmann, 1984, *Angew. Chem., Int. Ed. Engl.*, 23, 71; (b) M.H. Lyttle and D. Hudson, 1992, in *Peptides: Chemistry and Biology, Proc. of 12th Amer. Pept. Symp.* (J. A. Smith and J.E. Rivier, eds.), ESCOM, Leiden, p. 583; (c) A. Loffet and H.X. Zhang, 1993, *Int. J. Pept. Protein Res.*, 42, 346.

13. H. Kunz and C. Unverzagt, 1984, *Angew. Chem., Int. Ed. Engl.*, 23, 436.

14. B.W. Bycroft, W.C. Chan, S.R. Chhabra, and N.D. Hone, 1993, *J. Chem. Soc., Chem. Commun.*, 778.

15. S.R. Chhabra, B. Hothi, D.J. Evans, P.D. White, B.W. Bycroft, and W.C. Chan, 1998, *Tetrahedron Lett.*, 39, 1603.

16. W.C. Chan, B.W. Bycroft, D.J. Evans, and P.D. White, 1995, *J. Chem. Soc., Chem. Commun.*, 2209.

17. T. Johnson, M. Quibell, D. Owen, and R.C. Sheppard, 1993, *J. Chem. Soc., Chem. Commun.*, 369.

18. M. Quibell, W.G. Turnell, and T. Johnson, 1994, *Tetrahedron Lett.*, 35, 2237.

19. (a) M. Quibell, W.G. Turnell, and T. Johnson, 1994, *J. Org. Chem.*, 59, 1745; (b) M. Quibell, L.C. Packman, and T. Johnson, 1995, *J. Am. Chem. Soc.*, 117, 11656.

20. T. Wöhr, F. Wahl, A. Nefzi, B. Rohwedder, T. Sato, X. Sun, and M. Mutter, 1996, *J. Am. Chem. Soc.*, 118, 9218.

21. J. Coste, D. Le-Nguyen, and B. Castro, 1990, *Tetrahedron Lett.*, 31, 205.

22. (a) V. Dourtoglou, B. Gross, V. Lambropoulou, and C. Zioudrou, 1984, *Synthesis*, 572; (b) R. Knorr, A. Trzeciak, W. Bannwarth, and D. Gillessen, 1989, *Tetrahedron Lett.*, 30, 1927.

23. I. Abdelmoty, F. Albericio, L.A. Carpino, B.M. Foxman, and S.A. Kates, 1994, *Lett. Pept. Sci.*, 1, 57.

24. L.A. Carpino, 1993, *J. Am. Chem. Soc.*, 115, 4397.

25. L.A. Carpino, A. El-Faham, C.A. Minor, and F. Albericio, 1994, *J. Chem. Soc., Chem. Commun.*, 201.

26. L.A. Carpino, M. Beyermann, H. Wenschuh, and M. Bienert, 1996, *Acc. Chem. Res.*, 29, 268.

27. H. Wenschuh, M. Beyermann, E. Krause, M. Brudel, R. Winter, M. Schümann, L.A. Carpino, and M. Bienert, 1994, *J. Org. Chem.*, 59, 3275.

28. H. Wenschuh, M. Beyermann, A. El-Faham, S. Ghassemi, L.A. Carpino, and M. Bienert, 1995, *J. Chem. Soc., Chem. Commun.*, 669.

29. H. Wenschuh, M. Beyermann, H. Haber, J.K. Seydel, E. Krause, M. Bienert, L.A. Carpino, A. El-Faham and F. Albericio, 1995, *J. Org. Chem.*, 60, 405.

30. (a) L.A. Carpino, D. Sadat-Aalaee, H.G. Chao, and R.H. DeSelms, 1990, *J. Am. Chem. Soc.*, 112, 9651; (b) L.A. Carpino, E.-S.M.E. Mansour, and D. Sadat-Aalaee, 1991, *J. Org. Chem.*, 56, 2611.

31. C. Kaduk, H. Wenschuh, M. Beyermann, K. Forner, L.A. Carpino, and M. Bienert, 1995, *Lett. Pept. Sci.*, 2, 285.

32. L.A. Carpino and A. El-Faham, 1995, *J. Am. Chem. Soc.*, 117, 5401.

33. J.W. Perich, 1997, in *Methods in Enzymology* (G.B. Fields, ed.), vol. 289, Academic Press, San Diego, p. 245.

34. (a) E.A. Kitas, J.D. Wade, R.B. Johns, J.W. Perich, and G.W. Tregear, 1991, *J. Chem. Soc., Chem. Commun.*, 338; (b) E.A. Kitas, R. Knorr, A. Trzeciak, and W. Bannwarth, 1991, *Helv. Chim. Acta*, 74, 1314.

35. (a) J.W. Perich and E.C. Reynolds, 1991, *Int. J. Pept. Protein Res.*, 37, 572; (b) J.W. Perich, M. Ruzzene, L.A. Pinna, and E.C. Reynolds, 1994, *Int. J. Pept. Protein Res.*, 43, 39.

36. E.A. Ottinger, L.L. Shekels, D.A. Bernlohr, and G. Barany, 1993, *Biochemistry*, 32, 4354.

37. W.D.F. Meutermans and P.F. Alewood, 1996, *Tetrahedron Lett.*, 37, 4765.

38. P. White and J. Beythien, 1996, in *Innovation and Perspectives in Solid Phase Synthesis and Combinatorial Libraries, 4th Int. Symp.* (R. Epton, ed.), Mayflower Scientific, Birmingham, 557.

39. (a) R.M. Valerio, P.F. Alewood, R.B. Johns, and B.E. Kemp, 1989, *Int. J. Pept. Protein Res.*, 33, 428; (b) R.M. Valerio, J.W. Perich, E.A. Kitas, P.F. Alewood, and R.B. Johns, 1989, *Aust. J. Chem.*, 42, 1519; (c) E.A. Kitas, J.W. Perich, G.W. Tregear, and R.B. Johns, 1990, *J. Org. Chem.*, 55, 4181.

40. J.W. Perich, P.F. Alewood, and R.B. Johns, 1991, *Aust. J. Chem.*, 44, 233.

41. A. Otaka, K. Miyoshi, M. Kaneko, H. Tamamura, N. Fujii, M. Nomizu, T.R. Burke Jr, and P.P. Roller, 1995, *J. Org. Chem.*, 60, 3967.

42. T. Wakamiya, K. Sarutu, J.-I. Yasuoka, and S. Kusumoto, 1994, *Chem. Lett.*, 1099.

43. T. Johnson, L.C. Packman, C.B. Hyde, D. Owen, and M. Quibell, 1996, *J. Chem. Soc., Perkin Trans. 1*, 719.

44. P. Hormozdiari and D. Gani, 1996, *Tetrahedron Lett.*, 37, 8227.

45. (a) H. Kunz, 1987, *Angew. Chem., Int. Ed. Engl.*, 26, 294; (b) M. Meldal, 1994, *Curr. Opin. Struct. Biol.*, 4, 710; (c) J. Kihlberg, M. Elofsson, and L.A. Salvador, 1997, in *Methods in Enzymology* (G.B. Fields, ed.), vol. 289, Academic Press, San Diego, p. 221.

46. S.T. Anisfeld and P.T. Lansbury Jr, 1990, *J. Org. Chem.*, 55, 5560.

47. J. Offer, M. Quibell, and T. Johnson, 1996, *J. Chem. Soc., Perkin Trans. 1*, 175.

48. S. Peters, T. Bielfeldt, M. Meldal, K. Bock, and H. Paulsen, 1992, *J. Chem. Soc., Perkin Trans. 1*, 1163.

49. (a) E. Meinjohanns, M. Meldal, A. Schleyer, H. Paulsen, and K. Bock, 1996, *J. Chem. Soc., Perkin Trans. 1*, 985; (b) N. Mathieux, H. Paulsen, M. Meldal, and K. Bock, 1997, *J. Chem. Soc., Perkin Trans. 1*, 2359.

50. H. Paulsen, A. Schleyer, N. Mathieux, M. Meldal, and K. Bock, 1997, *J. Chem. Soc., Perkin Trans. 1*, 281.

51. E. Meinjohanns, M. Meldal, H. Paulsen, R.A. Dwek, and K. Bock, 1998, *J. Chem. Soc., Perkin Trans. 1*, 549.

52. (a) B.L. Karger and W.S. Hancock (eds.), 1996, *Methods in Enzymology*, vol. 270, Academic Press, San Diego; (b) B.L. Karger and W.S. Hancock (eds.), 1996, *Methods in Enzymology*, vol. 271, Academic Press, San Diego.

53. (a) R. Ramage and G. Raphy, 1992, *Tetrahedron Lett.*, 33, 385; (b) A.R. Brown, S.L. Irving and R. Ramage, 1993, *Tetrahedron Lett.*, 34, 7129.

54. M.A. Roggero, C. Servis, and G. Corradin, 1997, *FEBS Lett.*, 408, 285.

55. C. García-Echeverría, 1995, *J. Chem. Soc., Chem. Commun.*, 779.

56. (a) S. Funakoshi, H. Fukuda, and N. Fujii, 1991, *Proc. Natl. Acad. Sci. USA*, 88, 6981; (b) L.E. Canne, R.L. Winston, and S.B.H. Kent, 1997, *Tetrahedron Lett.*, 38, 3361.

57. C.T. Choma, G.T. Robillard, and D.R. Englebretsen, 1998, *Tetrahedron Lett.*, 39, 2417.

58. D.R. Englebretsen and P.F. Alewood, 1996, *Tetrahedron Lett.*, 37, 8431.

59. (a) H. Benz, 1994, *Synthesis*, 337; (b) P. Lloyd-Williams, F. Albericio, and E. Giralt, 1993, *Tetrahedron*, 49, 11065; (c) F. Albericio, P. Lloyd-Williams, and E. Giralt, 1997, in *Methods in Enzymology* (G.B. Fields, ed.), vol. 289, Academic Press, San Diego, p. 313.

60. J. Offer, T. Johnson, and M. Quibell, 1997, *Tetrahedron Lett.*, 38, 9047.

61. (a) T.W. Muir and S.B.H. Kent, 1993, *Curr. Opin. Biotechnol.*, 4, 420; (b) C.J.A. Wallace, 1995, *Curr. Opin. Biotechnol.*, 6, 403; (c) T.W. Muir, P.E. Dawson, and S.B.H. Kent, 1997, in *Methods in Enzymology* (G.B. Fields, ed.), vol. 289, Academic Press, San Diego, p. 266; (d) J. Wilken and S.B.H. Kent, 1998, *Curr. Opin. Biotechnol.*, 9, 412.

62. M. Schnölzer and S.B.H. Kent, 1992, *Science*, 256, 221.

63. M. Baca and S.B.H. Kent, 1993, *Proc. Natl. Acad. Sci. USA*, 90, 11638.

64. M.J. Williams, T.W. Muir, M.H. Ginsberg, and S.B.H. Kent, 1994, *J. Am. Chem. Soc.*, 116, 10797.

65. P.E. Dawson and S.B.H. Kent, 1993, *J. Am. Chem. Soc.*, 115, 7263.

66. D.R. Englebretsen, B.G. Garnham, D.A. Bergman, and P.F. Alewood, 1995, *Tetrahedron Lett.*, 36, 8871.

67. D.R. Englebretsen, C.T. Choma, and G.T. Robillard, 1998, *Tetrahedron Lett.*, 39, 4929.

68. T.W. Muir, M.J. Williams, M.H. Ginsberg, and S.B.H. Kent, 1994, *Biochemistry*, 33, 7701.

69. (a) H.K. Rau and W. Haehnel, 1998, *J. Am. Chem. Soc.*, 120, 468; (b) A.K. Wong, M.P. Jacobsen, D.J. Winzor, and D.P. Fairlie, 1998, *J. Am. Chem. Soc.*, 120, 3836.

70. (a) S.E. Cervigni, P. Dumy, and M. Mutter, 1996, *Angew. Chem., Int. Ed. Engl.*, 35, 1230; (b) E.C. Rodriguez, K.A. Winans, D.S. King, and C.R. Bertozzi, 1997, *J. Am. Chem. Soc.*, 119, 9905; (c) L.A. Marcauelle, E.C. Rodriguez, and C.R. Bertozzi, 1998, *Tetrahedron Lett.*, 39, 8417.

71. (a) K. Rose, 1994, *J. Am. Chem. Soc.*, 116, 30; (b) C. Rao and J.P. Tam, 1994, *J. Am. Chem. Soc.*, 116, 6975; (c) J. Shao and J.P. Tam, 1995, *J. Am. Chem. Soc.*, 117, 3893; (d) J.C. Spetzler and J.P. Tam, 1995, *Int. J. Pept. Protein Res.*, 45, 78; (e) T.D. Pallin and J.P. Tam, 1996, *J. Chem. Soc., Chem. Commun.*, 1345.

72. (a) C.-F. Liu and J.P. Tam, 1994, *Proc. Natl. Acad. Sci. USA*, 91, 6584; (b) C.-F. Liu and J.T. Tam, 1994, *J. Am. Chem. Soc.*, 116, 4149.

73. C.-F. Liu, C. Rao, and J.P. Tam, 1996, *J. Am. Chem. Soc.*, 118, 307.

74. L. Zhang, T.R. Torgerson, X.-Y. Liu, S. Timmons, A.D. Colosia, J. Hawiger, and J.P. Tam, 1998, *Proc. Natl. Acad. Sci. USA*, 95, 9184.

75. P. Botti, T.D. Pallin, and J.P. Tam, 1996, *J. Am. Chem. Soc.*, 118, 10018.

76. (a) M. Baca, T.W. Muir, M. Schnölzer, and S.B.H. Kent, 1995, *J. Am. Chem. Soc.*, 117, 1881; (b) L.E. Canne, A.R. Ferré-D'Amaré, S.K. Burley, and S.B.H. Kent, 1995, *J. Am. Chem. Soc.*, 117, 2998.

77. (a) P.E. Dawson, T.W. Muir, I. Clark-Lewis, and S.B.H. Kent, 1994, *Science*, 266, 776; (b) P.E. Dawson, 1997, in *Methods in Enzymology* (R. Horuk, ed.), vol. 287, Academic Press, San Diego, p. 34.

78. L.E. Canne, S.M. Walker, and S.B.H. Kent, 1995, *Tetrahedron Lett.*, 36, 1217.

79. P.E. Dawson, M.J. Churchill, M.R. Ghadiri, and S.B.H. Kent, 1997, *J. Am. Chem. Soc.*, 119, 4325.

80. T.M. Hackeng, C.M. Mounier, C. Bon, P.E. Dawson, J.H. Griffin, and S.B.H. Kent, 1997, *Proc. Natl. Acad. Sci. USA*, 94, 7845.

81. W. Lu, M.A. Qasim, and S.B.H. Kent, 1996, *J. Am. Chem. Soc.*, 118, 8518.

82. W. Lu, M.A. Starovasnik, and S.B.H. Kent, 1998, *FEBS Lett.*, 429, 31.

83. (a) H. Ueda, M.A. Siani, W. Gong, D.A. Thompson, G.G. Brown, and J.M. Wang, 1997, *J. Biol. Chem.*, 272, 24966; (b) C. Boshoff, Y. Endo, P.D. Collins, Y. Takeuchi, J.D. Reeves, V.L. Schweickart, M.A. Siani, T. Sasaki, T.J. Wiliams, P.W. Gray, P.S. Moore, Y. Chang, and R.A. Weiss, 1997, *Science*, 278, 290.

84. (a) K. Severin, D.H. Lee, J.A. Martinez, and M.R. Ghadiri, 1997, *Chem. Eur. J.*, 3, 1017; (b) S. Yao, I. Ghosh, R. Zutshi, and J. Chmielewski, 1997, *J. Am. Chem. Soc.*, 119, 10559; (c) S. Yao, I. Ghosh, R. Zutshi, and J. Chmielewski, 1998, *Angew. Chem., Int. Ed. Engl.*, 37, 478; (d) D.H. Lee, J.R. Granja, J.A. Martinez, K. Severin, and M.R. Ghadiri, 1996, *Nature*, 382, 525.

85. (a) J.A. Camarero and T.W. Muir, 1997, *J. Chem. Soc., Chem. Commun.*, 1369; (b) J.P. Tam and Y.-A. Lu, 1997, *Tetrahedron Lett.*, 38, 5599; (c) L. Zhang and J.P. Tam, 1997, *Tetrahedron Lett.*, 38, 4375; (d) L. Zhang and J.P. Tam, 1997, *J. Am. Chem. Soc.*, 119, 2363; (e) J.A. Camarero, J. Pavel, and T.W. Muir, 1998, *Angew. Chem., Int. Ed. Engl.*, 37, 347.

86. J.P. Tam, Y.-A. Lu, C.-F. Liu, and J. Shao, 1995, *Proc. Natl. Acad. Sci. USA*, 92, 12485.

87. C.-F. Liu, C. Rao, and J.P. Tam, 1996, *Tetrahedron Lett.*, 37, 933.

88. L.E. Canne, S.J. Bark, and S.B.H. Kent, 1996, *J. Am. Chem. Soc.*, 118, 5891.

89. L. Zhang and J.P. Tam, 1997, *Tetrahedron Lett.*, 38, 3.

90. Y. Shao, W. Lu, and S.B.H. Kent, 1998, *Tetrahedron Lett.*, 39, 3911.

91. S. Futaki, K. Sogawa, J. Maruyama, T. Asahara, M. Niwa, and H. Hojo, 1997, *Tetrahedron Lett.*, 38, 6237.

92. (a) T.W. Muir, D. Sondhi, and P.A. Cole, 1998, *Proc. Natl. Acad. Sci. USA*, 95, 6705; (b) K. Severinov and T.W. Muir, 1998, *J. Biol. Chem.*, 273, 16205.

93. (a) Á. Furka, F. Sebestyén, M. Asgedom, and G. Dibó, 1991, *Int. J. Pept. Protein Res.*, 37, 487; (b) K.S. Lam, S.E. Salmon, E.M. Hersh, V.J. Hruby, W.M. Kazmierski, and R.J. Knapp, 1991, *Nature*, 354, 82; (c) R.A. Houghten, C. Pinilla, S.E. Blondelle, J.R. Appel, C.T. Dooley, and J.H. Cuervo, 1991, *Nature*, 354, 84; (d) S.P.A. Fodor, J.L. Read, M.C. Pirrung, L. Stryer, A.T. Lu, and D. Solas, 1991, *Science*, 251, 767.

94. (a) J.N. Abelson (ed.), 1996, *Methods in Enzymology*, vol. 267, Academic Press, San Diego; (b) I.M. Chaiken and K.D. Janda (eds), 1996, *Molecular Diversity and Combinatorial Chemistry, Libraries and Drug Design*, American Chemical Society, Washington, DC; (c) A.W. Czarnik and S.H. DeWitt (eds), 1997, *A Practical Guide to Combinatorial Chemistry*, American Chemical Society, Washington, DC; (d) S.R. Wilson and A.W. Czarnik (eds), 1997, *Combinatorial Chemistry: Synthesis and Application*, Wiley, New York; (e) S. Cabilly (ed.), 1997, *Combinatorial Peptide Library Protocols*, Humana Press, New Jersey; (f) B.A. Bunin, 1998, *The Combinatorial Index*, Academic Press, San Diego; (g) N.K. Terrett, 1998, *Combinatorial Chemistry*, Oxford University Press, Oxford; (h) E.M. Gordon and J.F. Kerwin Jr (eds), 1998, *Combinatorial Chemistry and Molecular Diversity in Drug Discovery*, Wiley, New York.

95. (a) G. Jung and A.G. Beck-Sickinger, 1992, *Angew. Chem., Int. Ed. Engl.*, 31, 367; (b) M.A. Gallop, R.W. Barrett, W.J. Dower, S.P.A. Fodor, and E.M. Gordon, 1994, *J. Med. Chem.*, 37, 1233; (c) E.M. Gordon, R.W. Barrett, W.J. Dower, S.P.A. Fodor, and M.A. Gallop, 1994, *J. Med. Chem.*, 37, 1385; (d) K.D. Janda, 1994, *Proc. Natl. Acad. Sci. USA*, 91, 10779; (e) G. Lowe, 1995, *Chem. Soc. Rev.*, 24, 309; (f) N.K. Terrett, M. Gardner, D.W. Gordon, R.J. Kobylecki, and J. Steele, 1995, *Tetrahedron*, 51, 8135; (g) E.M. Gordon, 1995, *Curr. Opin. Biotechnol.*, 6, 624; (h) J.C. Chabala, 1995, *Curr. Opin. Biotechnol.*, 6, 632; (i) S.H. DeWitt and A.W. Czarnik, 1995, *Curr. Opin. Biotechnol.*, 6, 640; (j) P.H.H. Hermkens, H.C.J. Ottenheijm, and D. Rees, 1996, *Tetrahedron*, 52, 4527; (k) L.A. Thompson and

J.A. Ellman, 1996, *Chem. Rev.*, 96, 555; (l) J.S. Früchtel and G. Jung, 1996, *Angew. Chem., Int. Ed. Engl.*, 35, 17; (m) S.H. DeWitt and A.W. Czarnik, 1996, *Acc. Chem. Res.*, 29, 114; (n) R.W. Armstrong, A.P. Combs, P.A. Tempest, S.D. Brown, and T.A. Keating, 1996, *Acc. Chem. Res.*, 29, 123; (o) J.A. Ellman, 1996, *Acc. Chem. Res.*, 29, 132; (p) E.M. Gordon, M.A. Gallop, and D.V. Patel, 1996, *Acc. Chem. Res.*, 29, 144; (q) W.C. Still, 1996, *Acc. Chem. Res.*, 29, 155; (r) F. Balkenhohl, C. von dem Bussche-Hünnefeld, A. Lansky, and C. Zechel, 1996, *Angew. Chem., Int. Ed. Engl.*, 35, 2289; (s) C.D. Floyd, C.N. Lewis, and M. Whittaker, 1996, *Chem., Brit.*, 32 (March), 31; (t) M.J. Plunkett and J.A. Ellman, 1997, *Sci. Am.*, 276 (April), 55; (u) P.L. Meyers, 1997, *Curr. Opin. Biotechnol.*, 8, 701; (v) J.C. Hogan Jr, 1997, *Nature Biotechnol.*, 15, 328; (w) K.S. Lam, M. Lebl, and V. Krchnák, 1997, *Chem. Rev.*, 97, 411; (x) M.C. Pirrung, 1997, *Chem. Rev.*, 97, 473; (y) M. Lebl and V. Krchnák, 1997, in *Methods in Enzymology* (G.B. Fields, ed.), vol. 289, Academic Press, San Diego, p. 336; (z) P. Wentworth Jr and K.D. Janda, 1998, *Curr. Opin. Biotechnol.*, 9, 109 and references cited therein.

96. R.M. Williams, 1989, *Synthesis of Optically Active α-Amino Acids*, Pergamon Press, Oxford.

97. R.O. Duthaler, 1994, *Tetrahedron*, 50, 1540.

98. R.M. Williams and J.A. Hendrix, 1992, *Chem. Rev.*, 92, 889.

99. N.M. Kelly, A. Sutherland, and C.L. Willis, 1997, *Natural Product Reports*, 14, 205.

100. C.J. Easton, 1997, *Chem. Rev.*, 97, 53.

101. L.S. Hegedus, 1995, *Acc. Chem. Res.*, 28, 299.

102. (a) U. Schöllkopf, U. Groth, and C. Deng, 1981, *Angew. Chem., Int. Ed. Engl.*, 20, 798; (b) U. Schöllkopf, 1983, *Pure Appl. Chem.*, 55, 1799.

103. S.D. Bull, S.G. Davies, and W.O. Moss, 1998, *Tetrahedron: Asymmetry*, 9, 321.

104. D. Seebach, A.R. Sting, and M. Hoffmann, 1996, *Angew. Chem., Int. Ed. Engl.*, 35, 2709.

105. D. Seebach and A. Fadel, 1985, *Helv. Chim. Acta*, 68, 1243.

106. F. Alonso, S.G. Davies, A.S. Elend, and J.L. Haggitt, 1998, *J. Chem. Soc., Perkin Trans. 1*, 257.

107. D. Seebach and M. Hoffmann, 1998, *Eur. J. Org. Chem.*, 1337.

108. R.M. Williams and M.-N. Im, 1991, *J. Am. Chem. Soc.*, 113, 9276.

109. W. Oppolzer, R. Moretti, and C. Zhou, 1994, *Helv. Chim. Acta*, 77, 2363.

110. L. Lankiewicz, B. Nyassé, B. Fransson, L. Grehn, and U. Ragnarsson, 1994, *J. Chem. Soc., Perkin Trans. 1*, 2503.

111. S.D. Bull, S.G. Davies, S.W. Epstein, M.A. Leech, and J.V.A. Ouzman, 1998, *J. Chem. Soc., Perkin Trans. 1*, 2321.

112. A.G. Myers, J.L. Gleason, T. Yoon, and D.W. Kung, 1997, *J. Am. Chem. Soc.*, 119, 656.

113. J. Lee, W.-B. Choi, J.E. Lynch, R.P. Volante, and P.J. Reider, 1998, *Tetrahedron Lett.*, 39, 3679.

114. U. Schöllkopf, J. Nozulak, and M. Grauert, 1985, *Synthesis*, 55.

115. D.A. Evans and A.E. Weber, 1986, *J. Am. Chem. Soc.*, 108, 6757.

116. D. Blaser, S.Y. Ko, and D. Seebach, 1991, *J. Org. Chem.*, 56, 6230.

117. J.O. Durand and J.P. Genet, 1994, *Bull. Soc. Chim. Fr.*, 131, 612.

118. I.C. Baldwin, J.M.J. Williams, and R.P. Beckett, 1995, *Tetrahedron: Asymmetry*, 6, 1515.

119. E.J. Corey, M.C. Noe, and F. Xu, 1998, *Tetrahedron Lett.*, 39, 5347.

120. B. Lygo and P.G. Wainwright, 1997, *Tetrahedron Lett.*, 38, 8595.

121. B.M. Trost and C.B. Lee, 1998, *J. Am. Chem. Soc.*, 120, 6818.

122. (a) Y. Ito, M. Sawamura, and T. Hayashi, 1986, *J. Am. Chem. Soc.*, 108, 6405; (b) A. Togni and S.D. Pastor, 1990, *J. Org. Chem.*, 55, 1649; (c) V.A. Soloshonok and T. Hayashi, 1994, *Tetrahedron Lett.*, 35, 2713.

123. W. Oppolzer and R. Moretti, 1988, *Tetrahedron*, 44, 5541.

124. W. Oppolzer and O. Tamura, 1990, *Tetrahedron Lett.*, 31, 991.

125. D.A. Evans, T.C. Britton, J.A. Ellman, and R.L. Dorow, 1990, *J. Am. Chem. Soc.*, 112, 4011.

126. N. Zheng, J.D. Armstrong, J.C. McWilliams, and R.P. Volante, 1997, *Tetrahedron Lett.*, 38, 2817.

127. K.L. Reddy and K.B. Sharpless, 1998, *J. Am. Chem. Soc.*, 120, 1207.

128. F. Jeanjean, N. Pérol, J. Goré, and G. Fournet, 1997, *Tetrahedron Lett.*, 38, 7547.

129. Y.S. Park and P. Beak, 1997, *J. Org. Chem.*, 62, 1574.

130. C. Fehr, 1996, *Angew. Chem., Int. Ed. Engl.*, 35, 2567.

131. L. Duhamel, S. Fouquay, and J.-C. Plaquevent, 1986, *Tetrahedron Lett.*, 27, 4975.

132. M. Tabcheh, C. Guibourdenche, L. Pappalardo, M.-L. Roumestant, and P. Viallefont, 1998, *Tetrahedron: Asymmetry*, 9, 1493.

133. (a) W.S. Knowles, M.J. Sabacky, B.D. Vineyard, and D.J. Weikauff, 1975, *J. Am. Chem. Soc.*, 97, 2567; (b) W.S. Knowles, 1986, *J. Chem. Educ.*, 63, 222.

134. J.D. Morrison (ed.), 1985, *Asymmetric Synthesis*, vol. 5, Academic Press, London, Ch 1–3.

135. A.K. Ghosh and W. Liu, 1996, *J. Org. Chem.*, 61, 6175.

136. K.C. Nicolaou, G.-Q. Shi, K. Namoto, and F. Bernal, 1998, *J. Chem. Soc., Chem. Commun.*, 1757.

137. (a) H. Kunz, W. Sager, D. Schanzenbach, and M. Decker, 1991, *Liebigs Ann. Chem.*, 649; (b) H. Kunz, W. Pfrengle, K. Rück, and W. Sager, 1991, *Synthesis*, 1039; (c) H. Kunz and C. Rück, 1993, *Angew. Chem., Int. Ed. Engl.*, 32, 336.

138. (a) T.K. Chakraborty, G.V. Reddy, and K.A. Hussain, 1991, *Tetrahedron Lett.*, 32, 7597; (b) T. Inaba, I. Kozono, M. Fujita, and K. Ogura, 1992, *Bull. Chem. Soc. Jpn.*, 65, 2359; (c) T.K. Chakraborty, K.A. Hussain, and G.V. Reddy, 1995, *Tetrahedron*, 51, 9179; (d) J. Zhu, J.-P. Bouillon, G.P. Singh, J. Chastanet, and R. Beugelmans, 1995, *Tetrahedron Lett.*, 36, 7081.

139. C. Cativiela, M.D. Díaz-de-Villegas, J.A. Gálvez and J.I. García, 1996, *Tetrahedron*, 52, 9563.

140. (a) F.A. Davis, R.E. Reddy, and P.S. Portonovo, 1994, *Tetrahedron Lett.*, 35, 9351; (b) F.A. Davis, P.S. Portonovo, R.E. Reddy, and Y. Chiu, 1996, *J. Org. Chem.*, 61, 440.

141. F.A. Davis and D.L. Fanelli, 1998, *J. Org. Chem.*, 63, 1981.

142. M.S. Sigman and E.N. Jacobsen, 1998, *J. Am. Chem. Soc.*, 120, 5315.

143. M.S. Iyer, K.M. Gigstad, N.D. Namdev, and M. Lipton, 1996, *J. Am. Chem. Soc.*, 118, 4910.

144. (a) S.-H. Moon and Y. Ohfune, 1994, *J. Am. Chem. Soc.*, 116, 7405; (b) M. Horikawa, T. Nakajima, and Y. Ohfune, 1997, *Synlett*, 253.

145. (a) N.A. Petasis, A. Goodman, and I.A. Zavialov, 1997, *Tetrahedron*, 53, 16463; (b) N.A. Petasis and I.A. Zavialov, 1997, *J. Am. Chem. Soc.*, 119, 445.

146. C. Agami, D. Bihan, and C. Puchot-Kadouri, 1996, *Tetrahedron*, 52, 9079.

147. (a) R.F.W. Jackson, N. Wishart, A. Wood, K. James, and M.J. Wythes, 1992, *J. Org. Chem.*, 57, 3397; (b) M.J. Dunn, R.F.W. Jackson, J. Pietruszka, and D. Turner, 1995, *J. Org. Chem.*, 60, 2210.

148. R.F.W. Jackson, D. Turner, and M.H. Block, 1997, *J. Chem. Soc., Perkin Trans. 1*, 865.

149. C. Dugave and A. Ménez, 1996, *J. Org. Chem.*, 61, 6067.

150. C. Dugave and A. Ménez, 1997, *Tetrahedron: Asymmetry*, 8, 1453.

151. (a) L.D. Arnold, R.G. May, and J.C. Vederas, 1988, *J. Am. Chem. Soc.*, 110, 2237; (b) L.D. Arnold, J.C.G. Drover, and J.C. Vederas, 1987, *J. Am. Chem. Soc.*, 109, 4649.

152. J.E. Baldwin, C.N. Farthing, A.T. Russell, C.J. Schofield, and A.C. Spivey, 1996, *Tetrahedron Lett.*, 37, 3761.

153. (a) S.V. Pansare, G. Huyer, L.D. Arnold, and J.C. Vederas, 1991, *Org. Synth.*, 70, 1; (b) S.V. Pansare, L.D. Arnold, and J.C. Vederas, 1991, *Org. Synth.*, 70, 10.

154. M.A. Blaskovich, G. Evindar, N.G.W. Rose, S. Wilkinson, Y. Luo, and G.A. Lajoie, 1998, *J. Org. Chem.*, 63, 3631.

155. P. Garner and J.M. Park, 1991, *Org. Synth.*, 70, 18.

156. T. Laïb, J. Chastanet, and J. Zhu, 1998, *J. Org. Chem.*, 63, 1709.

157. A.F. Parsons, 1996, *Tetrahedron.*, 52, 4149.

158. (a) J.E. Baldwin, S.J. Bamford, A.M. Fryer, M.P.W. Rudolph, and M.E. Wood, 1997, *Tetrahedron*, 53, 5233; (b) J.E. Baldwin, A.M. Fryer, G.J. Pritchard, M.R. Spyvee, R.C. Whitehead, and M.E. Wood, 1998, *Tetrahedron*, 54, 7465.

159. R. Badorrey, C. Cativiela, M.D. Díaz-de-Villegas, and J.A. Gálvez, 1997, *Synthesis*, 747.

160. J.M. Jiménez, J. Rifé, and R.M. Ortuño, 1996, *Tetrahedron: Asymmetry*, 7, 537.

161. (a) D.C. Cole, 1994, *Tetrahedron*, 50, 9517; (b) E. Juaristi (ed.), 1997, *Enantioselective Synthesis of β-Amino Acids*, Wiley-VCH, New York.

162. (a) E. Arvanitis, H. Ernst, A.A. Ludwig (*née* D'Souza), A.J. Robinson, and P.B. Wyatt, 1998, *J. Chem. Soc., Perkin Trans. 1*, 521; (b) S. Azam, A.A. D'Souza, and P.B. Wyatt, 1996, *J. Chem. Soc., Perkin Trans. 1*, 621.

163. C. Gennari, M. Carcano, M. Donghi, N. Mongelli, E. Vanotti, and A. Vulpetti, 1997, *J. Org. Chem.*, 62, 4746.

164. E.N. Jacobsen, L. Deng, Y. Furukawa, and L.E. Martínez, 1994, *Tetrahedron*, 50, 4323.

165. G. Li and K.B. Sharpless, 1996, *Acta Chem. Scand.*, 1996, 50, 649.

166. G. Veeresha and A. Datta, 1997, *Tetrahedron Lett.*, 38, 5223.

167. P. Nebois and A.E. Greene, 1996, *J. Org. Chem.*, 61, 5210.

11 Synthetic pathways to naturally occurring cyclic peptides

A.B. Tabor

11.1 Introduction

The importance of cyclic peptides in the fields of medicinal, biological and synthetic chemistry cannot be overestimated. Cyclisation of a linear peptide is an established and successful strategy for constraining the peptide to adopt a reduced number of conformations. Often, with a biologically active peptide this may lead to greater specificity, tighter binding at the relevant receptor, better bioavailability, or resistance to proteolysis. Conversely, many naturally occurring cyclic peptides have been discovered with potent antibacterial, antiviral and antineoplastic properties.

Over the past five to ten years, major advances have been made in the development of reagents and strategies for complex cyclic peptide synthesis. Simultaneously, the discovery of novel, cyclic peptide natural products continues unabated, presenting new challenges for the synthetic chemist, as well as deeper insights into the design of biologically active nonnatural cyclic peptides. The aim of this chapter is to provide an overview of the exciting advances both in reagents and in strategy, and to describe highlights in the total synthesis of naturally occurring cyclic peptides that have occurred over the past decade, giving pointers to the primary literature. Several excellent recent reviews, referred to in the text, should be consulted for more comprehensive coverage of individual classes of cyclic peptides and novel reagents.

11.2 Synthesis of cyclic peptides by macrolactamisation and disulfide bond formation

11.2.1 General considerations

In nature, cyclisation of peptides can be achieved by the formation of a disulfide bond, typically between two cysteine residues, or by amide bond formation. The amide bond may be formed between the *N*- and *C*-termini, or between lysine and glutamic or aspartic acid side-chains, or between the *N*- or *C*-terminus and an appropriate side-chain.

The majority of simple cyclic peptides contain only amide linkages, and even with more complex cyclic peptides, cyclisation by amide bond formation is frequently the best strategy. The process of cyclisation is, of course, entropically unfavourable, and there are potential problems of competing dimerisation or oligomerisation of the linear precursor, all of which may lead to low yields. In addition, activation of the *C*-terminal—COOH group carries with it the risk of racemisation at an adjacent chiral centre. It is advisable, therefore, where there is a choice of positions for ring disconnection, to consider certain factors:[1-3]

- the disconnection should be chosen in such a way that the macrolactamisation step does not involve sterically hindered (β-branched or α,α-disubstituted) amino acids, or *N*-alkylated amino acids, as these generally give poor yields during coupling reactions;
- the cyclisation will be facilitated if the linear peptide precursor adopts a turn, rather than an extended conformation; peptides containing residues such as glycine or proline will often cyclise rapidly; it has been suggested that best results will be obtained if such residues are in the middle of the linear precursor;[4]
- it has been observed that cyclisation between a D and an L- residue is favourable.[5]

Macrolactamisation of large, linear peptides in solution generally presents no problems, provided the cyclisation step is carried out at high dilution. However, for medium and small cyclic peptides, of six or fewer residues, the ring-closure may require the linear peptide to adopt an unfavourable conformation. The ease of cyclisation therefore tends to be very sequence-dependent: if the linear peptide does not contain residues, such as glycine or proline, that induce a turn structure, cyclisation may be very slow and compete with racemisation or cyclodimerisation.

11.2.2 Solution-phase segment condensation and macrolactamisation

The choice of reagents for activation of the *C*-terminus of a protected linear precursor is often crucial to the success of a synthetic strategy, and the difficulties often encountered have led to a number of powerful reagents being developed to effect segment condensation and macrolactamisation.[6]

The combinations of reagents most commonly used for the formation of amide bonds and the synthesis of linear peptides, DCC/HOBt and DCC/DMAP, are particularly slow and prone to give epimerisation at the *C*-terminus.[7] Reliable cyclisations with low levels of racemisation may be achieved by the azide method[2], through use of diphenylphosphoryl azide (DPPA, **1**; Scheme 11.1) (which forms an azide *in situ*);[8] however, this method can be extremely slow. Conversion of the *C*-terminus to a pentafluorophenyl (Pfp) ester has become the method of choice for many solution-phase macrolactamisation strategies, it generally giving good results with unhindered amino acids.[9] BOP-Cl **2** has

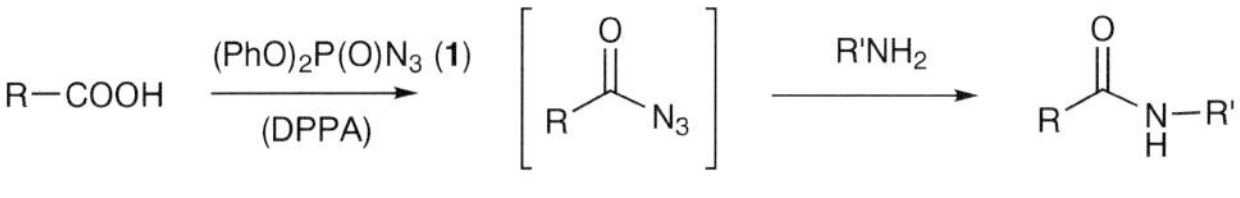

Scheme 11.1.

also proven an effective reagent for both coupling[10] and macrolactamisation;[3, 11] it has been postulated that this is because of intramolecular general base catalysis (Scheme 11.2).[12] The organophosphorus reagents diethylphosphonocyanidate [$(EtO)_2P(O)CN$, DEPC][13] and pentafluorophenyldiphenylphosphinate [$Ph_2P(O)OC_6F_5$, FDPP][14] have also been

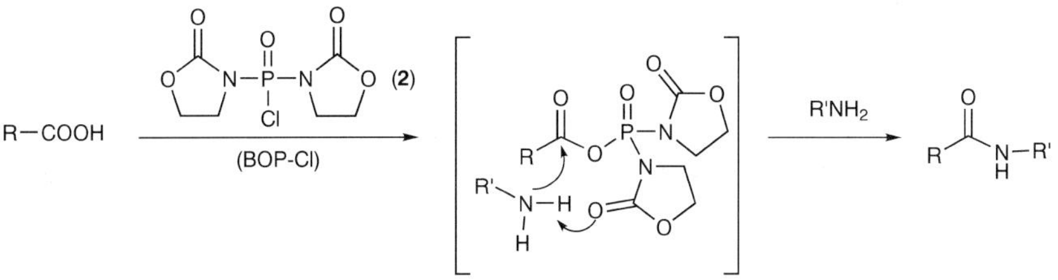

Scheme 11.2.

successfully used as coupling reagents.[15] FDPP, in particular, has given good results in macrolactamisations where other reagents have failed.[16]

Uronium reagents such as HBTU (**3**) and TBTU (**4**),[17] and phosphonium reagents such as BOP (**5**)[18] and PyBOP (**6**),[19] all of which form

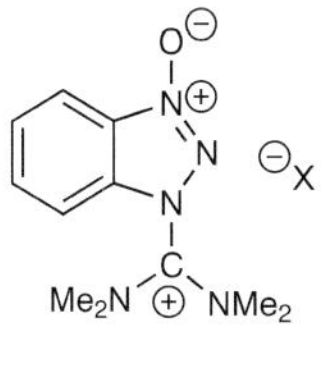

3, X⁻ = PF$_6$⁻, HBTU
4, X⁻ = BF$_4$⁻, TBTU

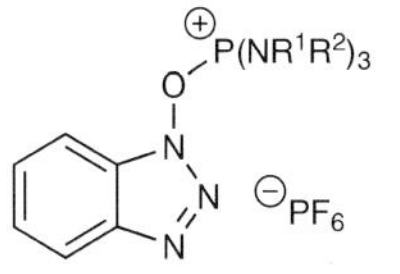

5, R^1 = R^2 = Me, BOP
6, R^1 = R^2 = -(CH$_2$)$_4$-, PyBOP

the activated HOBt ester *in situ*, have also been gainfully employed in linear and cyclic peptide syntheses. Although unacceptably high levels of racemisation can occasionally be encountered when using these reagents[7], such cyclisation reactions are usually rapid, and they have been extensively used for side-chain cyclisations.[20] However, in a recent comprehensive study by Carpino *et al.*, the azabenzatriazole-derived analogues, HATU (**7**), HAPyU (**8**) and PyAOP (**9**) were shown to be greatly superior in speed and yield of cyclisation, and generally gave

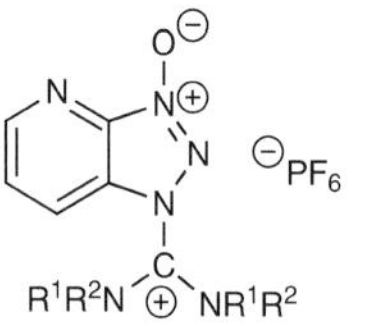

7, R¹ = R² = Me, HATU
8, R¹ = R² = -(CH₂)₄-, HAPyU

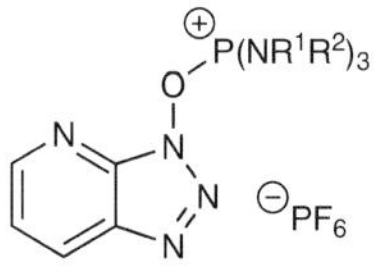

9, R¹ = R² = -(CH₂)₄- , PyAOP

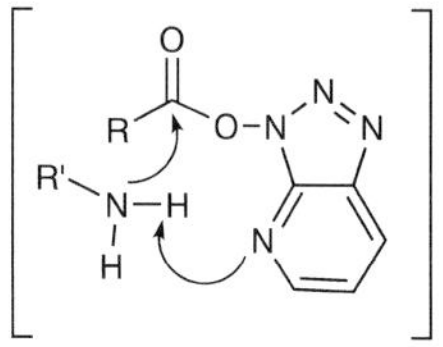

less than 10% *C*-terminal racemisation.[21] Again, this is thought to derive from intramolecular general base catalysis by the intermediate HOAt ester.[22]

A complementary approach to improving the head-to-tail cyclisation of small to medium length protected linear peptides is to induce a turn structure in the linear peptide by reversible amide bond alkylation. Cyclisation with HAPyU of a linear peptide with a backbone amide alkylated with the Hmb protecting group (Scheme 11.3)[21] gave double

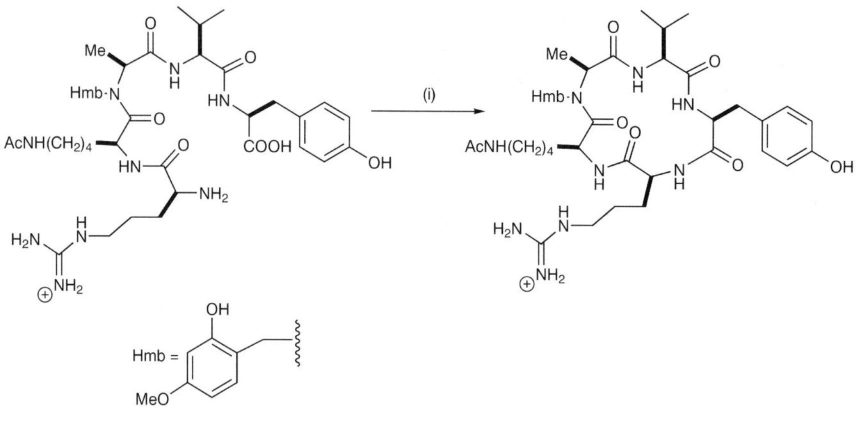

Scheme 11.3 (i) HAPyU (**8**), DIEA (3 equiv.).

the yield of cyclic peptide, compared with the linear peptide lacking the Hmb group. The Hmb group was subsequently removed by treatment with piperidine. In a similar approach,[23] Boc groups were added to the backbone amide positions of tetraphenylalanine, allowing the previously intractable linear peptide to cyclise.

During cyclisation of longer fully protected linear peptides, problems of limited solubility in aqueous or organic solvents may be encountered. These difficulties have been circumvented in two original approaches to the cyclisation of unprotected linear precursors in aqueous solution. Kent and co-workers[24] have used their native chemical ligation methodology to effect cyclisation of the binding loop region of Eglin c (Scheme 11.4). The unprotected linear peptide precursor was synthesised bearing an

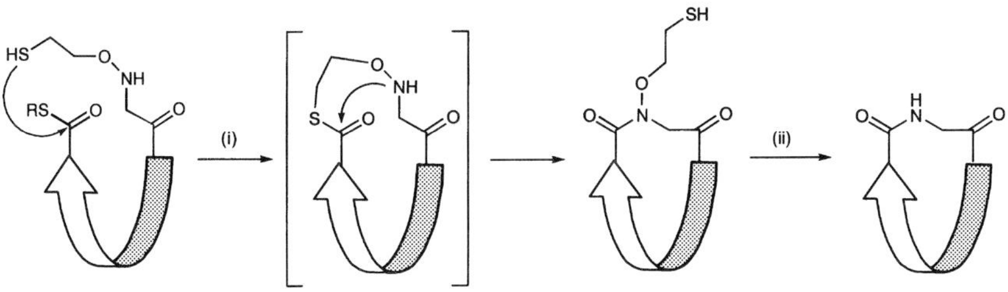

Scheme 11.4   (i) H_2O, pH 7.5; (ii) Zn, H^+.

oxyethanethiol group at the *N*-terminus; this then reacted at the *C*-terminus to form a cyclic thioester intermediate, which rearranged to the amide-linked cyclic peptide. The oxyethanethiol group was subsequently removed by reduction; no evidence of dimerisation or racemisation was seen. It should be noted that competition with the side-chains of aspartic acid or glutamic acid residues was not an issue. Zhang and Tam[25] have also exploited a transthioesterification reaction in their cyclisation of unprotected peptide fragments (Scheme 11.5). In this case, the unpro-

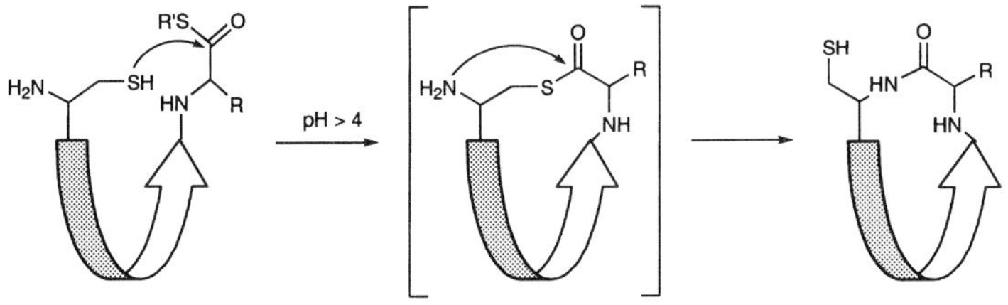

Scheme 11.5.

tected linear precursor was synthesised as a *C*-terminal thioester, with a Cys at the *N*-terminus; at pH > 4 the peptide cyclised via transthioesteri-fication followed by *S*- to *N*-acyl transfer. The process was highly regiospecific, with no side-reactions being observed with internal Cys, Lys or other side-chains. Significantly the method is free from racemisation risk, and it proceeds with no detectable oligomerisation. Another method, in which the C-terminal thioester is cyclised with the assistance of Ag^+ ions, has recently been reported,[26] allowing residues other than Cys to be used at the N-terminus.

11.2.3   On-resin cyclisation by amide bond formation

Until recently, the usual approach to simple cyclic peptides was to synthesise a fully protected linear peptide precursor, either stepwise in solution, or via a segment condensation approach or via solid-phase peptide synthesis, and then to carry out the macrolactamisation in solution. However, solid-phase synthesis of a fully protected linear precursor, followed by cyclisation on the solid phase, has the considerable advantage that competing dimerisation and oligomerisation are unlikely; it is also believed that a 'pseudo-dilution' effect may be operating in such cases.[27] Two recent approaches to solid-phase cyclic peptide synthesis via on-resin cyclisation have been especially successful. Ösapay et al.,[28] have developed a method based on the use of Kaiser oxime resin (Scheme 11.6). The linear peptide is assembled using Boc amino acids; removal of

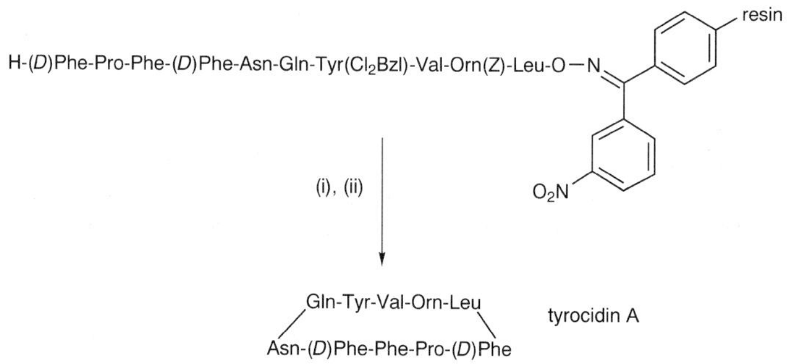

H-(D)Phe-Pro-Phe-(D)Phe-Asn-Gln-Tyr(Cl₂Bzl)-Val-Orn(Z)-Leu-O—N

resin

O_2N

(i), (ii)

Gln-Tyr-Val-Orn-Leu

tyrocidin A

Asn-(D)Phe-Phe-Pro-(D)Phe

Scheme 11.6   (i) DIEA; (ii) TMSOTf, TFA, thioanisole.

the Boc group from the N-terminus and neutralisation with DIEA is followed by head-to-tail cyclisation, either uncatalysed or in the presence of TMSOTf/TFA with simultaneous cleavage of the peptide from the resin. This method has been extensively used for the preparation of cyclic peptides, depsipeptides and fully protected cyclic peptide fragments for further segment synthesis[29] and has also been utilised for the formation of unnatural side-chain amide bridges.[30] A recent variation of this strategy, involving attachment of the first amino acid via a thioester linkage, synthesis of the linear peptide, and head-to-tail cyclisation, has been reported.[31]

A three-dimensional protecting-group strategy has also been developed by Alberico and co-workers (Scheme 11.7).[32] In this approach, the first

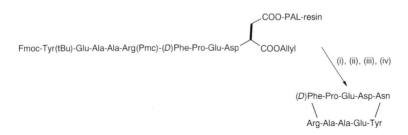

Scheme 11.7   (i) Pd(PPh₃)₄, DMSO/THF/0.5 M HCl/morpholine; (ii) piperidine; (iii) BOP (**5**), HOBt, DIEA; (iv) TFA, thioanisole, β-mercaptoethanol, anisole.

amino acid is attached to the solid support via the ω-carboxyl side-chain of Asp or Glu. Either *p*-alkoxylbenzyl alcohol (PAC) or tris(alkoxy)-benzylamide (PAL) supports can be used, with the α-carboxyl group protected as the allyl ester. Subsequent amino acids are coupled and deprotected using standard Fmoc-based chemistry, with the side-chains protected with *tert*-butyl and related acid-labile groups. With the cyclic precursor still attached to the resin, the *C*-terminus is selectively deprotected with Pd(0), the *N*-terminal Fmoc group removed and the peptide cyclised using BOP (**5**) or PyBOP (**6**). Cleavage of the peptide from the linker reveals the Asp/Glu (PAC support) or Asn/Gln (PAL support) side-chains. This strategy has been extended to allow the peptide to be anchored to the solid support by attaching the ε-NH₂ group of Lys to an active carbonate resin.[33]

Similar approaches to on-resin cyclisation have been reported by other groups,[34, 35] and the field has been reviewed.[36, 37] In a few cases, solid-phase and solution-phase syntheses of cyclic peptides have been compared[38, 39] and the solid-phase approach found to give greatly improved yields in the cyclisation step. In general, macrolactamisation reagents, such as HATU (**7**) and BOP (**5**), that work well for solution-phase cyclisations, also work well for on-resin cyclisations. In a recent study of on-resin head-to-tail cyclisations of pentapeptides, complete cyclisation was reported after only 30 min using HATU, with levels of racemisation below 10%.[40] Conversely, however, when HBTU (**3**) or TBTU (**4**) were used for on-resin cyclisations to give strained 10-membered lactams,[41] cyclisation failed and alkylguanidinium by-products were formed instead, arising from competing attack on the uronium reagents by the free amino group.[42] At present, until the factors influencing these reactions are better understood, each on-resin lactamisation should be carefully evaluated with several coupling reagents.

11.2.4   Synthesis of cyclic peptides containing disulfide linkages

Linear peptides containing two cysteine residues offer the opportunity for cyclisation by oxidative formation of an S–S bridge. However, many cyclic peptides contain several S–S bridges, or consist of two peptide chains joined by two or more such linkages, and in these cases it is vital to control the regiochemistry of disulfide bond formation. A number of different solutions to these problems have been reported, which fall into two categories. Two Cys residues bearing identical side-chain protection may be converted directly to the disulfide. The classical examples of this are the iodine oxidation of S-Trt- or S-Acm-protected Cys residues to give disulfide bridges, reported by Kamber et al.[43] More recent oxidising agents include thallium(III) trifluoroacetate on a range of protecting groups (Scheme 11.8)[44] and diphenylsulfoxide/methyl trichlorosilane on

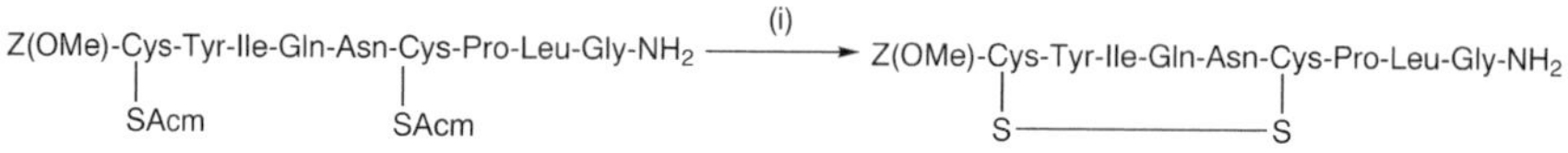

Z(OMe)-Cys-Tyr-Ile-Gln-Asn-Cys-Pro-Leu-Gly-NH$_2$ $\xrightarrow{\text{(i)}}$ Z(OMe)-Cys-Tyr-Ile-Gln-Asn-Cys-Pro-Leu-Gly-NH$_2$

  | | | |
SAcm   SAcm   S———————————S

Scheme 11.8   (i) $(CF_3COO)_3Tl$, TFA, 0°C.

S-Acm-protected Cys residues.[45] Alternatively, one Cys residue may be activated, for example as the 3-nitro-2-pyridinesulfenyl (Npys)[46] derivative, and reacted with a free thiol from another Cys residue; this strategy is particularly useful for the regiospecific synthesis of heterodimeric cyclic peptides joined by two disulfide bridges (Scheme 11.9).[47] Two recent comprehensive reviews[48, 49] detail all of the possible reagents and strategies for regiospecific disulfide bond formation and also give practical procedures.

11.3   Synthetic pathways to important classes of cyclic peptides

11.3.1   Cyclic peptides containing predominantly amide bonds

It is not possible to cover all the successful syntheses of cyclic peptides reported in the past decade in a review of this length. Many cyclic peptides of considerable biological importance have been synthesised, including the immunosuppressant cyclosporin,[50] and the field has been

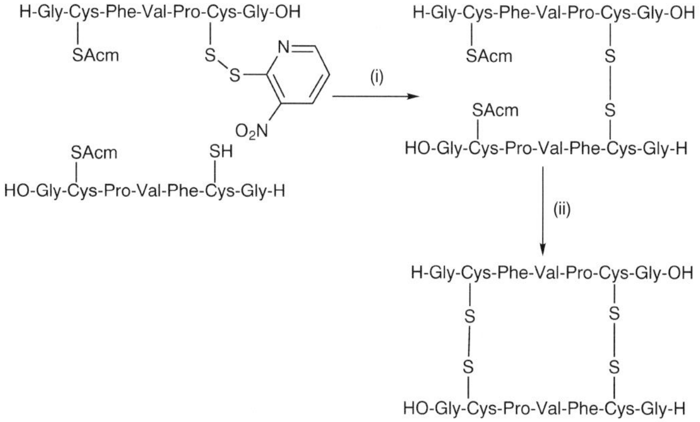

H-Gly-Cys-Phe-Val-Pro-Cys-Gly-OH

SAcm

O₂N

SAcm   SH

HO-Gly-Cys-Pro-Val-Phe-Cys-Gly-H

Scheme 11.9   (i) pH 6; (ii) I₂.

recently and comprehensively reviewed by Humphrey and Chamberlin.[1] In this section, syntheses of naturally occurring cyclic peptides containing predominantly amide bonds are discussed. The targets have been selected on the basis of the crucial biological interest of these peptides and the synthetic challenges presented by the array of nonproteinogenic amino acids contained therein. In the following sections, cyclic peptides which present special challenges arising from bridges formed between non-proteinogenic amino acids are presented.

The marine sponge cyclic peptides, cyclotheonamides A (**10**) and B (**11**), are potent inhibitors of serine proteases and have therefore aroused

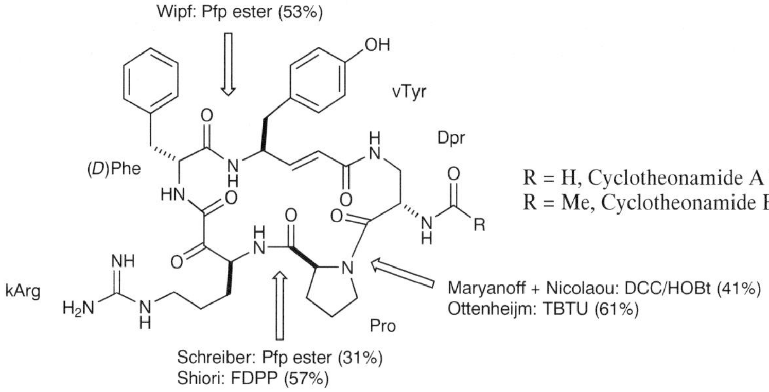

Wipf: Pfp ester (53%)

vTyr

Dpr

(D)Phe

R = H, Cyclotheonamide A **10**
R = Me, Cyclotheonamide B **11**

kArg

Maryanoff + Nicolaou: DCC/HOBt (41%)
Ottenheijm: TBTU (61%)

Pro

Schreiber: Pfp ester (31%)
Shiori: FDPP (57%)

considerable interest. Unusual features include the vinylogous tyrosine (vTyr) and α-ketoamide arginine homologue (kArg) residues. Most of the difficulties inherent in the reported syntheses derive from the necessity to protect the highly nucleophilic Arg side-chain and to carry through the kArg as an α-hydroxy acid, oxidising at the final step. Hagihara and Schreiber[51] reported the first synthesis of cyclotheonamide B via a stepwise coupling approach. The linear precursor 12 was synthesised with the kArg side-chain doubly protected and its α-hydroxy group unprotected (Scheme 11.10). Cyclisation was effected via the Pfp ester.

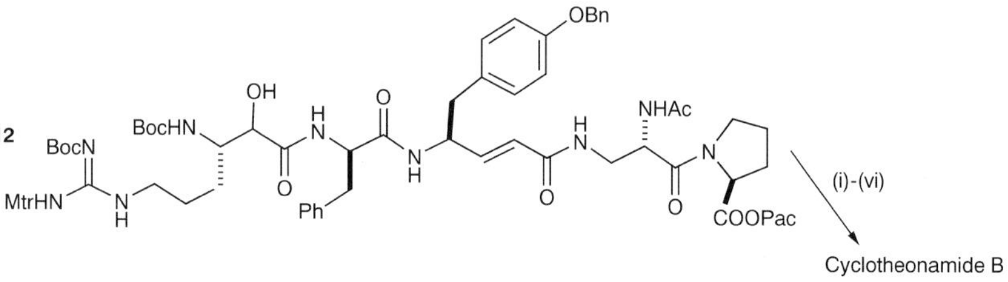

Scheme 11.10   (i) Zn, AcOH; (ii) Pfp-OH, DCC; (iii) TsOH, CH₂Cl₂; (iv) DIEA, DMAP (31%); (v) Dess–Martin periodinane (74%); (vi) TFA, C₆H₅SMe (58%). Pac = phenacyl.

Dess–Martin oxidation then provided the α-ketoamide in good yield, and the natural product was unmasked after side-chain deprotection. Wipf and Kim[52] approached cyclotheonamide A via a convergent [3+2] segment condensation strategy (Scheme 11.11). Condensation of fragments

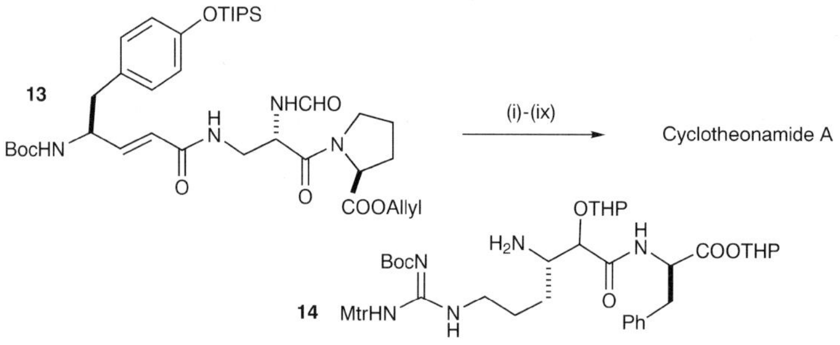

Scheme 11.11   (i) Pd(PPh₃)₄, dimedone (87%); (ii) NMM, ClOO-i-Bu, −20°C, then 14 (76%); (iii) 0.05 M HCl; (iv) Pfp-OH, DCC (80%); (v) sat. HCl, Et₂O/CH₂Cl₂; (vi) NMM (53%); (vii) Dess–Martin periodinane, 80°C (74%); (viii) HF/pyridine; (ix) TFA, C₆H₅SMe (36%).

13 and 14 was followed by selective hydrolysis of the THP ester, with cyclisation again achieved via the Pfp ester. Subsequent oxidation of the α-hydroxy acid proved more problematic, with elevated temperatures required for conversion to the ketoamide. Shiori and co-workers have reported two approaches to cyclotheonamide B.[16, 53] The second synthesis[16] is particularly interesting, as the problems associated with the kArg side-chain were circumvented by guanidination of an ornithine

analogue after the macrolactamisation step. Thus, fragments **15** and **16** (Scheme 11.12) were condensed using DEPC, the coupled product deprotected, and the precursor cyclised. FDPP was found to be greatly

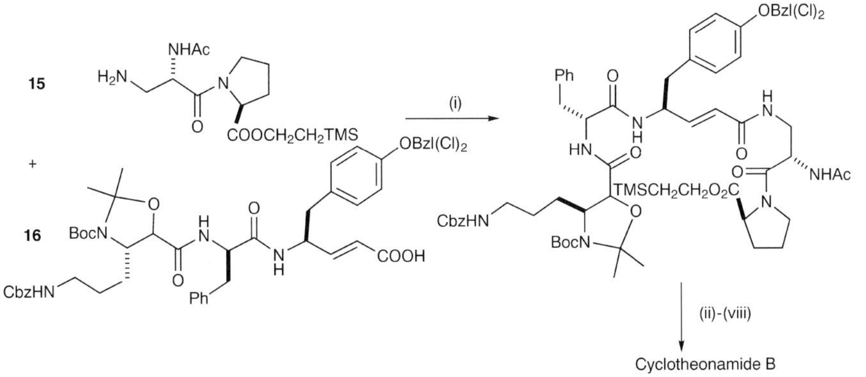

Scheme 11.12   (i) DEPC, DIEA; (ii) TBAF; (iii) TsOH; (iv) FDPP, DIEA; (v) 31% HBr-AcOH; (vi) N,N'-di(tert-butoxycarbonyl)thiourea, HgCl₂, Et₃N; (vii) Dess–Martin periodinane, 55°C; (viii) TFA, C₆H₅SMe.

superior for the cyclisation, with other methods such as DPPA, DEPC or formation of the Pfp ester giving very poor results. Selective deprotection of the ornithine side-chain was followed by guanidination; oxidation and deprotection completed the synthesis. The Maryanoff and Nicolaou groups[11] also utilised a [3+2] segment condensation strategy (Scheme 11.13), coupling fragments **17** and **18** to give linear peptide **19**. Careful

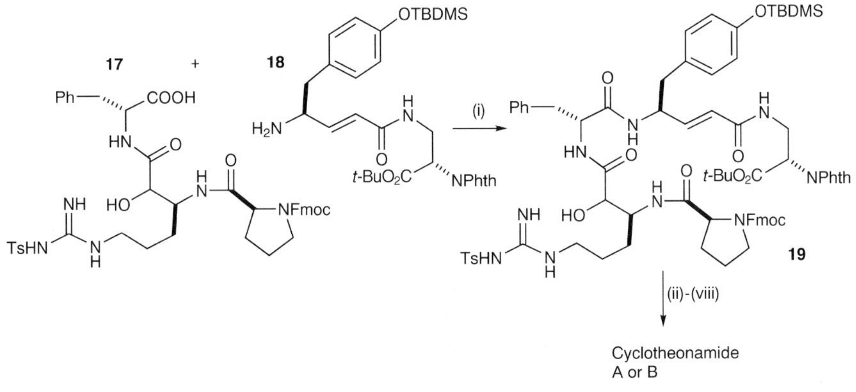

Scheme 11.13   (i) BOP-Cl, Et₃N (65%); (ii) Et₂NH; (iii) TFA; (iv) DCC, HOBt (41%); (v) hydrazine, 2-butenol (68%); (vi) acylation or formylation; (vii) Dess–Martin periodinane; (viii) HF, anisole (33%).

selection of the protecting group at the α-amino position of the di-aminopropionate (Dpr) residue allowed either cyclotheonamide A or B, or unnatural derivatives, to be synthesised from this advanced inter-mediate. Oxidation of the α-hydroxy acid in the final stages again proved problematic; this was attributed to unfavourable conformational features of the cyclic peptide. Ottenheijm and co-workers have also used a [3+2] segment condensation strategy,[54] utilising conventional benzyl, *t*-butyl and allyl protecting groups. The final cyclisation was carried out with TBTU (**4**) in high yield.

A number of natural products that disrupt cell signalling pathways by inhibiting Ser/Thr phosphatases have been reported.[55] These include the structurally related cyclic peptides motuporin (**20**), nodularin (**21**) and the microcystins (**22**). The complex amino acid Adda and the presence of a dehydroamino acid provide the main synthetic challenges of these

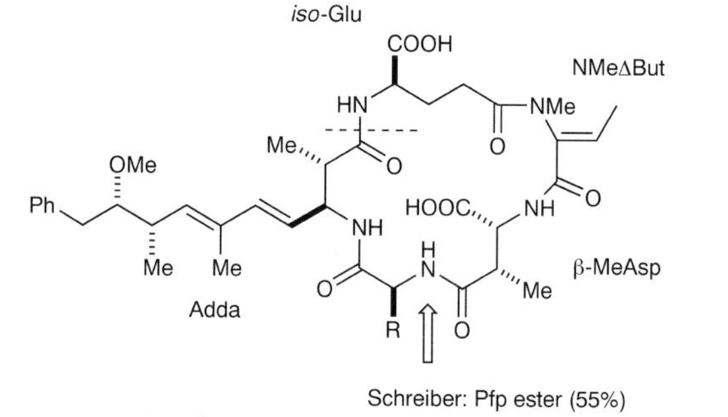

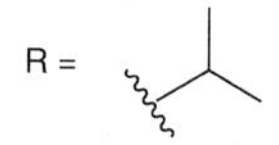

Schreiber: Pfp ester (55%)

R = [isopropyl/isobutyl group] motuporin **20**

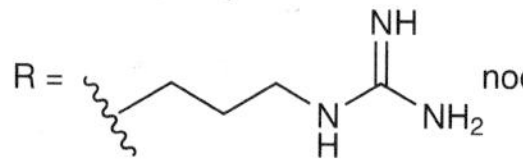

R = [chain terminating in guanidino group] nodularin **21**

Chamberlin: Pfp ester (56%)

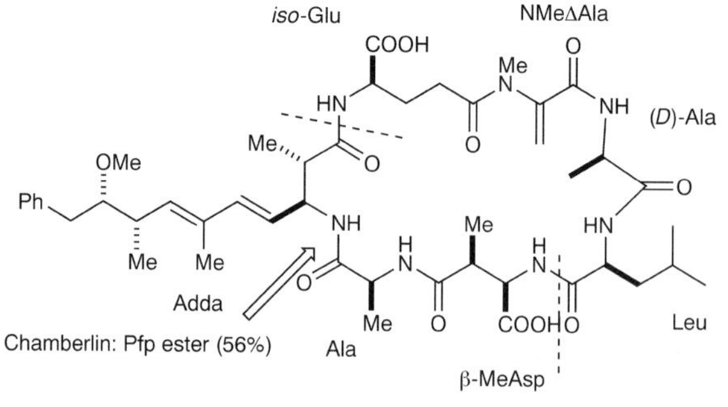

microcystin-LA **22**

peptides. Valentekovich and Schreiber[56] have synthesised motuporin via a [3+2] segment approach in which the disconnections were chosen such that segment condensation and subsequent cyclisation did not involve activation of an α-amido acid, to minimise racemisation. The Adda residue was synthesised in two sections. Crotylstannylation of BOM-protected mandelaldehyde, followed by elaboration, gave **23** (Scheme 11.14); this was coupled to aldehyde **25** (obtained via manipulation of the

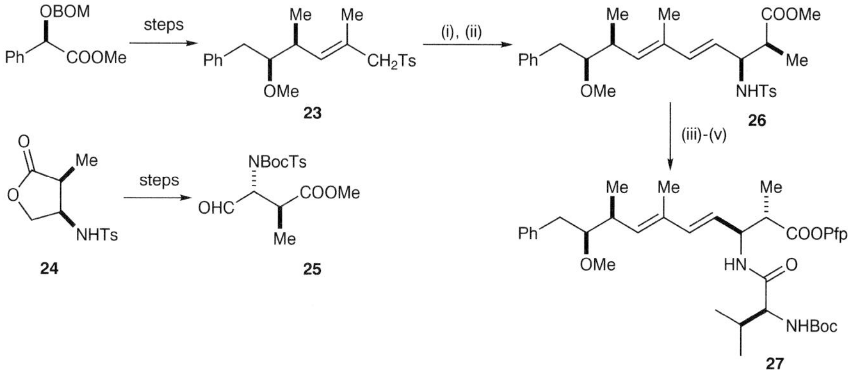

Scheme 11.14   (i) BuLi, then **25**; (ii) Na/naphthalene, −78°C (65%); (iii) LiOH; (iv) Na/naphthalene, −78°C; (v) Boc-Val-OPfp (5 equiv.); (vi) Pfp-OH, DCC (45%).

(D)-Thr-derived lactone **24**) by Julia olefination to give **26** as a single stereoisomer. Adjustment of the protecting groups, and coupling to Val-OPfp, gave fragment **27**. Lactone **24** also served as a precursor to the β-MeAsp residue (Scheme 11.15); standard coupling steps gave fragment **28**, in which the (D)-NMeThr residue serves as a masked form of NMeΔBut. Fragment condensation and macrolactamisation both utilised

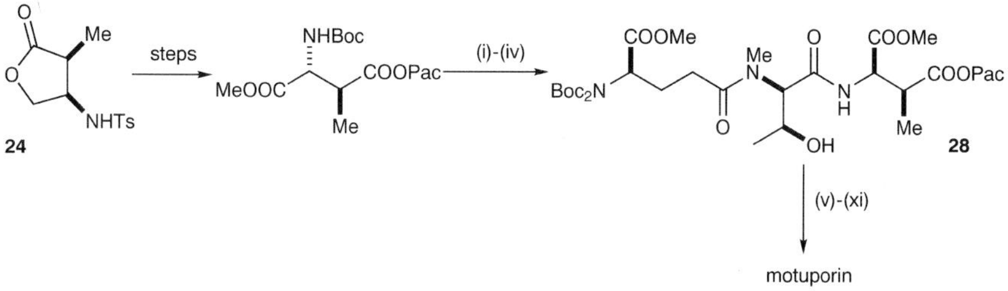

Scheme 11.15   (i) TFA/CH₂Cl₂; (ii) Boc-(D)N-MeThr, BOP, DIEA (88%); (iii) TFA/CH₂Cl₂; (iv) Boc₂-(D)Glu-OMe, BOP, DIEA (80%); (v) TFA/CH₂Cl₂; (vi) **27**, DIEA (82%); (vii) Zn, HOAc; (viii) Pfp-OH, DCC; (ix) TFA/CH₂Cl₂; (x) DIEA, DMAP (55%); (xi) Ba(OH)₂ (52%). Pac = phenacyl.

the Pfp ester methodology; saponification of the methyl esters and dehydration to the NMeΔBut residue were carried out simultaneously with Ba(OH)$_2$.

Chamberlin and co-workers have synthesised microcystin-LA via the coupling of three segments.[57] In this work, the Adda residue **32** was synthesised via Suzuki coupling of boronic acid **29** and iodoalkene **31** [derived from the protected β-MeAsp[58] (**30**); Scheme 11.16]; again,

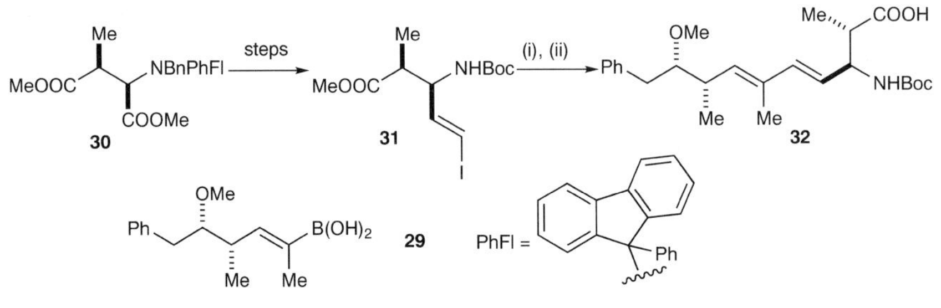

Scheme 11.16 (i) **29**, Pd(PPh$_3$)$_4$, TlOEt (82%); (ii) LiOH (88%).

only one stereoisomer was produced. In contrast to the synthesis of motuporin, the NMeΔAla residue was introduced at an earlier stage. The phosphonylsarcosine derivative **33** was prepared from methyl glyoxylate hemiacetal, and incorporated into peptide **34** (Scheme 11.17). A modified

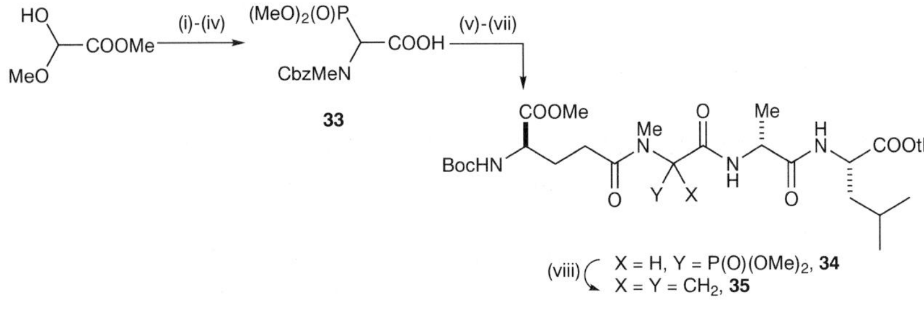

Scheme 11.17 (i) CbzNHMe, H$^+$, Et$_2$O (27%); (ii) MsCl, Et$_3$N; (iii) P(OMe)$_3$/NaI (75%); (iv) NaOH (98%); (v) (D)Ala-Leu-O-t-Bu, DCC, HOBt (91%); (vi) H$_2$, Pd/C; (vii) Boc-Glu-OMe, HATU (**7**), DIEA; (viii) CH$_2$O, K$_2$CO$_3$ (75%).

Schmidt reaction gave fragment **35**. The resulting dehydroamino acid residue proved stable to subsequent treatment with acids or nucleophiles in the later stages of the synthesis. Finally, adjustment of the protecting groups of **30**, followed by coupling to Ala-OCH$_2$CCl$_3$, gave fragment **36** (Scheme 11.18). HATU (**7**) was used for all of the coupling steps and also

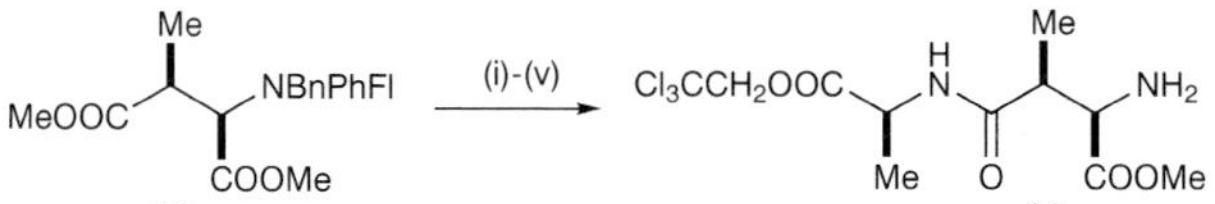

Scheme 11.18 (i) LiOH (70% after separation of epimers); (ii) H₂, Pd/C, TFA; (iii) Boc₂O (88%); (iv) Ala-OCH₂CCl₃, HATU (**7**), 88%; (v) TFA.

proved very effective for the segment condensation steps (Scheme 11.19). Macrolactamisation was again accomplished using the Pfp ester methodology; the ring-closure site was chosen in this case to place the

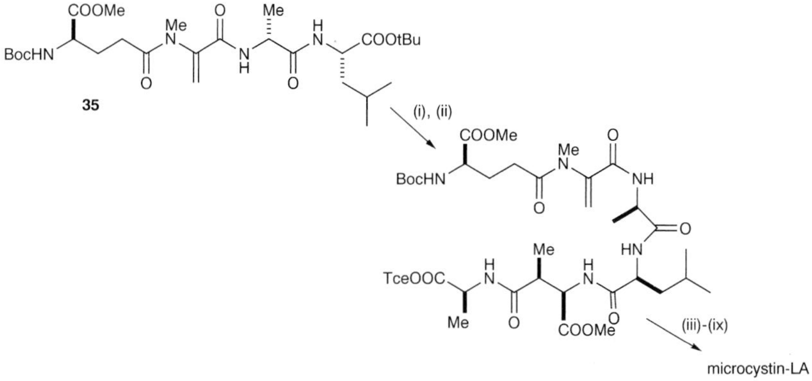

Scheme 11.19 (i) TFA, then Boc₂O (90%); (ii) **36**, HATU (**7**), collidine (80%); (iii) TFA; (iv) **32**, HATU, collidine (79%); (v) Zn/HOAc (64%); (vi) DCC, Pfp-OH (83%); (vii) TFA; (viii) NaHCO₃ (56%); (ix) LiOH (50%).

NMeΔAla residue, thought to induce a turn, in the centre of the linear precursor. Other syntheses of the Adda residue have also been reported,[59] and analogues of microcystin and nodularin have been prepared for studies of the structure–activity relationships of these inhibitors.[39, 60]

A third set of structurally interesting cyclic tetrapeptides are the family of fungal metabolites containing the (2S,9S)-2-amino-8-oxo-9,10-epoxydecanoic acid (Aoe) residue; these include chlamydocin (**37**) and trapoxin (**38**). The main challenge of these peptides lies in the Aoe residue itself, which is susceptible to nucleophilic attack. It is therefore necessary to introduce the epoxyketone at a late stage of the synthesis, as demonstrated by the first synthesis of chlamydocin, by Schmidt *et al.*, in 1984.[61] Baldwin and co-workers have reported a new synthesis of chlamydocin[62] in which the reactive side-chain is introduced at the end via radical coupling. Thus, tetrapeptide **39** was synthesised, containing

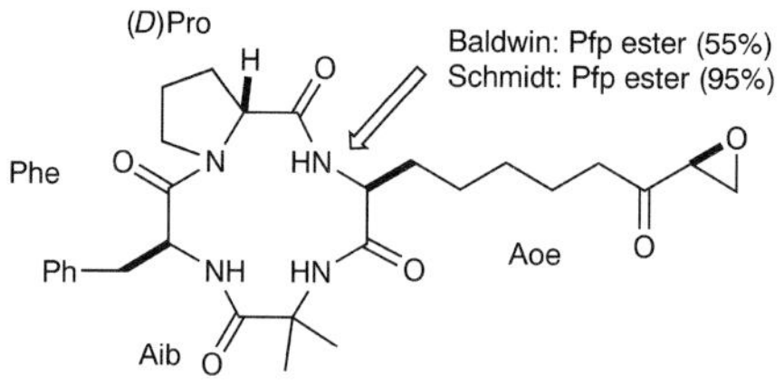

chlamydocin **37**

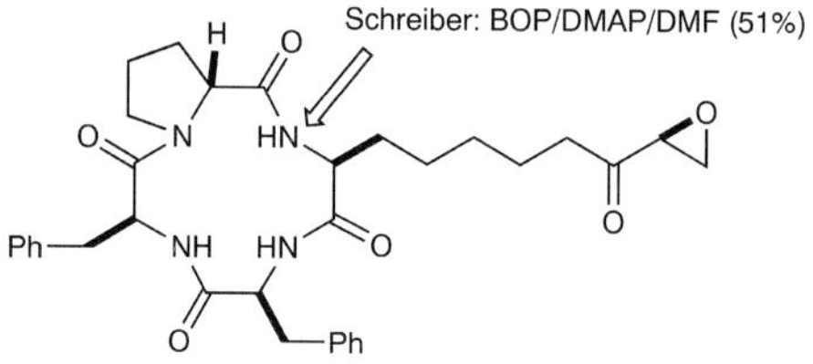

trapoxin **38**

(S)-2-amino-5-chloropentanoic acid as a precursor to Aoe, and cyclised via the Pfp ester (Scheme 11.20); no competing reaction of the amino group with the chloroalkane was seen. Free-radical homologation, followed by desilylation, gave chlamydocin. Schreiber and co-workers

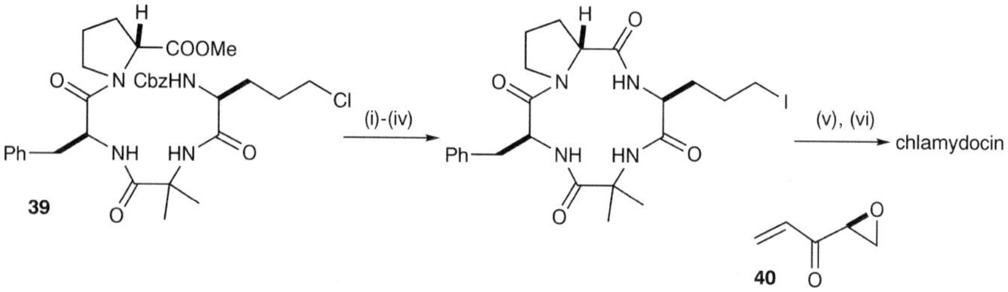

Scheme 11.20 (i) NaOH, MeOH; (ii) Pfp-OH, DCC; (iii) H_2, Pd/C, NMM, EtOH/dioxane, 95°C (55%); (iv) NaI (76%); (v) Bu_3SnH (3 equiv.) **40**, C_6H_6 (60%); (vi) TBAF (68%).

have reported the synthesis of trapoxin.[63] In this work, the Aoe precursor **41** was synthesised in moderate yield from (+)-2,3-O-isopropylidene-L-threitol and introduced into the linear tetrapeptide **42** (Scheme 11.21). In contrast to the previously reported syntheses of chlamydocin,[61, 62] cyclisation of **42** via the Pfp ester proved completely impossible: the macrolactamisation was only achieved in moderate yield after prolonged reaction with BOP. This provides a striking example of how the Thorpe–Ingold (gem-dialkyl) effect can greatly benefit a macrolactamisation reaction. Trapoxin lacks a gem-dialkyl residue which means that **42** will

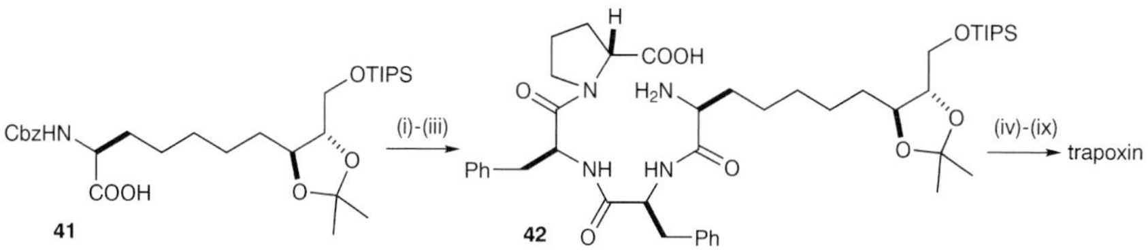

Scheme 11.21   (i) Phe-Phe-(D)-Pro-OMe, EDC, NMM, HOBt; (ii) LiOH; (iii) H₂, Pd/C (76%); (iv) BOP, DMAP, DMF, 25°C, 3 days; (v) TBAF (51%); (vi) TsCl, pyridine, DMAP; (vii) 5% HCl; (viii) DBU, 0°C (62%); (ix) Moffat oxidation (80%).

preferentially adopt a linear extended conformation, which keeps both reactive termini far removed from one another.

11.3.2   Cyclic peptides containing oxazoles, thiazoles and related heterocycles

A wealth of cyclic peptides containing oxazoles, thiazoles, oxazolines and *thiazolines* have been isolated, notably from marine tunicates.[64] Many possess cytotoxic, tumorigenic or other valuable bioactivities. Typically, these peptides consist of alternating heterocycle and amino-acid segments. Structural studies have shown that these architectural features give rise to cyclic peptides with well-defined conformations, sometimes with metal-chelating properties. However, the relationships between structure and biological activities in this series remain undefined. The pioneering synthetic work in this area was carried out by the groups of Shiori and Schmidt. In a series of elegant routes to peptides such as ulicyclamide[65, 66] and the patellamides[67, 68] procedures for the preparation and coupling of the oxazole and thiazole dipeptide segments, macro-cyclisation of the linear precursors and formation of the oxazoline moieties by dehydration of Thr or Ser residues were established. This work has been extensively reviewed by Wipf.[69]

Until recently, the synthesis of cyclic peptides in this series containing thiazoline moieties presented an insuperable challenge, as these residues proved extremely susceptible to racemisation even under mildly acidic or basic conditions.[70, 71] Pattenden and co-workers have reported an approach which circumvents this problem, forming the thiazoline ring in the final step of the synthesis by dehydration of a serine-derived thioamide. This strategy was first used in the synthesis of lissoclinamide 4 (**43**).[72] The appropriate linear precursor **45** was built up in stepwise fashion (Scheme 11.22) from the thiazole **44**, which was derived from condensation of the imino ester of D-Phe and Cys-OMe, followed by oxidation with MnO₂.[71] The key thioamide moiety was introduced with the thioacylating reagent **47**. Cyclisation with FDPP, at a site incapable of racemisation, afforded **46** in moderate yield. Dehydration was then accomplished using the Burgess reagent **48**[70, 73] to give lissoclinamide 4

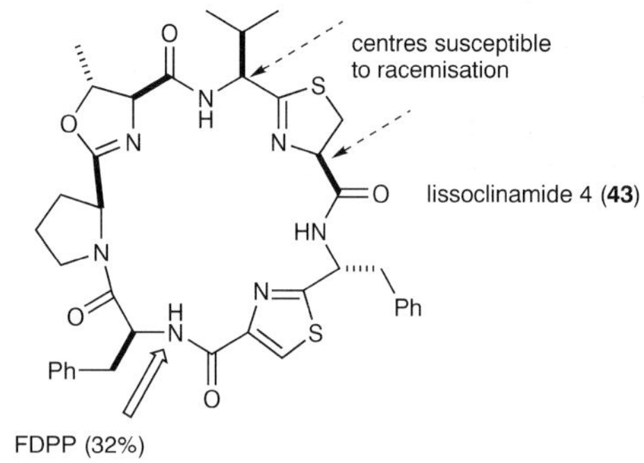

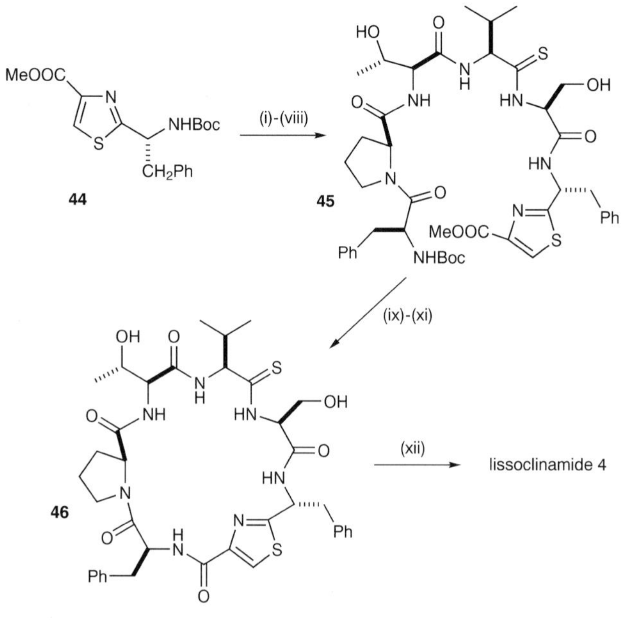

Scheme 11.22 (i) 50% TFA/CH$_2$Cl$_2$; (ii) Boc-Ser-OH, DCC, Et$_3$N (91%); (iii) 50% TFA/CH$_2$Cl$_2$, then NaHCO$_3$; (iv) **47**, DMF, 0°C−5°C (74%); (v) 50% TFA/CH$_2$Cl$_2$; (vi) Boc-*allo*-Thr-OH, DCC, HOBt, DIEA (71%); (vii) 50% TFA/CH$_2$Cl$_2$; (viii) Boc-Phe-Pro-OH, DCC, HOBt, DIEA (77%); (ix) NaOH/MeOH; (x) 50% TFA/CH$_2$Cl$_2$; (xi) FDPP, DIEA (32%); (xii) **48** (61%).

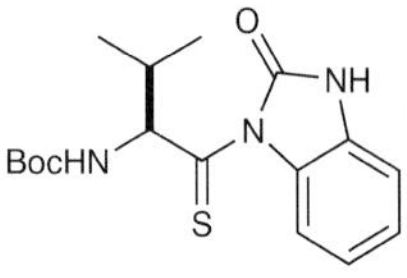

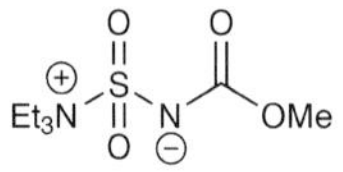

in good yield with little overall racemisation (5% of a second diastereoisomer was reported but not identified). This approach was also
used to form the oxazolidine ring in raocyclamide **49**,[74] and to form simultaneously the oxazolidine and thiazoline rings in cyclodidemnamide **50**.[75]

raocyclamide **49**   cyclodidemnamide **50**

In both of these cases, no racemisation was reported during the crucial
final dehydration step, indicating the mildness of the Burgess reagent
and the success of this strategy. A complimentary approach has been
used by Wipf and Fritch[76] in their synthesis of lissoclinamide 7 (**51**).

cyclisation (FDPP, 48%)

lissoclinamide 7 (**51**)

Their strategy was based on the observation that oxazolines may be
converted to thiazolines via a mild, two-step thiolysis–dehydration
sequence, and that thiolysis of Thr-derived oxazolines is generally much
slower than thiolysis of Ser-derived oxazolines.[77] Accordingly, the linear
precursor **54** was prepared from segments **52** and **53** (Scheme 11.23) and
transformed to the oxazoline precursor **55** via a sequence of cyclisation
and dehydration steps. The position for cyclisation was chosen on the
basis of the turn-inducing effect of the *allo*-Thr-derived oxazolidine,
which should facilitate cyclisation, although in practice this step proceeded in a modest yield. In the event, **55** proved to be more resistant
to thiolysis than expected on the basis of model studies on linear
oxazoline-bearing peptides, and the two-step sequence proceeded without

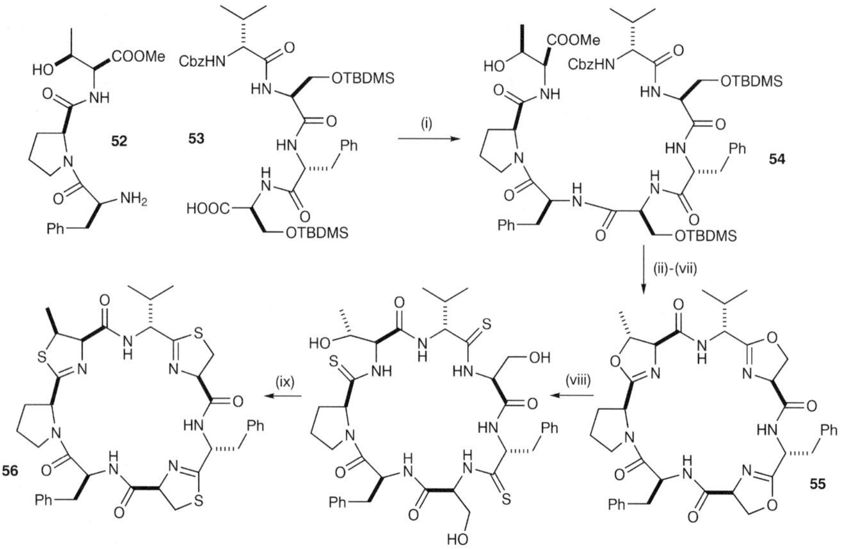

Scheme 11.23 (i) FDPP, Et$_3$N (81%); (ii) **48** (89%); (iii) H$_2$, Pd(OH)$_2$; (iv) NaOH, THF/MeOH; (v) FDPP; (vi) TBAF (21%); (vii) Ph$_3$P, DIAD, −60°C−0°C (85%); (viii) H$_2$S, Et$_3$N, 36°C, 30 h (90%); (ix) **48** (76%).

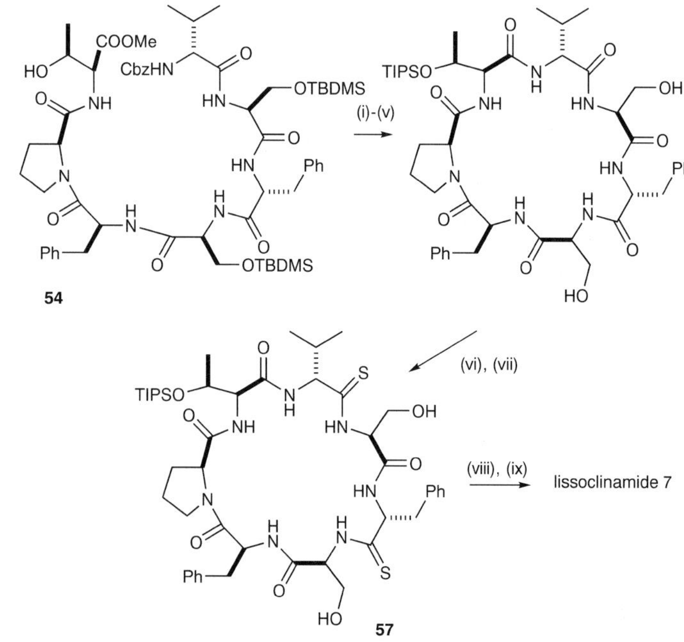

Scheme 11.24 (i) H$_2$, Pd/C; (ii) NaOH, MeOH/THF; (iii) FDPP, NaHCO$_3$ (48%); (iv) TIPS-OTf, 2,6-lutidine; (v) TsOH (61%); (vi) **48**, 65°C; (vii) H$_2$S, Et$_3$N, 4 days; (viii) TBAF (41%); (ix) **48** (90%).

selectivity to give **56**. In an alternative approach, **54** was cyclised without prior formation of the Thr-derived oxazoline (Scheme 11.24). Manipulation of the protecting groups followed by a sequence of dehydration, thiolysis and dehydration gave **57**, which was then selectively dehydrated to give the desired lissoclinamide 7. Yamada and co-workers have reported[78] a route to dolastatin E in which the sensitive thiazoline moiety is also formed by cyclodehydration, under Mitsunobu conditions, of the precursor thiopeptide in the last step. Contrary to previous reports,[70] no evidence of racemisation was seen, although the yield of this last step was highly dependent upon the stereochemistry of the precursor thiopeptide.

11.3.3   Cyclic peptides containing biaryl and biaryl ether linkages

Some of the most synthetically taxing cyclic peptides are the various families of antibiotics containing single or multiple biaryl and/or biaryl ether linkages. Of these, the greatest challenges have been presented by the multiply bridged vancomycin (**58**),[79] ereomycin (**59**) and teicoplanin (**60**). As well as presenting an irresistible challenge to the synthetic

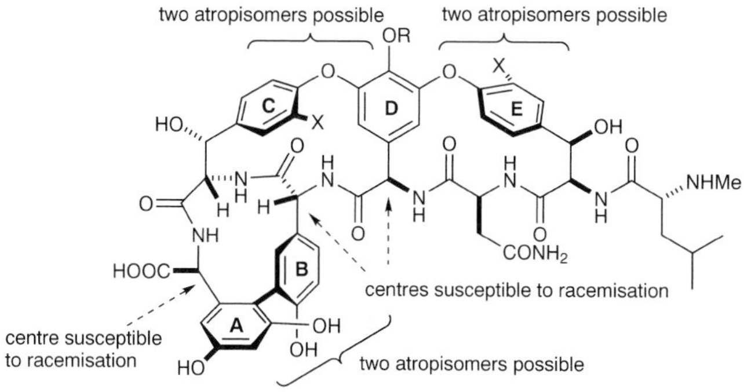

X = Cl, vancomycin **58**
X = H, ereomycin **59**
R = disaccharide

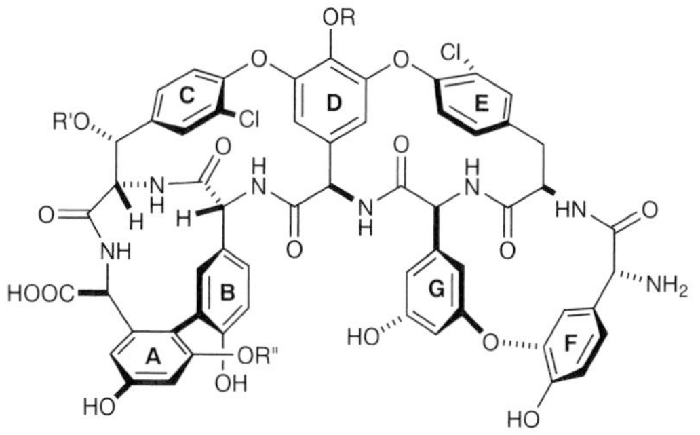

teicoplanin **60**
R, R', R" = sugars

chemist, these cyclic peptides are of crucial clinical importance. Vancomycin is the drug of choice for treatment of a variety of bacterial infections that are resistant to other classes of antibiotics. The recent emergence of vancomycin-resistant strains of methicillin-resistant *Staphylococcus aureus* (MRSA)[80] has encouraged renewed efforts towards a successful strategy for the synthesis of vancomycin and teicoplanin, which may in turn deliver novel and more potent analogues. The synthesis of the aglycone of vancomycin raises several major issues:

- three of the residues are derivatives of aryl glycine and are therefore highly susceptible to racemisation;
- restricted rotation about the biaryl and the biaryl ether bonds means that for each of the A-B, C-*O*-D and D-*O*-E rings, two noninterconverting rotational isomers (atropisomers) exist;
- new methodologies for the formation of biaryl- and biaryl -ether-linked amino acids, compatible with the easily racemised amino acid chiral centres, must be developed;
- a strategy for the formation of the biaryl-bridged A-B ring must be developed, capable of delivering the correct atropisomer;
- concurrently, an approach to the C-*O*-D-*O*-E bicyclic system is needed, either by formation of the biaryl ether bridges or via macrolactamisation. Again, it must be capable of delivering the correct atropisomers.

Many of these issues have been successfully addressed in synthetic approaches to less complex cyclic peptides containing single biaryl- or biaryl-ether linkages, such as deoxybouvardin and OF-4949-I, and in the synthesis of styrylamide-bearing peptide alkaloids such as frangulanine.[81] These approaches are discussed in an excellent review by Rama Rao, *et al.*, [82] in which the major synthetic advances in the quest for vancomycin, up to 1995, are also presented. Several other earlier reviews have also appeared, focusing on different approaches to the vancomycin aglycone.[83-85] Since then, the ultimate goal, the synthesis of the aglycone of vancomycin itself, has been achieved by the Evans[86] and Nicolaou[87] groups, with many other strategic advances being reported by other groups.

The strategy adopted by Evans and co-workers[86] is outlined in Scheme 11.25. The decision to start with A-B ring assembly, and to connect these rings by aryl coupling, was driven by early work in this field[88] which indicated that formation of the strained 12-membered ring by macrolactamisation would be impossible. The majority of the groups working in this area therefore developed intramolecular biaryl coupling methods

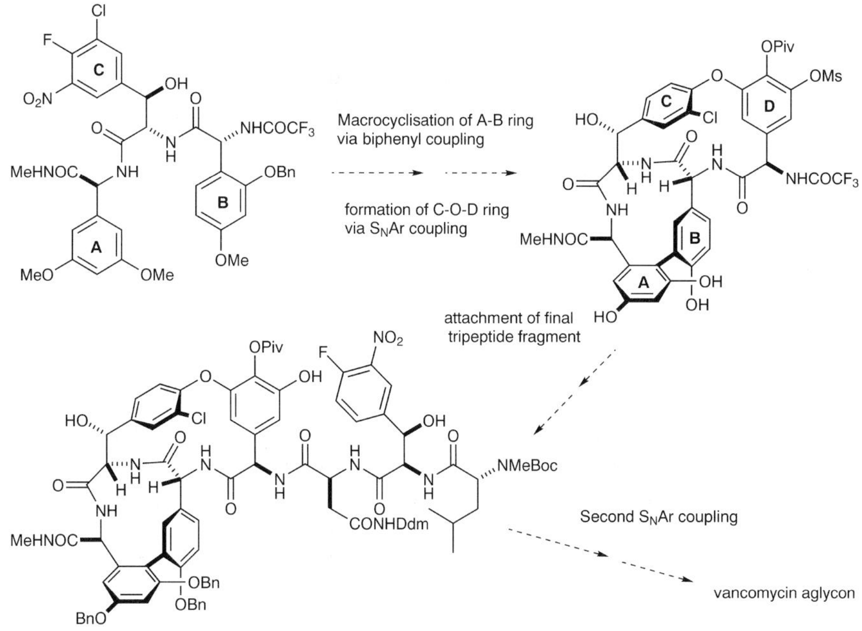

Scheme 11.25   Assembly strategy for Evans's synthesis of vancomycin aglycon.

for the formation of this bridge.[82] In his synthesis, Evans used his oxazolidinone methodology to prepare[89] the aryl glycine and β-hydroxy-tyrosine components, and he relied upon standard methods to assemble the precursor tripeptide **61** (Scheme 11.26). Oxidative coupling using VOF_3[90] closed the A-B ring, but gave the *incorrect* atropisomer **62** in > 95:5 diastereoisomeric excess [(de) > 95:5]. This was not unexpected;

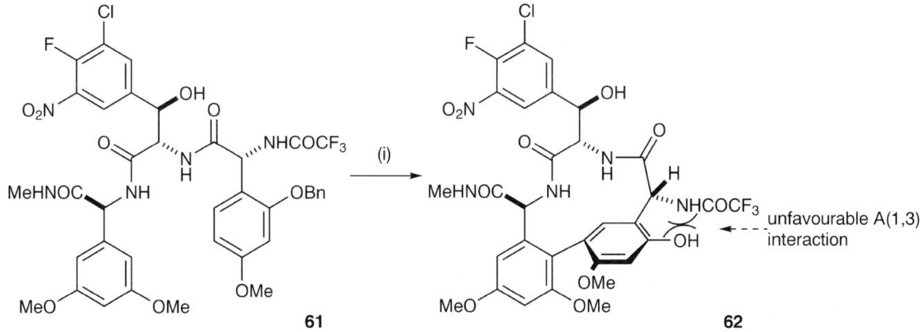

Scheme 11.26   (i) VOF_3, $BF_3 \cdot OEt_2$, $AgBF_4$, TFA/CH_2Cl_2 0°C; then $NaHB(OAc)_3$ (65%).

previous studies had shown that an unfavourable A(1,3) interaction between the *ortho*-hydroxyl and the adjacent Cα-stereocentre forced the A-B ring to adopt the unnatural atropisomeric configuration, and that removal of the *ortho*-hydroxyl, or inversion of the adjacent Cα-stereocentre, resulted in the formation of the correct atropisomer.[90, 91] However, it was anticipated that the C-*O*-D ring could be assembled regardless, and that the A-B ring stereochemistry could be adjusted at a later stage.

It was initially envisaged that the C-*O*-D and D-*O*-E rings would be formed via an oxidative coupling strategy. Methods such as the Ullmann diaryl ether condensation have been explored by Boger and co-workers[92] and Rama Rao,[93] and bromoquinone coupling has been suggested by Rama Rao and associates.[94] Tl(III)-mediated oxidative coupling (TTN) methods have been successfully used by Yamamura and co-workers[95] and Evans and co-workers,[96] notably in model studies of the C-*O*-D-*O*-E rings. However, the TTN method was found to be inadequate for ring-closure of **63** (Scheme 11.27) to form the D-*O*-E ring of the related orienticin C aglycone **64**,[97] owing to the unfavourable conformation of the preformed A-B and C-*O*-D rings.[98] A different coupling strategy was therefore required for the assembly of the C-*O*-D and D-*O*-E rings. Zhu and co-workers[84] have developed a highly successful intramolecular

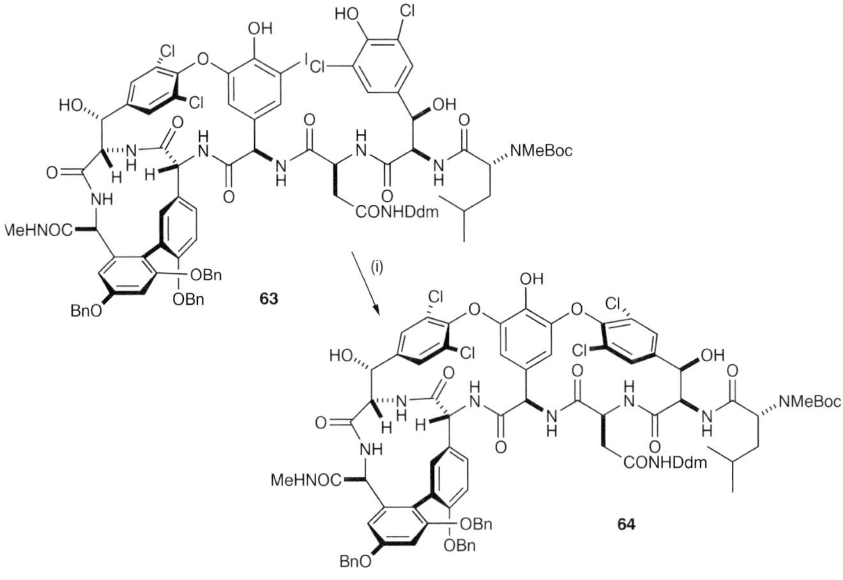

Scheme 11.27   (i) Tl(NO$_3$)$_3 \cdot$ 3H$_2$O, 3 Å sieves, 30:1 CH$_2$Cl$_2$:MeOH, r.t., then CrCl$_2$, 0°C (20%).

S_NAr strategy for the formation of the desired isodityrosine units, and have used it to access model vancomycin C-O-D-O-E rings[99] and model teicoplanin D-O-E-F-O-G rings.[100] Similar approaches have also been reported by the groups of Rama Rao,[101] Boger,[102] Rich[103] and Pearson.[104] This method was therefore used for the formation of both biaryl ether rings. Thus, aryl glycine **65** was coupled to **62**, giving tetrapeptide **66** (Scheme 11.28), which was cyclised to give **67** in a 5:1 ratio favouring the correct atropisomer, with the stereochemistry being controlled by the conformational preference of the nitro group. Removal of the nitro group and the *ortho*-hydroxyl on the A-B ring, and adjustment of the protecting groups, gave the bicyclic **68**. Gentle heating allowed the A-B ring to equilibrate to the desired atropisomer **69**, which, with the *ortho*-hydroxyl removed, is now thermodynamically favoured. Coupling of the tripeptide segment **70**, followed by a second S_NAr cyclisation, afforded **71**, in a 5:1 ratio again favouring the correct atropisomer. The stereoselectivity of this step compares favourably with the results from cyclisation of less-complex model systems (Scheme 11.29), in which little or no selectivity is seen:[105] it was demonstrated that the conformation of the naturally occurring A-B ring is responsible for the stereochemical bias of the cyclisation of **70**.[86] Adjustment of the protecting groups, notably the highly selective removal of the C-terminal amide group by nitrosation, led cleanly to the vancomycin aglycon.

The strategy adopted by Nicolaou and co-workers (Scheme 11.30) is conceptually quite different. Key to this approach was the preceding discovery[106] that the A-B ring of vancomycin could, after all, be cyclised by macrolactamisation, provided that the precursor peptide was pre-organised by an existing C-O-D biaryl ether (Scheme 11.31). Accordingly, the biaryl amino acid precursor **72** was synthesised (Scheme 11.32) via Suzuki coupling of **73** and **74**, giving a mixture of **72** and the undesired atropisomer **75**, which could be separated. Coupling of **72** to β-hydroxyl-tyrosine **76** and aryl glycine **77** gave the linear tripeptide **78**. The C-O-D biaryl ether linkage was then formed by a novel, Cu-mediated coupling

Scheme 11.28 (see overleaf) (i) NaHCO$_3$, MeOH, H$_2$O, 6 days; (ii) **65**, HATU, HOAt, collidine, $-20°C$ (65%); (iii) HF·pyridine, r.t., 1 h (85%); (iv) Na$_2$CO$_3$, DMSO, r.t., 1.5 h, then Tf$_2$NPh (79%: d.r. 5:1); (v) Zn, AcOH; (vi) NaNO$_2$, H$_3$PO$_4$, cat. Cu$_2$O, THF/H$_2$O; (vii) [Pd(dppf)Cl$_2$]·CH$_2$Cl$_2$, Et$_3$N, HCOOH, DMF, 75°C (77%); (viii) PivCl, Et$_3$N, THF; (ix) TFA, DMS, CH$_2$Cl$_2$; (x) (CF$_3$CO)$_2$O, 2,6-lutidine (88%); (xi) AlBr$_3$, then EtSH; (xii) MeOH, 55°C (54%); (xiii) BnBr, Cs$_2$CO$_3$, Bu$_4$NI; (xiv) LiSEt, 0°C; (xv) Allyl-Br, Cs$_2$CO$_3$; (xvi) LDA, THF, $-78°C$; (xvii) LiOH, THF/H$_2$O/MeOH (65%); (xviii) **70**, EDCI, HOAt (86%); (xix) CsF, DMSO, r.t. (95%): d.r. 5:1; (xx) Zn, AcOH; (xxi) HBF$_4$, *t*-BuONO, MeCN, then CuCl, CuCl$_2$, H$_2$O (65%); (xxii) N$_2$O$_4$, NaOAc, CH$_2$Cl$_2$/CH$_3$CN; (xxiii) H$_2$O$_2$, LiOH, THF/H$_2$O(68%); (xxiv) [Pd(PPh$_3$)$_4$], morpholine, THF (62%); (xxv) 10% Pd/C, 1,4-cyclohexadiene, EtOH (70%); (xxvi) TFA, DMS, CH$_2$Cl$_2$ (83%).

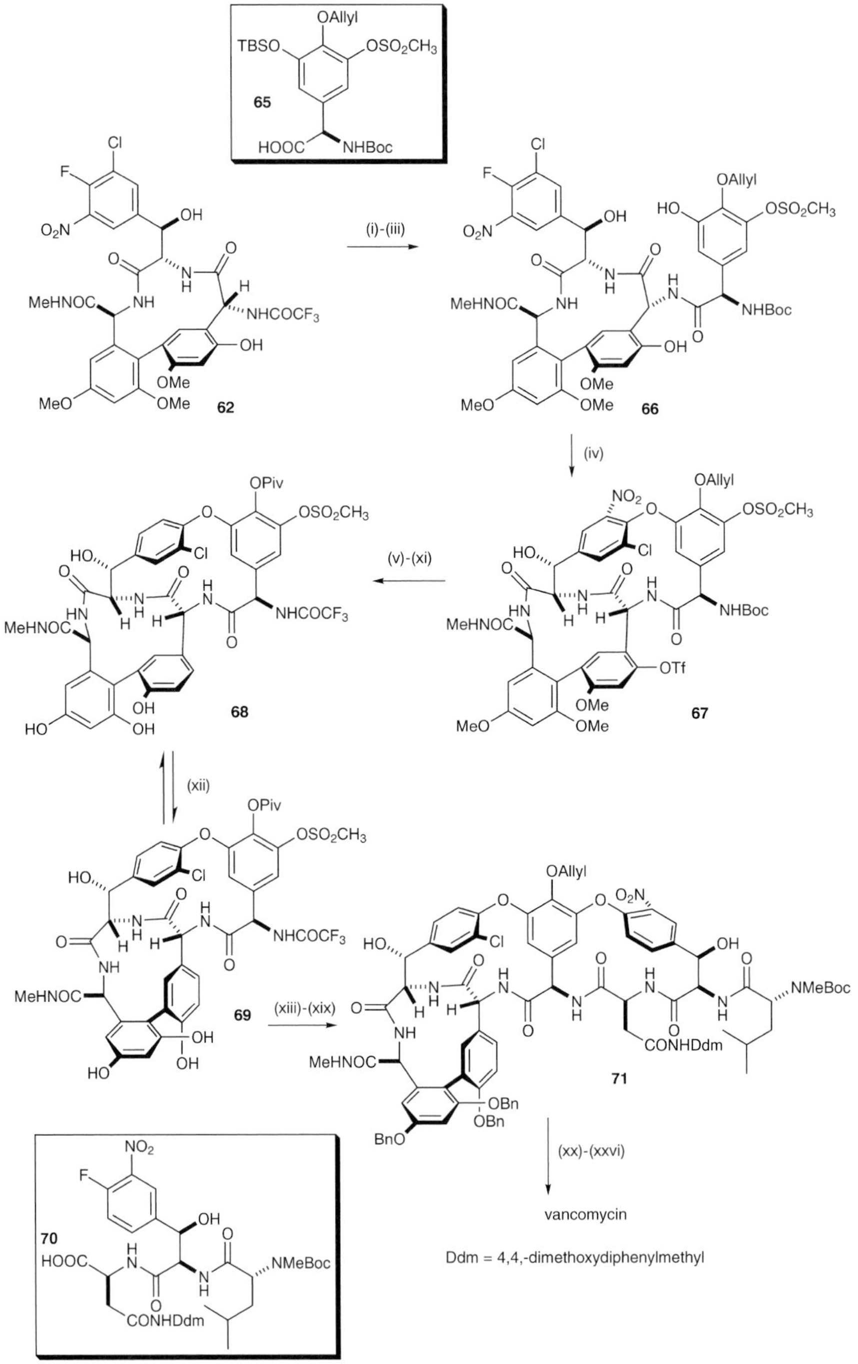
65
62
66
68
67
69
71
70
(i)-(iii)
(iv)
(v)-(xi)
(xii)
(xiii)-(xix)
(xx)-(xxvi)
vancomycin
Ddm = 4,4,-dimethoxydiphenylmethyl

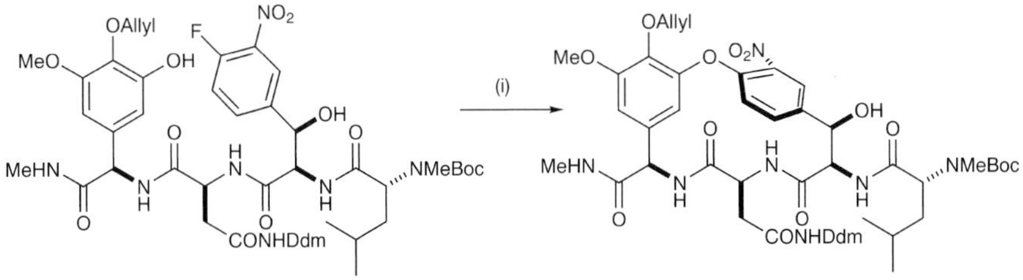

Scheme 11.29   (i) CsF, DMSO, r.t., 2.5 h (95%; diastereomeric ratio 1.4:1).

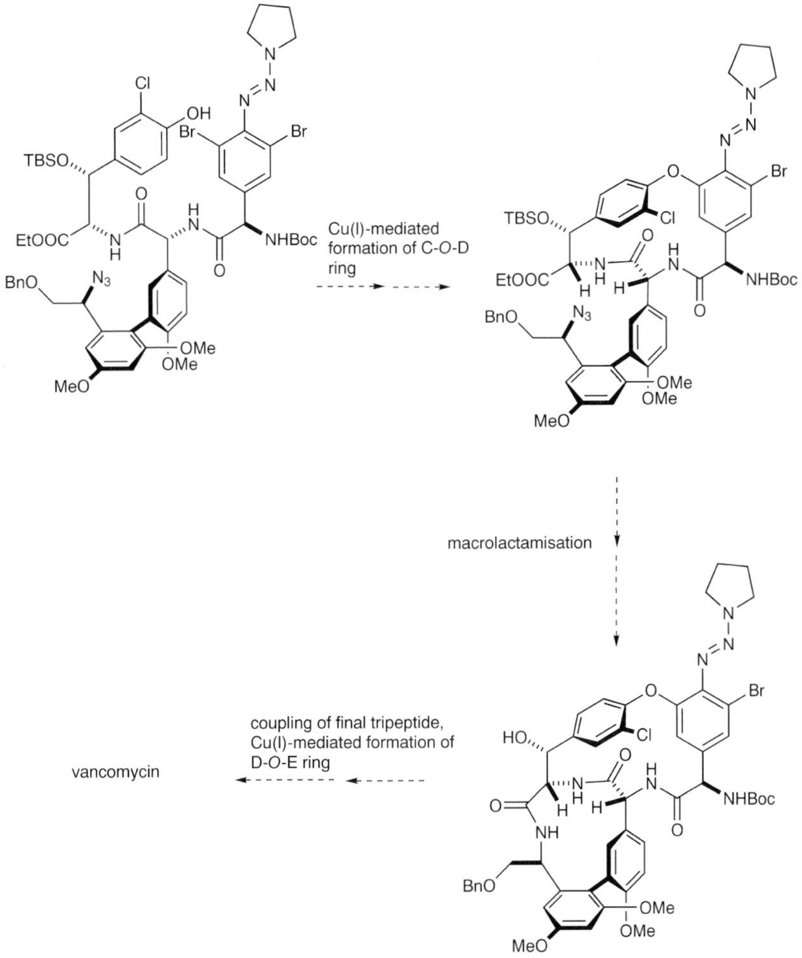

Scheme 11.30   Assembly strategy for Nicolaou synthesis of vancomycin algycon.

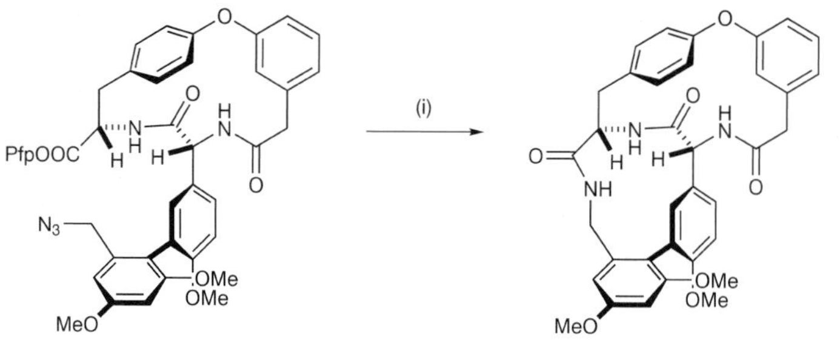

Scheme 11.31 (i) 4-pyrrolidinopyridine (3 equiv.) 10% Pd/C (30 mol%), dioxane/EtOH/ cyclohexene 90:8:2, high dilution, 90°C, 5 h, 30%.

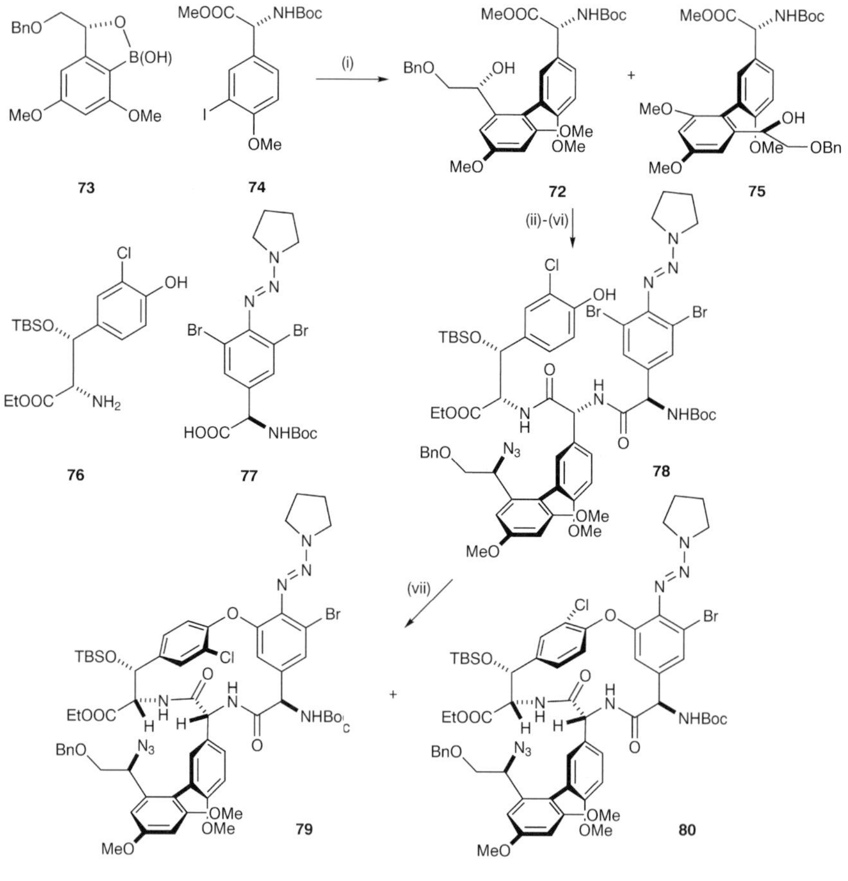

Scheme 11.32 (i) [Pd(PPh$_3$)$_4$] (0.2 equiv.) Na$_2$CO$_3$, PhMe/MeOH/H$_2$O (10:1:0.5) 90°C (84%, 2:1 **72:75**); (ii) DPPA, DEAD, Ph$_3$P, −20°C, (95%); (iii) LiOH, THF/H$_2$O (99%); (iv) EDC, HOBt, **76** (85%); (v) TMSOTf, 2,6-lutidine, CH$_2$Cl$_2$ (90%); (vi) EDC, HOAt, **77** (76%); (vii) CuBr.SMe$_2$, K$_2$CO$_3$, pyridine, MeCN, reflux (60%, 1:1 **79:80**).

protocol developed for this purpose by Nicolaou and co-workers,[107] in which the triazine acts as both a coordination centre for the Cu⁺ counterion and an electron sink for the attack of the incoming phenolate, to give **79** and the undesired **80** in a 1:1 ratio.

Compound **79** was then *O*-desilylated with TBAF, the azide reduced with Et₃P and the ethyl ester saponified. Macrolactamisation of the resulting amino acid using FDPP then proceeded, in good yield, to afford the bicyclic **81** (Scheme 11.33). It is notable that when this sequence was carried out using the wrong biaryl atropisomer, derived from **75**, the macrolactamisation reaction proceeded with racemisation at the Cα position shown (Scheme 11.34).

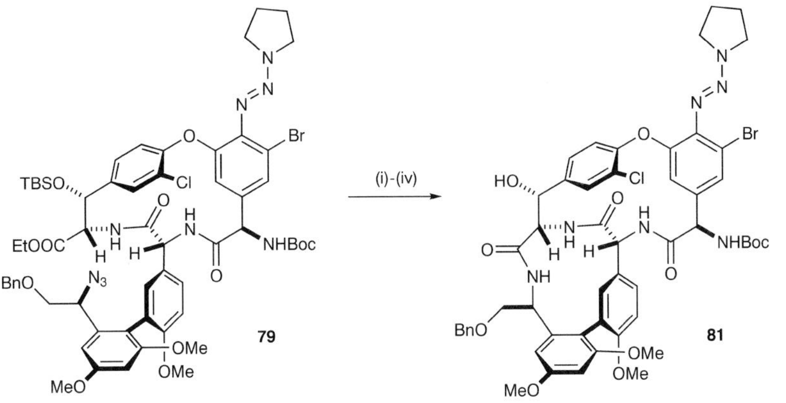

Scheme 11.33 (i) TBAF (80%); (ii) Et₃P (2 equiv.) H₂O (10 equiv.) MeCN (77%); (iii) LiOH, THF/H₂O (1:1) (68%); (iv) FDPP, DIEA (71%).

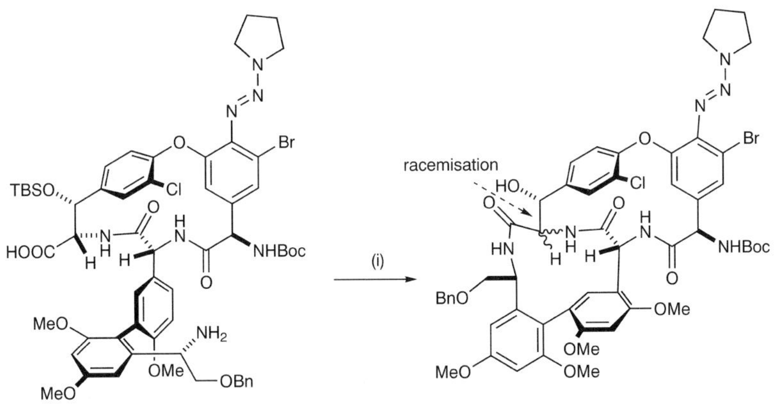

Scheme 11.34 (i) HATU, DIEA, DMF, 25°C, 8 h (50%).

With the A-B and C-*O*-D rings in place, tripeptide **82** was then attached to give **83** (Scheme 11.35) and the Cu-mediated cyclisation repeated to give **84** and **85** in a 1:3 ratio: the two atropisomers could be interconverted on heating. A sequence of functional group interconversion and deprotection steps then gave the vancomycin aglycon.

It is clear, from both of these synthetic routes, that control of the atropisomer ratio at each cyclisation step is crucial to the success of the synthesis. The results of these efforts,[86, 87] as well as the recent studies of Boger and co-workers[108] on the thermal equilibration of the various atropisomers, should shortly lead to a deeper understanding of the factors governing atropisomer selectivity.

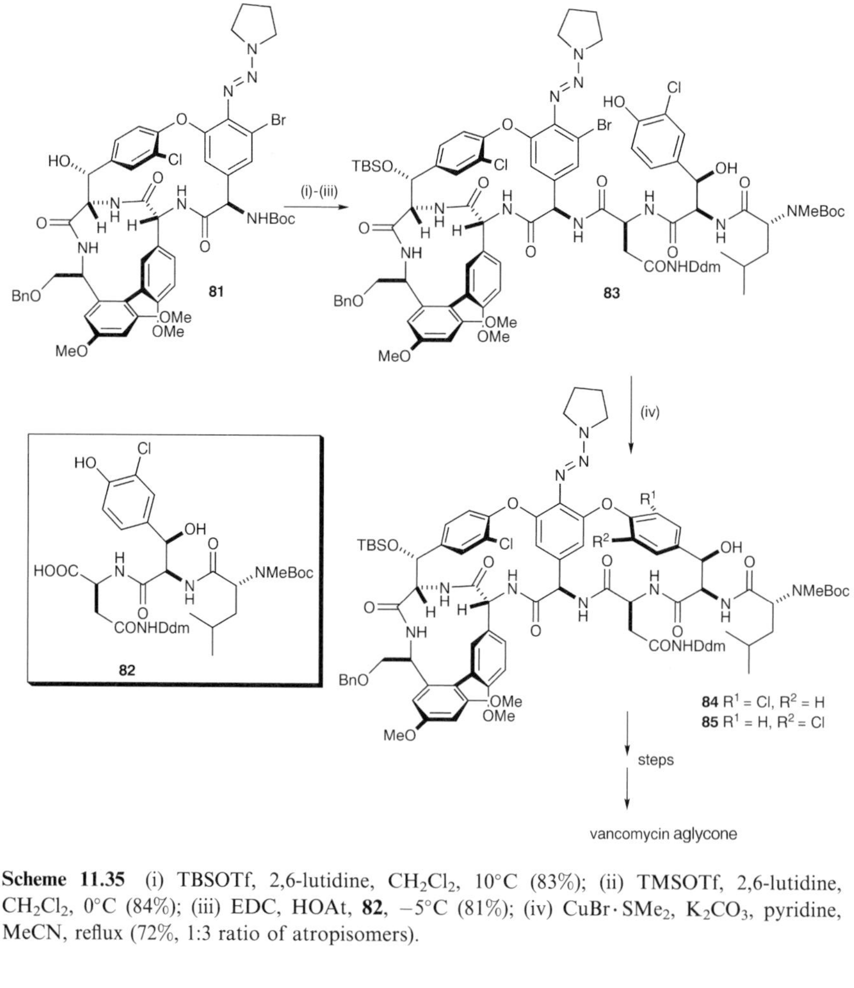

Scheme 11.35 (i) TBSOTf, 2,6-lutidine, CH_2Cl_2, 10°C (83%); (ii) TMSOTf, 2,6-lutidine, CH_2Cl_2, 0°C (84%); (iii) EDC, HOAt, **82**, −5°C (81%); (iv) CuBr·SMe_2, K_2CO_3, pyridine, MeCN, reflux (72%, 1:3 ratio of atropisomers).

11.3.4   Lantibiotics

The lantibiotic family of cyclic peptides are distinguished by one or more lanthionine or methyllanthionine bridges and often also contain dehydroamino acids. In spite of the intriguing structure of these peptides, only one total synthesis, of nisin (**86**), has so far been reported. Wakamiya and

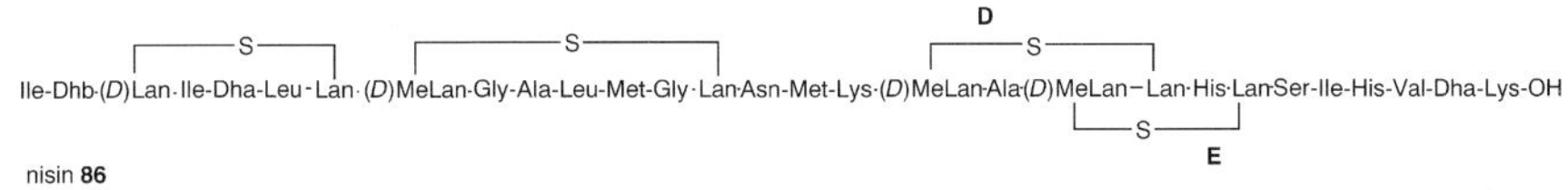

nisin **86**

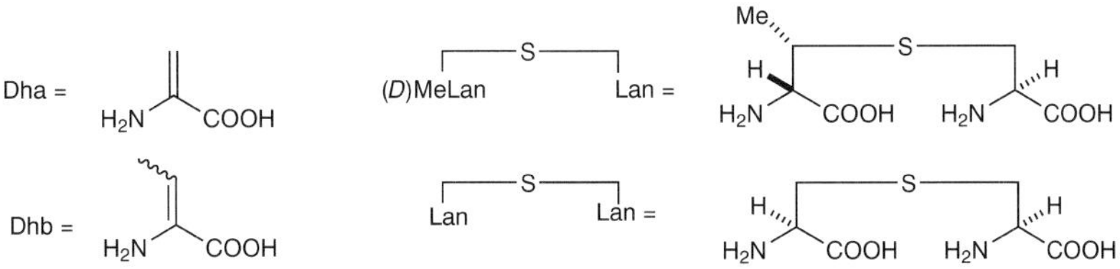

co-workers[109] approached the synthesis of the component rings of nisin by desulphurisation of suitable cystine-bridged precursors; the strategy is exemplified by the synthesis of the D-E rings of nisin (Scheme 11.36).[110]

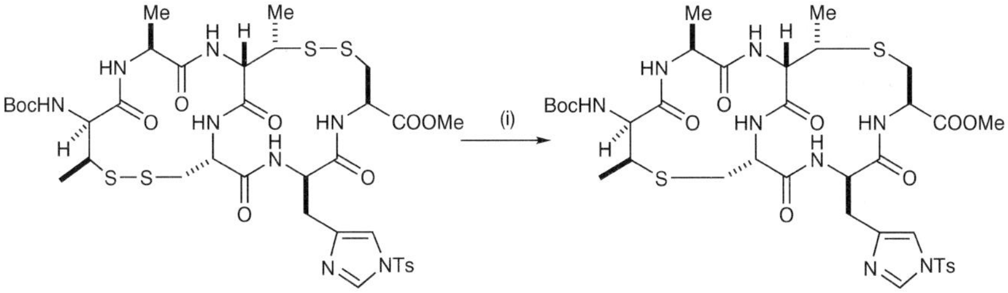

Scheme 11.36   Synthesis of rings D and E of nisin: (i) P(NEt$_2$)$_3$ (40%).

After synthesis of each ring, segment condensation was used to link the component rings together. The Dha residues were introduced as diaminopropyl (Dpr) groups, and the Dhb residues as Thr residues. Hoffman degradation of the Dpr residues and dehydration of the Thr residues in the final stages afforded nisin.

11.4   Future challenges

In addition to the completed syntheses described in this chapter, major advances towards the syntheses of other complex cyclic peptide architectures, such as diazonamide A,[111,112] keramamide F[113] and theonellamide F,[114] have been reported. Further naturally occurring cyclic peptides with intriguing architectures and novel biological properties continue to be discovered. These are regularly reviewed in the Royal Society of Chemistry Specialist Periodical Report *Amino Acids and Peptides*, published annually, and more detailed reviews of specific families of cyclic peptides are found in *Natural Product Reports*.[115] Although outside the scope of this chapter, the design and synthesis of nonnatural cyclic peptides as lead structures for drug discovery is also an exciting and expanding field, and good reviews of this area have been published.[116,117] The new synthetic challenges presented both by naturally occurring and by artificially designed cyclic and polycyclic peptide architectures will continue to inspire synthetic chemists for the foreseeable future.

References

1. J.M. Humphrey and A.R. Chamberlin, 1997, *Chem. Rev.*, 97, 2243.
2. K.D. Kopple, 1972, *J. Pharm. Sci.*, 61, 1345.
3. M.J.O. Anteunis and N.K. Sharma, 1988, *Bull. Soc. Chim. Belg.*, 97, 281.
4. F. Cavelier-Frontin, G. Pèpe, J. Verducci, D. Siri, and R. Jacquier, 1992, *J. Am. Chem. Soc.*, 114, 8885.
5. S.F. Brady, S.L. Varga, R.M. Freidinger, D.A. Schwenk, M. Mendlowski, F.W. Holly, and D.F. Veber, 1979, *J. Org. Chem.*, 44, 3101.
6. Q. Meng and M. Hesse, 1991, *Top. Curr. Chem.*, 161, 107.
7. N.L. Benoiton, Y.C. Lee, R. Steinauer, and F.M.F. Chen, 1992, *Int. J. Pept. Protein Res.*, 40, 559.
8. (a) T. Shiori and S. Yamada, 1974, *Chem. Pharm. Bull*, 22, 849; (b) T. Shiori, K. Ninomiya, and S. Yamada, 1972, *J. Am. Chem. Soc.*, 94, 6203.
9. (a) U. Schmidt, A. Lieberknecht, H. Griesser, and J. Talbiersky, 1982, *J. Org. Chem.*, 47, 3261; (b) U. Schmidt, 1986, *Pure Appl. Chem.*, 58, 295.
10. R.D. Tung, M.K. Dhaon, and D.H. Rich, 1986, *J. Org. Chem.*, 51, 3350.
11. B.E. Maryanoff, M.N. Greco, H.-C. Zhang, P. Andrade-Gordon, J.A. Kauffman, K.C. Nicolaou, A. Liu, and P.H. Brungs, 1995, *J. Am. Chem. Soc.*, 117, 1225.
12. C. Van der Auwera and M. Anteunis, 1987, *Int. J. Pept. Protein Res.*, 29, 574.
13. S. Takuma, Y. Hamada, and T. Shiori, 1982, *Chem. Pharm. Bull.*, 30, 3147.
14. S. Chen and J. Xu, 1991, *Tetrahedron Lett.*, 32, 6711.
15. J. Dudash, J.J. Jiang, S.C. Mayer, and M.M. Joullié, 1993, *Synth. Commun.*, 23, 349.
16. J. Deng, Y. Hamada, and T. Shiori, 1996, *Tetrahedron Lett.*, 37, 2261.
17. R. Knorr, A. Trzeciak, W. Bannwarth, and D. Gillessen, 1989, *Tetrahedron Lett.*, 30, 1927.
18. B. Castro, J.R. Dormoy, G. Elvin, and C. Selve, 1975, *Tetrahedron Lett.*, 14, 1219.
19. J. Coste, D. Le-Nguyen, and B. Castro, 1990, *Tetrahedron Lett.*, 31, 205.
20. A.M. Felix, C.-T. Wang, E.P. Heimer, and A. Fournier, 1988, *Int. J. Pept. Protein Res.*, 31, 231.
21. A. Ehrlich, H.-U. Heyne, R. Winter, M. Beyermann, H. Haber, L.A. Carpino, and M. Bienert, 1996, *J. Org. Chem.*, 61, 8831.
22. L.A. Carpino, 1993, *J. Am. Chem. Soc.*, 115, 4397.

23. F. Cavelier-Frontin, S. Achmad, J. Verducci, R. Jacquier, and G. Pèpe, 1993, *J. Mol. Struct.*, 286, 125.
24. Y. Shao, W. Lu, and S.B.H. Kent, 1998, *Tetrahedron Lett.*, 39, 3911.
25. L. Zhang and J.P. Tam, 1997, *J. Am. Chem. Soc.*, 119, 2363.
26. L. Zhang and J.P. Tam, 1997, *Tetrahedron Lett.*, 38, 4375.
27. S. Mazur and P. Jayalekshmy, 1978, *J. Am. Chem. Soc.*, 101, 677.
28. G. Ösapay, A. Profit, and J.W. Taylor, 1990, *Tetrahedron Lett.*, 31, 6121.
29. (a) A. Kapurniotu and J.W. Taylor, 1993, *Tetrahedron Lett.*, 34, 7031; (b) B.H. Lee, 1997, *Tetrahedron Lett.*, 38, 757.
30. G. Ösapay and J.W. Taylor, 1990, *J. Am. Chem. Soc.*, 112, 6046.
31. L.S. Richter, J.Y.K. Tom, and J.P. Burnier, 1994, *Tetrahedron Lett.*, 35, 5547.
32. (a) S.A. Kates, N.A. Solé, C.R. Johnson, D. Hudson, G. Barany, and F. Alberico, 1993, *Tetrahedron Lett.*, 34, 1549; (b) S.A. Kates, S.B. Daniels, and F. Alberico, 1993, *Anal. Biochem.*, 212, 303.
33. J. Alsina, F. Rabanal, E. Giralt, and F. Alberico, 1994, *Tetrahedron Lett.*, 35, 9633.
34. J.S. McMurray, C.A. Lewis, and N.U. Obeyesekere, 1994, *Pept. Res.*, 7, 195 and references cited therein.
35. (a) J.H. Lee, J.H. Griffin, and T.I. Nicas, 1996, *J. Org. Chem.*, 61, 3983; (b) I.R. Marsh, M. Bradley, and S.J. Teague, 1997, *J. Org. Chem.*, 62, 6199.
36. S.A. Kates, N.A. Solé, F. Alberico, and G. Barany, 1994, in *Peptides: Design, Synthesis and Biological Activity* (C. Basava, G.M. Anantharamaiah, eds.), Birkhauser, Boston, p. 39.
37. C. Blackburn, and S.A. Kates, 1997, in *Methods in Enzymology, Solid Phase Peptide Synthesis* (G.B. Fields, ed.), Academic Press, New York, vol. 289, p. 175.
38. K.L. Webster, T.J. Rutherford, and D. Gani, 1997, *Tetrahedron Lett.*, 38, 5716.
39. C. Taylor, R.J. Quinn, and P. Alewood, 1996, *Bioorg. Med. Chem. Lett.*, 6, 2107.
40. A.F. Spatola, K. Darlak, and P. Romanovskis, 1996, *Tetrahedron Lett.*, 37, 591.
41. S.C. Story and J.V. Aldrich, 1994, *Int. J. Pept. Protein Res.*, 43, 292.
42. For a further example, see: D. Delforge, M. Dieu, E. Delaive, M. Art, B. Gillon, B. Devreese, M. Raes, J. van Beeumen, and J. Remacle, 1996, *Lett. Pept. Sci.*, 3, 89.
43. B. Kamber, A. Hartmann, K. Eisler, B. Riniker, H. Rink, P. Sieber, and W. Rittel, 1980, *Helv. Chim. Acta*, 63, 899.
44. N. Fujii, A. Otaka, S. Funakoshi, K. Bessho, T. Watanabe, K. Akaji, and H. Yajima, 1987, *Chem. Pharm. Bull.*, 35, 2339.
45. K. Akaji, T. Tatsumi, M. Yoshida, T. Kimura, Y. Fujiwara, and Y. Kiso, 1992, *J. Am. Chem. Soc.*, 114, 4137.
46. M.S. Bernatowicz, R. Matsueda, and G.R. Matsueda, 1986, *Int. J. Pept. Protein Res.*, 28, 107.
47. M. Ruiz-Gayo, M. Royo, I. Fernández, F. Alberico, E. Giralt, and M. Pons, 1993, *J. Org. Chem.*, 58, 6319.
48. D. Andreu, F. Alberico, N.A. Solé, M.C. Munson, M. Ferrer, and G. Barany, 1994, in *Methods in Molecular Biology, Peptide Synthesis Protocols* (M.W. Pennington, B.M. Dunn, eds.), Humana Press, Totowa, NJ, vol. 35, p. 91, and other earlier reviews cited therein.
49. I. Annis, B. Hargittai, and G. Barany, 1997, in *Methods in Enzymology, Solid Phase Peptide Synthesis* (G.B. Fields, ed.), Academic Press, New York, vol. 289, p. 198.
50. W.J. Colucci, R.D. Tung, J.A. Petri, and D.A. Rich, 1990, *J. Org. Chem.*, 55, 2895.
51. M. Hagihara and S.L. Schreiber, 1992, *J. Am. Chem. Soc.*, 114, 6570.
52. P. Wipf and H. Kim, 1993, *J. Org. Chem.*, 58, 5592.
53. J. Deng, Y. Hamada, T. Shiori, S. Matsunaga, and N. Fusetani, 1994, *Angew. Chem., Int. Ed. Engl.*, 33, 1729.
54. H.M.M. Bastiaans, J.L. van der Baan, and H.C.J. Ottenheijm, 1997, *J. Org. Chem.*, 62, 3880.

55. C. MacKintosh and R.W. MacKintosh, 1994, *Trends Biol. Sci.*, 19, 444.
56. R.J. Valentekovich and S.L. Schreiber, 1995, *J. Am. Chem. Soc.*, 117, 9069.
57. J.M. Humphrey, J.B. Aggen, and A.R. Chamberlin, 1996, *J. Am. Chem. Soc.*, 118, 11759.
58. J.M. Humphrey, J.A. Hart, R.J. Bridges, and A.R. Chamberlin, 1994, *J. Org. Chem.*, 59, 2467.
59. (a) M. Namikoshi., K.L. Reinhart, A.M. Dahlem, V.R. Beasley, and W.W. Carmichael, 1989, *Tetrahedron Lett.*, 30, 4349; (b) T.K. Chakraborty and S.P. Joshi, 1990, *Tetrahedron Lett.*, 31, 2046; (c) M.F. Beatty, C. Jennings-White, and M.A. Avery, 1992, *J. Chem. Soc. Perkin Trans. I*, 1637; (d) F. D'Aniello, A. Mann, A. Schoenfelder, and M. Taddei, 1997, *Tetrahedron*, 53, 1447; (e) D.J. Cundy, A.C. Donohue, and T.D. McCarthy, 1998, *Tetrahedron Lett.*, 39, 5125.
60. (a) A.P. Mehrotra, K.L. Webster, and D. Gani, 1997, *J. Chem. Soc. Perkin Trans. I*, 2495; (b) A.B. Maude, A.P. Mehrotra, and D. Gani, 1997, *J. Chem. Soc. Perkin Trans. I*, 2513.
61. (a) U. Schmidt, A. Lieberknecht, H. Griesser, and F. Bartkowiak, 1984, *Angew. Chem., Int. Ed. Engl.*, 23, 318; (b) U. Schmidt, A. Lieberknecht, H. Griesser, R. Utz, T. Beuttler, and F. Bartkowiak, 1986, *Synthesis*, 361.
62. J.E. Baldwin, R.M. Adlington, C.R.A. Godfrey, and V.K. Patel, 1993, *Tetrahedron*, 49, 7837.
63. J. Taunton, J.L. Collins, and S.L. Schreiber, 1996, *J. Am. Chem. Soc.*, 118, 10412.
64. B.S. Davidson, 1993, *Chem. Rev.*, 93, 1771.
65. U. Schmidt and P. Gleich, 1985, *Angew. Chem., Int. Ed. Engl.*, 24, 569.
66. T. Sugiura, Y. Hamada, and T. Shiori, 1987, *Tetrahedron Lett.*, 28, 2251.
67. Y. Hamada, M. Shibata, and T. Shiori, 1985, *Tetrahedron Lett.*, 26, 5159.
68. U. Schmidt and H. Griesser, 1986, *Tetrahedron Lett.*, 27, 163.
69. P. Wipf, 1995, *Chem. Rev.*, 95, 2115.
70. P. Wipf and P. C. Fritch, 1994, *Tetrahedron Lett.*, 35, 5397.
71. C.D.J. Boden, G. Pattenden, and T. Ye, 1995, *Synlett*, 417.
72. C.D.J. Boden and G. Pattenden, 1995, *Tetrahedron Lett.*, 36, 6153.
73. G.M. Atkins and E.M. Burgess, 1968, *J. Am. Chem. Soc.*, 90, 4744.
74. D.J. Freeman and G. Pattenden, 1998, *Tetrahedron Lett.*, 39, 3251.
75. (a) C.D.J. Boden, M.C. Norley, and G. Pattenden, 1996, *Tetrahedron Lett.*, 37, 9111; (b) M.C. Norley and G. Pattenden, 1998, *Tetrahedron Lett.*, 39, 3087.
76. P. Wipf and P.C. Fritch, 1996, *J. Am. Chem. Soc.*, 118, 12358.
77. P. Wipf, C.P. Miller, S. Venkatraman, and P.C. Fritch, 1995, *Tetrahedron Lett.*, 36, 6395.
78. M. Nakamura, T. Shibata, K. Nakane, T. Nemoto, M. Ojika, and K. Yamada, 1995, *Tetrahedron Lett.*, 36, 5059.
79. D.H. Williams, 1996, *Nat. Prod. Rev.*, 13, 469 and references cited therein.
80. C.T. Walsh, S.L. Fisher, L.-S. Park, M. Prahalad, and Z. Wu., 1996, *Chem. Biol.*, 3, 21 and references cited therein.
81. (a) U. Schmidt, M. Zäh, and A. Lieberknecht, 1991, *J. Chem. Soc., Chem. Commun.*, 1002; (b) R.J. Heffner, J.J. Jiang, and M.M. Joullie, 1992, *J. Am. Chem. Soc.*, 114, 10181.
82. A.V. Rama Rao, M.K. Gurjar, K.L. Reddy, and A.S. Rao, 1995, *Chem. Rev.*, 95, 2135.
83. D.A. Evans and K.M. DeVries, 1994, in *Glycopeptide Antibiotics* (R. Nagarajan, ed.), Marcel Dekker, New York, p. 63.
84. J. Zhu, 1997, *Synlett*, 133.
85. R.M. Williams and J.A. Hendrix, 1992, *Chem. Rev.*, 92, 889.
86. (a) D.A. Evans, M.R. Wood, B.W. Trotter, T.I. Richardson, J.C. Barrow, and J.L. Katz, 1998, *Angew. Chem., Int. Ed. Engl.*, 37, 2700; (b) D.A. Evans, C.J. Dinsmore, P.S. Watson, M.R. Wood, T.I. Richardson, B.W. Trotter, and J.L. Katz, 1998, *Angew. Chem., Int. Ed. Engl.*, 37, 2704.

87. (a) K.C. Nicolaou, S. Natarajan, H. Li, N.F. Jain, R. Hughes, M.E. Solomon, J.M. Ramanjulu, C.N.C. Boddy, and M. Takayanagi, 1998, *Angew. Chem., Int. Ed. Engl.*, 37, 2708; (b) K.C. Nicolaou, N.F. Jain, S. Natarajan, R. Hughes, M.E. Solomon, H. Li, J.M. Ramanjulu, M. Takayanagi, A.E. Koumbis, and T. Bando, 1998, *Angew. Chem., Int. Ed. Engl.*, 37, 2714; (c) K.C. Nicolaou, M. Takayanagi, N.F. Jain, S. Natarajan, A.E. Koumbis, T. Bando, and J.M. Ramanjulu, 1998, *Angew. Chem., Int. Ed. Engl.*, 37, 2717.
88. A.G. Brown, M.J. Crimmin, and P.D. Edwards, 1992, *J. Chem. Soc. Perkin Trans. I*, 123.
89. (a) D.A. Evans and A.E. Weber, 1987, *J. Am. Chem. Soc.*, 109, 7151; (b) D.A. Evans, T.C. Britton, J.A. Ellman, and R.L. Dorow, 1990, *J. Am. Chem. Soc.*, 112, 4011; (c) D.A. Evans, D.A. Evrard, S.D. Rychnovsky, T. Früh, W.G. Whittingham, and K.M. DeVries, 1992, *Tetrahedron Lett.*, 33, 1189.
90. (a) D.A. Evans and C.J. Dinsmore, 1993, *Tetrahedron Lett.*, 34, 6029; (b) D.A. Evans, C.J. Dinsmore, D.A. Evrard, and K.M. DeVries, 1993, *J. Am. Chem. Soc.*, 115, 6426.
91. D.A. Evans, C.J. Dinsmore, A.M. Ratz, D.A. Evrard, and J.C. Barrow, 1997, *J. Am. Chem. Soc.*, 119, 3417.
92. (a) D.L. Boger and D. Yohannes, 1991, *J. Am. Chem. Soc.*, 113, 1427; (b) D.L. Boger, M.A. Patane, J. Zhou, 1994, *J. Am. Chem. Soc.*, 116, 8544.
93. A.V. Rama Rao, T.K. Chakraborty, K.L. Reddy, and A.S. Rao, 1992, *Tetrahedron Lett.*, 33, 4799.
94. A.V. Rama Rao, M.K. Gurjar, A.B. Reddy, and V.B. Khare, 1993, *Tetrahedron Lett.*, 34, 1657.
95. Y. Suzuki, S. Nishiyama, and S. Yamamura, 1990, *Tetrahedron Lett.*, 31, 4053.
96. D.A. Evans, J.A. Ellmann, and K.M. DeVries, 1989, *J. Am. Chem. Soc.*, 111, 8912.
97. D.A. Evans, J.C. Barrow, P.S. Watson, A.M. Ratz, C.J. Dinsmore, D.A. Evrard, K.M. DeVries, J.A. Ellman, S.D. Rychnovsky, and J. Lacour, 1997, *J. Am. Chem. Soc.*, 119, 3419.
98. D.A. Evans, C.J. Dinsmore, and A.M. Ratz, 1997, *Tetrahedron Lett.*, 38, 3189.
99. (a) R. Beugelmans, M. Bois-Choussy, C. Vergne, J.-P. Bouillon, and J. Zhu, 1996, *J. Chem. Soc., Chem. Commun.*, 1029; (b) C. Vergne, M. Bois-Choussy, and J. Zhu, 1988, *Synlett*, 1159.
100. M. Bois-Choussy, C. Vergne, L. Neuville, R. Beugelmans, and J. Zhu, 1997, *Tetrahedron Lett.*, 38, 5795.
101. A.V. Rama Rao, K.L. Reddy, and A.S. Rao, 1994, *Tetrahedron Lett.*, 35, 8465.
102. D.L. Boger, R.M. Borzilleri, and S. Nukui, 1995, *Bioorg. Med. Chem. Lett.*, 5, 3091.
103. J.W. Janetka and D.H. Rich, 1995, *J. Am. Chem. Soc.*, 117, 10585.
104. A.J. Pearson, P. Zhang, and G. Bignan, 1996, *J. Org. Chem.*, 61, 3940.
105. D.A. Evans and P.S. Watson, 1996, *Tetrahedron Lett.*, 37, 3251.
106. K.C. Nicolaou, J.M. Ramanjulu, S. Natarajan, S. Bräse, H. Li, C.N.C. Boddy, and F. Rübsam, 1997, *J. Chem. Soc., Chem. Commun.*, 1899.
107. K.C. Nicolaou, C.N.C. Boddy, S. Natarajan, T.Y. Yue, H. Li, S. Bräse, and J.M. Ramanjulu, 1997, *J. Am. Chem. Soc.*, 119, 3421.
108. (a) D.L. Boger, O. Loiseleur, S.L. Castle, R.T. Beresis, and J.H. Wu, 1997, *Bioorg. Med. Chem. Lett.*, 7, 3199; (b) D.L. Boger, R.T. Beresis, O. Loiseleur, J.H. Wu, and S.L. Castle, 1998, *Bioorg. Med. Chem. Lett.*, 8, 721; (c) D.L. Boger, S. Miyazaki, O. Loiseleur, R. T. Beresis, S.L. Castle, J.H. Wu, and Q. Jin, 1998, *J. Am. Chem. Soc.*, 120, 8920.
109. K. Fukase, M. Kitazawa, A. Sano, K. Shimbo, S. Horimoto, H. Fujita, A. Kubo, T. Wakamiya, and T. Shiba, 1992, *Bull. Chem. Soc. Jpn.*, 65, 2227 and references cited therein.
110. K. Fukase, Y. Oda, A. Kubo, T. Wakamiya, and T. Shiba, 1990, *Bull. Chem. Soc. Jpn.*, 63, 1758.
111. A. Boto, M. Ling, G. Meek, and G. Pattenden, 1998, *Tetrahedron Lett.*, 39, 8167.
112. C.J. Moody, K.J. Doyle, M.C. Elliott, and T.J. Mowlem, 1997, *J. Chem. Soc. Perkin Trans. I*, 2413.

113. J.A. Sowinski and P.L. Toogood, 1995, *Tetrahedron Lett.*, 36, 67.
114. (a) K. Tohdo, Y. Hamada, and T. Shiori, 1994, *Synlett,* 247; (b) K. Tohdo, Y. Hamada, and T. Shiori, 1994, *Synlett*, 250.
115. For example, see: (a) J.R. Lewis, 1998, *Nat. Prod. Rep.*, 15, 417; (b) D.J. Faulkner, 1997, *Nat. Prod. Rep.*, 14, 259.
116. V.J. Hruby and G.G. Bonner, 1994, in *Methods in Molecular Biology, Peptide Synthesis Protocols* (M.W. Pennington, B.M. Dunn, eds.), Humana Press, Totowa, NJ, vol. 35, p. 201.
117. D.P. Fairlie, G. Abbenante, and D.R. March, 1995, *Curr. Med. Chem.*, 2, 654.

12 The chemical synthesis of naturally occurring cyclodepsipeptides

K.J. Hale, G.S. Bhatia and M. Frigerio

12.1 Introduction

Cyclodepsipeptides are cyclic peptides possessing at least one ester linkage. Interest in such structures has risen steadily in recent years for several reasons. First, many cyclodepsipeptides elicit important pharmacological effects that make them medicinally valuable. Second, many cyclodepsipeptides serve as useful probe molecules for dissecting the regulatory pathways that control cell functioning. Third, many cyclodepsipeptides have bewilderingly complex and exquisitely ornate molecular architectures that pose a significant chemical challenge for total synthesis.

The scientific dividends that have arisen from synthetic progress in this area have been rich and bountiful. Not only has cyclodepsipeptide synthesis stimulated important breakthroughs in chemical reaction design, it has also significantly advanced our understanding of synthetic strategy and tactics, and has given biologists secure and predictable access to many high-purity low molecular weight biological probes. In many cases, the analogue structures that have been created have given remarkable insights into the critical biochemical events and signalling pathways that control cells. Thus, the chemical synthesis of cyclodepsipeptides should very much be seen as a cross-disciplinary endeavour, traversing many of the traditional boundaries that one would normally associate with mainstream synthetic organic chemistry, biochemistry, molecular biology and medicine.

In this chapter, we will attempt to give readers a flavour of how this important area of natural products research has progressed over the years 1988 to mid-1999. The selection criteria we have used for our coverage have been biological or medical significance, molecular complexity, unusualness of structure, or simply whether a synthesis has illustrated some important new methodological advance or breakthrough, or some novel strategy that would appear to be of more general synthetic significance. With this as background, we hope you will enjoy the journey of discovery on which we will now go.

12.2   Arenastatin A

Arenastatin A (cryptophycin 24) is a novel cyclodepsipeptide found in the Okinawan sponge *Dysidea arenia* as well as in the blue-green alga *Nostoc* sp. GSV224.[1] It and related family members have powerful anticancer properties[1,2] both *in vitro* and *in vivo*. Specifically, arenastatin A has an IC_{50} of 5 pg/ml when marshalled against human nasopharyngeal KB cancer cells,[1b] which makes it one of the most potent anticancer agents ever discovered. To date, only two total syntheses of arenastatin A have been recorded. The first was reported by Kobayashi in 1994,[3] the second by J.D White in 1998.[4] In White's strategy,[4] he opted to construct the 16-membered cyclodepsipeptide ring by macrolactamisation, which led to acid **5** and alcohol **11** being selected as intermediates. The synthesis of these two fragments is detailed in Schemes 12.1, 12.2 and 12.3.

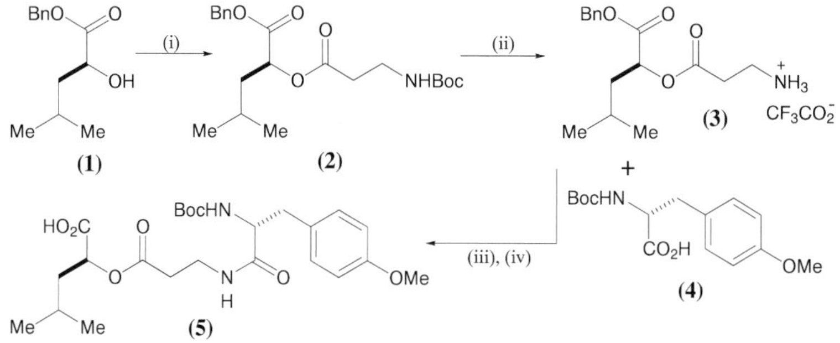

Scheme 12.1   (i) $HO_2CCH_2CH_2NHBoc$, EDCI, DMAP, 99%; (ii) TFA, CH_2Cl_2, 0°C; (iii) EDCI, HOBT, Et_3N, 99%; (iv) $Pd(OH)_2$, H_2, EtOAc.

A Boc-urethane/*O*-benzyl ester protecting group strategy was employed for the assembly of tripeptide **5**. The route developed exploited longstanding acid activation methods to establish the new ester and amide linkages (Scheme 12.1). In compound **11**, the *anti*-relationship between the two adjacent stereocentres was controlled through a Brown asymmetric allylboration reaction[5] between aldehyde **6** and (*E*)-crotyldiisopinylcampheylborane, available from the (+)-methoxyborane (Scheme 12.2). The *anti* product **7** was isolated with greater than 50:1 diastereoselectivity and in 85% enantiomeric excess (ee). A series of Wittig olefinations were then used to anchor stereoselectively the two (*E*)-disubstituted olefins within stryrene **11**.

An alternative route to **11** was also put in place (Scheme 12.3). In this, the readily available aldehyde **12**[6] was exploited as the chiral starting intermediate. It underwent a highly stereoselective *C*-allylation reaction with allyltri-*n*-butylstannane[7] to provide the *anti*-homoallylic alcohol **13**

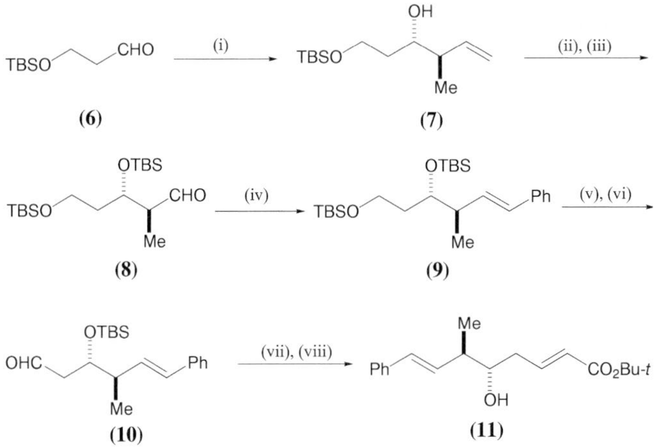

Scheme 12.2 (i) *Trans* CH$_3$CH=CHCH$_3$, *t*-BuOK, *n*-BuLi, −75°C to −45°C, (+)-(Ipc)$_2$-BOMe, BF$_3$·OEt$_2$, −78°C then **6**, −78°C to RT, 71% (dr > 50:1, er >1 2:1); (ii) TBSCl, imid, DMF, RT, 93%; (iii) O$_3$, CH$_2$Cl$_2$, −78°C, then Me$_2$S, −78°C to RT, 62%; (iv) *n*-BuLi, THF, PhCH$_2$P(O)(OEt)$_2$, −78°C to RT, 79%; (v) HF-Py, THF, RT, 99%; (vi) Dess–Martin periodinane, RT, CH$_2$Cl$_2$, 98%; (vii) Ph$_3$P=CHCO$_2$Bu-*t*, RT, 85%; (viii) TBAF, THF, 75% (20% overall yield for 8 steps).

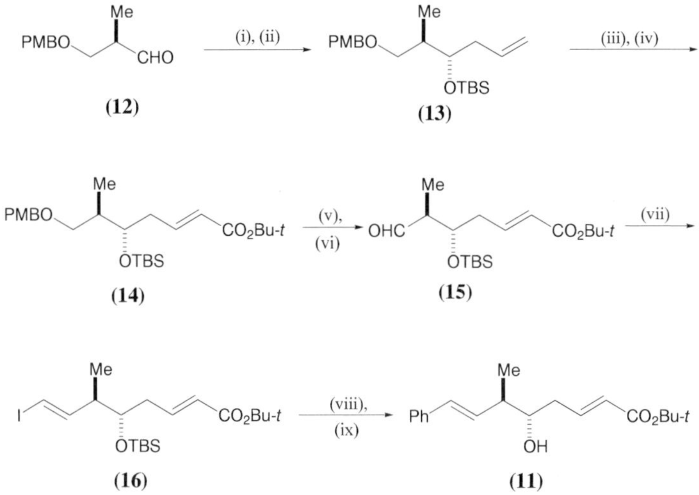

Scheme 12.3 (i) Bu$_3$SnCH$_2$CH=CH$_2$, SnCl$_4$, CH$_2$Cl$_2$, −100°C, 76% (20:1); (ii) TBSCl, imid, DMF, RT, 91%; (iii) OsO$_4$ (cat.), NaIO$_4$, THF-H$_2$O, 76%; (iv) Ph$_3$P=CHCO$_2$Bu-*t*, CH$_2$Cl$_2$, RT, 95%; (v) DDQ, CH$_2$Cl$_2$-H$_2$O, 0°C to RT, 92%; (vi) Dess–Martin periodinane, CH$_2$Cl$_2$, 92%; (vii) CHI$_3$, CrCl$_2$, THF, 0°C, 62%; (viii) PhSnMe$_3$, DMF, PdCl$_2$(MeCN)$_2$, RT, 67%; (ix) TBAF, THF, 75%, (12% overall yield for 9 steps).

with greater than 20:1 diastereoselectivity. A Wittig olefination again featured for installation of the α,β-enoate unit, and a Takai iodoolefination[8]/Stille cross-coupling[9] tactic was relied upon for introduction of the styryl moiety.

Having secured pathways to acid **5** and alcohol **11**, only four more steps were required for completion of the arenastatin A synthetic venture (Scheme 12.4). These included an esterification reaction to form **17**, a simultaneous cleavage of the Boc urethane and *t*-butyl ester groups with trifluoroacetic acid, a macrolactamisation of the resulting amino acid via the acid azide method and, finally, a stereoselective epoxidation of the more electron-rich double bond with dimethyldioxirane. The viability of the latter transformation was established by Kobayashi during his first total synthesis of the target in 1994.[3]

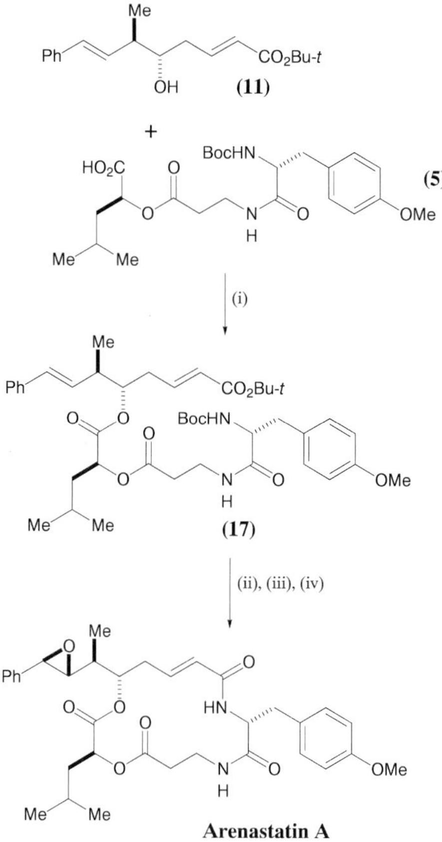

Scheme 12.4 (i) DIC, DMAP, CH$_2$Cl$_2$, RT, 93%; (ii) TFA, CH$_2$Cl$_2$, 0°C; (iii) DPPA, NaHCO$_3$, DMF, 0°C, 57% (2 steps); (iv) DMDO, CH$_2$Cl$_2$, −30°C, 24 h, 80%, (dr 3:1).

Clearly, the advent of both these syntheses has significantly expedited the future synthetic assembly of novel analogues for this biologically interesting class of natural products, which are thought to exert their anticancer effects by an irreversible inhibition of tubulin polymerisation.

12.3   Dolastatin D

In 1993, Yamada and co-workers reported[10] their discovery of the markedly cytotoxic cyclodepsipeptide, dolastatin D, in aqueous methanol extracts of the Japanese sea hare *Dolabella auricularia*. Its structure was assigned by chemical degradation and spectroscopic analysis. These synergistic efforts revealed a novel pentadepsipeptidal array in which the unusual β-amino acid, (2*R*,3*R*)-3-amino-2-methylbutanoic acid, was present. Observation of the latter amino acid was of particular significance, for it had never previously been encountered in nature, yet it had been synthesised chemically.[11] Dolastatin D shows moderate cytotoxicity against HeLa-S3 cells, its IC_{50} being 2.2 µg/ml.

In order to place future biological work on a more secure footing, Yamada and co-workers developed the total synthesis of dolastatin D presented in Scheme 12.5. Items of particular merit in their pathway include the high-yielding *N*-methylation of tridepsipeptide **22** with sodium hydride and methyl iodide, which did not interfere with the stereochemical integrity of other labile stereocentres, and the BOP-mediated[12] macrolactamisation which created the 16-membered ring of dolastatin D. The latter reaction proceeded in a very respectable 66% yield, in sharp contrast to the rather meagre 13% yield of natural product obtained using BOP-Cl.[13]

The substantial variations in product yield that are frequently recorded in macrocyclisations mediated by different dehydrating reagents illustrates one important and quite general point widely appreciated by the cognoscenti of cyclodepsipeptide ring synthesis, and that is, one should always be prepared to evaluate a *very wide range* of carboxyl activation reagents to obtain a good yield of a particular cyclodepsipeptide product. In essence, one must very much be prepared to adopt a 'combinatorial' approach when tackling such macrocyclic ring closures.

12.4   Enopeptin B

The awesome combination of a cyclodepsipeptide, a Michael acceptor polyene and an *N*-acyl-2-aminocyclopentane-1,3-dione moiety had never previously been encountered in a natural product prior to Isono and

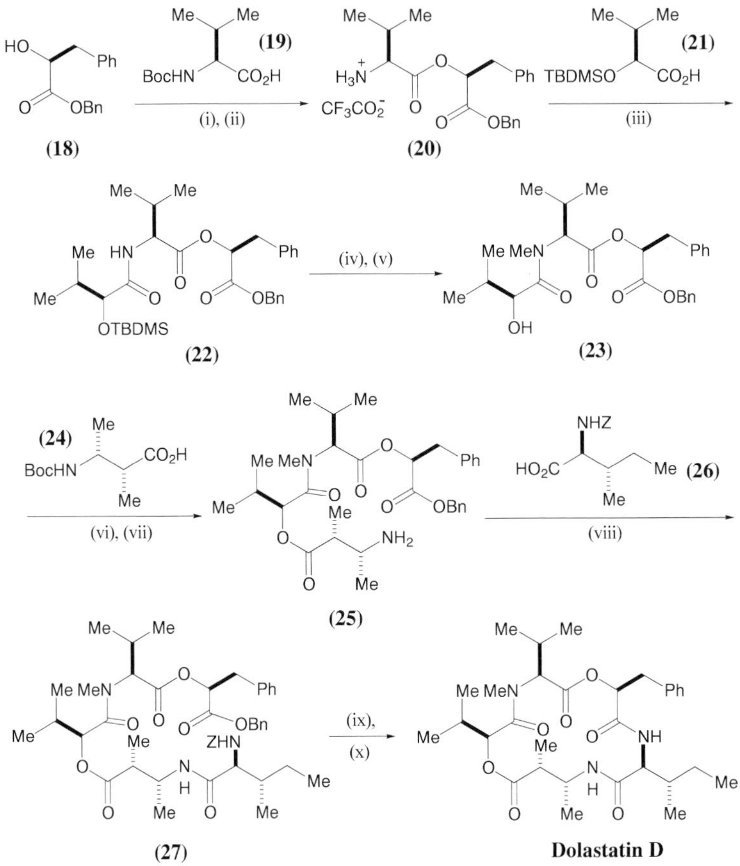

Scheme 12.5 (i) **19**, DCC, DMAP, CH$_2$Cl$_2$, 0°C then RT, 88%; (ii) CF$_3$CO$_2$H, CH$_2$Cl$_2$, 0°C; (iii) **21**, DEPC, Et$_3$N, DMF, 0°C then RT, 89% (2 steps); (iv) NaH, MeI, DMF, 0°C, 75%; (v) 12% HF, H$_2$O, MeCN, RT, 98%; (vi) **24**, DCC, DMAP, CSA, CH$_2$Cl$_2$, 0°C then RT, 98%; (vii) CF$_3$CO$_2$H, CH$_2$Cl$_2$, 0°C; (viii) **26**, DEPC, Et$_3$N, DMF, 0°C then RT, 82% (2 steps); (ix) H$_2$, 10% Pd/C, MeOH, AcOH, H$_2$O, RT, 91%; (x) regime A: BOPCl, Et$_3$N, CH$_2$Cl$_2$, 0°C–4°C, 13%; regime B: HOSu, DCC, DMF, CH$_2$Cl$_2$, 41%; regime C: DPPA, Et$_3$N, DMF, 0°C then RT, 47%; regime D: BOP, NaHCO$_3$, DMF, RT, 66%.

co-workers' discovery of enopeptins A and B in the culture filtrates of *Streptomyces* sp. RK-1051.[14] Both molecules exhibit significant antibiotic effects *in vitro* against Gram-positive strains of bacteria, with *Staphylococcus aureus* being especially mortal in their presence. The chemical challenges posed by enopeptins A and B cannot be overemphasised. Two issues that must be grappled with in any total synthesis are the low nucleophilicity of 2-aminocyclopentanediones in *N*-acylation reactions and the high sensitivity of the enopeptins towards acids, bases and strong nucleophiles. Notwithstanding these difficulties, Schmidt and co-workers[15] synthesised enopeptin B by the strategy outlined in Schemes 12.6 and 12.7. Early effort focused on the preparation of pentadespsipeptide **28** by conventional peptide coupling technology. The phenacyl ester was then detached from **28** with zinc dust in acetic acid, and a pentafluorophenyl ester introduced at that carbonyl. The Boc group was excised from **29** by treatment with HCl in dioxane, and the amine fully unveiled under biphasic conditions with aqueous sodium bicarbonate and chloroform. This regime brought about a very successful cyclisation to compound **30**. Catalytic hydrogenolysis then completed the synthesis of subtarget **31**.

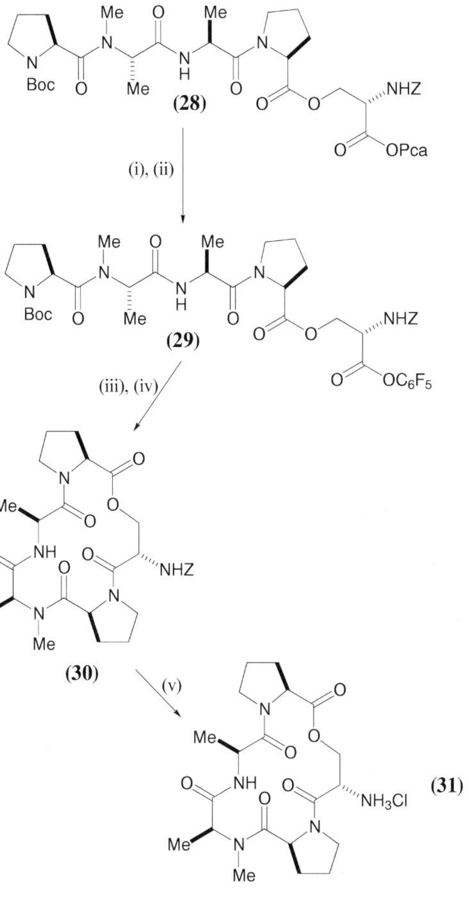

Scheme 12.6   (i) Zn, AcOH, RT, 4 h; (ii) EDC, CH$_2$Cl$_2$, pentafluorophenol, −20°C to RT, 20 h; (iii) HCl, dioxane, RT, 2 h; (iv) NaHCO$_3$, CHCl$_3$, H$_2$O, RT, 6 h, 68% (4 steps); (v) MeOH, HCl, Pd/C, H$_2$, RT, 6 h.

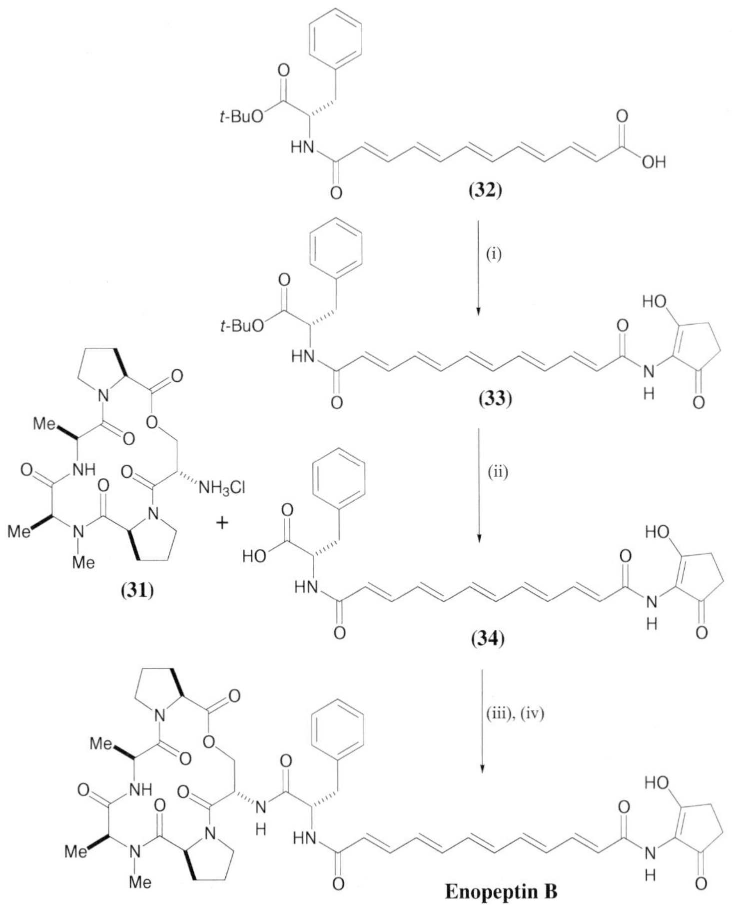

Scheme 12.7 (i) i-BuOCOCl, NMM, THF, 0°C, 2 h then 2-aminocyclopentane-1,3-dione hydrochloride, NMM, 0°C, 3 h, 49%; (ii) TFA, CH$_2$Cl$_2$, RT, 2 h; (iii) PvCl, **34**, NMM, THF, 0°C, 2 h; (iv) **31**, NMM, DMF, RT, 12 h, 8% (3 steps).

Mixed anhydride coupling technology proved optimal for connecting the 2-aminocyclopentane-1,3-dione component to the partially-protected polyene acid **32**. The success of this coupling is noteworthy as the 2-aminocyclopentanedione was left completely unprotected. Unfortunately, the remaining sequence to enopeptin B was low-yielding (8% over two steps). Since amine **31** underwent N-acylation with Boc-(S)-phenylalanine and HATU[16] in high yield, this suggested that intramolecular O- to N-acyl rearrangement[17] (to give a ring-contracted macrolactam) was not problematic in this endgame. The low yield of natural product more likely reflects the lability of the polyene under the strongly acidic conditions used to deprotect the *t*-butyl ester. Future investigations in this area might benefit from the use of an *O*-allyl ester in the final stages. The latter protecting group could potentially be removed under neutral conditions via Pd(0) catalysis.

12.5   L-156,602

In 1990, Hensens and co-workers[18] at Merck Sharp and Dohme Research Laboratories announced their structure determination of L-156,602, a naturally occurring competitive inhibitor of C5a anaphylatoxin binding to its receptor on human neutrophils. C5a antagonists are of interest on account of their significant therapeutic potential for the treatment of inflammatory disorders and allergic disease states. L-156,602 has been synthesised in optically pure form by a Merck chemistry team[19] headed by Charles Caldwell and Phillipe Durette. Their synthetic sequence is depicted fully in Schemes 12.9 to 12.14.

Strategically it was considered prudent to assemble a partially protected linear cyclodepsipeptide precursor **35** (Scheme 12.8) in which the piperazic acid nitrogens were masked by Z-groups, the hydroxamic acid NOH components were blocked as *O*-benzyl ethers and the anomeric hydroxyl was protected as a methyl glycoside. Macrolactamisation was considered most favourable between the *N*-benzyloxy-L-alanine and glycine residues as this mode of ring-closure would allow the most nucleophilic and least sterically encumbered nitrogen to be used for the attack on the activated carbonyl. The hexapeptide forbear of **35** would be assembled by a [2+2+2] fragment condensation between **40**, **41** and **42**. The pyran activated ester **36** would be constructed via a Seebach enantioretentive Claisen condensation reaction.[20]

Caldwell's route[19b] to the tetrahydropyranylpropionate sector of L-156,602 set off with a diastereoselective Frater–Seebach alkylation reaction[21] between the enolate derived from methyl (*R*)-3-hydroxy-butanoate **43** and (*S*)-1-iodo-2-methylbutane **44** (Scheme 12.9), which furnished alcohol **45** in 46% yield. After ester reduction and Swern oxidation to obtain aldehyde **47**, a Wittig condensation, hydrogenation and *O*-desilylation provided lactone **38**. The latter was then subjected to a highly diastereoselective Claisen condensation with the lithium enolate formed from (2*R*,5*R*)-2-*t*-butyl-5-methyl-1,3-dioxolan-4-one **39**. The desired product **49** was isolated in 75% yield as a single diastereoisomer; a 16% yield of **38** was also recovered. A Fischer glycosidation was next used to block the hemiketal functionality so as to prevent a retro-Claisen reaction from occurring during the base-mediated cleavage of the lactone in **49**. Unfortunately, saponification of the glycosidated lactone took

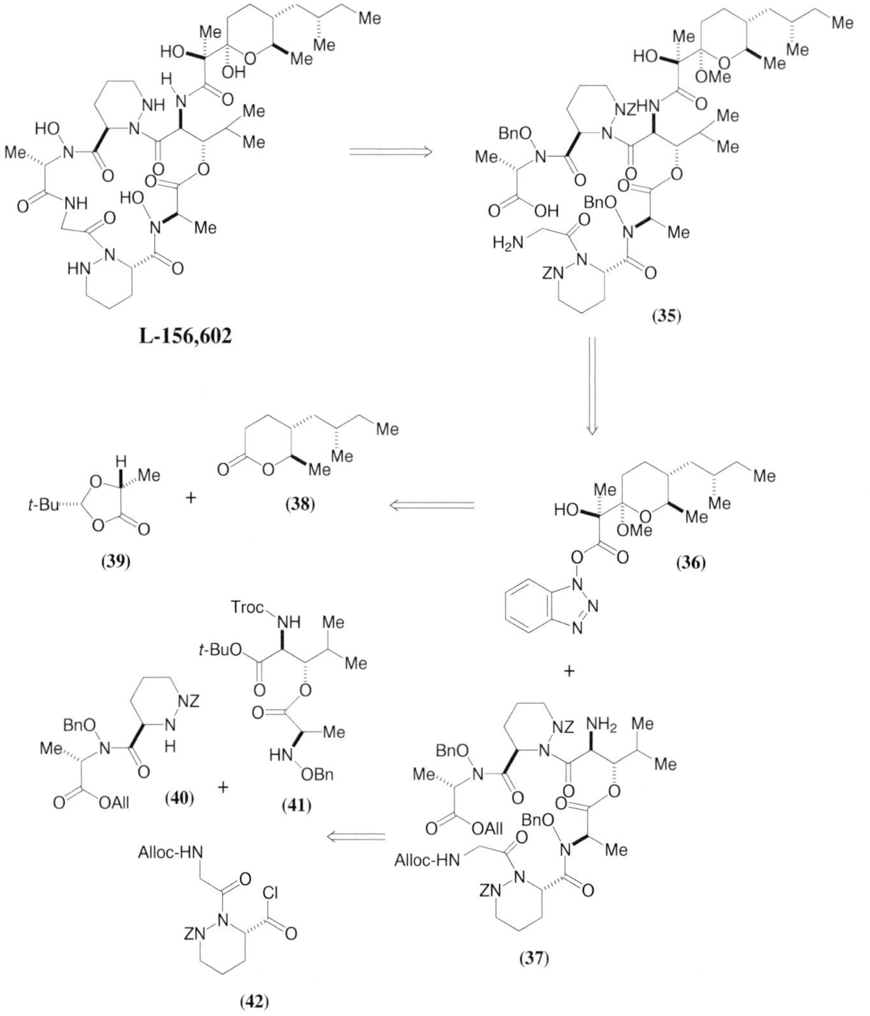

Scheme 12.8 The Caldwell–Durette retrosynthetic analysis of L-156,602.

place only slowly. The product acid also had a tendency to decompose when attempts were made to isolate it. The best method for cleaving the *t*-butyl-1,3-dioxolanone moiety from the glycoside was to transesterify with NaOMe in methanol and then saponify with aqueous KOH in ethanol. For stability reasons, the potassium carboxylate salt was the intermediate actually isolated from this cleavage, and it was used for the subsequent reaction with Castro's BOP reagent[12] to obtain activated ester **36**.

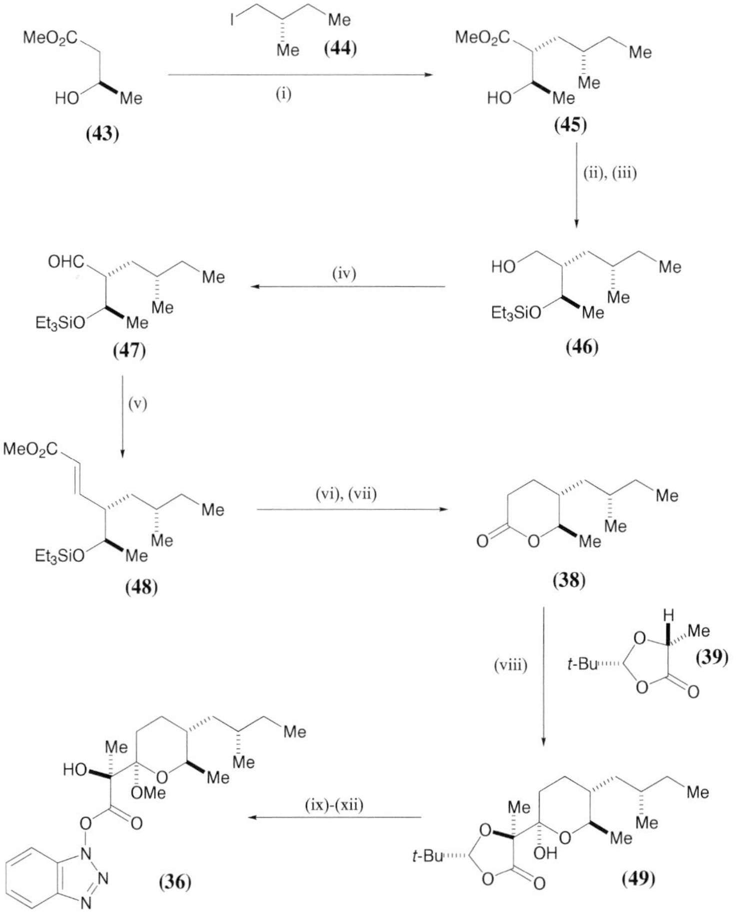

Scheme 12.9 (i) $i\text{-Pr}_2\text{NLi}$, THF, $-50°\text{C}$; cool to $-78°\text{C}$ then add **44**, warm to RT, 46%; (ii) Et_3SiOTf, CH_2Cl_2, 2,6-lutidine, $-40°\text{C}$ to RT; (iii) DIBAH, PhMe, $-70°\text{C}$ to $-40°\text{C}$, 52% (2 steps); (iv) Me_2SO, $(\text{COCl})_2$, CH_2Cl_2, $-70°\text{C}$, Et_3N, $-20°\text{C}$, 98%; (v) THF, Δ, $\text{Ph}_3\text{P}{=}\text{CHCO}_2\text{Me}$, 70%; (vi) H_2, Pd/C, EtOAc, 98%; (vii) $n\text{-Bu}_4\text{NF}$, THF; (viii) **39**, $i\text{-Pr}_2\text{NLi}$, THF, $-70°\text{C}$, add **38**, warm to $-25°\text{C}$, 75%; (ix) HCl, MeOH, $0°\text{C}$ to RT, 89%; (x) NaOMe, MeOH, 78%; (xi) Aq. KOH, EtOH; (xii) $\text{BtOP}^+(\text{NMe}_2)_3\text{PF}_6^-$, N-methylmorpholine, 4 Å sieves, DMF.

Implementing the aforementioned [2+2+2]-fragment condensation strategy required Durette and co-workers[19a] initially to gain access to the dipeptides **40** and **42**, and the depsipeptide **41** (Scheme 12.8). The enantiomeric Z-piperazic acid derivatives **52** and **57** were prepared by a route which exploited a Diels–Alder cycloaddition between di-t-butylazodicarboxylate **51** and methyl 2,4-pentadienoate **50**

(Scheme 12.10). After saturation of the double bond in the cycloadduct, saponification and trifluoroacetic-acid-mediated deprotection of the Boc groups furnished (±)-piperazic acid trifluoroacetic acid salt in 61% overall yield from **50** and **51**. The *N(1)*-Z derivative was then prepared and an optical resolution performed with (+)- and (−)-ephedrines according to the procedure described by Hassall, Johnson and Theobald.[22] Typically, this protocol delivered **52** (Scheme 12.10) and **57** (Scheme 12.11) in 18% and 15% yield, respectively. Acid **52** was next converted to acid chloride **53** (Scheme 12.10) and a Carpino biphasic coupling[23] executed with the partially protected hydroxamic acid ester **56**. Recourse to acid chloride activation was required here because of attenuated nucleophilicity in the *O*-benzylated hydroxamate partner, as well as steric shielding around its *N*-atom. Diethylamine in acetonitrile finally excised the Fmoc urethane from the coupled product to secure **40**.

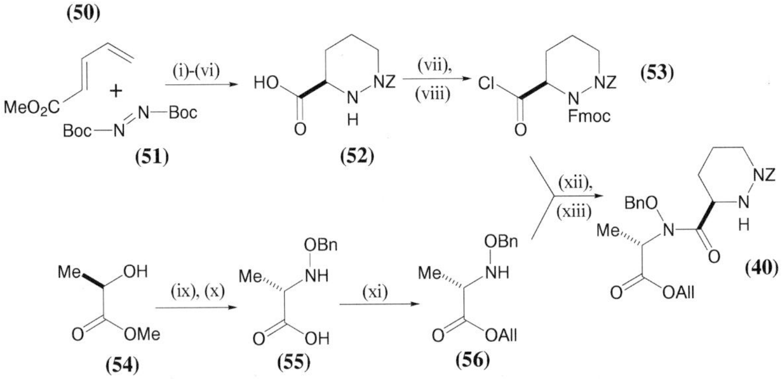

Scheme 12.10 (i) CCl$_4$, reflux, 36 h, 65%; (ii) H$_2$, Pd/C, MeOH; (iii) aq. KOH, MeOH, 93%, 2 steps; (iv) TFA, CH$_2$Cl$_2$, 100%; (v) ZCl, aq. NaOH, 70%; (vi) Optical resolution with (−)- and (+)-ephedrines, 15% of (3*S*)-Z-Piz, 18% of (3*R*)-Z-Piz; (vii) Me$_3$SiCl, *i*-Pr$_2$NEt, 0°C, CH$_2$Cl$_2$, then add Fmoc-Cl, 92%; (viii) (COCl)$_2$, cat. DMF; (ix) Tf$_2$O, CH$_2$Cl$_2$ then BnONH$_2$, 2,6-lutidine; (x) NaOH, THF, 58%, 3 steps; (xi) CDI, CH$_2$Cl$_2$ then allyl alcohol, 83%; (xii) 10% aq. NaHCO$_3$, CH$_2$Cl$_2$; (xiii) Et$_2$NH, MeCN, 62% (3 steps).

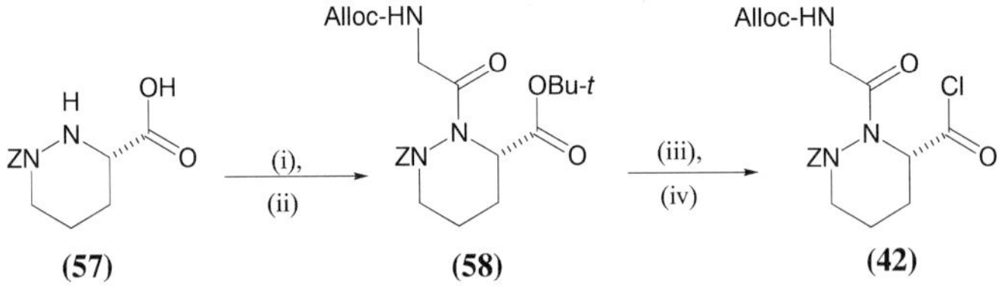

Scheme 12.11 (i) Isobutene, cat. H$_2$SO$_4$, 1,4-dioxane; (ii) Alloc-HNCH$_2$COCl, 10% aq. NaHCO$_3$, CH$_2$Cl$_2$, 97%; (iii) CF$_3$CO$_2$H, 98%; (iv) (COCl)$_2$, cat DMF.

A Carpino biphasic amino-acid chloride coupling[23] tactic again featured in the synthesis of **58** (Scheme 12.11). The latter was converted to acid chloride **42** in a further two steps.

Caldwell and Bondy[24] had earlier reported a convenient synthesis of (2*S*,3*S*)-3-hydroxyleucine **59** utilising Sharpless asymmetric epoxidation chemistry. After regioselective protection of **59** in the manner shown (Scheme 12.12), **60** was coupled to **62** with 1,1'-carbonyldiimidazole to obtain **41**. The synthesis proceeded with a silver-cyanide-mediated amidation reaction between acid chloride **42** and depsipeptide **41** (Scheme

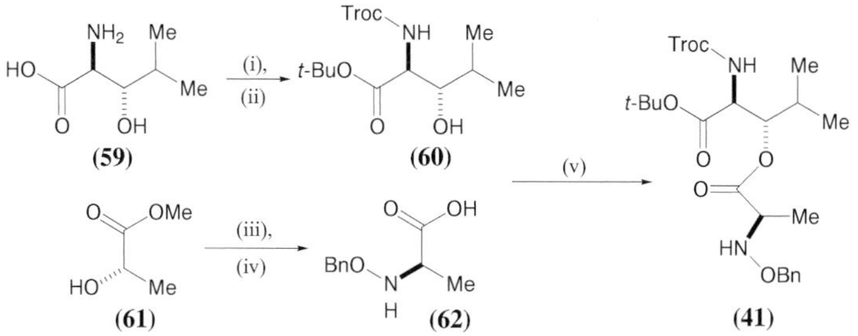

Scheme 12.12   (i) Cl₃CCH₂OC(O)Cl, aq. NaHCO₃, CH₂Cl₂; (ii) *i*-PrN=C(OBu-*t*)NPr-*i*, 63%; (iii) Tf₂O, CH₂Cl₂, 0°C then BnO-NH₂, 2,6-lutidine; (iv) aq. NaOH, THF, 58%, 3 steps; (v) CDI, CH₂Cl₂, 67%.

12.13). The *t*-butyl ester was then detached from **63** by acidolysis, and the liberated acid converted to acid chloride **64**. A Carpino amidation between **64** and **40** next ensued to assemble **65** (Scheme 12.13) and its Troc urethane reductively excised with zinc dust (Scheme 12.14). The product amine **37** underwent *N*-acylation with activated ester **36** in 56% yield. The terminal *O*-allyl protecting groups were then removed from **66** with Pd(0) and tri-*n*-butylstannane[25] to give the macrolactamisation precursor **35**. It transpired that the best conditions for effecting 19-membered ring closure activated the acid with *n*-propane phosphonic anhydride[26] and DMAP. Significantly, this macrolactamisation procedure did not cause epimerisation at the α-centre of the activated ester residue. The final step of the Caldwell–Durette route was a hydrogenolytic deprotection of the macrolactamisation product to obtain L-156,602 in 53% yield and complete this landmark synthesis.

12.6   A83586C

The story of A83586C[27] began in Eli Lilly's laboratories in 1988. There, a team of isolation chemists led by Tim Smitka came across an

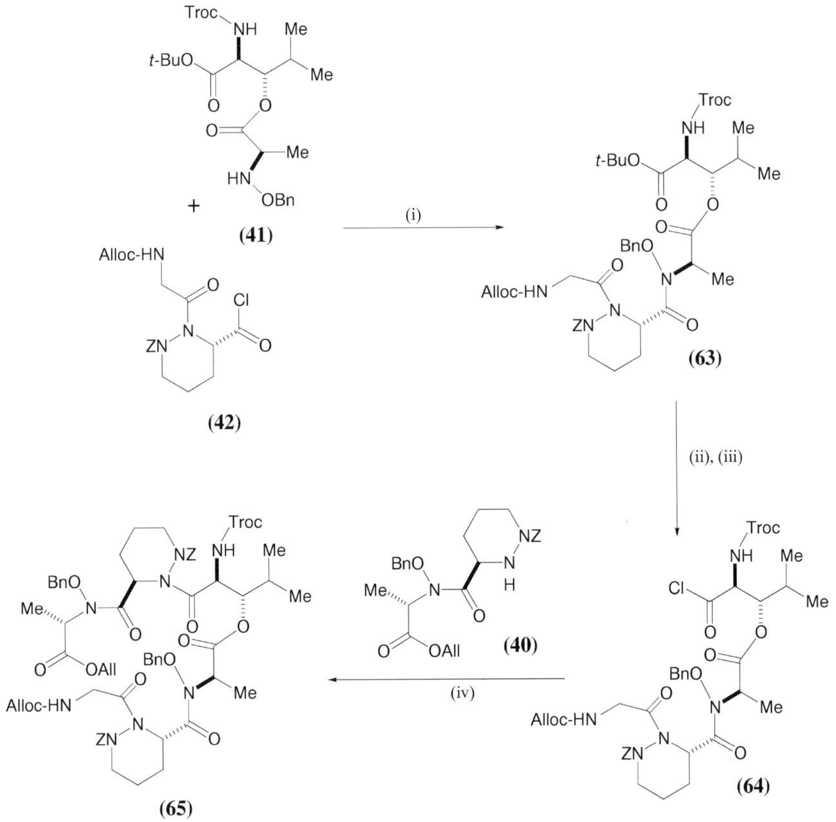

Scheme 12.13 (i) AgCN, PhMe, 90°C, 77%; (ii) CF_3CO_2H; (iii) $(COCl)_2$; (iv) 10% aq. $NaHCO_3$, CH_2Cl_2, 70% (3 steps).

exceedingly potent antitumour antibiotic in the culture filtrates of *Streptomyces karnatakensis*.[27] They termed this molecule A83586C (Scheme 12.15) and they deduced its structure through a combination of single-crystal X-ray analysis and chemical degradation. Smitka recognised that A83586C closely resembled azinothricin (see structure), a related antibacterial agent discovered two years earlier by Hubert Maehr and co-workers at Hoffmann La-Roche.[28] In fact, both compounds differ only at their C(37)-stereocentres and in the nature of their hydroxamic acid residues.

Despite the fact that both molecules are remarkably effective at combating Gram-positive strains of bacteria *in vitro*, their clinical development as antibacterial drugs was never considered viable because of their high toxicity in mice at very high drug concentrations (a 9.3 mg/kg dosage

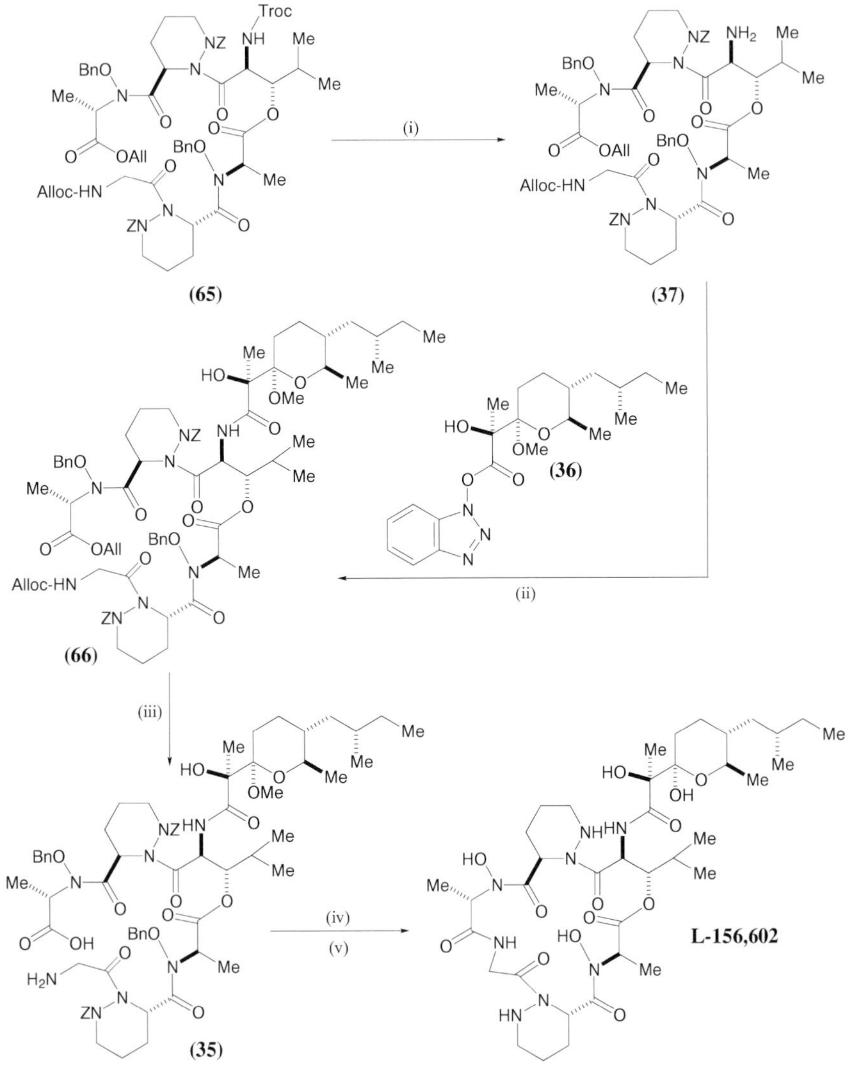

Scheme 12.14   (i) Zn, AcOH; (ii) **36**, DMF, 56% (from C_{14} acid); (iii) Bu_3SnH, H_2O, CH_2Cl_2, $(Ph_3P)_2PdCl_2$; (iv) $[n\text{-PrP(O)O}]_3$, CH_2Cl_2, DMAP, 57% (2 steps); (v) H_2, Pd/C, MeOH, 53%.

of A83586C is lethal to a mouse). As a consequence, synthetic interest in this class quickly dissipated. However, it was reawakened in 1990 when it emerged that citropeptin (see structure)[29] could confer a 120% life extension on mice with P388 lymphocytic leukaemia when administered at the *non-lethal* dosage of 2 mg/kg/day. Interest was further

The A83586C family of cyclodepsipeptides

heightened in 1997 soon after it was discovered that another member of this class, GE3 (see structure),[30] could bring about a 47% reduction in tumour size when given as a single 2 mg/kg dosage to mice with 'incurable' PSN-1 human pancreatic carcinoma. When combined with mechanism-of-action studies which suggested that GE3 could halt cell-cycle progression in A431 lung cancer cells at the G1 to S phase, by preventing unregulated E2F transcription factors from binding to their recognition sites, synthetic and biological curiosity in the A83586C family was further piqued.

In late 1997 our group reported the first asymmetric total synthesis of A83586C by a highly chemoselective coupling strategy which deliberately avoided the use of heteroatom protecting groups at the final stages.[31] The route was predicated upon the successful union of a fully decorated activated ester **68** with a preassembled, nonprotected, cyclodepsipeptide **69** (Scheme 12.15). It was further envisioned that the resulting glycal **67** would then be capable of being chemoselectively hydrated under very mild acid conditions to complete the synthesis. The plan evolved out of

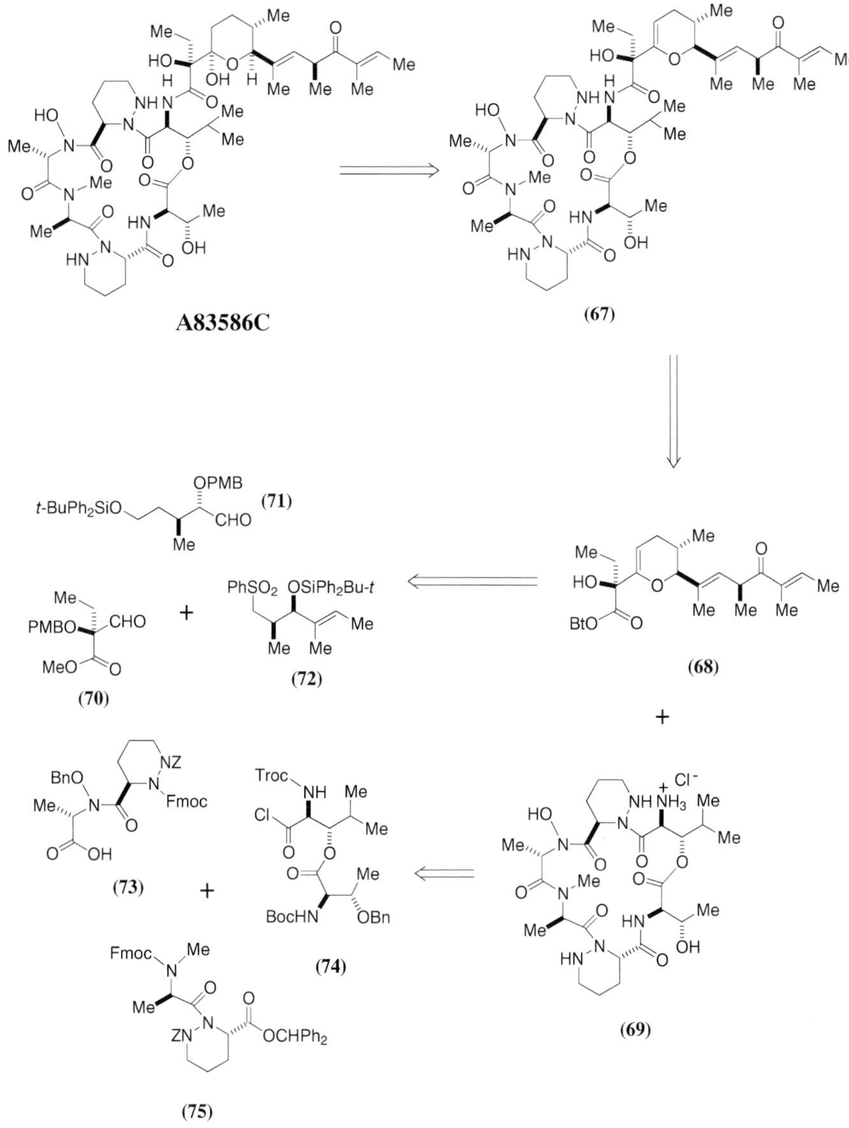

Scheme 12.15   The Hale retrosynthetic analysis of A83586C.

a gradual recognition that virtually all protecting-group strategies for this target would falter at the endgame stages, given the target's lability towards acids, bases, oxidants, many reducing protocols, and the majority of strong nucleophiles.

Synthetic planning for the pyran sector centred around the preparation of intermediates **70**, **71** and **72**. Sulfone **72** was assembled in six steps from the enantiopure propionamide **76** (Scheme 12.16). An Evans asym-

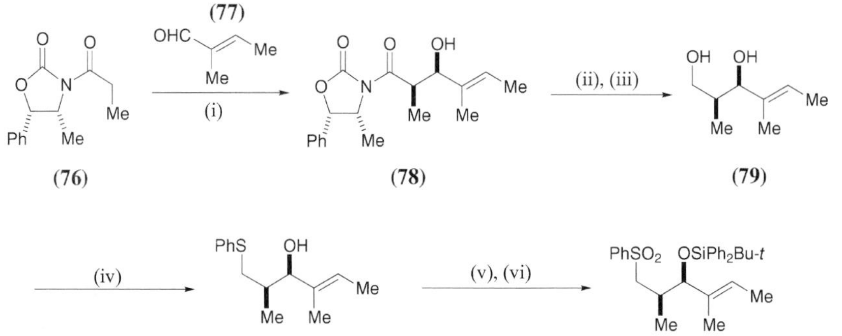

Scheme 12.16 (i) n-Bu$_2$BOTf, Et$_3$N, CH$_2$Cl$_2$, $-5°$C to $-10°$C, cool to $-78°$C, add **77**, warm to RT, 85%; (ii) NaOMe, MeOH, CH$_2$Cl$_2$, $-15°$C, 10 min, 85%; (iii) DIBAL-H, CH$_2$Cl$_2$ $-78°$C to $-15°$C, 88%; (iv) (PhS)$_2$, Bu$_3$P, DMF, RT, 82%; (v) t-BuPh$_2$SiCl, DMF, imid, 85°C, 96%; (vi) oxone, THF, MeOH, H$_2$O, RT, 99%.

metric aldol reaction[32] was used to install the two stereocentres and the (E)-trisubstituted alkene present in **72** in a single step. The aldol adduct **78** was taken forward to **80** by transesterification, methyl ester reduction and selective thioetherification with tributylphosphine and phenyldisulfide. O-Silylation and Trost–Curran oxidation[33] with oxone then completed the synthetic sequence to **72**, which proceeded in a very healthy 50% overall yield.

Efficient procurement of the *anti*-relationship that existed between the two adjacent stereocentres in **71** was considered the main priority in the preparation of this fragment. This stereochemical arrangement was fashioned by a chelation-controlled epoxide ring-opening reaction on the chiral 2,3-epoxy alcohol **83**[34] [itself obtained by Sharpless asymmetric epoxidation (AE)] with trimethylaluminium,[35] which proceeded with 20:1 selectivity in favour of the desired C(3)-ring-opened product (Scheme 12.17). After separation of the minor undesired component by chromatography, the major 1,2-diol was protected as *p*-methoxybenzylidene acetal **84**, and the latter reductively cleaved[36] to position regioselectively a *p*-methoxybenzyl group on the secondary oxygen atom. The primary alcohol was then oxidised by the Swern method to obtain aldehyde **71**.

Having charted a successful course to subtargets **71** and **72**, the group effected unification as shown in Scheme 12.18. Ketone **86** was obtained

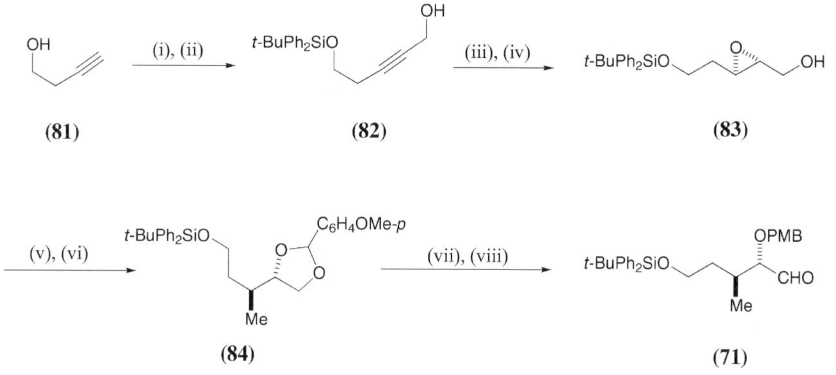

Scheme 12.17 (i) *t*-BuPh$_2$SiCl, imid, DMF, RT, 76%; (ii) *n*-BuLi, THF, −30°C, (HCHO)$_n$, warm to RT, 73%; (iii) REDAL, Et$_2$O, 0°C, 73%; (iv) *t*-BuO$_2$H, Ti(O-*i*-Pr)$_4$, CH$_2$Cl$_2$, 4 Å MS, −20°C, (−)-DET, 84%, 93% ee; (v) Me$_3$Al, Hex, 0°C, 75%; (vi) *p*-MeO-C$_6$H$_4$-CH(OMe)$_2$, DMF, *p*-TsOH, 55°C, 84%; (vii) DIBAL-H, CH$_2$Cl$_2$, −78°C to RT, 78%; (viii) Me$_2$SO, (COCl)$_2$, CH$_2$Cl$_2$, −78°C, Et$_3$N, warm to RT, 83%.

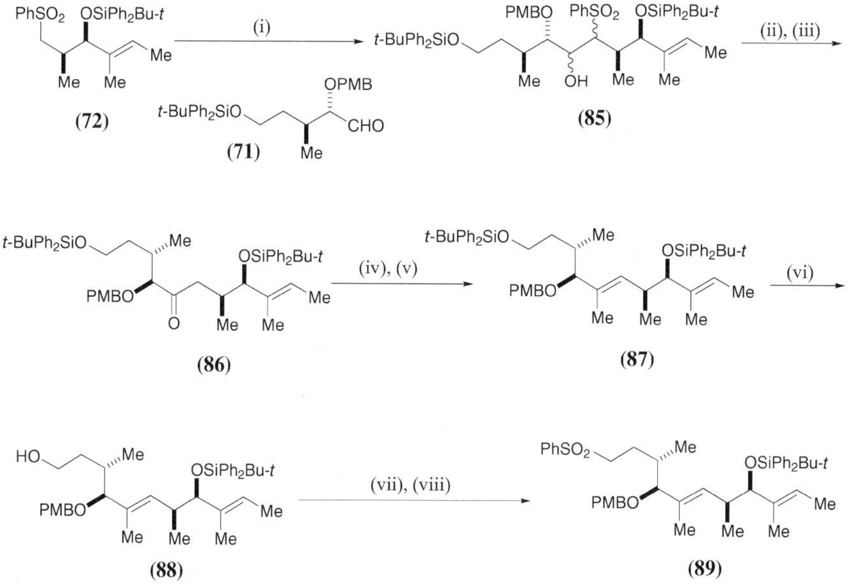

Scheme 12.18 (i) *n*-BuLi, THF, −78°C, then add **71**, 64%–75%; (ii) (CF$_3$CO$_2$)O, Me$_2$SO, CH$_2$Cl$_2$, −78°C; Et$_3$N, warm to RT, 92%; (iii) Bu$_3$SnH, AIBN, PhMe, reflux, 82%; (iv) MeMgBr, THF, −78°C to RT, 98%; (v) POCl$_3$, Py, 55°C, 97%; (vi) HF-Pyridine, THF, 86%; (vii) Bu$_3$P, DMF, (PhS)$_2$, 94%; (viii) oxone, THF, MeOH, H$_2$O, 96%.

after the coupled β-hydroxysulfones **85** were subjected to a Swern oxidation with trifluoroacetic anhydride and DMSO, and the resulting β-ketosulfone mixture was desulfonylated via Smith's free-radical tri-*n*-butylstannane method.[37] Although **86** could be converted to an appropriate (*Z*)-enol triflate in high yield, all efforts at transforming this intermediate into **87** were unsuccessful using the Stille[9] or McMurry–Scott[38] cross-coupling protocols. A Grignard addition/POCl₃-pyridine-mediated dehydration tactic was therefore implemented on ketone **86** to access **87**. The latter was the predominant isomer of the 2.6:1 mixture of alkenes that formed, the other component being the 1,1-disubstituted alkene. Owing to the difficulties in separating these alkenes at this stage, their fractionation was attempted after the primary TBDPS groups had been selectively removed with HF–py complex. Alcohol **88** was purified in 55% overall yield from ketone **86** after flash chromatography. A further two steps were required to convert it to phenylsulfone **89**.

An efficient entry into the chiral β-aldehydo ester **70** was made possible through use of the Sharpless catalytic asymmetric dihydroxylation reaction (Scheme 12.19).[39] This performed admirably on **90**, delivering diol

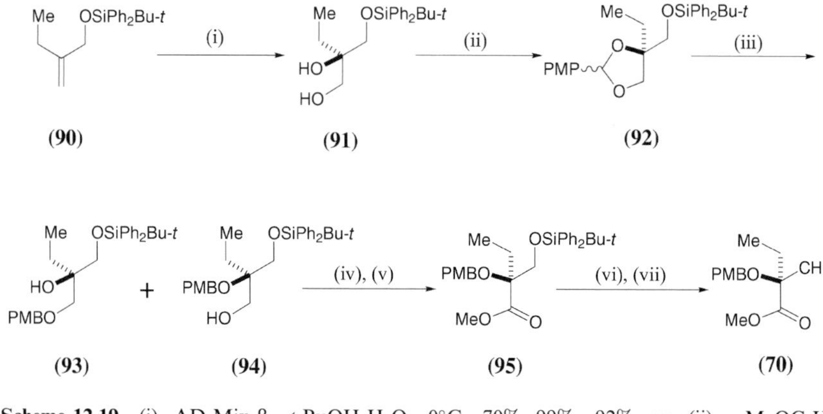

Scheme 12.19 (i) AD-Mix-β, *t*-BuOH-H₂O, 0°C, 70%–99%, 92% ee; (ii) *p*-MeOC₆H₄-CH(OMe)₂, cat. PPTS, DMF, 55°C, 98%; (iii) *i*-Bu₂AlH, CH₂Cl₂, −78°C, 2 h, then RT, 5 min, 63%–69%, regioselectivity **93**:**94** = 1:2; (iv) PDC, DMF, RT, 4 d; (v) CH₂N₂, Et₂O, 64%, 2 steps; (vi) *n*-Bu₄NF, THF, 74%; (vii) Me₂SO, (COCl)₂, CH₂Cl₂, −78°C, Et₃N, 0°C, 82%.

91 in 99% yield and greater than 92% ee. A *p*-methoxybenzylidenation and acetal reduction tactic was again used to position a PMB group on the more hindered hydroxyl of this product. Unfortunately, however, the regioselectivity of this reduction was not high, a 2:1 mixture of the alcohols **94** and **93** being formed. Fortunately, these could be readily separated from one another by chromatography. An oxidation–esterification sequence on **94** next delivered methyl ester **95** which was converted to **70** by *O*-desilylation and Swern oxidation.

Union of **89** and **70** was effected by adding the sulfone anion onto aldehyde **70** at low temperature (Scheme 12.20). The ensuing mixture of

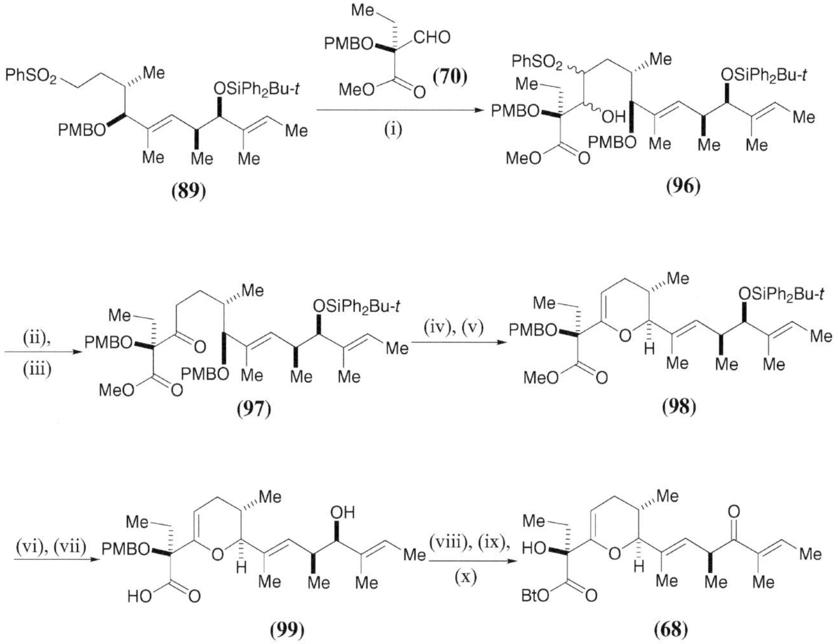

Scheme 12.20 (i) *n*-BuLi, THF, −78°C, then add sulfone anion to **70**, 82%; (ii) Me$_2$SO, (CF$_3$CO)$_2$O, CH$_2$Cl$_2$, −78°C, then Et$_3$N, 0°C, 93%; (iii) Al-Hg, 10% aq. THF, reflux, 96%; (iv) DDQ, CH$_2$Cl$_2$, H$_2$O, 0°C, 3 h; (v) PPTS, MeOH, 60°C, 1 h, 77%, 2 steps; (vi) *n*-Bu$_4$NF, DMF, RT, 48 h, 84%; (vii) EtSLi, HMPA, THF, RT, 2 h; (viii) *i*-Pr$_2$NEt, BtOP(NMe$_2$)$_3$PF$_6$, CH$_2$Cl$_2$, 99%, 2 steps; (ix) (CF$_3$CO)$_2$O, Me$_2$SO, CH$_2$Cl$_2$, −78°C; Et$_3$N, to 0°C, 97%; (x) DDQ, wet CHCl$_3$, RT, 3 h, 71%.

β-hydroxysulfones **96** were then oxidised, and desulfonylation accomplished via the aluminium amalgam reduction method of Corey–Chaykovsky.[40] A chemoselective deprotection of the more electron-rich and less-hindered secondary PMB ether in **97** was next attempted, using a limited quantity of DDQ[41] (1.2 equiv.) in ice-cold aqueous CH$_2$Cl$_2$. This provided a mixture of three products: the linear δ-hydroxy ketone and the two α- and β-ring-closed hemiketals. Owing to the great difficulties encountered in working with an equilibrating three-component mixture, a dehydrative ring-closure was effected with methanol and PPTS to obtain glycal **98**, and this carried forward to acid **99**. The

N-hydroxybenzotriazole ester was prepared by treatment with Castro's BOP reagent. Of special note here was the compatibility of this esterification method with the presence of the allylic secondary hydroxyl. Another noteworthy set of reactions, conducted in the presence of the HOBT ester, were the Swern oxidation used to obtain the terminal enone, and the deprotection of the tertiary PMB ether with DDQ[41] in wet chloroform to gain access to the activated ester **68**. The success of these synthetic methods in this demanding environment is particularly gratifying.

As part of our effort to improve significantly the synthetic technology available for preparing functionalised and nonfunctionalised homochiral piperazic acid derivatives, our group introduced the novel method of tandem electrophilic hydrazination–nucleophilic cyclisation during the synthesis of A83586C.[42] In this protocol, a bromovaleric acid derivative is tethered onto an Evans chiral oxazolidinone and an asymmetric electrophilic hydrazination[43] performed at −78°C. Then, rather than protically quenching the aza-anion at low temperature, as one would do in a normal Evans hydrazination,[43] the tethered bromovaleryl aza-anion is allowed to warm up to room temperature in the presence of DMPU, whereupon a facile tandem ring-closure[42] takes place to give the piperazic acid derivative. In the case of **100**, the ring-closed product **101** is obtained with high overall efficiency (Scheme 12.21); its enantiomer is also readily available by this approach. The development of this protocol proved critical to the overall success of the A83586C synthetic venture.

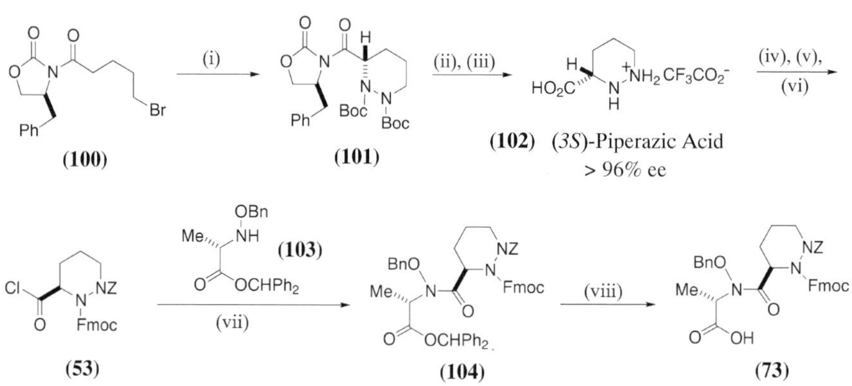

Scheme 12.21   (i) LDA, THF, −78°C; then *t*-BuO$_2$C—N=N—CO$_2$Bu-*t*; then add DMPU and warm to RT; (ii) aq. LiOH, THF, 0°C; (iii) CF$_3$CO$_2$H, CH$_2$Cl$_2$, 68% over 3 steps; (iv) ZCl, 1 M NaOH, PhMe, 0°C, 93%; (v) Me$_3$SiCl, *i*-Pr$_2$NEt, CH$_2$Cl$_2$, reflux; Fmoc-Cl, 0°C, then RT, 18 h, 85%; (vi) (COCl)$_2$, C$_6$H$_6$, 60°C, 1.5 h; (vii) 10% aq. NaHCO$_3$, CH$_2$Cl$_2$, 0°C, 81%, 2 steps; (viii) CF$_3$CO$_2$H, CH$_2$Cl$_2$, 99%.

As well as allowing dipeptide **73** to be readily prepared on large scale (Scheme 12.21), the method also allowed an expedient route to be developed to tetrapeptide **110** (Scheme 12.22). Starting from known Fmoc-N(Me)-D-Ala **105**, acid chloride **106** was formed, and its coupling

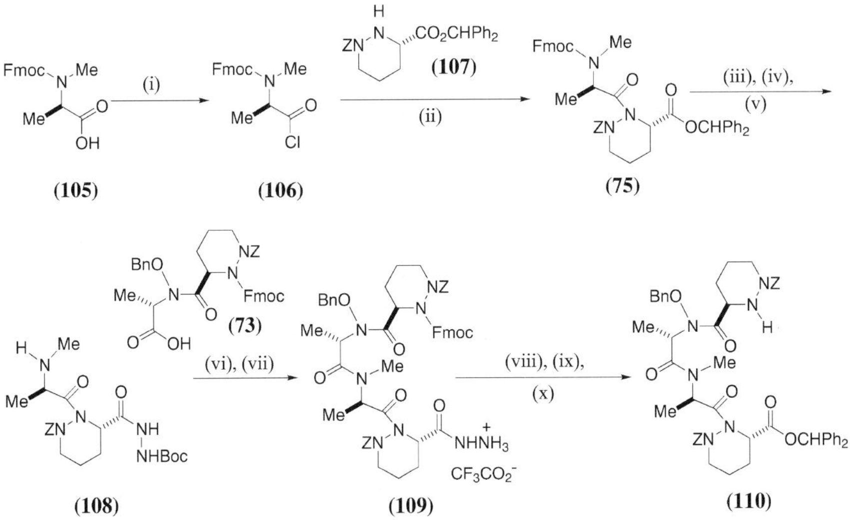

Scheme 12.22 (i) (COCl)$_2$, CH$_2$Cl$_2$, RT, 2 h; (ii) AgCN, PhMe, 70°C, 92%, 2 steps; (iii) CF$_3$CO$_2$H, PhOH, CH$_2$Cl$_2$; (iv) Boc-NHNH$_2$, DCC, HOBt, THF, 0°C to RT, 93%, 2 steps; (v) Et$_2$NH-MeCN (1:1); (vi) Et$_3$N, −10°C, then add BOP-Cl, warm to 0°C, 4 h (75%, 2 steps); (vii) CF$_3$CO$_2$H, CH$_2$Cl$_2$; (viii) NBS, THF, H$_2$O, 0°C to RT; (ix) Ph$_2$C=N$_2$, Me$_2$CO, RT, 12 h, 68%, 3 steps; (x) Et$_2$NH, MeCN.

to **107** effected with silver cyanide. Fmoc amino-acid chloride coupling technology was used here primarily because these intermediates have very good stability, in contrast to their α-Boc- or α-Z-protected counterparts, which are often exceedingly difficult to prepare and handle. Fmoc-protected α-amino-acid chlorides are also highly reactive, coupling efficiently to donors of low nucleophilicity, with minimal epimerisation at the α-stereocentre. Acid chloride activation was essential for the preparation of **75**, since activated ester and mixed anhydride methods had both failed to mediate this particular amidation. The N(2)-acylation of N(1)-acylated α-hydrazino acids is frequently unsuccessful with mixed anhydrides or activated esters, usually because the N(2)-atom is electronically deactivated by the electron-withdrawing N(1)-acyl group and also because there is substantial steric hindrance around the N(2) atom in a partially acylated hydrazine.

Unfortunately, conditions could not be established for removing the Fmoc protecting group from **75** without also causing exclusive formation of the diketopiperazine. In order to circumvent this problem, the diphenylmethyl ester group of **75** was replaced with a *t*-butylcarb-azide[44] function (Scheme 12.22). This tactic now permitted an efficient deprotection of the Fmoc group to obtain **108**. Happily, the [2+2] fragment condensation between **108** and **73** proceeded satisfactorily when the acid component was activated with BOP-Cl. After removal of the Boc group with TFA, and conversion of the acylhydrazine unit in **109** to an acid (using *N*-bromosuccinimide[45] in THF and water), the carboxyl was esterified with diphenyldiazomethane and the Fmoc group cleaved to obtain **110**.

One of the best large-scale methods currently available for obtaining (2*S*,3*S*)-3-hydroxyleucine is the six-step procedure of Manaviazar and Delisser,[46] which relies upon Sharpless AD and Sharpless cyclic sulfate chemistry to install the *anti*-aminoalcohol motif in **59** (Scheme 12.23). (2*S*,3*S*)-3-Hydroxyleucine, so obtained, was transformed into **119** in a further three steps, the depsipeptide linkage being forged through a DMAP-assisted DCC coupling between **115** and **118**. *O*-Deallylation of **119** with Pd(0) and morpholine, followed by chlorination with oxalyl chloride, subsequently provided the depsipeptide acid chloride **74**.

The amidation between **110** and **74** took place efficiently (86% yield) if silver cyanide was used as the promoter, and the reaction mixture was heated at 60°C for *only* two minutes (Scheme 12.24). More prolonged heating invariably resulted in decomposition of **121**. After Z for Troc interchange in **121**, mild acidolysis with TFA selectively removed the Boc and diphenylmethyl ester groups, to provide the key macrolactamisation precursor **122**. Of all the reagents evaluated for securing ring-closure, only Carpino's HATU system[16] delivered the desired product in 25%–40% yield. Deprotection of this material was best accomplished by catalytic hydrogenation over a Pd-C catalyst in methanolic HCl. One equivalent of acid was added to the reaction mixture to protonate the hydroxyleucine amine as soon as it was liberated, and thereby prevent *O*- to *N*-acyl migration[17] from occurring in this residue. The finale of the synthesis was a rapid (room temperature for 10 min) chemo- and regio-selective coupling between **69** and **68** mediated by Et$_3$N which furnished **67** in 31% overall yield from the protected macrolactam. The glycal unit of **67** was then coaxed into undergoing a very clean and highly regioselective hydration reaction[31a,b] in wet CDCl$_3$ to deliver A83586C in quantitative yield.

Our group has also demonstrated that its synthesis is amenable to the construction of quite elaborate synthetic analogues such as 4-*epi*-A83586C,[47] and some of these are currently being used to probe the role of unregulated E2F transcription factors in the development of certain cancers.

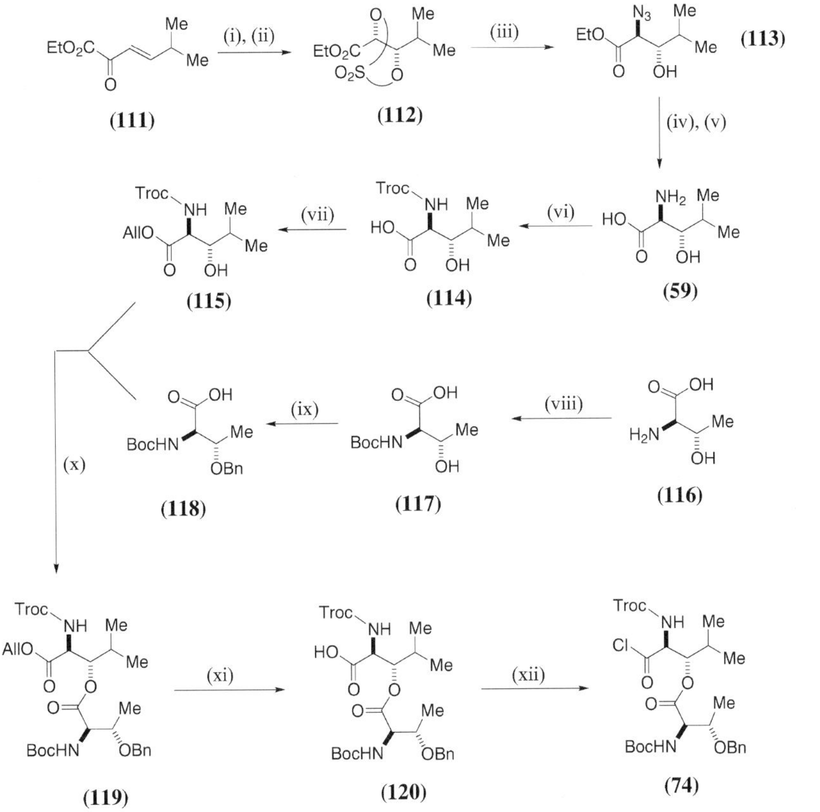

Scheme 12.23 (i) AD-mix-α, MeSO$_2$NH$_2$, t-BuOH-H$_2$O (1:1), 0°C, 96 h, 95%, 97% ee; (ii) SOCl$_2$, CCl$_4$, reflux, 2 h; add RuCl$_3$.xH$_2$O, NaIO$_4$, MeCN, H$_2$O, 92%; (iii) NaN$_3$, H$_2$O-Me$_2$CO (1:10), 1.5 h; 20% aq. H$_2$SO$_4$, Et$_2$O, 24 h, 92%; (iv) 1 M aq. NaOH, THF, 0°C, 93%; (v) H$_2$, Pd(OH)$_2$, MeOH, 93%; (vi) Troc-Cl, aq. NaOH, THF, RT, 90%; (vii) NaHCO$_3$, DMF, RT, AllBr, 92%; (viii) Boc$_2$O, aq. NaOH, 1,4-dioxane, 0°C, 90%; (ix) NaH, DMF, BnBr, 32%; (x) DCC, DMAP, CH$_2$Cl$_2$, 0°C to RT, 83–88%; (xi) (Ph$_3$P)$_4$Pd, morpholine, THF, 0°C to RT, 59%; (xii) (COCl)$_2$, C$_6$H$_6$, RT, 2.5 h.

12.7   Himastatin

From a purely biosynthetic and structural standpoint, the dimeric architecture of himastatin can only be described as fascinating. However, when combined with impressive antibiotic effects against a range of Gram-positive organisms, and a good antitumour profile in several animal tumour models, the inherent attractiveness of this molecule as a synthetic target gains in stature. First isolated in 1990 from the actinomycete strain ATCC 53653 by a group of scientists at Bristol Myers Squibb,[48]

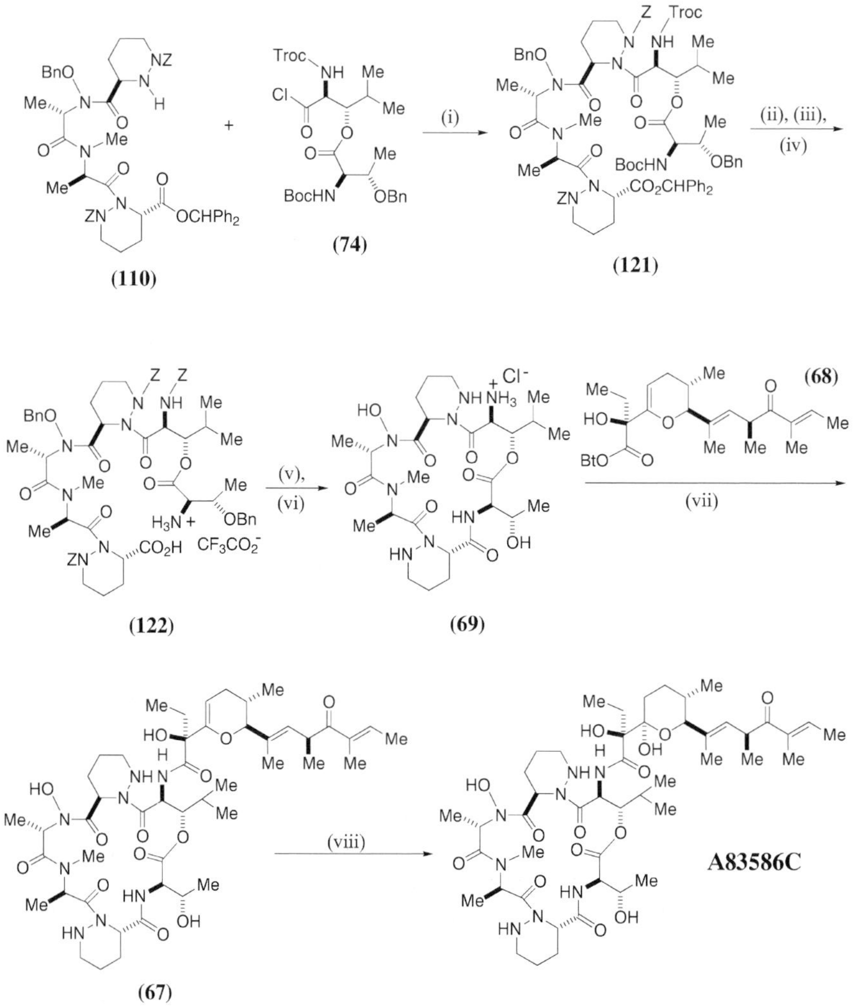

Scheme 12.24 (i) AgCN (1.3 equiv.), C_6H_6, 60°C, 2 min, 73%–86%; (ii) Zn, AcOH, H_2O, RT, 3 h; (iii) Z-Cl, 10% aq. $NaHCO_3$, CH_2Cl_2, RT, 2 h, 78%, 2 steps; (iv) CF_3CO_2H, CH_2Cl_2, PhOH, 0°C, 1 h, 100%; (v) mix with NEM (13.5 equiv.), add to HATU (10 equiv.) in CH_2Cl_2 at 0°C over 3 h, warm to RT, HIGH DILUTION, 30 h, 25%–40%; (vi) H_2 (1 atm), 10% Pd-C, MeOH, dry HCl (1 equiv.), 24 h; (vii) mix together in CH_2Cl_2, cool to −78°C, add Et_3N (9.3 equiv.) then warm to RT for 10 min., 31%, 2 steps; (viii) $CDCl_3$ (wet), 72 h, 0°C, 100%.

six years elapsed before a reasonably confident (but incorrect) structure could be proposed for himastatin by Leet and co-workers. The structure they originally put forward was actually that of 2-*epi*-himastatin.[49] However, a subsequent total synthesis of 2-*epi*-himastatin by Kamenecka

2-*epi*-HIMASTATIN

HIMASTATIN

and Danishefsky[50] showed that the original structural assignment for himastatin was incorrect, and it led these workers to propose that the true structure probably had the C(2)-and C(2′)-carboxamido groups disposed in an L- rather than a D-configuration (compare structures illustrated).[50] It will be appreciated that the new himastatin structure proposed by Danishefsky and Kamenecka also had these same carboxamido groupings bearing a *syn* rather than a *trans* relationship to the tertiary hydroxyls present in the pyrrolindoline ring systems. In order to confirm their new stereochemical proposal, as well as open up a synthetic pathway to novel congeners of the natural product, Danishefsky and Kamenecka carried out a total synthesis of the revised stereostructure. Their elegant route (shown in Schemes 12.25–12.28)[51] is notable for its efficient and stereoselective construction of the (3*R*,5*R*)-5-hydroxypiperazic acid and *syn-cis*-pyrrolo[2,3-*b*]indoline units, and for its quite spectacular use of a Stille cross-coupling reaction to create the biaryl C—C linkage found in the target.

Initial synthetic effort was marshalled towards obtaining the key bis(pyrroloindoline) subunit **131** (Scheme 12.25). It was to be coupled to the pentadepsipeptide **145**, whose synthesis is presented in Schemes 12.26 and 12.27. Danishefsky envisioned that the requisite *syn-cis*-pyrrolo[2,3-*b*]indoline skeletal arrangement in **131** could be built up through a stereospecific oxidative cyclisation[52] of an appropriately functionalised L-tryptophan derivative. After screening a wide range of partially protected

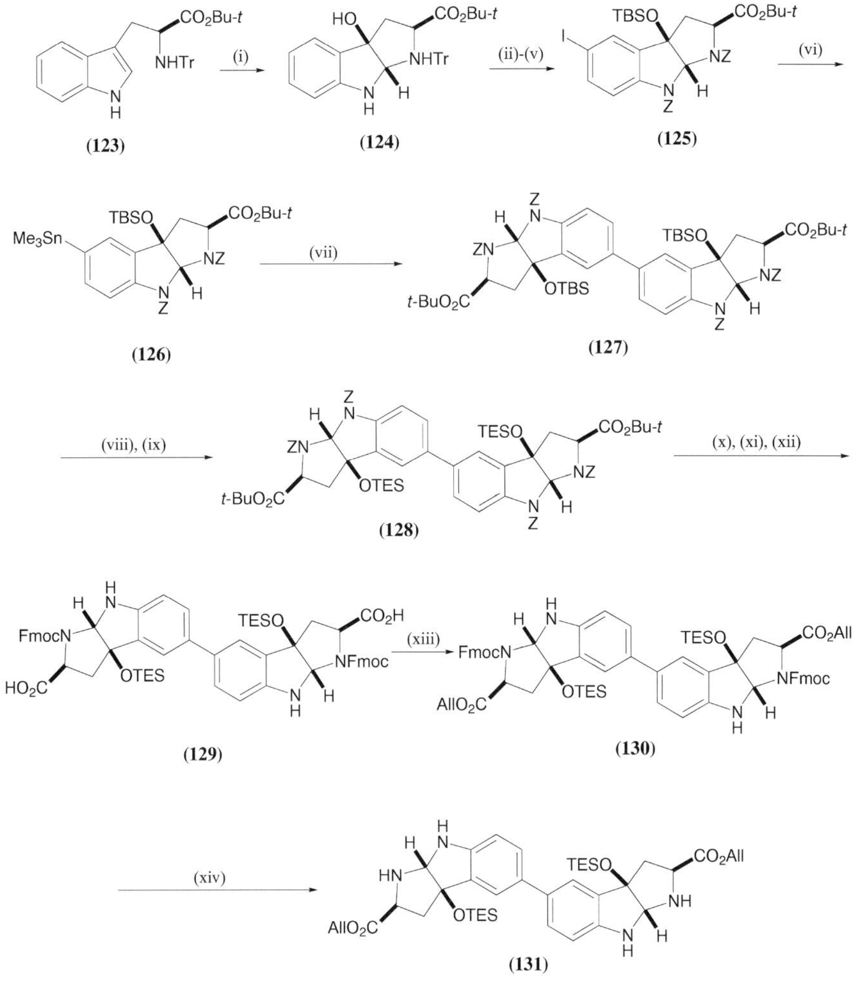

Scheme 12.25 (i) DMDO, CH_2Cl_2, $-78°C$; (ii) AcOH, MeOH, CH_2Cl_2; (iii) ZCl, py, CH_2Cl_2; (iv) TBSCl, DBU, MeCN; (v) ICl, 2,6-di-*tert*-butylpyridine, CH_2Cl_2, 75%; (vi) Me_6Sn_2, $[Pd(Ph_3)_4]$, THF, 86%; (vii) **125**, $[Pd_2dba_3]$, $AsPh_3$, DMF, 45°C, 79%; (viii) TBAF, THF, 91%; (ix) TESCl, DBU, DMF, 92%; (x) H_2, Pd/C, EtOAc, 100%; (xi) Fmoc-HOSu, py, CH_2Cl_2, 95%; (xii) TESOTf, 2,6-lutidine, CH_2Cl_2; (xiii) allyl alcohol, DBAD, PPh_3, THF, (90%, 2 steps); (xiv) piperidine, MeCN, 74%.

L-tryptophan derivatives in such reactions, it was eventually demonstrated that oxidation of the *t*-butyl ester of N_b-trityl-L-tryptophan **123** with dimethyldioxirane could produce the desired tricycle **124** in 70% yield, unaccompanied by any of the *anti-cis* stereoisomer (Scheme 12.25). The ease with which this formidable-looking ring system could be

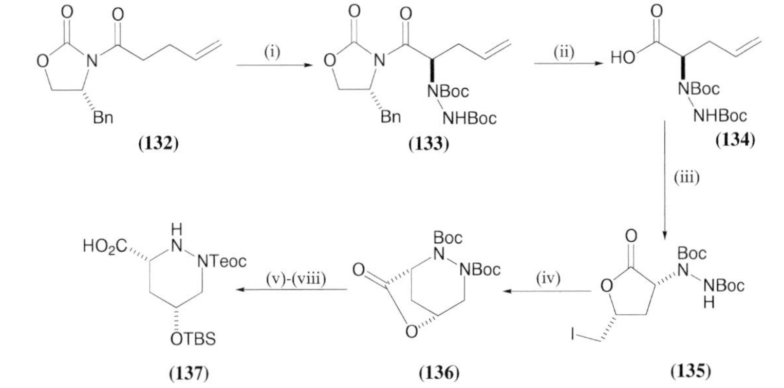

Scheme 12.26 (i) NaHMDS, THF, $-78°C$, Boc—N=N—Boc, 80%; (ii) LiOOH, THF, 89%; (iii) NIS, Ti(OiPr)$_4$, 80%; (iv) NaH, DMF, 74%; (v) TFA, MeOH; (vi) TeocCl, py; (vii) TBSOTf, 2,6-lutidine; (viii) aq. LiOH, 82%.

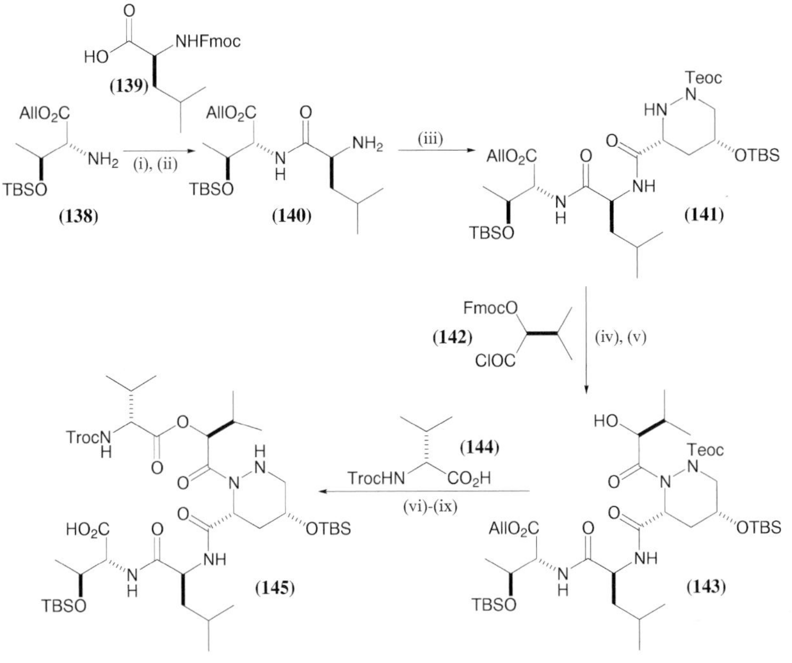

Scheme 12.27 (i) **139**, EDCl, DMAP, CH$_2$Cl$_2$; (ii) piperidine, CH$_3$CN, 76%; (iii) **137**, HATU, HOAt, collidine, CH$_2$Cl$_2$, 95%; (iv) collidine, CH$_2$Cl$_2$, **142**; (v) piperidine, CH$_3$CN, 96%; (vi) **144**, IPCC, Et$_3$N, DMAP, CH$_2$Cl$_2$; (vii) ZnCl$_2$, CH$_3$NO$_2$; (viii) TBSOTf, 2,6-lutidine, CH$_2$Cl$_2$; (ix) PhSiH$_3$, [Pd(PPh$_3$)$_4$], THF, 72% (4 steps).

constructed by this methodology can only be described as breathtaking. Following removal of the N-trityl group, and Z-urethane protection of the exposed nitrogen atoms, a TBS group was anchored on the tertiary hydroxyl, and a regiospecific iodination performed at C(5) with iodine monochloride. The entire four-step sequence proceeded in 75% overall yield. A Pd(0) cross-coupling reaction[9] was next effected between the aryl iodide **125** and hexamethyldistannane to obtain arylstannane **126**. A second Stille cross-coupling[9] was then performed between **125** and **126** to obtain **127** in high yield. Since model studies had revealed that deprotection of the angular TBS protecting group in related macrocyclic systems could be problematic, this grouping was replaced with a more labile TES function whilst still at this early stage in the proceedings. The four Z-groups were next jettisoned from **128** and several other protecting-group adjustments made until eventually subtarget **131** emerged.

Danishefsky's requirement for a stereoselective pathway to **137** led him to introduce a new method for building protected 5-hydroxypiperazic acid derivatives (Scheme 12.26), which nicely complements the earlier tandem electrophilic hydrazination–nucleophilic cyclisation approach to these molecules, reported by our group.[42c] The key steps in the Danishefsky–Kamenecka pathway to **137** (Scheme 12.26) were a stereo-specific Evans hydrazination[43] on **132** to obtain **133**, an iodolactonisation on acid **134** to access **135** (with 6:1 *cis:trans* selectivity) and an intra-molecular N-alkylation mediated by base to form the protected piperazic acid derivative **136**. Protecting-group adjustment finally provided acid **137**, appropriately adorned for coupling to dipeptide amine **140** (Scheme 12.27). Yet again, acid chloride amidation technology was needed for N(2)-acylation of the partially protected piperazic acid residue in tripeptide **141**. Attachment of the fifth residue to **143** proceeded along fairly conventional lines. Two items of particular note in this sequence, however, were the deprotection of the Teoc group with zinc chloride in nitromethane, and the Pd(0)/phenylsilane reductive cleavage of the O-allyl ester unit in the presence of the Troc group to access **145**.

As a result of earlier model studies, Danishefsky and Kamenecka were confident that **131** would undergo N-acylation with **145** at its more nucleophilic pyrrolo[2,3-b]indoline N_b-atom. In the event, this coupling proceeded as anticipated (Scheme 12.28), it leading to **146** after successive deprotection of the O-allyl ester and Troc functionalities. Having arrived at the *seco* bis-amino-acid structure **146**, the only obstacles that remained were macrolactamisation and global O-desilylation. These transforma-tions were accomplished with HATU, and TBAF/aq. acetic acid, respec-tively, himastatin being isolated in a respectable 35% yield. Significantly, this magnificent synthesis confirmed the new stereostructure proposed for himastatin by Danishefsky and Kamenecka.

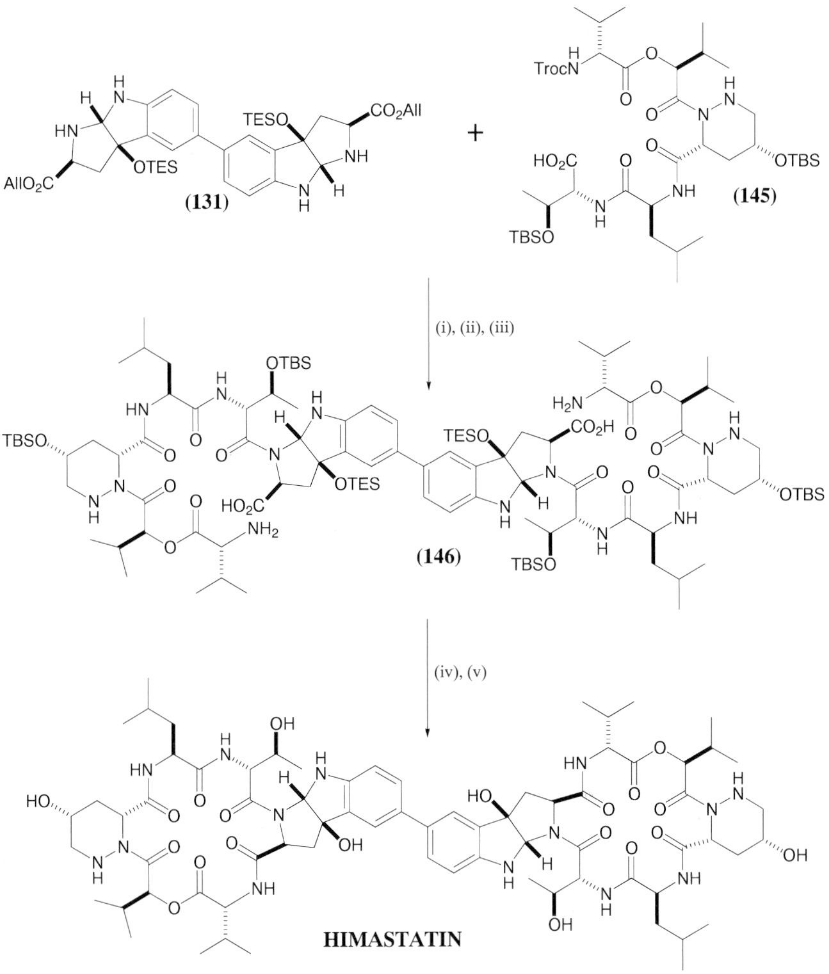

Scheme 12.28 (i) HATU, HOAt, collidine, CH_2Cl_2, $-10°C$ to RT, 60%; (ii) [Pd(PPh$_3$)$_4$], PhSiH$_3$, THF; (iii) Pb/Cd, NH$_4$OAc, THF, 56%; (iv) HOAt, HATU, i-Pr$_2$NEt, DMF; (v) TBAF, THF, AcOH, 35% (2 steps).

12.8   Luzopeptin C

Continuing with the theme of dimeric cyclodepsipeptides, Boger's group at Scripps have recently announced the first total synthesis[53] of luzo-peptin C. Luzopeptin C[54] is a cyclic decadepsipeptide (see structure), isolated from *Actinomadura luzonensis*, which has the ability to suppress human immunodeficiency virus (HIV) replication in infected MT-4 cells

Luzopeptin C

at noncytotoxic drug concentrations (IC$_{100}$ = 2.5–5.0 µg/mL).[55] It operates by inhibiting HIV reverse transcriptase enzymes, which also include strains of the single and double mutant variety. In addition, luzopeptin C has a significant antitumour profile in several model systems. Prominent amongst its many unusual features are the functionalised quinoline-2-carboxylate units that are appended to the depsipeptide ring via two D-serine α-nitrogen atoms. A pair of tetrahydropyridazine 3-carboxylic acid and N-methyl-β-hydroxy-L-valine residues are also apparent in the cyclic peptidal sequence.

A number of experiments have confirmed that the luzopeptins can function as DNA bis-intercalating agents. In other words, the luzopeptins not only have the capacity to bind intramolecularly to single DNA molecules, they also can strongly bind to two DNA molecules. In this regard, members of the luzopeptin class biologically resemble other C(2)-symmetrial decadepsipeptides such as sandramycin, the quinoxapeptins and quinaldopeptin.

The synthetic challenges propounded by luzopeptin C are plentiful. Although early synthetic work in the antrimycin area by Shin[56] (Scheme 12.29) had actually provided the fundamental technology needed to assemble the homochiral 2,3,4,5-tetrahydro-3-pyridazinecarboxylic acid units found in the luzopeptin sequence, it was only after Ciufolini and Xi[57] applied the Shin methodology to more relevant model systems that such an approach looked feasible for use on the luzopeptin problem itself (Scheme 12.30). Given these encouraging antecedents, in their route to luzopeptin C, Boger and associates decided to assemble a cyclodepsipeptide precursor that would allow the tetrahydro-4-hydroxy-pyridazine-3-carboxamide rings to be constructed through the aforementioned trifluoroacetic-acid-induced cyclocondensation strategy.[53b] Specifically, Boger envisioned that the desired cyclodepsipeptide core would be available from the macrolactamisation of *seco* amino acid **147**, and that this in turn would be accessible from **148** and **149** by a [5+5]-fragment condensation (Scheme 12.31).

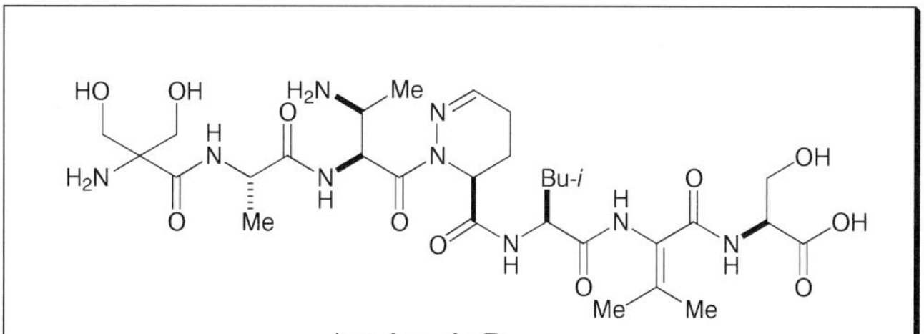

Antrimycin Dv

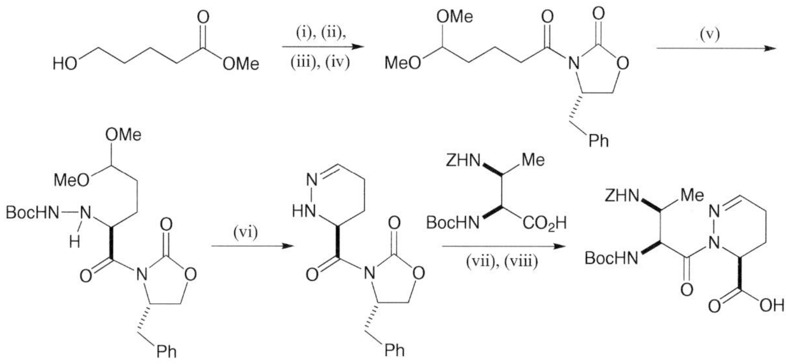

Scheme 12.29 (i) PCC, CH$_2$Cl$_2$, 78%; (ii) p-TsOH, MeOH, 74%; (iii) 1 M aq. LiOH, MeOH, 65%; (iv) PivCl, Et$_3$N, THF, −78°C then N-lithium (S)-4-benzyl-2-oxazolidinone, 96%; (v) LDA, THF, −78°C, DBAD, 93% (> 90% d.e.); (vi) CF$_3$CO$_2$H, CH$_2$Cl$_2$, 83%; (vii) DCC, DMAP, CH$_2$Cl$_2$, 0°C, 1 h, then RT, 5 h, 83%; (viii) 1 M aq. LiOH, 30% aq. H$_2$O$_2$, THF-H$_2$O (3:1), 0°C, 3 h, 89%.

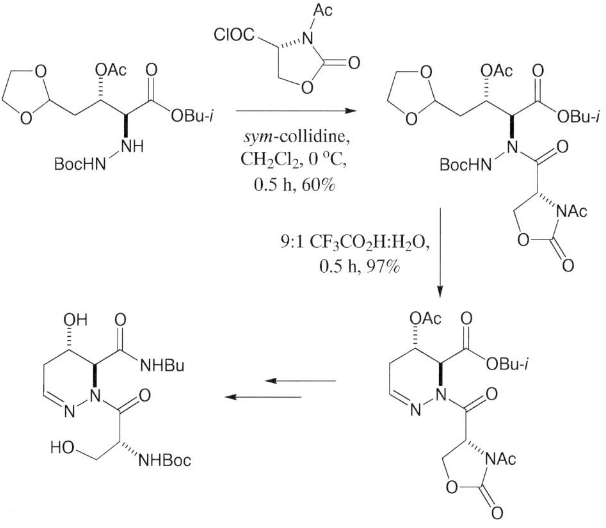

Scheme 12.30 Ciufolini's model work on the luzopeptin problem.

Scheme 12.32 summarises the elegant chemistry used by Boger and co-workers to arrive at fragments **148** and **149**. The lone stereocentre found in tripeptide **158** was forged through a Sharpless asymmetric epoxidation/base-mediated intramolecular epoxide ring-opening sequence on allylic

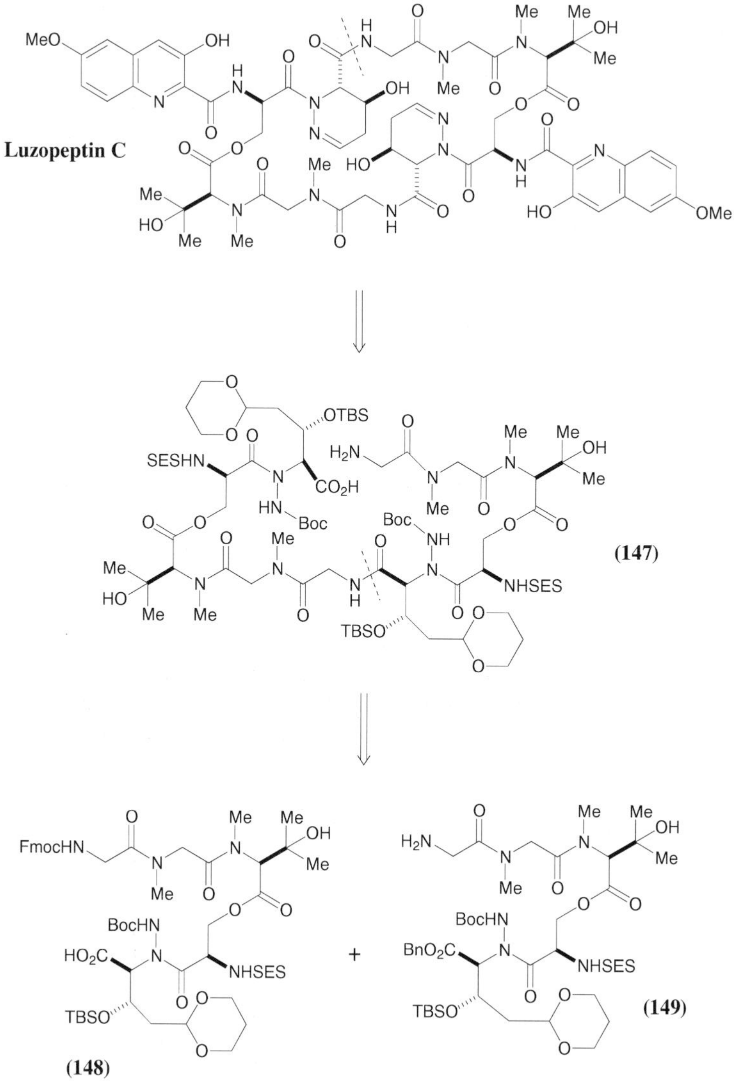

Scheme 12.31   Boger's retrosynthetic analysis of luzopeptin C.

alcohol **150**. This provided the oxazolidinone **153** which was converted to **157** in five more steps. Following an Fmoc for Boc protecting-group interchange, esterification of **158** with alcohol **159** and DCC-DMAP yielded depsipeptide **160**, the common precursor needed for both **148** and **149**.

A most impressive set of reactions was used to hew the acyclic 4-hydroxy-2,3,4,5-tetrahydropyridazine-3-carboxylic acid precursor **166** required for the preparation of hydrazinodipeptide **159** (Scheme 12.33). The venerable Sharpless AD reaction[39] was again deployed on enoate **162** for generating the appropriately functionalised homochiral diol needed for obtention of the protected α-amino-ester **164**. The latter was synthesised by azide inversion of the α-nitrobenzenesulfonyl ester **163**, protection with TBS triflate and azide reduction with triphenylphosphine. Treatment of **165** with Collet's oxaziridine reagent[58] in DMF next positioned an NBoc group on the primary amine, to give the hydrazine intermediate **166**. An acid chloride coupling with **167** subsequently provided the cyclic carbamate **168** which was selectively transesterified in the presence of the *O*-benzyl ester to complete **159**.

With **148** and **149** in hand only six more steps were needed to complete the first total synthesis of luzopeptin C. The optimal conditions for performing the [5+5]-fragment condensation activated the acid component **148** with EDCI and HOAt. Transfer hydrogenation simultaneously removed the Fmoc and Bn protecting groups from the product decadepsipeptide. *Seco* amino acid **147** cyclised smoothly with the EDCI/HOAt reagent combination to give **169** (Scheme 12.34). Subjection of this product to the Shin–Ciufolini cyclocondensation conditions (TFA) fashioned the desired tetrahydro-4-hydroxypyridazine-3-carboxamide ring systems. Deprotection of the product was accomplished with neat *anhydrous* HF in anisole at 0°C. Significantly, this deprotection regime was compatible with the delicate hydrazone functionality present in the cyclodepsipeptide core. The resulting diamine salt was finally coupled to 3-hydroxy-6-methoxyquinoline-2-carboxylic acid to provide luzopeptin C in excellent yield for the last two steps.

Scheme 12.32   (see overleaf) (i) Sharpless epoxidation; (ii) MeNCO, CH_2Cl_2, 23°C, 2 h, 94%; (iii) NaH, THF, 25°C, 24–72 h, 66%–85%; (iv) DHP, PPTS, CH_2Cl_2, 23°C, 17 h, 99%; (v) KOH, (CH_2OH)-H_2O, 150°C, 25 h, 92%–94%; (vi) Boc-Gly-Sar-OH, EDCI, HOAt, DMF, 23°C, 83%; (vii) *p*-TsOH, MeOH, 23°C, 24 h, 84%; (viii) RuO_4-$NaIO_4$, CCl_4-MeCN-H_2O (2:2:3), 23°C, 24 h, 87%; (ix) 4 M HCl-dioxane, 23°C, 0.5 h, then FmocCl, 10% aq. $NaHCO_3$-dioxane, 23°C, 8 h, 79%; (x) DCC-DMAP (3:2), CH_2Cl_2, −20°C to 0°C, 17 h, 73%; (xi) H_2, 10% Pd/C, MeOH, 10°C–12°C, 3 h, 76%–78%; (xii) Et_2NH-MeCN (1:2), 23°C, 20 min, 100%.

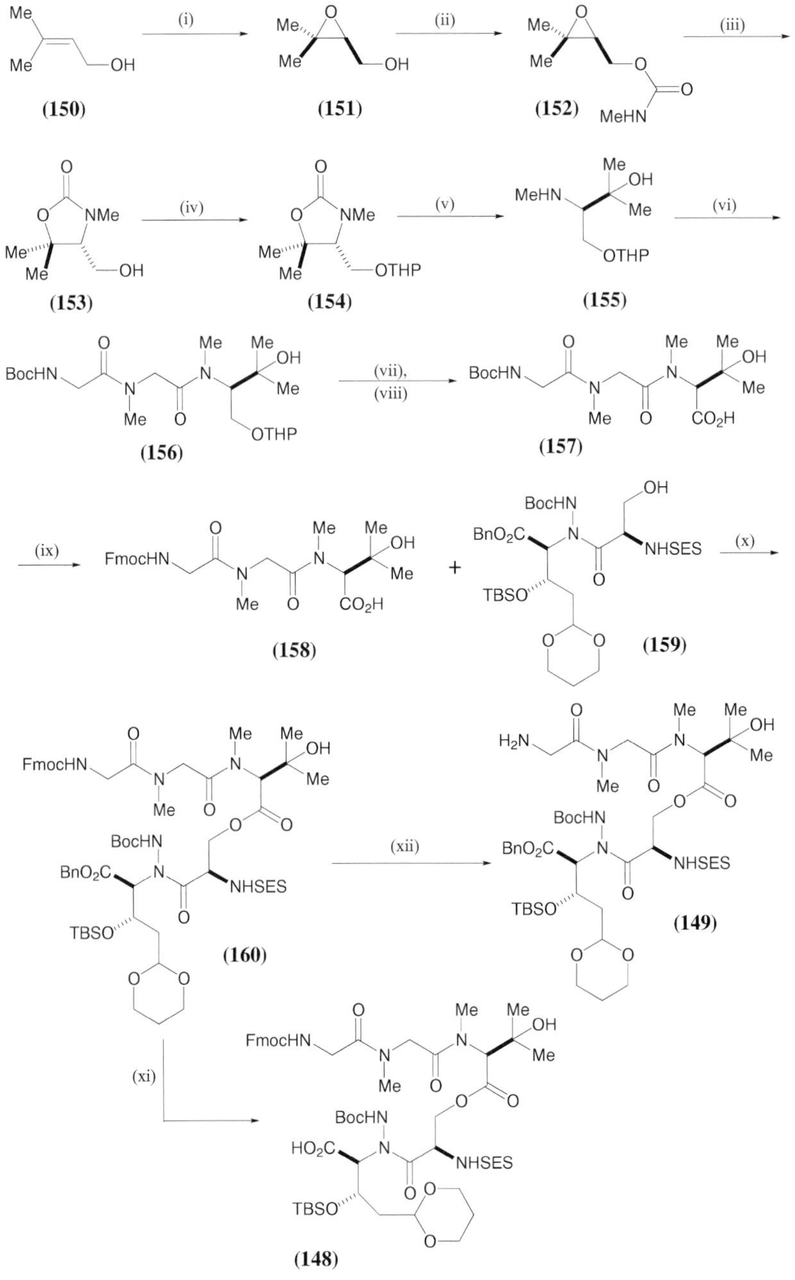

Me
Me
OH
(i)
(150)
Me
Me
O
OH
(ii)
(151)
O
(iii)
(152)
MeHN
O
NMe
(iv)
(153)
OTHP
(v)
(154)
MeHN
OH
(vi)
(155)
BocHN
(vii),
(viii)
(156)
CO2H
(157)
(ix)
FmocHN
(158)
BnO2C
NHSES
(x)
TBSO
(159)
(160)
(xii)
H2N
(149)
(xi)
HO2C
(148)

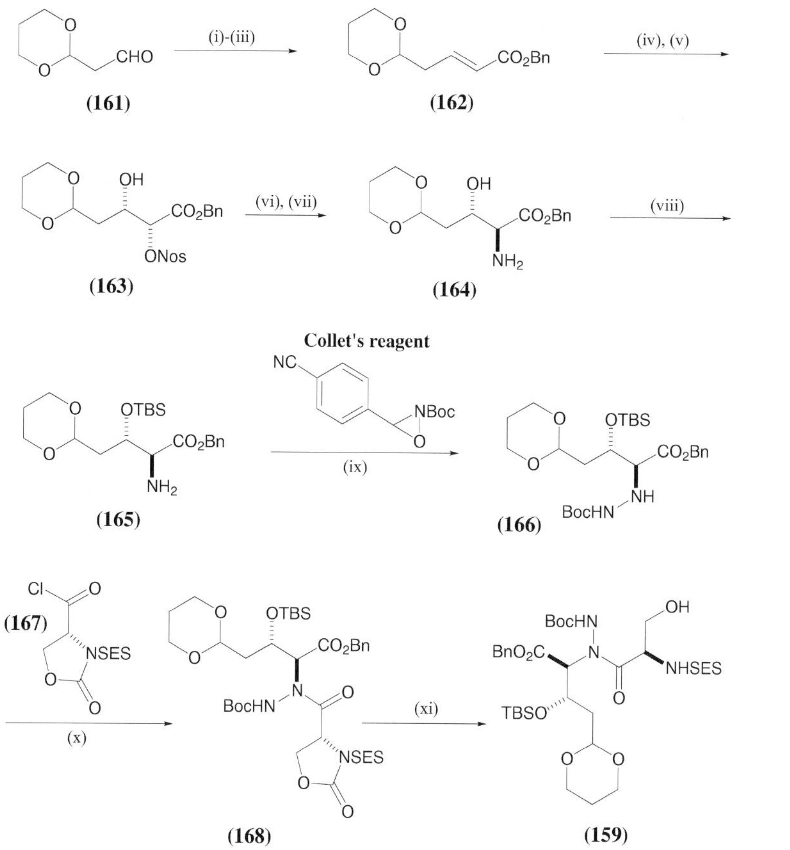

Scheme 12.33 (i) $(MeO)_2P(=O)CH_2CO_2Me$, KOBu-t, MeOH, 0°C to RT, 1 h, 88%; (ii) LiOH, THF, H_2O, RT, 18 h, conc. *in vacuo*; (iii) BnBr, DMF, RT, 92%; (iv) AD-Mix-α, $MeSO_2NH_2$, t-BuOH-H_2O (1:1), 4°C, 24 h, 80% (99% ee); (v) 4-nitrobenzenesulfonyl chloride, Et_3N, CH_2Cl_2, 4°C, 12 h, 68%; (vi) NaN_3, DMF, 50°C, 14 h, 87%; (vii) Ph_3P, THF, H_2O, 50°C, overnight, 93%; (viii) TBSOTf, 2,6-lutidine, 0°C, 3.5 h, 98%; (ix) Collet's reagent, DMF, RT, 24 h, 72%; (x) CH_2Cl_2, K_2CO_3, 0°C, 2 h, 85%; (xi) Cs_2CO_3, MeOH, 35 min, RT, 64%.

12.9 Sanglifehrin A

Without doubt, one of the most structurally intriguing cyclodepsipeptides to have been identified in recent years is sanglifehrin A. Isolated by a group of Novartis scientists[59] from the Malawian soil microorganism *Streptomyces* sp. A92-308110, sanglifehrin A captured immediate biological attention on account of its impressive range of immunosuppressant actions. The great breadth of novel structural features present in

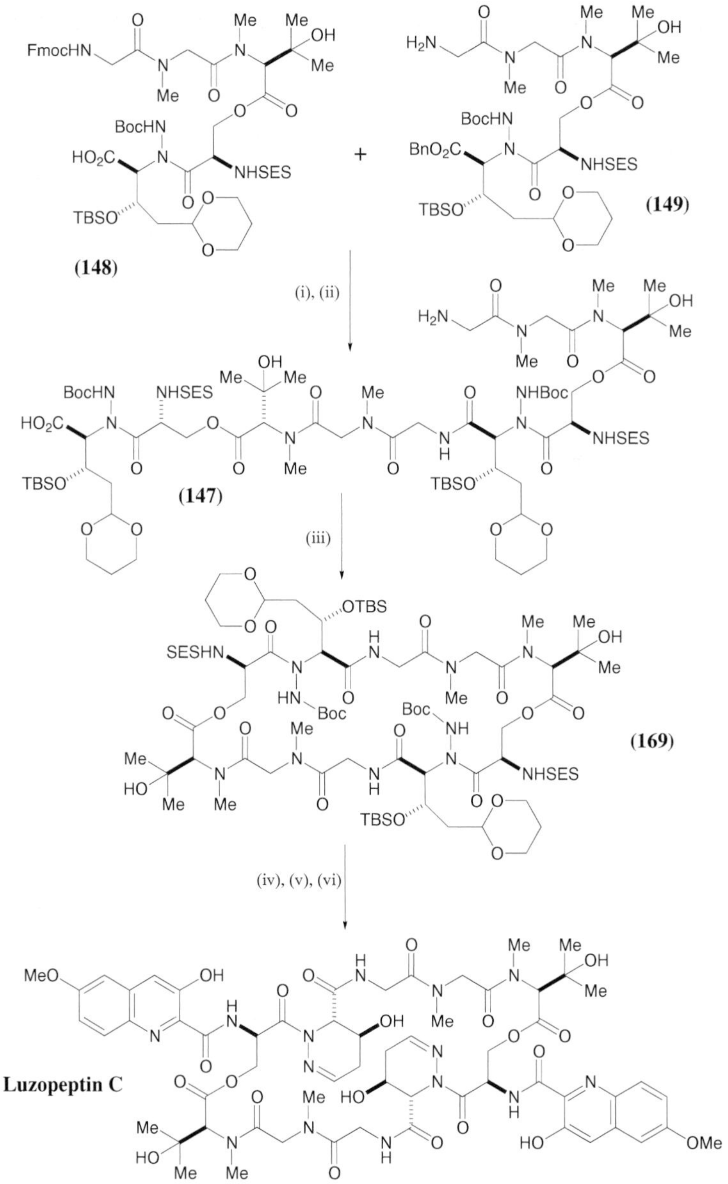

Scheme 12.34   (i) EDCI-HOAt (1:1), CH$_2$Cl$_2$, 0°C, 2 h, 64% (from **160**); (ii) 25% aq. HCO$_2$NH$_4$, 10% Pd/C, EtOH-H$_2$O, 23°C, 4 h, 98%; (iii) EDCI-HOAt (1:1.1), CH$_2$Cl$_2$, 0°C, 16 h, 66%; (iv) TFA-CH$_2$Cl$_2$-anisole (1:1:0.4), 0°C for 2 h then 0°C to 23°C for 1 h, 68%; (v) HF, anisole, 0°C, 1.5 h; (vi) EDCI-HOBt, NaHCO$_3$, 3-hydroxy-6-methoxyquinoline-2-carboxylic acid, DMF, 23°C, 11 h 80% (2 steps).

sanglifehrin A, which include a highly functionalised [5.5]spirolactam ring system, two stereodefined diene arrays and an *N(1)*-acylated piperazic acid unit, as well as 17 asymmetric centres, make this a prime candidate for total synthesis.[60]

To date, only one total synthesis of sanglifehrin A has been achieved and this is due to the Nicolaou group.[61] Their retrosynthetic strategy dissected the target at the central σ-bond of both diene motifs and also severed the C(13)-amide linkage (Scheme 12.35). As a consequence,

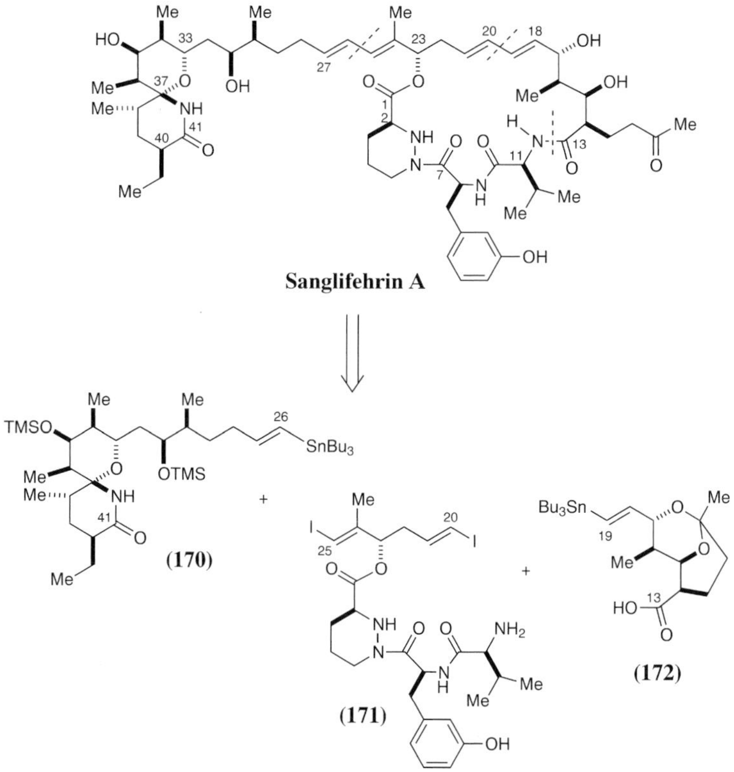

Scheme 12.35   Nicolaou's retrosynthetic strategy for sanglifehrin A.

stannanes **170** and **172**, and the bis(vinyl iodide) **171**, emerged as key intermediates. Nicolaou firstly planned to establish the C(13)-amide bond between **171** and **172** and then to effect an intramolecular Stille cross-coupling[9] reaction between the less-hindered vinyl iodide unit of this product and its internally tethered vinylstannane component. A second intermolecular Stille cross-coupling[9] would then be performed between

170 and the ring-closed cyclodepsipeptide to give the natural product after removal of the protecting groups.

3-Pentanone 173 was considered the optimal starting material for the synthesis of spirolactam 170 (Scheme 12.36). A Paterson asymmetric aldol reaction[62] between the (Z)-(+)-isopinocampheylboron enolate derived from 173 and methacrolein furnished the syn-aldol adduct 174 after O-silylation. A second Paterson substrate-controlled anti-aldol reaction[63] and an in situ lithium borohydride reduction next formulated the anti–anti relationship between the newly-induced stereocentres in 175. After processing towards the butyrate ester 176, an Ireland ester–enolate Claisen rearrangement[64] was called upon to install stereospecifically the C(40)-ethyl stereocentre. A regio- and stereoselective hydroboration reaction on 177 next set the C(38)-methyl stereochemistry. This process also reduced the carboxylic acid grouping to a primary alcohol. A chemo-selective TPAP oxidation[65] on the primary alcohol, in the presence of the newly installed secondary hydroxyl, yielded an intermediary hemiacetal that underwent immediate oxidation to the δ-valerolactone derivative 178. A Weinreb amidation reaction with Me_2AlNH_2 next appended the terminal amide functionality needed for constructing the [5.5]spirolactam unit.[66] The actual intermediate employed for spirolactam ring assembly was aldehyde 179. It collapsed to the desired product when treated with aqueous HF in acetonitrile. After O-silylation with TESOTf, a Brown asymmetric crotylboration[67] on aldehyde 180 with [(+)-Ipc$_2$B(Z-crotyl)] afforded a 3:1 mixture of alcohols enriched in the desired syn-isomer, which was purified and O-silylated to obtain 181. The terminal alkene in 181 was now ozonolytically degraded and the product aldehyde homologated to the enal with $TMSCH_2CH{=}NBu\text{-}t$/LDA under Corey's conditions.[68] A Lindlar hydrogenation subsequently removed the unsaturation from this enal to give 182. In situ generation of di-methyl (diazomethyl)phosphonate, from the basic hydrolysis of [$MeCOC({=}N_2)P({=}O)(OMe)_2$],[69] and subsequent reaction with 182, efficiently converted the latter into the terminal alkyne in 94% yield. Terminal acetylene bromination to obtain 183 and protecting-group interchange next set the stage for a Guibé Pd(0)-mediated reductive hydrostannation reaction,[70] which proceeded with excellent regio- and stereocontrol to furnish the desired vinyltin fragment 170 in good yield.

Depsipeptide 192 was assembled by the strategy enunciated in Scheme 12.37. Iodo-enal 184 was asymmetrically allylborated to give 185 and the alkene oxidatively degraded. After temporary protection, the product aldehyde 186 was stereoselectively iodoolefinated under Takai conditions[8] with $CHI_3{-}CrCl_2$ to obtain 187. O-Desilylation with TBAF and an EDC-promoted esterification with the known piperazic acid derivative 188[42a, b] then followed. Deprotection of both Boc groups from

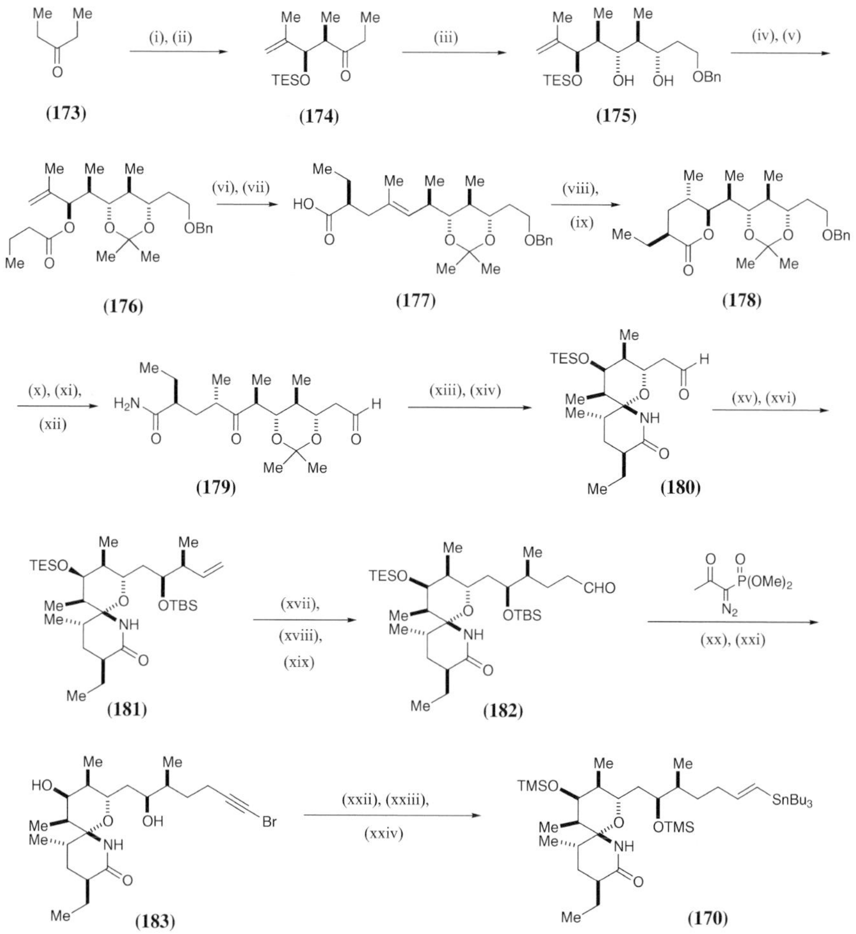

Scheme 12.36 (i) (+)-Ipc₂BOTf, *i*-Pr₂NEt, THF, −78°C, 2 h then methacrolein, 10 h; 30% aq. H₂O₂; (ii) TESCl, imid, CH₂Cl₂, 0°C, 2 h, 97% (2 steps); (iii) Cy₂BCl, Et₃N, Et₂O, 0°C, 1.5 h, then BnO(CH₂)₂CHO (2 equiv.), −78°C to −10°C, 4 h; then LiBH₄, −78°C to RT, 12 h, then NaBO₃·4H₂O, THF-H₂O, (3:2), RT, 12 h, 72%; (iv) CSA (0.1 equiv.), Me₂C(OMe)₂, Me₂CO, 72 h, 95%; (v) (PrCO)₂O, Et₃N, CH₂Cl₂, 24 h, 98%; (vi) LDA, TBSCl, THF, −78°C, then HMPA:THF (1:5), warm to 0°C, 1 h; (vii) PhMe, 70°C, 2 h, 84%; (viii) BH₃, THF, −20°C, 17 h, then NaBO₃·4H₂O, THF-H₂O (3:2), RT, 12 h; (ix) TPAP, NMO, CH₂Cl₂, RT, 8 h, 51% (2 steps); (x) Me₂AlNH₂, (20 equiv.), CH₂Cl₂, RT, 2 h; (xi) H₂, 10% Pd(OH)₂-C, EtOH, 12 h, 90% (2 steps); (xii) DMP, Py, CH₂Cl₂, RT, 0.5 h, 55%; (xiii) aq. HF, MeCN, 36 h, 90%; (xiv) TESOTf, 2,6-lutidine, CH₂Cl₂, −40°C to RT, 88%; (xv) (Z)-crotyl(+)Ipc₂B, THF, −78°C, then NaBO₃·4H₂O, THF-H₂O (3:2), RT, 12 h, 67%; (xvi) TBSOTf, 2,6-lutidine, CH₂Cl₂, −40°C to RT, 88%; (xvii) O₃, CH₂Cl₂, −78°C, then Ph₃P, warm to RT, 92%; (xviii) LDA (5 equiv.), TMSCH₂CH=NBu-*t*, THF, −78°C to −20°C, 2 h, 87%; (xix) H₂, Lindlar cat., MeOH, 3.5 h, 94%; (xx) Diazoketone (2 equiv.), K₂CO₃ (2.5 equiv.), MeOH, RT, 3.5 h, 94%; (xxi) NBS (1.2 equiv.), AgNO₃ (0.3 equiv.), Me₂CO, 1 h, 93%; (xxii) aq. HF, MeCN, RT, 3 h; then TBAF (1.2 equiv.), THF, 15 min, 79%; (xxiii) CF₃C(=O)N(Me)SiMe₃ (excess), RT, 1 h, 98%; (xxiv) [Pd₂(dba)₃]·CHCl₃ (0.1 equiv.), Ph₃P (0.8 equiv.), Bu₃SnH (2.2 equiv.), RT, 0.5 h, 70%.

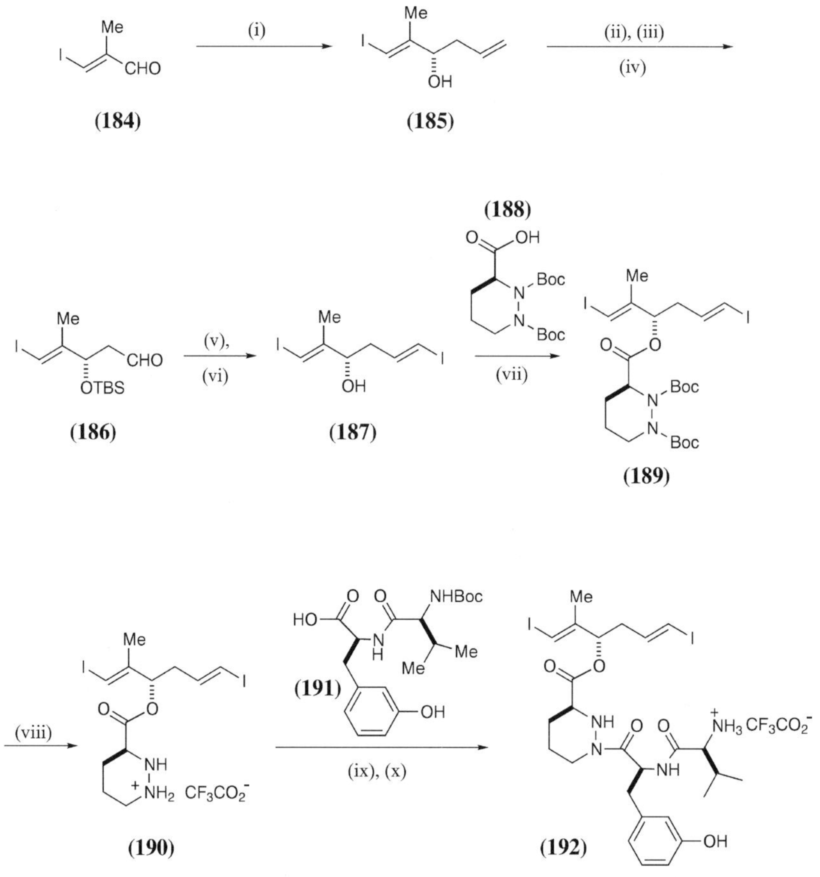

(184)   (185)

(188)

(186)   (187)   (189)

(190)   (191)   (192)

Scheme 12.37 (i) (−)-(Ipc)$_2$BOMe, AllMgBr, Et$_2$O, 0°C to RT, then −100°C add **184**, 80%; (ii) TBSCl, imid., DMF, 84%; (iii) OsO$_4$, NMO, t-BuOH-H$_2$O, 89%; (iv) NaIO$_4$, MeOH, RT, 92%; (v) CHI$_3$ (2 equiv.), CrCl$_2$ (6 equiv.), dioxane-THF (9:1), −5°C to RT, 12 h, 57%; (vi) TBAF, THF, 0°C to RT, 15 min, 88%; (vii) **188** (2 equiv.), EDC (2 equiv.), 4-PPy (0.1 equiv.), i-Pr$_2$NEt (1 equiv.), CH$_2$Cl$_2$, 0°C to RT, 64%; (viii) CF$_3$CO$_2$H-CH$_2$Cl$_2$ (1:1), 0°C to RT, 2 h; (ix) **191** (1 equiv.), HOAt (1 equiv.), i-Pr$_2$NEt, EDC (1.2 equiv.), CH$_2$Cl$_2$, 66% (2 steps); (x) CF$_3$CO$_2$H-CH$_2$Cl$_2$ (1:9), 0°C to RT, 4 h.

189 unmasked the hydrazine grouping allowing a regioselective acylation to be implemented at the less-hindered and more nucleophilic nitrogen atom of **190**. A second Boc deprotection with TFA finally delivered **192**. The route devised for dipeptide acid **191** is presented in Scheme 12.38. It featured a Duphos-rhodium-mediated asymmetric hydrogenation reaction[71] on the dehydroamino acid ester **195** to access the tyrosine analogue **196** in high ee.

The three key steps which hallmarked the synthetic pathway to vinylstannane **172** (Scheme 12.39) were the regiocontrolled C-homo-allylation of 2,3-epoxyalcohol **198** used to obtain diol **199**, the Wacker

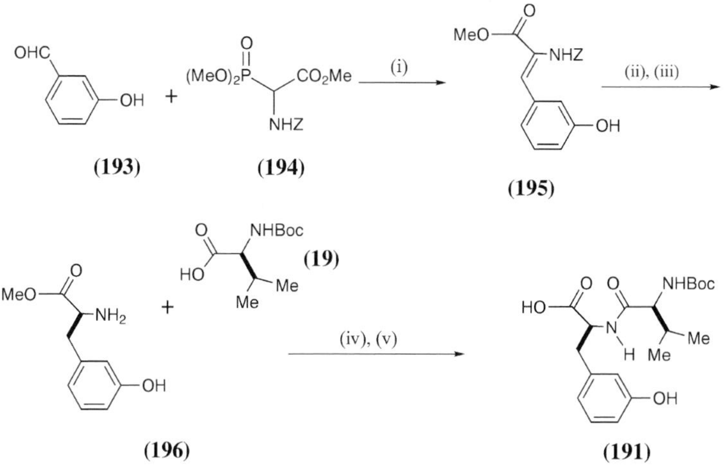

Scheme 12.38 (i) DBU, CH$_2$Cl$_2$, 25°C, 2 h, 90%; (ii) H$_2$, 60 psi, [(S, S)-Et-DuP-Rh]$^+$TfO$^-$, MeOH, 96 h, 98% ee, 90%; (iii) 10% Pd/C, H$_2$, MeOH, 25°C, 12 h, 96%; (iv) EDC, HOAt, CH$_2$Cl$_2$, 0°C to 25°C, 3 h, 78%; (v) LiOH, THF-H$_2$O (3:1), 0°C to 25°C, 1.5 h, 89%.

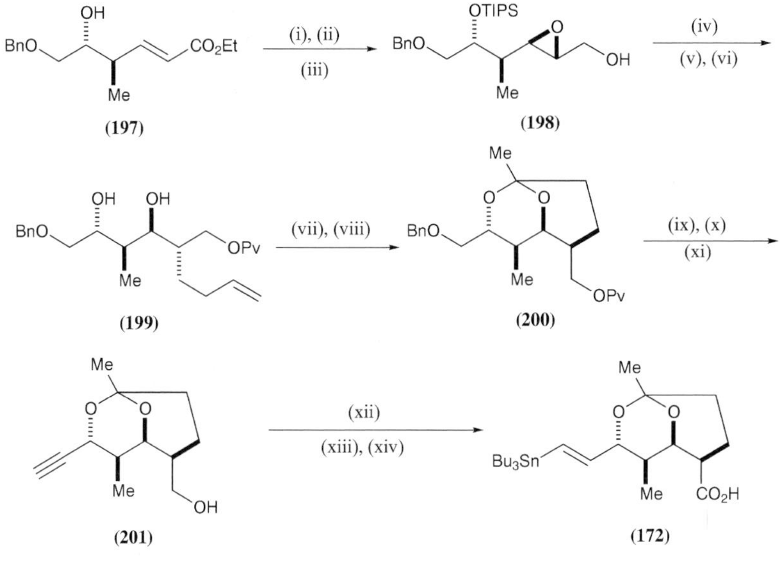

Scheme 12.39 (i) TIPSCl, imid., DMF, 60°C, 24 h, 95%; (ii) DIBAL-H, CH$_2$Cl$_2$, −78°C, 2 h, 87%; (iii) m-CBPA, CH$_2$Cl$_2$, −25°C, β:α (6:1), 100%; (iv) H$_2$C=CHCH$_2$CH$_2$MgBr, CuI, Et$_2$O-THF (1:1), −40°C to −20°C, 18 h, 80%; (v) PvCl, py, 25°C, 24 h, 95%; (vi) TBAF, THF, 25°C, 1 h, 81%; (vii) PdCl$_2$, benzoquinone, DMF-H$_2$O (7:1), 25°C, 3 h; (viii) TsOH · H$_2$O, C$_6$H$_6$, reflux, 88% (2 steps); (ix) 10% Pd/C, H$_2$, EtOH, 25°C, 1 h; (x) TPAP, NMO, 4 Å MS, CH$_2$Cl$_2$, 25°C, 0.3 h, 83% (2 steps); (xi) MeCOC(=N$_2$)PO(OMe)$_2$, K$_2$CO$_3$, MeOH, 0°C to 25°C, 13 h, then K$_2$CO$_3$, 25°C, 24 h, 98%; (xii) n-Bu$_3$SnH, i-Pr$_2$NEt, P(o-Tol)$_3$, PdCl$_2$(PhCN)$_2$, CH$_2$Cl$_2$, −20°C, 1 h, 79%; (xiii) TPAP, NMO, 4 Å MS, CH$_2$Cl$_2$, 25°C, 15 min; (xiv) NaClO$_2$, NaH$_2$PO$_4$, 2-methylbut-2-ene, THF-t-BuOH-H$_2$O (1:5:1), 25°C, 15 min.

oxidation of this product to secure the methyl ketone and the internal ketalisation of this ketone to fashion ketal **200**. The latter was processed towards acetylene **201** by a multistep sequence that again exploited the Ohira dimethyl (diazomethyl)phosphonate alkynylation reaction.[69] A Pd(0)-catalysed hydrostannylation of this acetylene berthed the (*E*)-vinylstannane unit in **172**, while a TPAP/sodium chlorite double oxidation sequence finally introduced its acid functionality.

A HATU coupling successfully connected the two fragments **172** with **192** together in 45% yield (Scheme 12.40). The critical regioselective intramolecular Stille ring-closure proceeded in similar yield. The unification of **170** and **203** again exploited Stille Pd(0)-mediated cross-coupling chemistry for the stereocontrolled installation of the second (*E*,*E*)-diene motif. After a series of deprotections, sanglifehrin A was

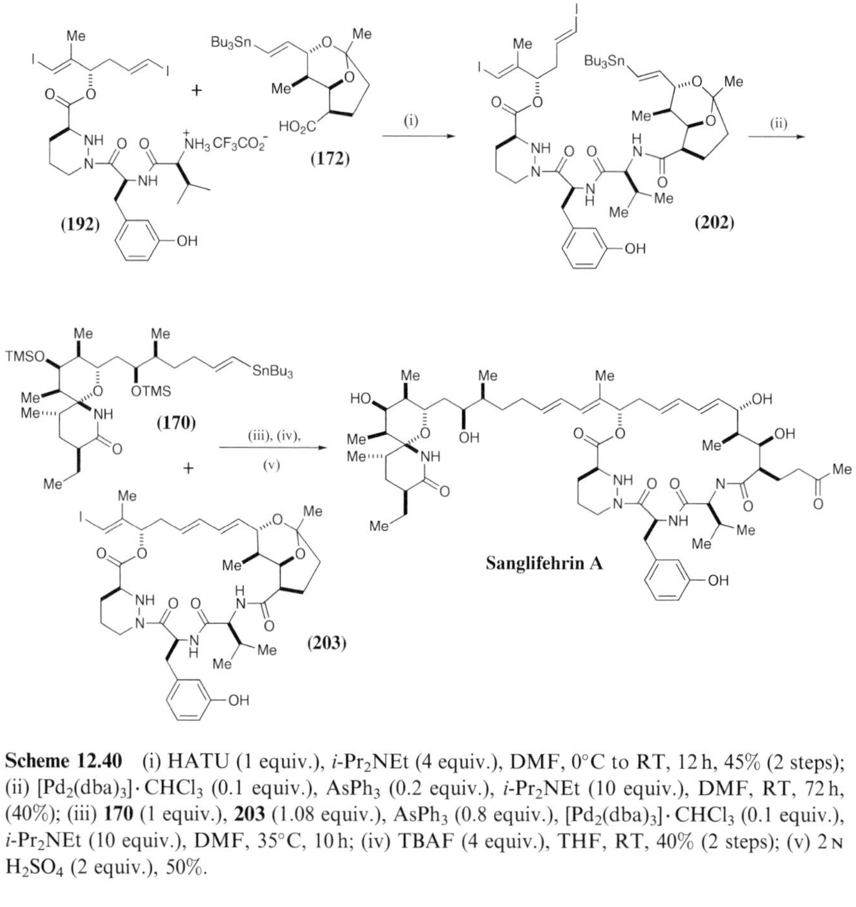

Scheme 12.40 (i) HATU (1 equiv.), *i*-Pr$_2$NEt (4 equiv.), DMF, 0°C to RT, 12 h, 45% (2 steps); (ii) [Pd$_2$(dba)$_3$]·CHCl$_3$ (0.1 equiv.), AsPh$_3$ (0.2 equiv.), *i*-Pr$_2$NEt (10 equiv.), DMF, RT, 72 h, (40%); (iii) **170** (1 equiv.), **203** (1.08 equiv.), AsPh$_3$ (0.8 equiv.), [Pd$_2$(dba)$_3$]·CHCl$_3$ (0.1 equiv.), *i*-Pr$_2$NEt (10 equiv.), DMF, 35°C, 10 h; (iv) TBAF (4 equiv.), THF, RT, 40% (2 steps); (v) 2 N H$_2$SO$_4$ (2 equiv.), 50%.

finally isolated in 20% overall yield for the last three steps, completing this fully stereocontrolled synthesis which stands out for its exceptional synthetic beauty.

12.10   (−)-Pateamine A

Yet another cyclodepsipeptide showing promise as a drug for transplant surgery is the marine natural product (−)-pateamine A.[72] It displays powerful immunosuppressive effects in the mixed lymphocyte culture response (MLR) assay ($IC_{50} = 2.6\,\text{nM}$) at concentrations that are not particularly cytotoxic [LCV (lymphocyte viability) assay/MLR ratio > 1000]. A measure of its efficacy as an immunosuppressant, when compared with cyclosporin A, can be gauged from the results obtained in the mouse-skin graft-rejection assay. (−)-Pateamine A resulted in a 15-day survival period, whereas cyclosporin A elicited only a 10-day survival time when administered under identical conditions. In an effort to elucidate the mechanism by which (−)-pateamine A induces immunosuppression, which is currently unknown, Romo and co-workers embarked on a total synthesis[73, 74] with a view to obtaining synthetic analogues that could be used to assist in the isolation and identification of the putative cellular receptor. Scheme 12.41 depicts the key elements of Romo's synthetic planning for this target. His intention was to stereoselectively introduce the C(10)–C(17)-trienyl side by a Stille cross-coupling reaction between **204** and **205**. A base-induced β-lactam ring-opening was further envisaged on **206** for bringing about macro-lactonisation. A Mitsunobu reaction[75] between **207** and **208** was proposed for establishment of the ester linkage in **206**, and a Lindlar reduction on the eneyne macrocyclisation product was anticipated for controlling the (*E*,*Z*)-dienoate geometry required in **205**.

The synthesis of eneyne acid **208** commenced with a Taber-modified Noyori reduction[76] on ethyl acetoacetate to obtain (*S*)-2-hydroxybutyrate (Scheme 12.42). The Taber-modified methodology[76] generally gives an improved enantiomeric purity and higher yield for this product than the yeast reduction methods which also can be used to access this material (94% ee compared with 84%–90% ee). After *O*-silylation and low-temperature reduction to obtain aldehyde **210**, a two-step Corey–Fuchs alkynylation[77] led to **211**. A Negishi zirconocene dichloride-catalysed carboalumination reaction[78] on **211,** stereo- and regioselectively converted it to **212**, after trapping of the intermediary vinylalane with iodine. A Sonogashira cross-coupling[79] with propargyl alcohol subsequently provided alcohol **213**, which was readily transformed into acid **208** in two more steps.

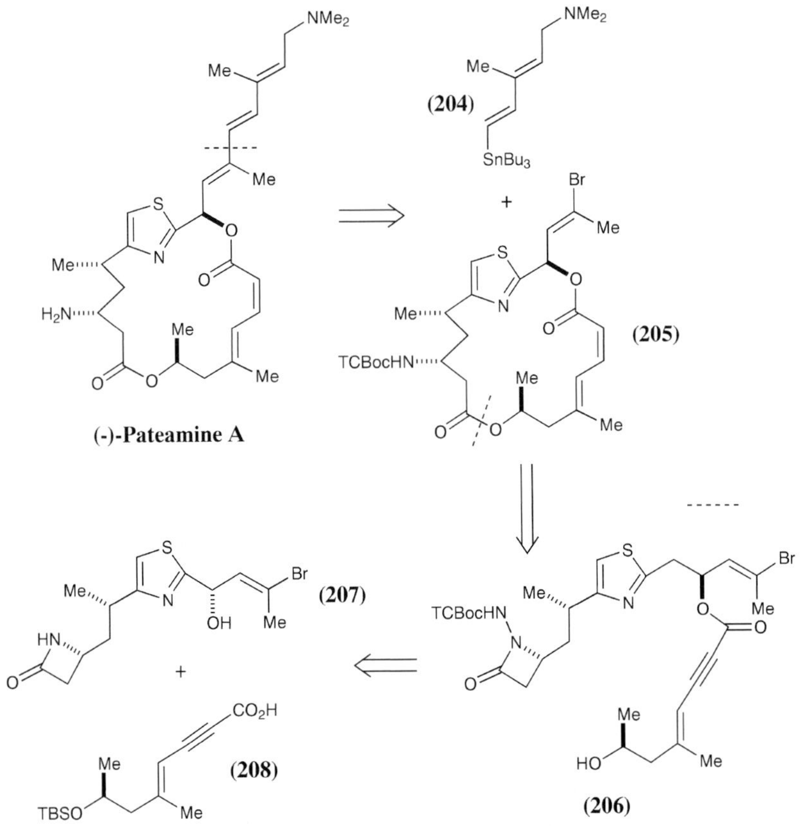

Scheme 12.41   Romo's retrosynthetic planning for (−)-pateamine A.

A Nagao acetate aldol reaction[80] between known aldehyde **215** and the N-acetylthiazolidinethione **214** set the C(10)-allylic alcohol stereochemistry present in aldehyde **224** (Scheme 12.43). After hydroxyl protection, ammonolytic cleavage of the auxiliary and thionation of the resulting amide with Belleau's reagent [(PhOPh)$_2$P$_2$S$_4$][81] yielded the thioamide precursor **217** needed for the modified Hantzsch thiazole synthesis.[82] Further elaboration of **217** into **222** set the ground for a Hruby asymmetric conjugate addition reaction[83] with excess MeMgBr and CuBr·Me$_2$S to obtain **223** with a 6.4:1 level of selectivity. A Weinreb transamidation/ DIBAL reduction protocol subsequently afforded aldehyde **224**. A further bout of Nagao acetate aldol chemistry on **224** ultimately yielded the N-benzyloxyamide **226** after removal of the auxiliary (Scheme 12.44). It was transformed into the C(3)-inverted β-lactam **207** by an intramolecular Mitsunobu lactamisation and a samarium

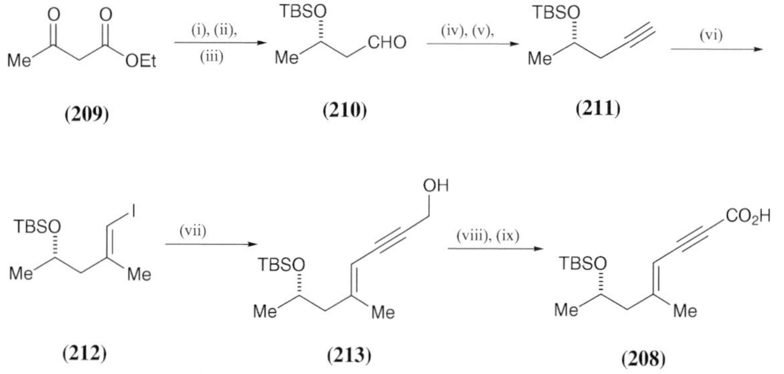

(209)   (210)   (211)

(212)   (213)   (208)

Scheme 12.42   (i) H_2 (200 psi), RuCl$_2$-(S)-BINAP, EtOH, 130°C, 10 h, 97% (94% ee); (ii) TBSCl, imid, DMF, RT, 12 h, 97%; (iii) i-Bu$_2$AlH, CH$_2$Cl$_2$, −90°C; (iv) Zn dust, Ph$_3$P, CBr$_4$, CH$_2$Cl$_2$, 48 h, then add **210** at 0°C, 1 h, 92% (2 steps); (v) n-BuLi, THF, −78°C, 2 h to RT, 10 h, 95%; (vi) Cp$_2$ZrCl$_2$, Me$_3$Al, hex, CH$_2$Cl$_2$, 0°C then RT, 20 h, then I$_2$ in Et$_2$O at 0°C, 60%; (vii) (Ph$_3$P)$_4$Pd (1.7 mol%), C$_6$H$_6$, 0°C, then propargyl alcohol (1.3 equiv.), n-PrNH$_2$ (5.2 equiv.), then CuI (20 mol%), warm to RT, 20 h, 94%; (viii) Me$_2$SO, (COCl)$_2$, CH$_2$Cl$_2$, −78°C, 20 min., then Et$_3$N, 76%; (ix) MeCN, NaH$_2$PO$_4$ pH 4.26 buffer, t-BuOH, then 30% H$_2$O$_2$, NaClO$_2$ (20 equiv.), 0°C, 2.5 h, 61%.

diiodide N—O bond cleavage. Romo settled for a trichloro-t-butoxy carbamate (TCBoc) protecting group strategy for blocking the β-lactam nitrogen; this simultaneously activated the lactam carbonyl for the ensuing O-acylation. However, before the latter macrocyclisation could be implemented, the eneyne acid **208** first had to be positioned at O(10) with inversion of configuration. The TIPS group was therefore removed from **207** with buffered TBAF, and a Mitsunobu esterification performed on the resulting allylic alcohol at low temperature with acid **208**. The key macrocyclisation intermediate was subsequently unveiled by treatment of the Mitsunobu product with HF-pyridine; this cleaved the TBS group from O(24). After a substantial amount of experimentation, the macrolactonisation of **206** was eventually accomplished in good yield (68%) by employing the reaction conditions of Palomo and co-workers[84] for the intermolecular alcoholysis of β-lactams, except now Romo operated under conditions of very high dilution. Having successfully negotiated the major obstacles presented by this novel ring-closure, Romo and co-workers next set about sculpting the (E,Z)-enoate extending from C(19)–C(22) in the target. The latter feature was readily installed by a Lindlar reduction on **227**. A Stille reaction[9] was next used to elaborate the C(10)–C(17)-trienylamine side-chain stereoselectively; the appropriate stannane **204** was prepared according to Scheme 12.45. The final step of Romo's synthesis (Scheme 12.44) was removal of the TCBoc protecting

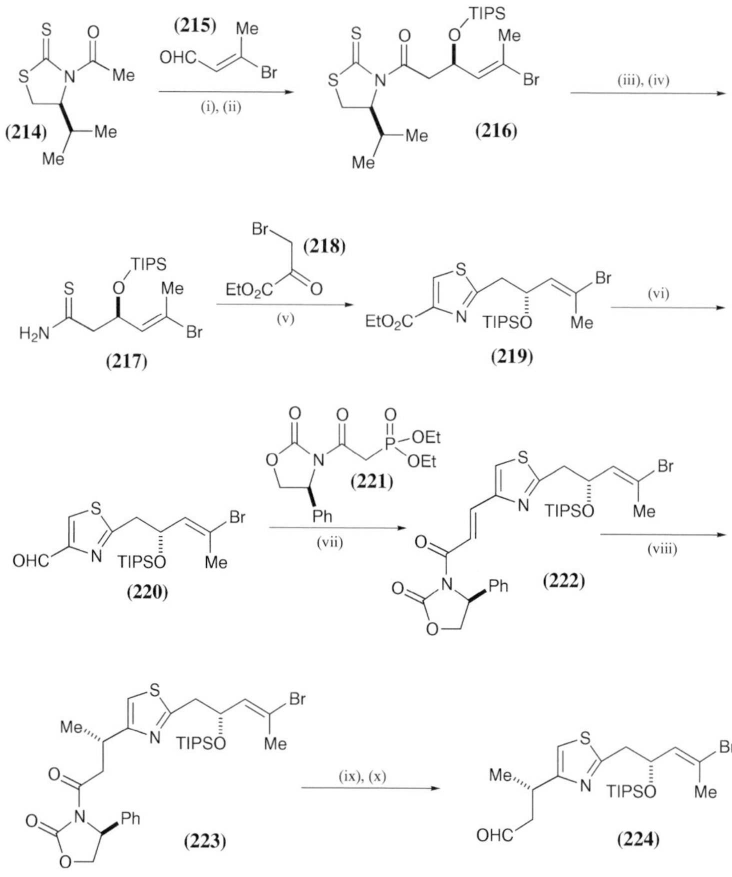

Scheme 12.43 (i) Sn(OTf)$_2$, EtN-piperidine, CH$_2$Cl$_2$, 4.5 h, $-50°$C, then add **215** at $-78°$C, 3 h, 84% ($>$19:1 dr); (ii) TIPSOTf, 2,6-lutidine, CH$_2$Cl$_2$, 0°C to RT, 40 min, 99%; (iii) NH$_3$ (g), CH$_2$Cl$_2$, 0°C, 90%; (iv) Belleau's reagent, THF, RT, 0.5 h, 72%; (v) ethyl bromopyruvate **218** (3 equiv.), KHCO$_3$ (5 equiv.), DME, $-30°$C to RT, then $-30°$C, 2,6-lutidine, (CF$_3$CO)$_2$O, 1 h, 72%; (vi) i-Bu$_2$AlH, CH$_2$Cl$_2$, $-90°$C, 3 h, then MeOH, 89%; (vii) **221**, THF, NaN(SiMe$_3$)$_2$, RT, 1.5 h, 90%; (viii) CuBr-Me$_2$S, THF, Me$_2$S, $-78°$C then MeMgBr (3 M in Et$_2$O), 15 min, warm to 0°C 15 min, add **222** at $-78°$C, warm to $-30°$C, 3.5 h, 77%; (ix) MeON(Me)AlMe$_2$, CH$_2$Cl$_2$, PhMe, $-20°$C, 8 h, 82%; (x) i-Bu$_2$AlH, CH$_2$Cl$_2$, $-78°$C, 1.3 h, then MeOH, 95%.

group from the C(3)-amine using the Cd–Pb deprotection methodology introduced by Ciufolini[85] for removing Troc groups from sensitive molecules. Significantly, these very mild reducing conditions did not perturb or damage the delicate (E,Z)-enoate domain that is present in $(-)$-pateamine A, allowing the natural product to be isolated in a stalwart 78% yield.

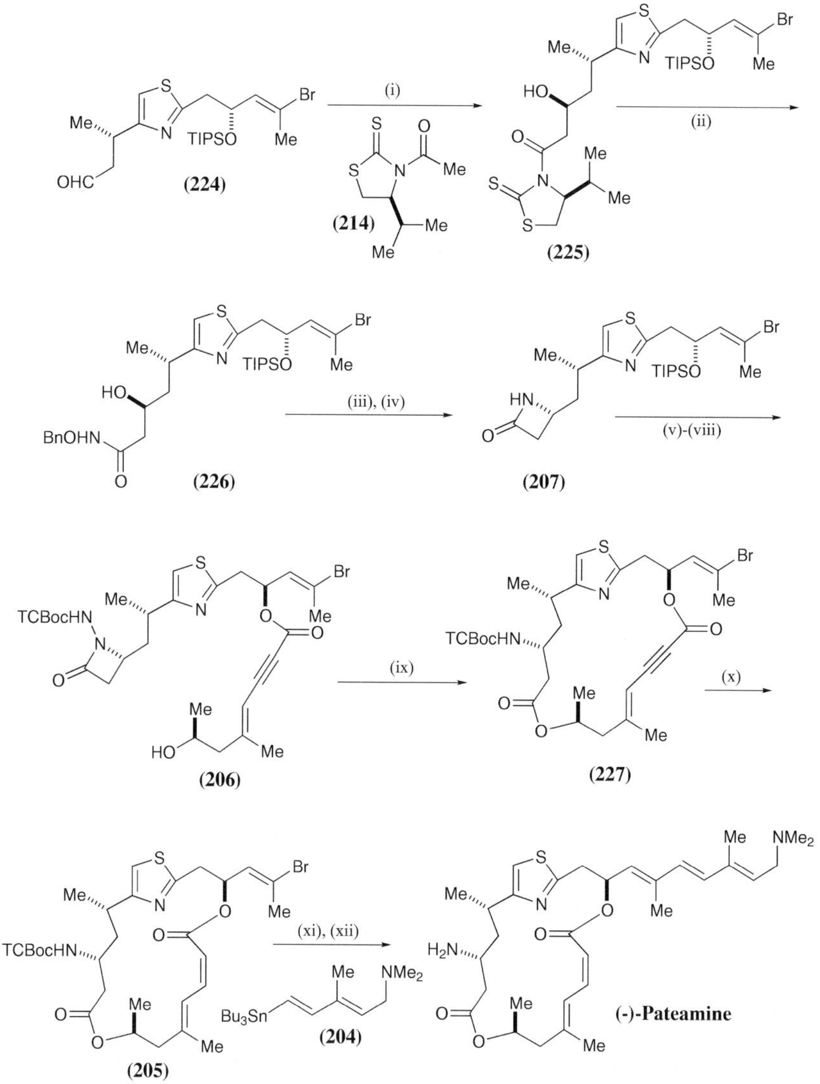

Scheme 12.44 (i) **214**, Sn(OTf)₂, EtN-pip, CH₂Cl₂, 4 h, −50°C, then add **224** at −78°C, 1 h, 90% (> 19:1 dr); (ii) BnONH₃Cl, CH₂Cl₂, Et₃N, 0°C, then RT, 27 h, 90%; (iii) Ph₃P, DIAD, THF, 92%; (iv) SmI₂, THF-H₂O (20:1), 0°C, 1 h, 96%; (v) TCBocCl, CH₂Cl₂, DMAP, Et₃N, RT, (vi) TBAF, THF, AcOH (20 mol%), 78%; (vii) Acid **208**, Ph₃P, DIAD, THF, −20°C, 1.5 h, 57%; (viii) HF-py, py, THF, 0°C to RT, then 5 h, 78%; (ix) Et₄NCN (9 equiv.), CH₂Cl₂ (0.0018 M), 0°C to RT, 6 h, 68%; (x) H₂, Pd/CaCO₃/Pb, MeOH, 17 h, ca. 50%; (xi) Pd₂(dba)₃·CHCl₃ (15 mol%), AsPh₃ (60 mol%), THF, stannane **204**, 33 h, RT, 27% (57% based on SM); (xii) 1 N aq. NH₄OAc, THF, Cd-Pb couple, 3.5 h, 78%.

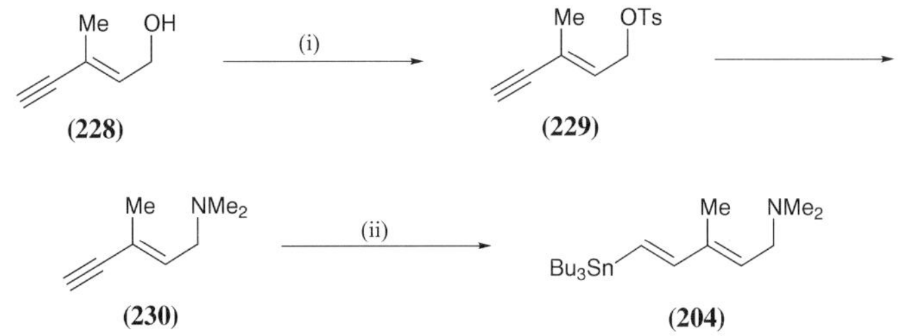

Scheme 12.45 (i) *n*-BuLi, THF, hex, −78°C, 1 h, then *p*-TsCl; bubble Me$_2$NH into reaction mixture for 15 min, transfer to sealed tube, warm to RT, 1.5 h, 69%; (ii) CuCN (0.9 equiv.), THF, −60°C, then add *n*-BuLi (1.8 equiv.) over 15 min, warm to −40°C, add Bu$_3$SnH (1.8 equiv.), 1 h, then add amine **230** in THF, stir 1 h, 71%.

12.11   FR-901,228

Numerous studies have correlated the presence of *ras* oncogenes with human tumour development.[86] It is believed that activated *ras* expression is essential for maintenance of the transformed phenotype and that the latter is highly tumourigenic. Agents that can selectively reverse the phenotype of an oncogenic *ras*-transformant usually have the ability to 'de-transform' tumourigenic cell lines, and often this normalisation process is associated with significant *in vivo* antitumour activity. Recently, the structurally unique antitumour bicyclic depsipeptide FR-901,228 (Scheme 12.48) was isolated from culture broths of *Chromobacterioum violaceum* no. 968.[87] It can restore to normal the morphology of a Ha-*ras* transformant to normal in NIH-3T3 cells; it is also highly active an antitumour drug in a number of mouse-based assays. Given its ability to target components of the *ras*-mediated signal transduction pathway, FR-901,228 is a molecule of great medical interest, and, as a consequence, a number of synthetic endeavours have been sparked towards it. Simon and co-workers[88] have recently announced a very impressive total synthesis of FR-901,228 by a strategy that delays disulfide ring assembly until the final step of the synthesis.

Cyclodepsipeptide ring construction began with the synthesis of the protected mercaptohydroxyheptenoic acid **236** (Scheme 12.46). The transformation of most interest in this sequence was the Carreira catalytic asymmetric aldol reaction[89] executed between the silyl ketene acetal **233** and aldehyde **232** to obtain **235**, which was converted into acid **236** by simple base hydrolysis.

An Fmoc protecting-group strategy was exploited for the assembly of **240** (Scheme 12.47) with Castro's BOP reagent being used iteratively

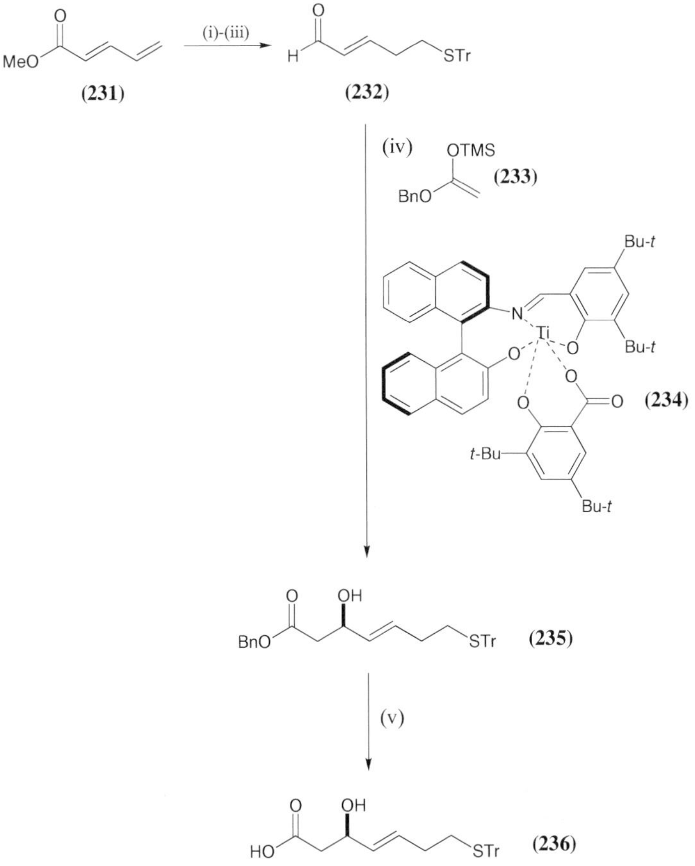

Scheme 12.46   (i) TrSH, Cs$_2$CO$_3$, 78%; (ii) DIBAL-H, 91%; (iii) (COCl)$_2$, DMSO, Et$_3$N, 74%; (iv) **233**, Ti(IV) catalyst **234**, Et$_2$O, $-10°$C; n-Bu$_4$NF, 99%; (v) LiOH, MeOH, 100%.

for each amidation step. The fact that all these couplings were carried out without protection of the threonine β-hydroxyl is especially notable.

The best method for closing the macrolactone ring of **243** generated a phosphonium ion from the *seco*-acid hydroxyl of **244** by Mitsunobu activation (Scheme 12.48); this tactic brought about ring-closure in 62% yield. Interestingly, the Mitsunobu macrolactonisation process greatly benefited from the addition of p-TsOH·H$_2$O to the reaction mixture; the latter additive completely suppressed elimination of the activated allylic alcohol. By way of contrast, the macrolactonisation of *seco* acid **242** (prepared by a similar route) by acid activation (Scheme 12.48) afford-ed the desired product **243** only in very low yield. To the best of our

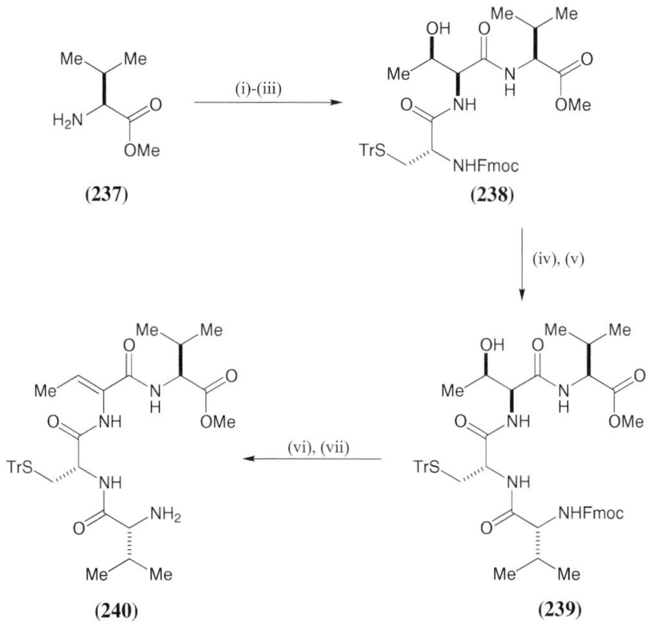

Scheme 12.47 (i) Fmoc-L-Thr, BOP, DIEA, 95%; (ii) Et$_2$NH; (iii) Fmoc-D-Cys(STr), BOP, DIEA, 92%; (iv) Et$_2$NH; (v) Fmoc-D-Val, BOP, DIEA, 86%; (vi) Ts$_2$O, 97%; (vii) DABCO, Et$_2$NH, 96%.

knowledge, this is the first example of a Mitsunobu macrolactonisation being used successfully to form a cyclodepsipeptide ring.

Disulfide ring assembly was achieved in 84% yield by oxidation of the (S-trityl)lactone with iodine in dilute methanol.[90] Significantly, this procedure did not damage the potentially sensitive dehydroamino acid functionality also present within FR-901,228.

12.12   Doliculide

Doliculide is an exceedingly powerful antineoplastic agent isolated[91] from the Japanese sea hare *D. auricularia*. From a synthetic perspective, the 1,3,5-*syn*-trimethylalkane substructure embedded within its 15-carbon polyketide domain provides its most significant challenge. Yamada's solution[92] to the 1,3,5-*syn*-trimethylalkane issue was to employ a series of iterative Evans *syn*-aldol[32]/Barton–McCombie deoxygenation[93] reactions to build up this motif, the process commencing from aldehyde **245** (Scheme 12.49). The *anti*-relationship found between C(6) and C(7) in

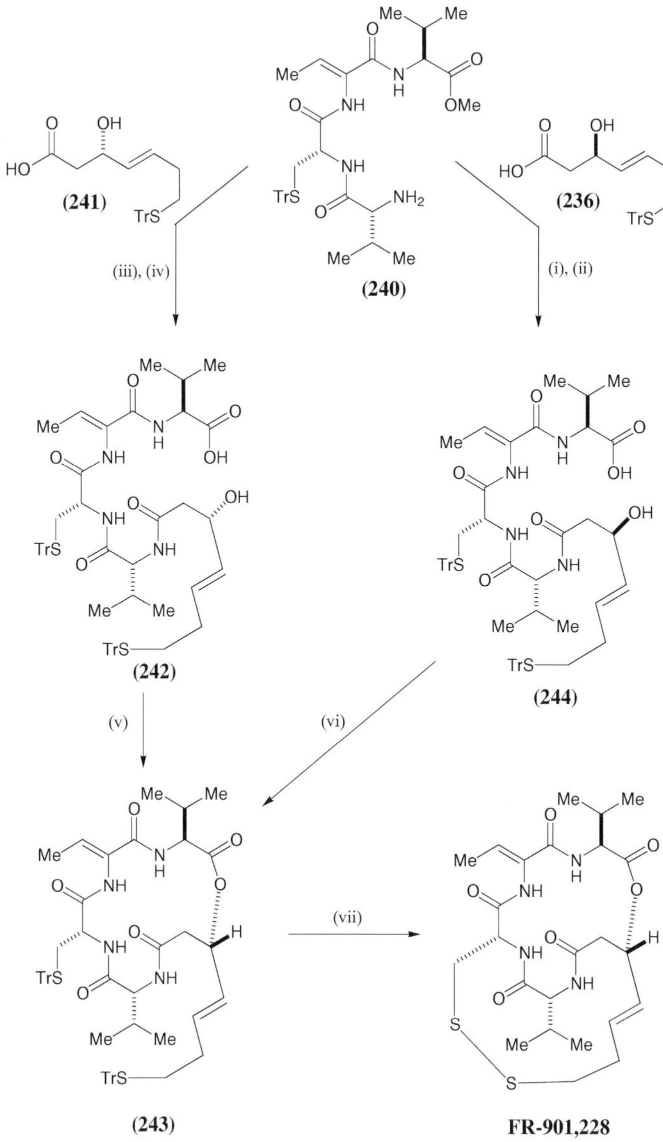

Scheme 12.48   (i) **236**, BOP, DIEA, 95%; (ii) LiOH, 98%; (iii) **241**, BOP, DIEA, 98%; (iv) LiOH, 93%; (v) EDCI, DMAP, p-TsOH, < 5%; (vi) DIAD (20 equiv.), PPh$_3$ (25 equiv.), THF, p-TsOH (5 equiv.), add **244** over 2 h at 0°C, then stir 1 h, 62%; (vii) I$_2$, MeOH, 84%.

subtarget **255** was established by a Mitsunobu inversion on alcohol **253** to give **254**. A routine series of protecting-group manipulations then converted **254** into **255**.

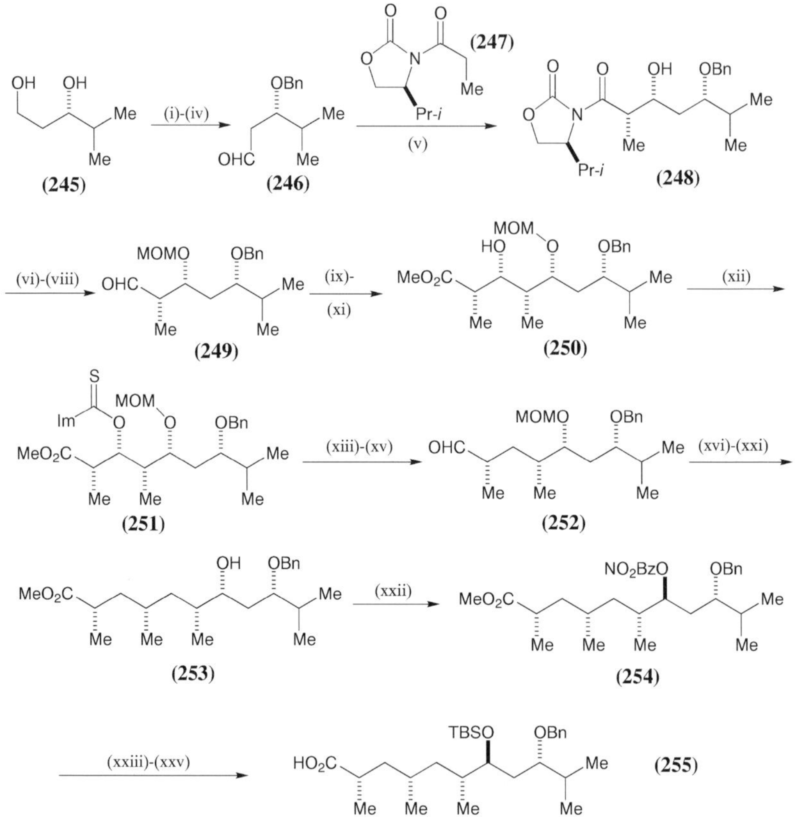

Scheme 12.49 (i) TrCl, Et$_3$N, DMAP, CH$_2$Cl$_2$, 20°C, 13 h; (ii) BnBr, NaH, DMF, RT, 2.5 h; (iii) conc. HCl, MeOH, THF, 30°C, 4 h; (iv) DMSO, (COCl)$_2$, Et$_3$N, CH$_2$Cl$_2$, −78°C for 20 min then 0°C for 15 min, 86% (4 steps); (v) **247**, Bu$_2$BOTf, Et$_3$N, CH$_2$Cl$_2$, −78°C for 30 min, 0°C 1 h, 95%; (vi) Me(MeO)NH.HCl, Me$_3$Al, THF, −19°C to −6°C, 2 h; (vii) MeOCH$_2$Cl, i-Pr$_2$NEt, 0°C to RT, 2 h; (viii) DIBAL-H, THF, −78°C, 10 min, 80% (3 steps); (ix) **247**, Et$_3$N, Bu$_2$BOTf, CH$_2$Cl$_2$, −78°C for 30 min then 0°C for 1 h; (x) LiOH, H$_2$O$_2$, THF, H$_2$O, 0°C, 1.5 h; (xi) CH$_2$N$_2$, Et$_2$O, CHCl$_3$, RT, 5 min, 85% (3 steps); (xii) Im$_2$CS, THF, Δ, 10 h, 93%; (xiii) Bu$_3$SnH, PhMe, Δ, 13 min, 79%; (xiv) LiAlH$_4$, THF, 0°C, 10 min; (xv) DMSO, (COCl)$_2$, Et$_3$N, −78°C for 10 min then 0°C for 10 min, 95% (2 steps); (xvi) **247**, Bu$_2$BOTf, Et$_3$N, CH$_2$Cl$_2$, −78°C for 30 min then 0°C for 1 h; (xvii) LiOH, H$_2$O$_2$, H$_2$O, 0°C, 1.5 h; (xviii) CH$_2$N$_2$, Et$_2$O, CHCl$_3$, RT, 5 min; (xix) Im$_2$CS, THF, Δ, 10 h; (xx) Bu$_3$SnH, PhMe, Δ, 13 min, 58% (5 steps); (xxi) conc. HCl, MeOH, 50°C, 2 h; (xxii) Ph$_3$P, p-NO$_2$C$_6$H$_4$CO$_2$H, DEAD, Et$_2$O, RT, 17.5 h, 69% (2 steps); (xxiii) NaOH, MeOH, H$_2$O, 45°C, 2 h; (xxiv) TBSOTf, Et$_3$N, CH$_2$Cl$_2$, 0°C, 15 min; (xxv) K$_2$CO$_3$, MeOH, THF, H$_2$O, 40°C, 1 h, 99% (3 steps).

Following elaboration of acid **255** into the *seco N*-methylamino acid **259** by successive attachment of the glycine and protected 3-iodo-*N*-methyl tyrosine residues, ring closure was effected with BOP-Cl and Et$_3$N

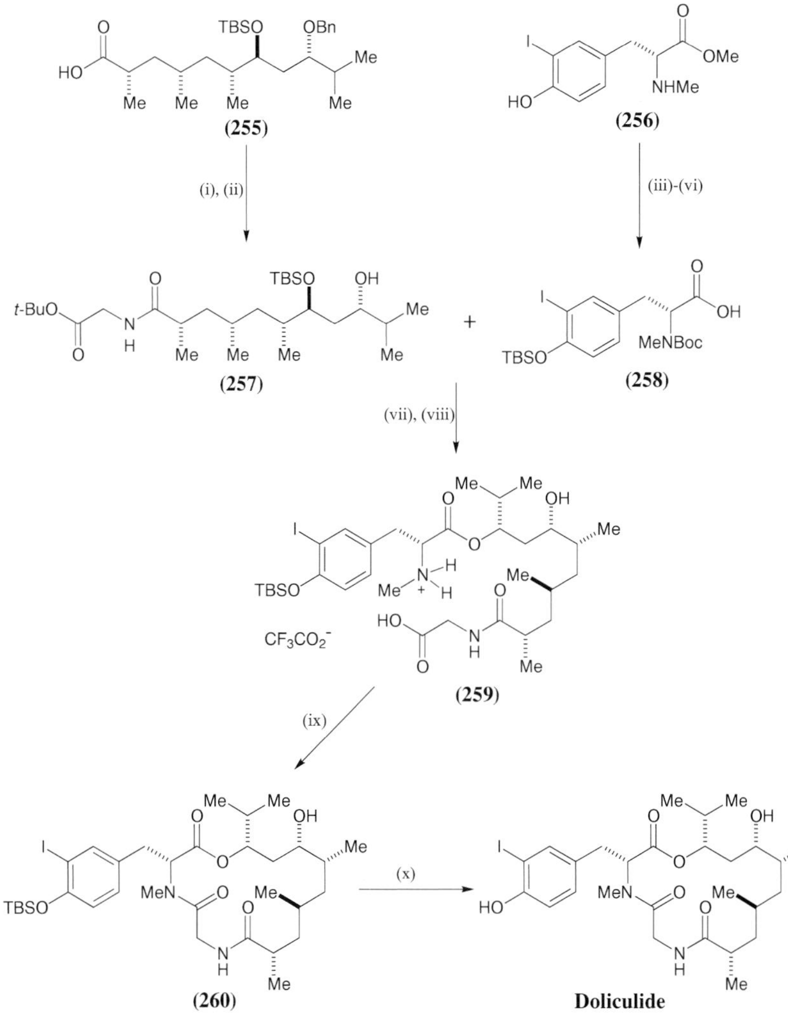

Scheme 12.50 (i) Glycine *tert*-butyl ester hydrochloride, DEPC, Et_3N, DMF, 0°C, 0.5 h, 98%; (ii) H_2, 20% $Pd(OH)_2$/C, 4°C, 1.5 h, 95%; (iii) Boc_2O, Et_3N, CH_2Cl_2, 5°C, 14 h; (iv) LiOH, THF, H_2O, RT, 1 h; (v) TBSCl, imidazole, DMF, 50°C, 1 h; (vi) K_2CO_3, H_2O, MeOH, THF, RT, 0.5 h, 67% (4 steps); (vii) DCC, DMAP, CH_2Cl_2, −20°C, 2 h, 94%; (viii) CF_3CO_2H, CH_2Cl_2, RT, 3 h; (ix) BOP-Cl, Et_3N, CH_2Cl_2, 0°C to 25°C, 19 h, 74%; (x) Bu_4NF, THF, 0°C, 5 min, 99%.

(Scheme 12.50). Macrolactamisation was attempted at this site since this avoided problems with loss of stereochemistry in the macrocyclisation precursor. In this regard, when a macrolactonisation strategy was investigated for ring-closure, a total epimerisation of the 3-iodotyrosine moiety was recorded; such occurrences are frequently problematical during the activation of N-methylamino acid residues. Completion of the doliculide synthetic venture was accomplished by a TBAF-induced O-desilylation which was complete within five minutes.

12.13   Didemnin A

In 1981, Rinehart and co-workers reported the isolation of a new class of cyclodepsipeptides, the didemnins,[94] from a Caribbean tunicate of the family Didemnidae. All of these peptides had a common macrocyclic framework and differed from one other only in the nature of the side-chains they had attached to the L-threonine residue. All are potent inhibitors of L1210 leukaemia cell growth, and all show activity *in vivo* against murine P388 lymphocytic leukaemia and B16 melanoma.[94] Originally, the didemnins had their structures assigned by degradative and spectroscopic methods, but this led to some misassignments which were later corrected by Rinehart[94] and other workers.[95, 96]

Although there have been several very elegant syntheses of the didemnins, the Joullié routes[97] to didemnins A, B and C have the greatest aesthetic appeal as they proceed through a common macrocyclic intermediate which is assembled in a fully stereocontrolled manner. All the other didemnin syntheses[94c, 98, 99] prepare the macrocycle in a nonstereocontrolled fashion. Joullié's retrosynthetic workplan for didemnin A[97] envisaged cyclisation between the L-leucine and (2S,4S)-Hip residues (Scheme 12.51), which was in contrast to all previously reported syntheses, which accomplished ring closure at the Pro-Leu,[98] the Ile-Thr[94c] and the N,O-Me₂Tyr-Pro[99] termini, respectively. The Joullié strategy[97] also severed the amide linkage connecting the L-Thr and Hip-isostatine residues. As a consequence, the early stages of her route focused on the preparation of fragments **263** (Scheme 12.54) and **264** (Schemes 12.52 and 12.53).

The Hip-isostatine unit **264** was built up from **271** and **272**. The approach used to access **271** (Scheme 12.52) exploited a Gennari chelation-controlled aldol condensation[100] between the (E)-silylketene thioacetal **266** and the aldehyde **267** to set up the stereotriad present in **268**. A series of protecting-group adjustments followed thereafter, the primary objective being to position an O-acetate group at the carbon adjacent to the isopropyl group to access **271**. Enolisation of this acetate and trapping with the Z-protected D-alloisoleucine imidazolide **272** next provided the β-ketoester **273** (Scheme 12.53) which was reduced with very high *anti*-selectivity (12.6:1), to give **264** after O-silylation, and N-debenzylation. The avenue used to obtain tetrapeptide **263** proceeded

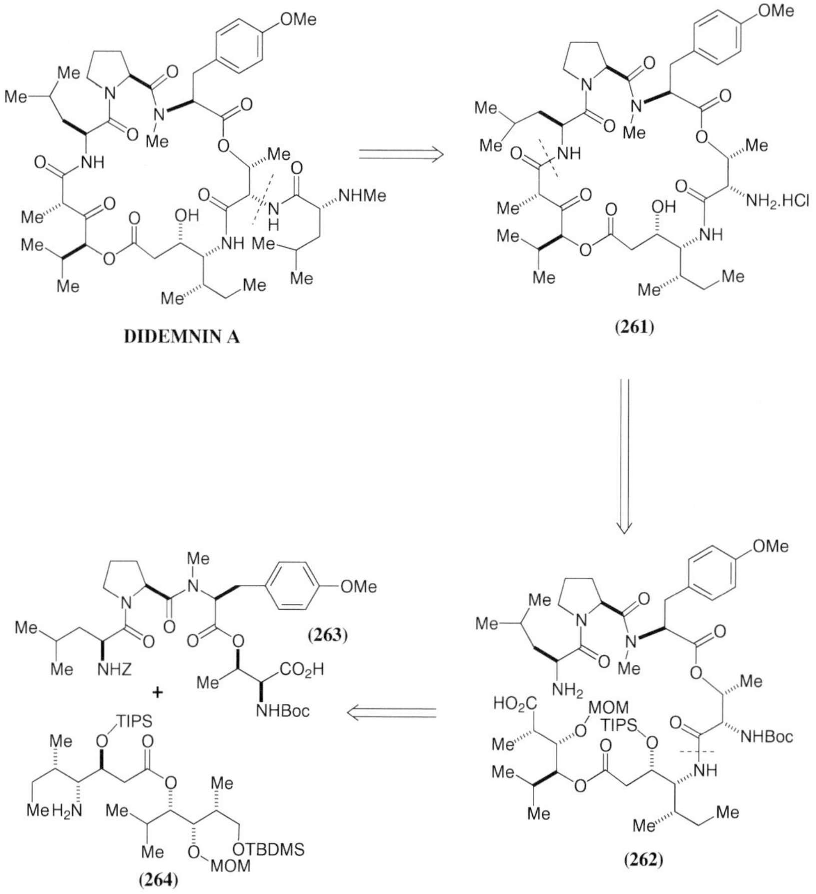

Scheme 12.51   Joullie's strategic analysis of didemnin A.

along quite traditional lines (Scheme 12.54), depsipeptide formation being accomplished by a mixed anhydride coupling protocol.[101]

Fragments **263** and **264** were united via Benoiton's isopropenyl chloroformate methodology[101] (Scheme 12.55). This reagent was found to be superior to all others for acid activation. The advance towards *seco* amino acid **262** required cleavage of the TBS group from **281** with mild acid, implementation of a two-step oxidation procedure to convert the primary alcohol to the carboxylic acid and a hydrogenolysis to unmask the amine. Diphenylphosphoryl azide generally gave the highest yield of macrolactamisation product **282**. The next task was selectively to strip off the various protecting groups from **282** to obtain **261**. Removal of the

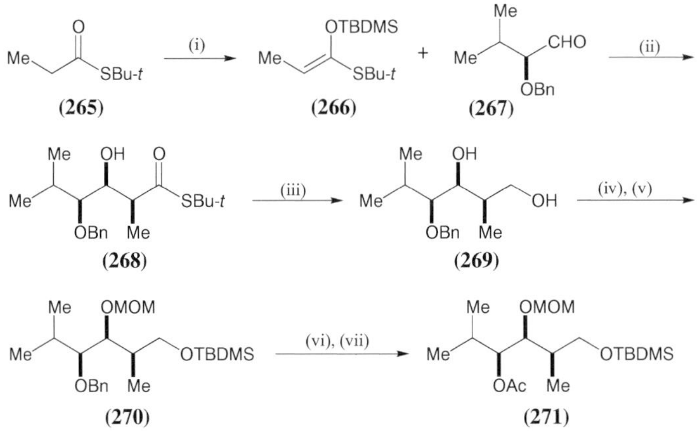

Scheme 12.52 (i) LDA, THF, HMPA, t-BuMe$_2$SiCl; (ii) SnCl$_4$, CH$_2$Cl$_2$, $-78°$C, 0.5 h, 74% (2 steps); (iii) LiAlH$_4$, Et$_2$O, 0°C to RT, 92%; (iv) t-BuMe$_2$SiCl, DMAP, CH$_2$Cl$_2$, Et$_3$N, 6 h, 99%; (v) CH$_2$Cl$_2$, MeOCH$_2$Cl, i-Pr$_2$NEt, 0°C, 24 h, 96%; (vi) Na, NH$_3$, THF, $-78°$C, 0.5 h, 96%; (vii) Ac$_2$O, Et$_3$N, CH$_2$Cl$_2$, DMAP, RT, 12 h, 90%.

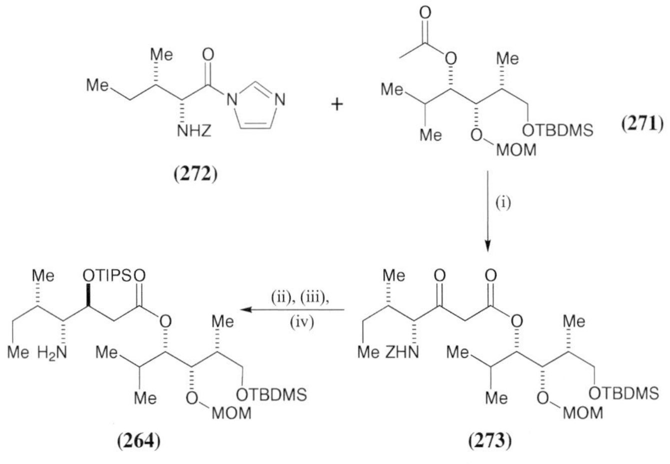

Scheme 12.53 (i) **271**, LDA, THF, $-78°$C, 1 h, 76%; (ii) KBH$_4$, EtOH, 0°C, 1 h then RT, 24 h, 83%; (iii) i-Pr$_3$SiOTf, 2,6-lutidine, CH$_2$Cl$_2$, 0°C, 3 h, 93%, 12.6:1 $anti$:syn; (iv) H$_2$, Pd/C, MeOH, 100%.

MOM group from the Hip residue initially proved troublesome, but was eventually accomplished very cleanly in 93% yield with dimethylboron bromide at low temperature. Swern oxidation of the product alcohol with trifluoroacetic anhydride and DMSO next provided the desired ketone without epimerising the adjacent 2-methyl group. Finally, the Boc and

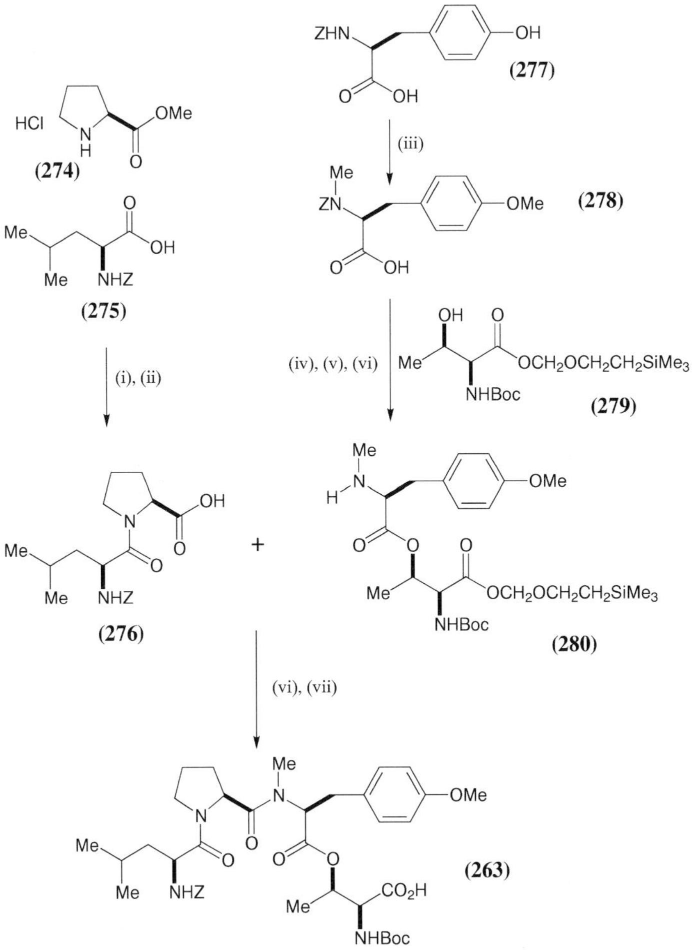

Scheme 12.54 (i) DCC, NMM, CH$_2$Cl$_2$, HOBT, 0°C, 24 h, 96%; (ii) LiOH·H$_2$O, THF, MeOH, H$_2$O (1:1:1), 0°C, 6 h, 94%; (iii) Powdered KOH (10 equiv.), Bu$_4$NHSO$_4$ (10% by wt.), THF, RT, 15 min then (MeO)$_2$SO$_2$, 0.5 h, 82%; (iv) isopropenyl chloroformate (1.1 equiv.), DMAP (0.2 equiv.), CH$_2$Cl$_2$, 0°C, 1 h, (87%); (v) H$_2$ 40 psi, 10% Pd-C, EtOAc-MeOH (1:1), 3 h; (vi) acid **276**, CH$_2$Cl$_2$, −15°C, add BOP-Cl (1.2 equiv.), then NMM (1.2 equiv.), stir −15°C for 0.5 h, add **280** and NMM (1.2 equiv.), 0°C, 6 h, 85%; (vii) 15% soln of 48% aq. HF in MeCN, −20°C, then −10°C, 5 h, 88%.

TIPS groups were simultaneously detached with hydrogen chloride in ethyl acetate to afford **261**. Didemnin A was synthesised in a further two steps by a BOP-promoted coupling with Z-NMe-D-Leu and a hydrogenolysis.

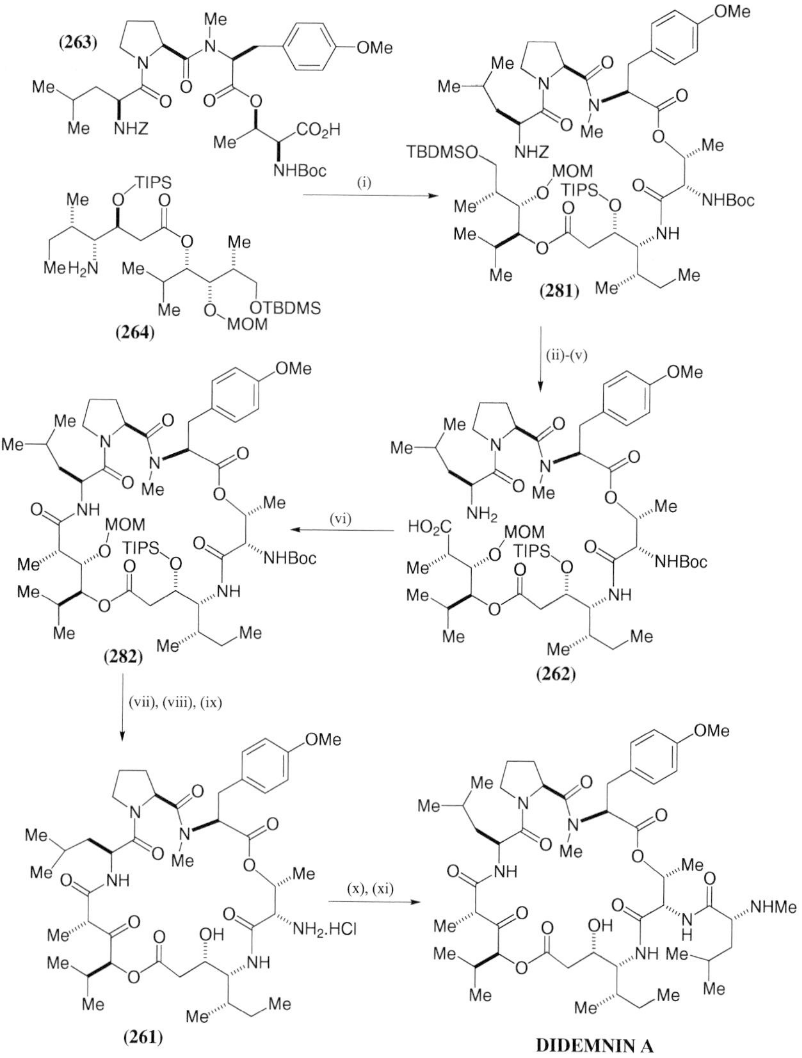

Scheme 12.55 (i) Isoprenyl chloroformate, N-methylmorpholine, THF, $-15°C$ to $0°C$, 60%; (ii) AcOH, THF, H_2O (3:1:1), 12 h, RT, 86%; (iii) TFAA, DMSO, CH_2Cl_2, Et_3N, $-78°C$; (iv) $KMnO_4$, t-BuOH, 5% NaH_2PO_4, 0.5 h, RT; (v) H_2, 10% Pd/C, MeOH, EtOAc (1:1), 40 psi, 24 h; (vi) $(PhO)_2P(O)N_3$, DMF, $NaHCO_3$, $0°C$, 72 h, 40% 4 steps; (vii) Me_2BBr, CH_2Cl_2, $-78°C$, 1 h, 93%; (viii) $(CF_3CO)_2O$, DMSO, Et_3N, CH_2Cl_2, $-78°C$, 1.5 h, 92%; (ix) HCl, EtOAc, $-30°C$ to $0°C$, 90%; (x) N-Z-methyl-D-leucine, BOP, CH_2Cl_2, N-methylmorpholine, 75%; (xi) H_2, 10% Pd/C, MeOH, EtOAc (1:1), 40 psi, 18 h, 85%.

12.14   PF1022A

In closing, we would finally like to highlight Scherkenbeck's total synthesis[102] of the potent anthelmintic agent PF1022A (see structure) which is outlined in Schemes 12.56 and 12.57. One particularly novel

PF1022A

aspect of the synthesis is the efficient construction of **288** and **289** by the S_N2 displacement of **284** and **285** with the caesium carboxylate salt of partially protected amino acid **287** (Scheme 12.56). The fact that these reactions were not overshadowed by the spectre of competing elimination is most exciting, and indicates that due consideration should be given to this novel methodology for establishing the ester linkages of depsipeptides in future years. Another item of note in this synthesis is the excellent yield recorded for the BOP-Cl mediated lactamisation reaction of the linear precursor **293** (Scheme 12.57). The authors attribute the great facility of this macrocyclisation to the existence of *cis*-amide rotamers in the *seco* amino-acid cyclisation intermediate. This confers a U-type turn conformation on the compound which forces the terminal amino and carboxylic acid termini into proximity to favour the cyclisation process.

12.15   Epilogue

All the synthetic endeavours we have outlined in this chapter superbly illustrate the tremendous power of modern-day asymmetric organic synthesis. Many of the syntheses also showcase the potentially enormous contribution that synthetic organic chemistry can make to the fields of biology and medicine. The syntheses discussed reflect strongly the latest methods and strategies used by the modern practitioners of synthetic organic chemistry and they all demonstrate how man's desire to

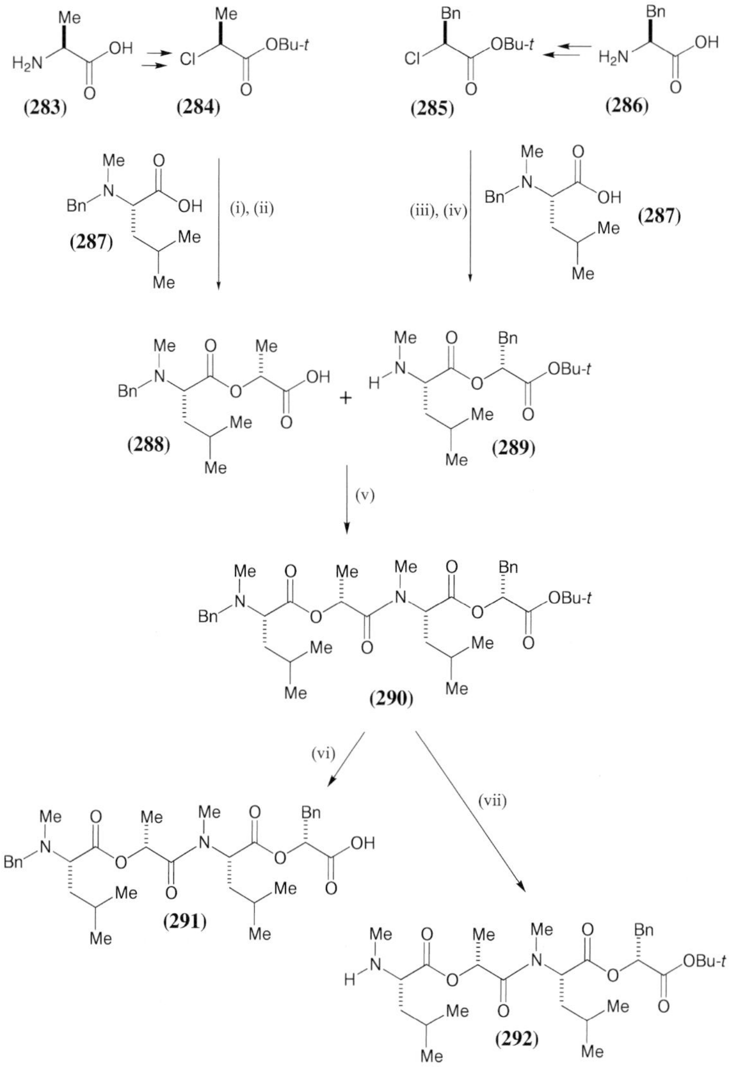

Scheme 12.56 (i) **287**, Cs$_2$CO$_3$, DMSO, RT, 20 h, 82%; (ii) HCl (g), CH$_2$Cl$_2$, 0°C for 2 h then RT for 12 h, 95%; (iii) **287**, Cs$_2$CO$_3$, DMSO, RT, 20 h, 29%; (iv) Pd(OH)$_2$/C, H$_2$, EtOH, RT, 88%; (v) BOP-Cl, i-Pr$_2$NEt, CH$_2$Cl$_2$, 0°C for 1 h then RT for 1 h, 80%; (vi) HCl (g), CH$_2$Cl$_2$, 0°C for 2 h then RT for 12 h, 80%; (vii) Pd(OH)$_2$/C, H$_2$, EtOH, RT, 74%.

build nature's most synthetically challenging and complex architectures stimulates, develops, and nurtures the discipline of organic chemistry as a whole.

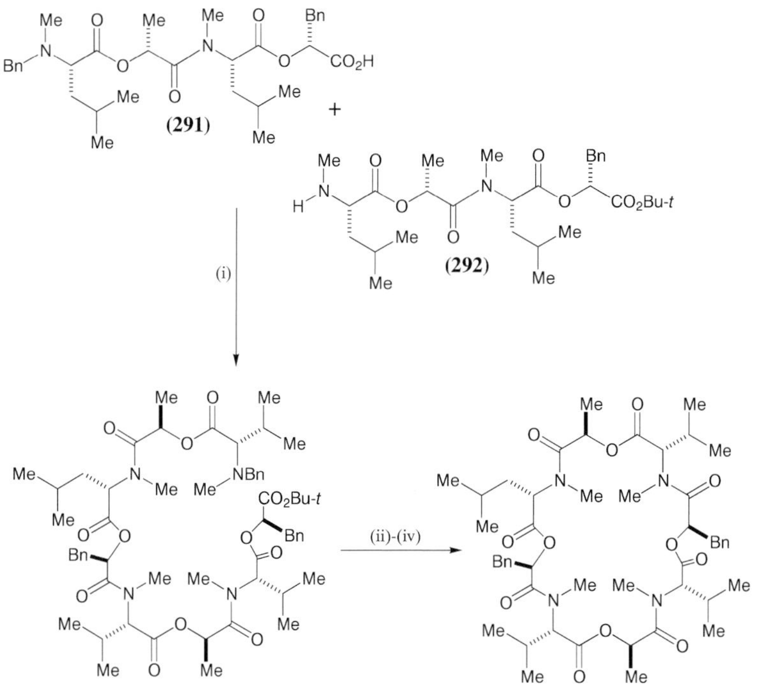

Scheme 12.57 (i) BOP-Cl, i-Pr$_2$NEt, CH$_2$Cl$_2$, 0°C, 1.5 h, 79%; (ii) Dowex$^®$ 1-X4, MeOH-H$_2$O (1:1), RT, 3 h, 95%; (iii) Pd(OH)$_2$/C, H$_2$, EtOH, 2 h, 92%; (iv) BOP-Cl, i-Pr$_2$NEt, CH$_2$Cl$_2$, 0°C to RT, 48 h, 87%.

Acknowledgements

We thank Professor K.C. Nicolaou of Scripps for very kindly supplying us with a preprint of his communication on the total synthesis of sanglifehrin A. We also thank Drs. C. Caldwell and P.L. Durette of Merck, Sharp & Dohme Research Laboratories in Rahway for supplying us with extra details of their L-156,602 synthesis.

References

1. (a) M. Kobayashi, S. Aoki, N. Ohyabu, M. Kurosu, W. Wang, and I. Kitagawa, 1994, *Tetrahedron Lett.*, 35, 7969; (b) M. Kobayashi, M. Kurosu, N. Ohyabu, W. Wang, S. Fujii, and I. Kitagawa, 1994, *Chem. Pharm. Bull.*, 42, 2196.
2. (a) Y. Koiso, K. Morita, M. Kobayashi, W. Wang, N. Ohyabu, and S. Iwasaki, 1996, *Chem.-Biol. Interact.*, 102, 183; (b) K. Morita, Y. Koiso, Y. Hasimoto, M. Kobayashi, W. Wang, N. Ohyabu, and I. Iwasaki, 1997, *Biol. Pharm. Bull. (Japan)*, 20, 171.

3. (a) M. Kobayashi, M. Kurosu, W. Wang, and I. Kitagawa, 1994, *Chem. Pharm. Bull.*, 42, 2394; (b) M. Kobayashi, W. Wang, N. Ohyabu, M. Kurosa, and I. Kitagawa, 1995, *Chem. Pharm. Bull.*, 43, 1598.

4. J.D. White, J. Hong, and L. Robarge, 1998, *Tetrahedron Lett.*, 39, 8779.

5. P.K. Jadhav, K.S. Bhat, P.T. Perumal, and H.C. Brown, 1986, *J. Org. Chem.*, 51, 432.

6. S. Shimizu, S. Nakamura, M. Nakada, and M. Shibasaki, 1996, *Tetrahedron*, 52, 13363.

7. (a) R.J. Linderman, K.P. Cusack, and M.R. Taber, 1996, *Tetrahedron Lett.*, 37, 6649; (b) G.E. Keck, M. Park, and D. Krishnamurthy, 1993, *J. Org. Chem.*, 58, 3787.

8. K. Takai, K. Nitta, and K. Utimoto, 1986, *J. Am. Chem. Soc.*, 108, 7408.

9. (a) V. Farina, V. Krishnamurthy, and W. Scott, 1997, *Org. React.*, 50, 1; (b) J.K. Stille, 1986, *Angew. Chem., Int. Edn. Engl.*, 25, 508.

10. H. Sone, T. Nemoto, H. Ishiwata, M. Ojika, and K. Yamada, 1993, *Tetrahedron Lett.*, 34, 8449.

11. (a) D. Seebach and H. Estermann, 1987, *Tetrahedron Lett.*, 28, 3103; (b) idem. 1988, *Helv. Chim. Acta*, 71, 1824; (c) E. Juaristi, J. Escalante, B. Lamatsch, and D. Seebach, 1992, *J. Org. Chem.*, 57, 2396.

12. B. Castro, J.R. Dormoy, G. Evin, and C. Selve, 1975, *Tetrahedron Lett.*, 1219.

13. R.D. Tung and D.H. Rich, 1985, *J. Am. Chem. Soc.*, 107, 4342.

14. (a) H. Osada, T. Yano, H. Koshino, and K. Isono, 1991, *J. Antibiotics*, 44, 1463; (b) H. Koshino, H. Osada, T. Yano, J. Uzawa, and K. Isono, 1991, *Tetrahedron Lett.*, 32, 7707.

15. U. Schmidt, K. Neumann, A. Schumacher, and S. Weinbrenner, 1997, *Angew. Chem., Int. Edn. Engl.*, 36, 1110.

16. L.A. Carpino, 1993, *J. Am. Chem. Soc.*, 115, 4937.

17. For examples of *O*- to *N*-acyl shifts in cyclodepsipeptides see: (a) A.B. Mauger and O.A. Stuart, 1987, *Int. J. Peptide Protein Res.*, 30, 481; (b) A.B. Mauger and O.A. Stuart, 1989, *Int. J. Peptide Protein Res.*, 34, 196 and the references cited therein. See also reference 19a below.

18. O.D. Hensens, R.P. Borris, L.R. Koupal, C.G. Caldwell, S.A. Currie, A.A. Haidri, C.F. Homnick, S.S. Honeycutt, S.M. Lindenmayer, C.D. Schwartz, B.A. Weissberger, H.B. Woodruff, D.L. Zink, L. Zitano, J.M. Fieldhouse, T. Rollins, M.S. Springer, and J.P. Springer, 1991, *J. Antibiotics*, 44, 249.

19. (a) P.L. Durette, F. Baker, P.L. Barker, J. Boger, S.S. Bondy, M.L. Hammond, T.J. Lanza, A.A. Pessolano, and C.G. Caldwell, 1990, *Tetrahedron Lett.*, 31, 1237; (b) C.G. Caldwell, K.M. Rupprecht, S.S. Bondy, and A.A. Davis, 1990, *J. Org. Chem.*, 55, 2355.

20. D. Seebach, R. Naef, and G. Calderari, 1984, *Tetrahedron*, 40, 1313.

21. (a) G. Frater, 1979, *Helv. Chim. Acta*, 62, 2825; (b) D. Seebach and D. Wasmuth, 1980, *Helv. Chim. Acta*, 63, 197; (c) D. Seebach, H.-F. Chow, R.F.W. Jackson, M.A. Sutter, S. Thaisrivongs, and J. Zimmerman, 1986, *Justus Liebigs Ann. Chem.*, 1281.

22. C.H. Hassall, W.H. Johnson, and C.J. Theobald, 1979, *J. Chem. Soc., Perkin Trans. 1*, 1451.

23. L.A. Carpino, B.J. Cohen, K.E. Stephens Jr, S.Y. Sadat-Aalaee, J.-H. Tien, and D.C. Langridge, 1986, *J. Org. Chem.*, 51, 3734.

24. C.G. Caldwell and S.S. Bondy, 1990, *Synthesis*, 34.

25. O. Dangles, F. Guibé, G. Balavoine, S. Lavielle, and A. Marquet, 1987, *J. Org. Chem.*, 52, 4984.

26. H. Wissmann and H.J. Kleiner, 1980, *Angew. Chem., Int. Edn. Engl.*, 19, 133.

27. T.A. Smitka, J.B. Deeter, A.H. Hunt, F.P. Mertz, R.M. Ellis, L.D. Boeck, and R.C. Yao, 1988, *J. Antibiotics*, 41, 726.

28. H. Maehr, C. Liu, N.J. Palleroni, J. Smallheer, L. Todaro, T.H. Williams, and J.F. Blount, 1986, *J. Antibiotics*, 39, 17.

29. Y. Hayakawa, M. Nakagawa, Y. Toda, and H. Seto, 1990, *Agric. Biol. Chem.*, 54, 1007.

30. Y. Sakai, T. Yoshida, T. Tsujita, K. Ochiai, T. Agatsuma, Y. Saitoh, F. Tanaka, T. Akiyama, S. Akinaga, and T. Mikuzumi, 1997, *J. Antibiotics*, 50, 659.

31. (a) K.J. Hale and J. Cai, 1997, *J. Chem. Soc., Chem. Commun.*, 2319; (b) K.J. Hale, J. Cai, and V.M. Delisser, 1996, *Tetrahedron Lett.*, 37, 9345; (c) K.J. Hale and J. Cai, 1996, *Tetrahedron Lett.*, 37, 4233; (d) K.J. Hale, J. Cai, S. Manaviazar, and S.A. Peak, 1995, *Tetrahedron Lett.*, 36, 6965; (e) K.J. Hale, V.M. Delisser, L.-K. Yeh, S.A. Peak, S. Manaviazar, and G.S. Bhatia, 1994, *Tetrahedron Lett.*, 35, 7685; (f) K.J. Hale, G.S. Bhatia, S.A. Peak, and S. Manaviazar, 1993, *Tetrahedron Lett.*, 34, 5343.

32. (a) D.A. Evans, J. Bartroli, and T.L. Shih, 1981, *J. Am. Chem. Soc.*, 103, 2127; R. Baker and J.L. Castro, 1990, *J. Chem. Soc., Perkin Trans. 1*, 47.

33. B.M. Trost and D.P. Curran, 1981, *Tetrahedron Lett.*, 22, 1287.

34. The enantiomer of **83** had previously been prepared: H.A. Vaccaro, D.E. Levy, A. Sawabe, T. Jaetsch, and S. Masamune, 1992, *Tetrahedron Lett.*, 33, 1937.

35. T. Suzuki, H. Saimoto, H. Tomioka, K. Oshima, and H. Nozaki, 1982, *Tetrahedron Lett.*, 23, 3597.

36. S. Takano, M. Akiyama, S. Sato, and K. Ogasawara, 1983, *Chem. Lett.*, 1539.

37. A.B. Smith III, K.J. Hale, and J.P. McCauley Jr, 1989, *Tetrahedron Lett.*, 30, 5579.

38. (a) J.E. McMurry and W.J. Scott, 1983, *Tetrahedron Lett.*, 24, 979; (b) A.B. Smith III, K.J. Hale, L.M. Laakso, K. Chen, and A. Riera, 1989, *Tetrahedron Lett.*, 30, 6963.

39. K.B. Sharpless, W. Amberg, Y.L. Bennani, G.A. Crispino, J. Hartung, K.-S. Jeong, H.-L. Kwong, K. Morikawa, Z.-W. Wang, D. Xu, and X.-L. Zhang, 1992, *J. Org. Chem.*, 57, 2768.

40. (a) E.J. Corey and M. Chaykovsky, 1964, *J. Am. Chem. Soc.*, 86, 1639; ibid. 1965, 87, 1345.

41. K. Horita, T. Yoshioka, T. Tanaka, Y. Oikawa, and O. Yonemitsu, 1986, *Tetrahedron*, 42, 3021.

42. (a) (3*S*)- and (3*R*)-Piperazic acids: (a) K.J. Hale, V.M. Delisser, and S. Manaviazar, 1992, *Tetrahedron Lett.*, 33, 7613; (b) K.J. Hale, J. Cai, V. Delisser, S. Manaviazar, S.A. Peak, G.S. Bhatia, T.C. Collins, and N. Jogiya, 1996, *Tetrahedron*, 52, 1047; (3*S*,5*S*)-5-Hydroxypiperazic acid: (c) K.J. Hale, N. Jogiya, and S. Manaviazar, 1998, *Tetrahedron Lett.*, 39, 7163.

43. (a) D.A. Evans, T.C. Britton, R.L. Dorow, and J.F. Dellaria, 1986, *J. Am. Chem. Soc.*, 108, 6395; (b) D.A. Evans, T.C. Britton, R.L. Dorow, and J.F. Dellaria, 1987, *Tetrahedron*, 43, 3803.

44. R.A. Boissonnas, St. Guttmann, and P.-A. Jaquenoud, 1960, *Helv. Chim. Acta*, 43, 1349.

45. H.T. Cheung and E.R. Blout, 1965, *J. Org. Chem.*, 30, 315.

46. K.J. Hale, S. Manaviazar, and V.M. Delisser, 1994, *Tetrahedron*, 50, 9181.

47. K.J. Hale, J. Cai, and G. Williams, 1998, *Synlett*, 149.

48. (a) K.S. Lam, G.A. Hesler, J.M. Mattel, S.W. Mamber, and S. Forenza, 1990, *J. Antibiotics*, 43, 956; (b) J.E. Leet, D.R. Schroeder, B.S. Krishnan, and J.A. Matson, 1990, *J. Antibiotics*, 43, 961.

49. J.E. Leet, D.R. Schroeder, J. Golik, J.A. Matson, T.W. Doyle, K.S. Lam, S.E. Hill, M.S. Lee, J.L. Whitney, and B.S. Krishnan, 1996, *J. Antibiotics*, 49, 299.

50. T.M. Kamenecka and S.J. Danishefsky, 1998, *Angew. Chem., Int. Edn. Engl.*, 37, 2993.

51. T.M. Kamenecka and S.J. Danishefsky, 1998, *Angew. Chem., Int. Edn. Engl.*, 37, 2995.

52. M. Ohno, T.F. Spande, and B. Witkop, 1968, *J. Am. Chem. Soc.*, 90, 6521.

53. (a) D.L. Boger, M.W. Ledeboer, and M. Kume, 1999, *J. Am. Chem. Soc.*, 121, 1098; (b) D.L. Boger and G. Schule, 1998, *J. Org. Chem.*, 63, 6421.

54. (a) M. Konishi, H. Ohkuma, F. Sakai, T. Tsuno, H. Koshiyama, T. Naito, and H. Kawaguchi, 1981, *J. Am. Chem. Soc.*, 103, 1241; (b) E. Arnold and J. Clardy, 1981, *J. Am. Chem. Soc.*, 103, 1243; (c) M. Konishi, H. Ohkuma, F. Sakai, T. Tsuno, H. Koshiyama, T. Naito, and H. Kawaguchi, 1981, *J. Antibiotics*, 34, 148.

55. (a) Y. Inouye, Y. Take, K. Oogose, A. Kubo, and S. Nakamura, 1987, *J. Antibiotics*, 40, 105; (b) Y. Take, Y. Inouye, S. Nakamura, H.S. Allaudeen, and A. Kubo, 1989, *J. Antibiotics*, 42, 107.

56. (a) Y. Nakamura and C. Shin, 1991, *Chem. Lett.*, 1953; (b) Y. Nakamura, A. Ito, and C. Shin, 1994, *Bull. Chem. Soc. Jpn.*, 67, 2151.

57. (a) M.A. Ciufolini and N. Xi, 1994, *J. Chem. Soc., Chem. Commun.*, 1867; (b) M.A. Ciufolini and N. Xi, 1997, *J. Org. Chem.*, 62, 2320.

58. (a) J. Vidal, L. Guy, S. Sterin, and A. Collet, 1993, *J. Org. Chem.*, 58, 4791; (b) J. Vidal, S. Damestoy, L. Guy, J.-C. Hannachi, A. Aubry, and A. Collet, 1997, *Chem. Eur. J.*, 3, 1691.

59. (a) T. Fehr, L. Oberer, V. Quesniaux Ryffel, J.-J. Sanglier, W. Schuler, and R. Sedrani, 1997, *PCT Int. Appl.*, WO 9702285A/970123, Sandoz Ltd, Switzerland; (b) J.-J. Sanglier, V. Quesniaux, T. Fehr, H. Hofmann, M. Mahnke, K. Memmert, W. Schuler, G. Zenke, L. Gscwind, C. Mauer, and W. Schilling, 1999, *J. Antibiotics*, 52, 466; (c) T. Fehr, J. Kallen, L. Oberer, J.-J. Sanglier, and W. Schilling, 1999, *J. Antibiotics*, 52, 474.

60. R. Banteli, I. Brun, P. Hall, and R. Metternich, 1999, *Tetrahedron Lett.*, 40, 2109.

61. (a) K.C. Nicolaou, J. Xu, F. Murphy, S. Barluenga, O. Baudoin, H. Wei, D.L.F. Gray, and T. Ohshima, 1999, *Angew. Chem., Int. Edn. Engl.*, 38, 2447; (b) K.C. Nicolaou, T. Ohshima, F. Murphy, S. Barluenga, J. Xu, and N. Winssinger, 1999, *J. Chem. Soc., Chem. Commun.*, 809.

62. (a) I. Paterson, M.A. Lister, and C.K. McClure, 1986, *Tetrahedron Lett.*, 27, 4787; (b) I. Paterson, 1988, *Chem. Ind. (Lond.)*, 390; (c) I. Paterson and J. Goodman, 1989, *Tetrahedron Lett.*, 30, 997; (d) I. Paterson, J.M. Goodman, M.A. Lister, R.C. Schumann, C.K. McClure, and R. Norcross, 1990, *Tetrahedron*, 46, 4663; (e) I. Paterson and A.N. Hulme, 1995, *J. Org. Chem.*, 60, 3288.

63. (a) H.C. Brown, R.K. Dhar, R.K. Bakshi, P.K. Pandiarajan, and B. Singaram, 1989, *J. Am. Chem. Soc.*, 111, 3441; (b) A. Vulpetti, A. Bernadi, C. Gennari, J.M. Goodman, and I. Paterson, 1993, *Tetrahedron*, 49, 685; (c) I. Paterson and M.V. Perkins, 1992, *Tetrahedron Lett.*, 33, 801.

64. (a) R.E. Ireland, P.G.M. Wuts, and B. Ernst, 1981, *J. Am. Chem. Soc.*, 103, 3205; (b) R.E. Ireland and M.G. Smith, 1988, *J. Am. Chem. Soc.*, 110, 854.

65. S.V. Ley, J. Norman, W.P. Griffith, and S.P. Marsden, 1994, *Synthesis*, 639.

66. A. Basha, M. Lipton, and S.M. Weinreb, 1977, *Tetrahedron Lett.*, 4171.

67. H.C. Brown and K.S. Bhat, 1986, *J. Am. Chem. Soc.*, 108, 293.

68. E.J. Corey, D. Enders, and M.G. Bock, 1976, *Tetrahedron Lett.*, 7.

69. S. Ohira, 1989, *Syn. Comm.*, 19, 561.

70. (a) H.X. Zhang, F. Guibé, and G. Balavoine, 1990, *J. Org. Chem.*, 55, 1857; (b) C.D.J. Boden, G. Pattenden, and T. Ye, 1996, *J. Chem. Soc., Perkin Trans. 1*, 2417.

71. M.J. Burk, O.E. Feaster, W.A. Nugent, and R.L. Harlow, 1993, *J. Am. Chem. Soc.*, 115, 10125.

72. (a) P.T. Northcote, J.W. Blunt, and M.H.G. Munro, 1991, *Tetrahedron Lett.*, 32, 6411; (b) R. Rzasa, D. Romo, D.J. Stirling, J.W. Blunt, and M.H.G. Munro, 1995, *Tetrahedron Lett.*, 36, 5307.

73. (a) R.M. Rzasa, H.A. Shea, and D. Romo, 1998, *J. Am. Chem. Soc.*, 120, 591; (b) D. Romo, R.M. Rzasa, H.A. Shea, K. Park, J.M. Langenhan, L. Sun, A. Akhiezer, and J.O. Liu, 1998, *J. Am. Chem. Soc.*, 120, 12237.

74. For other synthetic studies, see: D.J. Critcher and G. Pattenden, 1996, *Tetrahedron Lett.*, 37, 9107.

75. (a) O. Mitsunobu, 1981, *Synthesis*, 1; (b) D.L. Hughes, 1992, *Org. React.*, 42, 335.

76. (a) R. Noyori, T. Ohkuma, M. Kitamura, H. Takaya, N. Sato, H. Kumobayashi, and S. Akutagawa, 1987, *J. Am. Chem. Soc.*, 109, 5856; (b) D.F. Taber and L.J. Silverberg, 1991, *Tetrahedron Lett.*, 32, 4227; (c) D.F. Taber, P.B. Deker, and L.J. Silverberg, 1992, *J. Org. Chem.*, 57, 5990.

77. E.J. Corey and P.L. Fuchs, 1972, *Tetrahedron Lett.*, 3769.
78. (a) D.E. Van Horn and E. Negishi, 1978, *J. Am. Chem. Soc.*, 100, 2252; (b) T. Yoshida and E. Negishi, 1981, *J. Am. Chem. Soc.*, 103, 4985; (c) E. Negishi and D. Choueiry, 1995, in *Encyclopedia of Organic Reagents* (L.A. Paquette, ed.), John Wiley and Sons, Chichester, 5195.
79. K. Sonagashira, Y. Tohda, and N. Hagihara, 1975, *Tetrahedron Lett.*, 4467.
80. (a) Y. Nagao, Y. Hagiwara, T. Kumagai, M. Ochiai, T. Inoue, K. Hashimoto, and E. Fujita, 1986, *J. Org. Chem.*, 51, 2391; (b) V. Calo, V. Fiandanese, A. Nacci, and A. Scilimati, 1994, *Tetrahedron*, 50, 7283.
81. G. Lajoie, F. Lepine, L. Maziak, and B. Belleau, 1983, *Tetrahedron Lett.*, 24, 3815.
82. E. Aguilar and A.I. Meyers, 1994, *Tetrahedron Lett.*, 35, 2473.
83. (a) E. Nicolas, K.C. Russell, and V.J. Hruby, 1993, *J. Org. Chem.*, 58, 766; (b) G. Li, D. Patel, and V.J. Hruby, 1993, *Tetrahedron: Asymmetry*, 4, 2315.
84. C. Palomo, H.M. Aizpurua, C. Cuevas, A. Mielgo, and R. Galarza, 1995, *Tetrahedron Lett.*, 36, 9027.
85. Q. Dong, C.E. Anderson, and M.A. Ciufolini, 1995, *Tetrahedron Lett.*, 36, 5681.
86. G.L. Bolton, J.S. Sebolt-Leopold, and J.C. Hodges, 1994, *Ann. Rep. Med. Chem.*, 29, 165.
87. (a) H. Ueda, H. Nakajima, Y. Hori, T. Fujita, M. Nishimura, T. Goto, and M. Okuhara, 1994, *J. Antibiotics*, 47, 301; (b) N. Shigematsu, H. Ueda, S. Takase, H. Tanaka, K. Yamamoto, and T. Tada, 1994, *J. Antibiotics*, 47, 311; (c) H. Ueda, T. Manda, S. Matsumoto, S. Mukumoto, F. Nishigaki, I. Kawamura, K. Shimomura, 1994, *J. Antibiotics*, 47, 315; (d) H. Ueda, H. Nakajima, T. Goto, and M. Okuhara, 1994, *Biosci. Biotechnol. Biochem.*, 58, 1579.
88. W.L. Khan, J. Wu, W. Xing, and J.A. Simon, 1996, *J. Am. Chem. Soc.*, 118, 7237.
89. E.M. Carreira, R.A. Singer, and W. Lee, 1994, *J. Am. Chem. Soc.*, 116, 8837.
90. B. Kamber, A. Hartmann, K. Eisler, B. Riniker, H. Rink, P. Sieber, and W. Rittel, 1980, *Helv. Chim. Acta*, 63, 899.
91. H. Ishiwata, T. Nemoto, M. Ojika, and K. Yamada, 1994, *J. Org. Chem.*, 59, 4710.
92. H. Ishiwata, H. Sone, H. Kigoshi, and K. Yamada, 1994, *J. Org. Chem.*, 59, 4712.
93. D.H.R. Barton and S.W. McCombie, 1975, *J. Chem. Soc., Perkin Trans. 1*, 1574.
94. (a) K.L. Rinehart Jr, J.B. Gloer, J.C. Cook, S.A. Mizsak, and T.A. Scahill, 1981, *J. Am. Chem. Soc.*, 103, 1857; (b) K.L. Rinehart Jr, J.B. Gloer, R.G. Hughes Jr, H.E. Renis, J.P. McGovren, E.B. Swynenberg, D.A. Stringfellow, S.L. Kuentzel, and L.H. Li, 1981, *Science*, 212, 933; (c) K.L. Rinehart Jr, V. Kishore, S. Nagarajan, R.J. Lake, J.B. Gloer, F.A. Bozich, K.-M. Li, R.E. Maleczka Jr, W.L. Todsen, M.H.G. Munro, D.W. Sullins, and R. Sakai, 1987, *J. Am. Chem. Soc.*, 109, 6846, and the references cited therein.
95. W.R. Ewing, K.L. Bhat, and M.M. Joullié, 1986, *Tetrahedron*, 42, 5863.
96. (a) B.D. Harris, K.L. Bhat, and M.M. Joullié, 1987, *Tetrahedron Lett.*, 28, 2837; (b) B.D. Harris and M.M. Joullié, 1988, *Tetrahedron*, 44, 3489.
97. (a) W.R. Ewing, B.D. Harris, W.-R. Li, and M.M. Joullié, 1989, *Tetrahedron Lett.*, 30, 3757; (b) W.-R. Li, W.R. Ewing, B.D. Harris, and M.M. Joullié, 1990, *J. Am. Chem. Soc.*, 112, 7659.
98. (a) U. Schmidt, M. Kroner, and H. Griesser, 1988, *Tetrahedron Lett.*, 29, 4407; (b) U. Schmidt, M. Kroner, and H. Griesser, 1988, *Tetrahedron Lett.*, 29, 3057; (b) U. Schmidt, M. Kroner, and H. Griesser, 1991, *Synthesis*, 294.
99. Y. Hamada, Y. Kondo, M. Shibata, and T. Shioiri, 1989, *J. Am. Chem Soc.*, 111, 669.
100. C. Gennari, M.G. Beratta, A. Bernardi, G. Moro, C. Scolastico, and R. Todeshini, 1986, *Tetrahedron*, 42, 893.
101. F.M.F. Chen, Y. Lee, R. Steinaver, and N.L. Benoiton, 1987, *Can. J. Chem.*, 65, 613.
102. (a) J. Scherkenbeck, A. Plant, A. Harder, and N. Mencke, 1995, *Tetrahedron*, 51, 8459; (b) J. Scherkenbeck, A. Plant, A. Harder, and H. Dyker, 1998, *Biorg. Med. Chem. Lett.*, 8, 1035.

Index